The Role of Tropics in Climate Change

The Role of Tropics in Climate Change
Global Case Studies

Edited by

Neloy Khare
Ministry of Earth Sciences Government of India,
New Delhi, India

Elsevier
Radarweg 29, PO Box 211, 1000 AE Amsterdam, Netherlands
50 Hampshire Street, 5th Floor, Cambridge, MA 02139, United States

Copyright © 2024 Elsevier Inc. All rights reserved, including those for text and data mining, AI training, and similar technologies.

No part of this publication may be reproduced or transmitted in any form or by any means, electronic or mechanical, including photocopying, recording, or any information storage and retrieval system, without permission in writing from the publisher. Details on how to seek permission, further information about the Publisher's permissions policies and our arrangements with organizations such as the Copyright Clearance Center and the Copyright Licensing Agency, can be found at our website: www.elsevier.com/permissions.

This book and the individual contributions contained in it are protected under copyright by the Publisher (other than as may be noted herein).

Notices

Knowledge and best practice in this field are constantly changing. As new research and experience broaden our understanding, changes in research methods, professional practices, or medical treatment may become necessary.

Practitioners and researchers must always rely on their own experience and knowledge in evaluating and using any information, methods, compounds, or experiments described herein. In using such information or methods they should be mindful of their own safety and the safety of others, including parties for whom they have a professional responsibility.

To the fullest extent of the law, neither the Publisher nor the authors, contributors, or editors, assume any liability for any injury and/or damage to persons or property as a matter of products liability, negligence or otherwise, or from any use or operation of any methods, products, instructions, or ideas contained in the material herein.

ISBN: 978-0-323-99519-1

For information on all Elsevier publications visit our website at https://www.elsevier.com/books-and-journals

Publisher: Candice G. Janco
Acquisitions Editor: Jessica Mack
Editorial Project Manager: Teddy A. Lewis
Production Project Manager: Rashmi Manoharan
Cover Designer: Victoria Pearson Esser

Typeset by TNQ Technologies

Contents

Contributors xv
About the editor.. xix
Foreword xxi
Preface xxiii
Acknowledgments... xxxi

CHAPTER 1　The importance of tropics in the changing climate............................1
　　　　　　　D. Rajan and S. Gautam
　　1 Introduction 1
　　2 Organization of the chapter 1
　　　　2.1　Crisis of human-caused climate change 2
　　　　2.2　General atmospheric circulation 3
　　　　2.3　The impact of climate on the tropical region 4
　　　　2.4　The impact of the tropical belt on the climate 4
　　　　2.5　Implications for the future climate 6
　　　　2.6　Climate change in the Indian context 7
　　　　2.7　Climate change and national security 7
　　　　2.8　Climate change and South Asia 8
　　　　2.9　Impact of climate change on Indian agriculture.............................8
　　　　2.10　Tropical teleconnection impacts on the Antarctic climate changes....9
　　　　2.11　Projection/prediction of precipitations and teleconnection due
　　　　　　　to climate change 11
　　　　2.12　Lesson from pandemic COVID 11
　　　　2.13　Adaptation to the heat wave and climate change........................... 12
　　　　2.14　Recent observations of greenhouse gas, temperature, and ocean
　　　　　　　heat in the warming climate situation.. 12
　　3 Conclusion and pathways 14
　　References.. 14
　　Further reading 15

**CHAPTER 2　Climate change impacts on water resources and agriculture
　　　　　　　in Southeast Asia with a focus on Thailand, Myanmar,
　　　　　　　and Cambodia** 17
　　　　　　　G. Srinivasan, Anshul Agarwal and Upeakshika Bandara
　　1 Introduction 17
　　2 Climate change over Southeast Asia 18
　　　　2.1　Myanmar.. 20
　　　　2.2　Thailand... 22
　　　　2.3　Cambodia.. 23

v

    3 Impacts on agriculture and water resources 24
        3.1 Myanmar: impacts 27
        3.2 Thailand: impacts 27
        3.3 Cambodia: impacts 28
    4 Practices for adaptive actions 28
    5 Priorities for climate-resilient agriculture and water sectors 29
        5.1 Enhancing climate information and its use 29
        5.2 National institutional framework 30
        5.3 Capacities for research and development in the country context 30
    References...30

CHAPTER 3 Interrelationship between climate justice and migration: A case of south western coastal region of Bangladesh 33
    M. Ashrafuzzaman
    1 Introduction 33
        1.1 Theoretical framework: Climate justice and migration................................35
        1.2 Description of migration 35
        1.3 When society and climate experience transformation..................................36
        1.4 Study area 36
    2 Methodology 37
        2.1 Determining sample size using quantitative methods 37
        2.2 Methods of gathering information 39
        2.3 Qualitative approach...39
    3 Results: Quantitative analysis 40
        3.1 Migration behavior of the respondents 40
        3.2 Why people initiate migration?..41
        3.3 Period of migration 42
        3.4 Diverse migration related to climate change 43
        3.5 Types of migration 44
        3.6 Reasons for migration 45
    4 Qualitative analysis 46
        4.1 Methods used included participatory rural appraisal, case study, focus group discussions, workshops, interviews, key informant interviews, and in-depth interviews 46
    5 Discussions 50
        5.1 The relationship between climate justice and climate-induced migration..........53
        5.2 Example of adaptation in global migration 54
    6 Conclusions 55
    References...56

CHAPTER 4 Climate variability, observed climate trends, and future climate projections for Sri Lanka 61
I.M. Shiromani Priyanthika Jayawardena, D.W.T.T. Darshika, H.M.R.C. Herath and H.A.S.U. Hapuarachchi
  1 Introduction 61
    1.1 Spatial and temporal variations of rainfall 63
    1.2 Temperature variations 66
    1.3 Spatial and temporal variations of lightning 67
  2 Interannual and intraseasonal rainfall variability in Sri Lanka 71
    2.1 Impact of El Nino and La Nina on interannual variability of seasonal rainfall in Sri Lanka 72
    2.2 Impact of Indian Ocean Dipole on interannual variability of monthly rainfall from June to December in Sri Lanka 76
    2.3 Impact of the Madden-Julian oscillation (MJO) on intra-seasonal rainfall variability in Sri Lanka 76
    2.4 Impact of the boreal summer intra-seasonal oscillation (BSISO) on intra-seasonal SWM rainfall variability in Sri Lanka 80
  3 Observed local trends 82
    3.1 Recent observed trends in climatic extremes in Sri Lanka (1981–2015) 84
  4 Climate change—future climate change projections 88
    4.1 Future climate projections of seasonal rainfall and annual rainfall for Sri Lanka 88
    4.2 Future climate projections of annual temperature for Sri Lanka and projected heatwave occurrence in Colombo 92
  5 Conclusions 96
  References 100
  Further reading 103

CHAPTER 5 A review study on the artifacts, vertebrate remains, and volcanic ash of peninsular Malaysia and their significance to reconstruct paleoclimate 105
Ajab Singh and Neloy Khare
  1 Introduction 105
  2 History of human occupation in the Peninsula 106
  3 Origin, migration, and settlement of hominids in Malaysia 108
  4 Evolution and migration of quaternary faunas and their detailed studies 110
    4.1 Animal remains from the alluvial deposits 110
    4.2 Animal remains from the cave deposits 111
  5 Occurrence and source of volcanic ash 112
  6 Controversies over the impact of YTT ash on the ecosystem 113

	7 Conclusions	114
	Acknowledgments	115
	References	115

CHAPTER 6 Bhutan and the geography of climate change ... 121
Jeetendra Prakash Aryal, Medha Bisht and Dil Bahadur Rahut

	1 Introduction	121
	2 The discourse on climate change in Bhutan	122
	3 Hydropower in Bhutan in context to climate change	122
	4 Climate change, global warming, and glaciers in Bhutan	124
	5 Agriculture in the context of climate change	125
	6 Geographies of climate change	126
	References	127
	Further reading	129

CHAPTER 7 Study of spatio-temporal climate variability and changes in South America using climate data tools ... 131
Amba Shalishe, Anirudh Bhowmick, Subodh Kumar Chaturvedi and Jai Ram Ojha

	1 Introduction	131
	2 Data and methodology	132
	3 Results and discussions	132
	3.1 Precipitation variability	132
	3.2 Temperature variability	136
	4 Conclusions	141
	References	141
	Further reading	143

CHAPTER 8 Climate change scenario over Japan: trends and impacts ... 145
Sridhara Nayak and Tetsuya Takemi

	1 Introduction	145
	2 Present climate over Japan	146
	2.1 Precipitation	147
	2.2 Temperature	149
	3 Projected climate change over Japan	153
	3.1 Precipitation	153
	3.2 Temperature	157
	4 Impacts	161
	5 Summary	166
	6 Declarations	166
	6.1 Availability of data and material	166

Funding 167
Authors' contributions 167
Acknowledgments 167
References 167

CHAPTER 9 A brief review of climate in Taiwan 171
Mahjoor Ahmad Lone, Kweku Afrifa Yamoah and Tsai-Wen Lin
1 Introduction: hydroclimate of Taiwan 171
2 Contemporary climate in Taiwan 171
3 Hydroclimate variability in Taiwan during the late Holocene 172
4 Critical issues remaining 175
    4.1 Dynamical relationship between ENSO and typhoons 175
    4.2 Unavailability of hydroclimate datasets over the LGM in Taiwan 175
    4.3 Dating uncertainties for proxy records 175
References 176

CHAPTER 10 Potential of the lake sediments study of Ethiopia in understanding the Holocene climatic conditions: a case study of Chamo Lake, Southern Nations, Nationalities, and People's Region, Ethiopia 177
Subodh Kumar Chaturvedi, Anirudh Bhowmick, Gosaye Berhanu, Geremu Gecho and Jai Ram Ojha
1 Introduction 177
2 Description of the study area 178
    2.1 Physiography of Ethiopia 178
    2.2 Chamo Lake 181
    2.3 Geomorphology 181
    2.4 Climate 183
    2.5 Geology 183
3 Material and methods 188
    3.1 Sediment texture analysis 190
    3.2 XRF analysis 191
    3.3 Organic matter estimation 191
4 Results and discussion 191
    4.1 Sediment texture analysis 191
    4.2 Geochemical analysis of lake sediments 196
    4.3 Statistical analysis of elements in sediments 197
5 Conclusions and recommendation 203
Acknowledgments 203
References 203

CHAPTER 11 High-resolution paleoclimatic records from the tropical delta shelf off Ayeyarwady, Myanmar.......... 207
Rajani Panchang and Rajiv Nigam
1 Preamble..........207
2 Why foraminifera?..........208
3 Status of foraminiferal studies from the Bay of Bengal..........209
4 Significance of studying benthic foraminiferal distributions on the Ayeyarwady shelf..........211
5 The Indo-Myanmar joint oceanographic studies..........211
6 Objectives of the present study..........212
7 Myanmar Ayeyarwady delta shelf..........212
    7.1 Andaman Sea..........212
    7.2 Location and extent of the Hinterland..........215
    7.3 Coastline..........215
    7.4 Drainage..........215
    7.5 The Ayeyarwady Delta..........217
    7.6 Climate..........218
    7.7 Coastal geomorphology..........218
    7.8 Sediment distribution on the Ayeyarwady Shelf..........219
    7.9 Coastal hydrography..........219
    7.10 Tectonic setting of the study area..........225
8 Data collection and analysis: the systematic approach to paleoclimatic studies..........227
    8.1 On board sampling and data collection..........227
    8.2 Laboratory analysis..........230
    8.3 Geochronology..........234
    8.4 Coral sclerites: associated microfauna with environmental significance..........236
9 Comparison of the foraminiferal assemblages from Ayeyarwady delta shelf with adjoining coasts: insights into global ocean-atmospheric linkages of the study area..........236
10 Relict benthic foraminifera and sea level history along west coast of Myanmar..........238
    10.1 Relict benthic foraminifera..........238
    10.2 Larger benthic foraminifera..........238
    10.3 Associated relict fauna..........240
    10.4 Integration of relict faunal proxy data..........245
    10.5 Proposition of a regional sea-level curve..........247
    10.6 Comparison with other sea-level curves..........247
    10.7 Significance of sea level fluctuations off Myanmar..........251

- 11 Subsurface foraminiferal signatures on Ayeyarwady Delta shelf: intradecadal paleoclimatic records for the past five centuries 251
  - 11.1 Sub-surface distribution of foraminifera 252
  - 11.2 Asterorotalia trispinosa as an indicator of paleomonsoons 254
- 12 Periodicity in paleoclimatic records 258
  - 12.1 Asian signatures of the Little Ice Age (LIA) 259
  - 12.2 Geochemical proxies as modern reinforcement to traditional micropaleontology 260
- 13 Conclusions 262
- Acknowledgments 264
- References 264
- Further reading 271

CHAPTER 12 Sedimentation model for the Pinjor Formation of the Upper Siwalik, Dehradun Sub-basin, Garhwal Himalaya, India: a response of climate and tectonics 273

Bijendra Pathak and U.K. Shukla

- 1 Introduction 273
- 2 Geology of the area 274
- 3 Methodology 275
- 4 Lithofacies analysis 275
  - 4.1 Horizontal and low-angle planar gravelly sandstone 275
  - 4.2 Planar cross-bedded gravelly sandstone 281
  - 4.3 Matrix-to-clast-supported conglomerate 282
  - 4.4 Clast-supported conglomerate 282
  - 4.5 Ripple cross-bedded sandstone 282
  - 4.6 Laminated mudstone 283
- 5 Lithofacies association 283
- 6 Palaeocurrent pattern 284
- 7 Petrography 285
- 8 Depositional model 287
- 9 Discussion 289
  - 9.1 Sedimentation 289
  - 9.2 Climate versus tectonics control on sedimentation 290
- 10 Conclusions 291
- Acknowledgments 292
- References 292

CHAPTER 13 A catalogue of Quaternary diatoms from the Asian tropics with their environmental indication potential for paleolimnological applications **295**
Mital Thacker and Balasubramanian Karthick
1 Introduction 295
2 Materials and methods 297
3 Result 298
    3.1 Freshwater taxa 299
    3.2 Brackish environment 351
    3.3 Marine environment 352
4 Conclusions 361
Acknowledgments 361
References 361

CHAPTER 14 Environmental disaster in Pithoragarh district, Uttarakhand, India: an analogy to climate change and anthropogenic interference **377**
Akhouri Bishwapriya, Prakash K. Gajbhiye, Milind Vasantrao Dakate, Santosh Kumar Tripathi and Abhishek Kumar Chaurasia
1 Introduction 377
    1.1 Study area 379
    1.2 Slope stability assessment 382
    1.3 Geotechnical assessment 386
2 Conclusions 400
Acknowledgments 401
References 401
Further reading 401

CHAPTER 15 Geological/paleontological applications in marine archeology: few examples from Indian waters **403**
Rajiv Nigam
1 Ground penetrating radar for subsurface information 404
2 Marine sediments as a source of reconstruction of the past 405
3 Microfossils with special reference to foraminifera 405
4 Foraminifera 406
5 Reconstruction of past sea-level changes 407
6 Sea-level fluctuations and marine archeology 409
    6.1 Lothal 409
    6.2 Unknown city off Surat, Gujarat 411
    6.3 Submerged Dwarka 412
    6.4 Dholavira 414

	6.5 Ramsetu	417
	6.6 Other areas	418
7	Conclusions	418
	Acknowledgements	419
	References	419

CHAPTER 16 Introspecting contribution and preparedness of tropical agriculture against climate change: an Indian perspective 423

S. Suresh Ramanan, M. Prabhakar, Mohammed Osman and A. Arunachalam

1	Background	423
2	Methodology	425
3	Results	426
4	Discussion	427
5	Conclusions	432
	Acknowledgments	432
	References	432

CHAPTER 17 Mechanisms and proxies of solar forcing on climate and a peek into Indian paleoclimatic records 437

Rajani Panchang, Mugdha Ambokar, Kalyani Panchamwar and Neloy Khare

1	Introduction	437
2	Sun as the eternal driver of the Earth's climate: an historical perspective	437
3	Sun: the external driver of Earth's climate	440
4	The climatic manifestations of solar activity: obvious correlations of the inquisitive	441
5	Solar cycles	442
	5.1 Sunspots: structure and nature	442
	5.2 Grand Minima and Grand Maxima	443
	5.3 ENSO and Solar cycles	446
	5.4 Schwabe cycle (the 11-year cycle)	447
	5.5 Hale cycle (the 22-year cycle)	450
	5.6 Brückner-Egeson-Lockyer Cycle (the 30- to 40-year cycle)	451
	5.7 Gleissberg cycle (70- to 100-year cycle)	452
	5.8 Suess Cycle or de Vries Cycle (~210-year cycle)	453
	5.9 Eddy Cycle (~1000-year cycle)	455
	5.10 Hallstatt/Bray cycle (the 2500-year cycle)	456
	5.11 Milankovitch cycles—Earth's orbital variations	459
6	Proxies and paleoclimatic records of solar cycles	463
	6.1 Short-term solar activity	463

6.2 Long-term solar activity 465
6.3 Strontium isotopic ratios 470
6.4 Magnetic susceptibility 471
6.5 Foraminiferal signatures 473
6.6 Speleothem records 475
6.7 Tree rings 476
6.8 Ice cores 479
7 Prediction of solar cycles 479
8 Conclusions 481
References 482

Glossary 491
Author index 497
Index 519

Contributors

Anshul Agarwal
Regional Integrated Multi-Hazard Early-Warning System for Africa and Asia (RIMES), Klong Luang, Pathum Thani, Thailand

Mugdha Ambokar
Marine Bio-Geo Research-Lab for Climate and Environment (MBReCE), Department of Environmental Science, Savitribai Phule Pune University, Pune, Maharashtra, India

A. Arunachalam
ICAR-Central Agroforestry Research Institute, Jhansi, Uttar Pradesh, India

Jeetendra Prakash Aryal
International Center for Biosaline Agriculture, Dubai, United Arab Emirates

M. Ashrafuzzaman
Climate Change and Sustainable Development Policies, University of Lisbon & Nova University of Lisbon, Lisbon, Portugal; Institute of Social Sciences, University of Lisbon, Lisbon, Portugal; Norwich Research Park, University of East Anglia, Norwich, United Kingdom; Department of Geography, University of Valencia, Valencia, Spain; Department of Anthropology, University of Chittagong, Chattogram, Bangladesh

Upeakshika Bandara
Regional Integrated Multi-Hazard Early-Warning System for Africa and Asia (RIMES), Klong Luang, Pathum Thani, Thailand

Gosaye Berhanu
Department of Geology, College of Natural Sciences, Arba Minch University, Arba Minch, Ethiopia

Anirudh Bhowmick
Faculty of Meteorology and Hydrology, Arba Minch Water Technology Institute, Arba Minch University, Arba Minch, Ethiopia

Medha Bisht
South Asian University, New Delhi, India

Akhouri Bishwapriya
Geological Survey of India, Patna, Bihar, India

Subodh Kumar Chaturvedi
Institute of Hydrocarbon, Energy and Geo-Resources, ONGC Center for Advanced Studies, University of Lucknow, Lucknow, Uttar Pradesh, India

Abhishek Kumar Chaurasia
Geological Survey of India, Patna, Bihar, India

Milind Vasantrao Dakate
Geological Survey of India, Nagpur, Maharashtra, India

D.W.T.T. Darshika
Department of Meteorology, Colombo, Sri Lanka

Prakash K. Gajbhiye
Geological Survey of India, Pune, Maharashtra, India

S. Gautam
Karunya Institute of Technology and Sciences (Deemed to be University), Coimbatore, Tamil Nadu, India

Geremu Gecho
Department of Geology, College of Natural Sciences, Arba Minch University, Arba Minch, Ethiopia

H.A.S.U. Hapuarachchi
Department of Meteorology, Colombo, Sri Lanka

H.M.R.C. Herath
Department of Meteorology, Colombo, Sri Lanka

Balasubramanian Karthick
Biodiversity & Paleobiology Group, Agharkar Research Institute, Pune, Maharashtra, India; Department of Botany, Savitribai Phule Pune University, Pune, Maharashtra, India

Neloy Khare
Ministry of Earth Sciences, New Delhi, India

Tsai-Wen Lin
Department of Geosciences, National Taiwan University, Taipei, Taiwan; Alfred Wegener Institute of Polar and Marine Research, Bremerhaven, Germany

Mahjoor Ahmad Lone
Department of Geography and Environmental Sciences, Northumbria University, Newcastle upon Tyne, United Kingdom

Sridhara Nayak
Research and Development Center, Japan Meteorogical Corporation, Osaka, Japan; Disaster Prevention Research Institute, Kyoto University, Kyoto, Japan

Rajiv Nigam
CSIR-National Institute of Oceanography, Dona Paula, Goa, India

Jai Ram Ojha
Department of Geology, College of Natural Sciences, Arba Minch University, Arba Minch, Ethiopia

Mohammed Osman
ICAR-Central Research Institute for Dryland Agriculture, Hyderabad, Telangana, India

Kalyani Panchamwar
Marine Bio-Geo Research-Lab for Climate and Environment (MBReCE), Department of Environmental Science, Savitribai Phule Pune University, Pune, Maharashtra, India

Rajani Panchang
Marine Bio-Geo Research-Lab for Climate and Environment (MBReCE), Department of Environmental Science, Savitribai Phule Pune University, Pune, Maharashtra, India; CSIR-National Institute of Oceanography, Dona Paula, Goa, India

Bijendra Pathak
Centre of Advanced Study in Geology, Banaras Hindu University, Varanasi, Uttar Pradesh, India

M. Prabhakar
ICAR-Central Research Institute for Dryland Agriculture, Hyderabad, Telangana, India

I.M. Shiromani Priyanthika Jayawardena
Department of Meteorology, Colombo, Sri Lanka

Dil Bahadur Rahut
Asian Development Bank Institute, Tokyo, Japan

D. Rajan
Karunya Institute of Technology and Sciences (Deemed to be University), Coimbatore, Tamil Nadu, India

Amba Shalishe
Faculty of Meteorology and Hydrology, Arba Minch Water Technology Institute, Arba Minch University, Arba Minch, Ethiopia

U.K. Shukla
Centre of Advanced Study in Geology, Banaras Hindu University, Varanasi, Uttar Pradesh, India

Ajab Singh
Department of Geology, Faculty of Science, University of Malaya, Kuala Lumpur, Malaysia

G. Srinivasan
Regional Integrated Multi-Hazard Early-Warning System for Africa and Asia (RIMES), Klong Luang, Pathum Thani, Thailand

S. Suresh Ramanan
ICAR-Central Agroforestry Research Institute, Jhansi, Uttar Pradesh, India

Tetsuya Takemi
Disaster Prevention Research Institute, Kyoto University, Kyoto, Japan

Mital Thacker
Biodiversity & Paleobiology Group, Agharkar Research Institute, Pune, Maharashtra, India; Department of Botany, Savitribai Phule Pune University, Pune, Maharashtra, India

Santosh Kumar Tripathi
Geological Survey of India, Hyderabad, Telangana, India

Kweku Afrifa Yamoah
Department of Archaeology, University of York, York, United Kingdom

About the editor

Dr. Neloy Khare, presently Adviser/Scientist "G" to the Government of India at MoES, has a very distinctive acumen not only in administration but also in quality science and research in his areas of expertise covering a large spectrum of geographically distinct locations like the Antarctic, Arctic, Southern Ocean, Bay of Bengal, Arabian Sea, Indian Ocean, etc. Dr. Khare has over 30 years of experience in the field of paleoclimate research using paleobiology (palaeontology)/teaching/science management/administration/coordination for scientific programs (including the Indian Polar Program), etc. Having completed his doctorate (Ph.D.) in tropical marine regions and Doctor of Science (DSc) in Southern High latitude marine regions toward environmental/climatic implications using various proxies including foraminifera (microfossil), he has made significant contributions in the field of paleoclimatology of southern high latitude regions (the Antarctic and the Southern Ocean) using micropaleontology as a tool. These studies coupled with his paleoclimatic reconstructions from tropical regions helped understand causal linkages and teleconnections between the processes taking place in Southern high latitudes with that of climate variability occurring in tropical regions. Dr. Khare has been conferred Honorary Professor and Adjunct Professor by many Indian universities.

He has a very impressive list of publications to his credit. He has over 150 research articles in national and international scientific journals. He has also edited the special issue of *Polar Science* (Elsevier), *Journal of Asian Earth Sciences* (Elsevier), *Quaternary International* (Elsevier), Journal of Applied Geophysics, and *Frontiers in Marine Science* as its Managing/Guest Editor. Dr. Khare has authored/edited many books, popular science articles, and technical reports. The Government of India and several professional bodies have bestowed him with many prestigious awards for his humble scientific contributions to past climate changes/oceanography/polar Science and southern oceanography. One of his books has received the most coveted award "Rajiv Gandhi National Award-2013" conferred by the Honorable President of India. Others include the ISCA Young Scientist Award, BOYSCAST Fellowship, CIES French Fellowship, Krishnan Gold Medal, Best Scientist Award, Eminent Scientist Award, ISCA Platinum Jubilee Lecture, and IGU Fellowship, besides many. Dr. Khare has made tremendous efforts to popularize ocean science and polar science across the country by way of delivering many invited lectures, radio talks and publishing popular science articles.

Dr. Khare sailed in the Arctic Ocean as a part of "Science PUB" in 2008 during the International Polar Year campaign for scientific exploration and earned the distinction of being the first Indian to sail in the Arctic Ocean.

Foreword

It gives me great pleasure in writing this Foreword that presents diverse perspectives in a focused way. The present book titled *The Role of Tropics in Climate Change—Global Case Studies* very aptly describes the understanding of climate change and its scenario over the tropical regions. It delves into a tropical record of climate change, covering geographically distinct archives of diversified unique proxies and locations of the tropics. It encompasses chapters on climate change assessment in space and time over tropical regions involving various tropical countries in an integrated manner in 18 distinctive chapters.

The Importance of Tropics in the Changing Climate by **Rajan and Gautam**, whereas a detailed account of Climate Change Impacts on Water Resources and Agriculture in Southeast Asia with a focus on Thailand, Myanmar, and Cambodia has been provided by **Srinivasan,** followed by a presentation of a case study of south western coastal region of Bangladesh highlighting the Interrelationship between climate justice and migration by **Ashrafuzzaman** and climate change projection. Precipitation concentration changes over Bangladesh have been put forth by **Towfiqul Islam et al.** On the contrary, **Priyanthika Jayawardena et al.** have focused on the climate variability, observed climate trends, and future climate projections for *Sri Lanka.* Similarly, **Singh et al.** made a thorough review of the artifacts, vertebrate remains and volcanic ash of Peninsular *Malaysia*, and their probable potential in gleaning past climate signals and **Aryal et al.** emphasized the climate change in *Bhutan* in their chapter.

This book also devotes dedicated chapters to the climate variability and change of *South America* by **Shalishe et al.** and the climate change scenario over *Japan* by **Nayak and Takemi.** While climate change trends and patterns over the *Taiwan* region have been highlighted by **Lone et al.** the potential of the lacustrine sediments in understanding the Holocene climatic conditions over *Ethiopia* has been explored by **Chaturvedi et al.** In a bid to reconstruct past climatic conditions over Myanmar region, **Panchang and Nigam** generated high-resolution palaeoclimatic records from the tropical delta shelf of Ayeyarwady, Myanmar.

Interestingly, from the Indian perspective, the geological response of climate and tectonics for the Pinjor Formation of the Upper Siwalik, Dehradun Subbasin, Garhwal Himalaya, *India,* through sediment model has been addressed by **Pathak and Shukla.** Whereas, Quaternary diatoms from the Asian tropics with their environmental indication potential for paleolimnological applications has been catalogued by **Thacker and Balasubramanian** and an assessment of the environmental disaster in Pithoragarh district, Uttarakhand, owing to climate change and anthropogenic interference has been made by **Akhouri et al.** On the contrary, **Nigam** explored the utility of the geological/paleontological applications in marine archaeology from Indian waters and **Ramanan et al.** estimated the contribution and preparedness of tropical agriculture against climate change from the Indian perspective. **Panchang et al.** have provided a detailed review of the mechanisms and proxies of solar forcing on climate with special reference to the Indian palaeoclimatic records.

Greenhouse gas—triggered climate change in future will have repercussions on global mean climate and sea level with contrasting regional manifestations, such as inconsistent temperatures rising for several decades, change in the rainfall pattern, and altered tropical cyclone activity. Undoubtedly, the changes in climate variability are inevitable irrespective of the fact whether the outcomes of climate model experiments can foresee the regional climate change prediction. It is therefore a fact that future climate scenarios attain significance based on certain articulated assumptions on climate change impact assessments, though these may be difficult to detect on natural and human systems.

The tropical regions are known to play an important role in the dynamics of global climate change. The climate is changing in the tropics, as it is in the rest of the world. The tropical ecosystems appear to be more sensitive to climate change and less able to store carbon. These tropical ecosystems are probably responding to global warming more energetically than anyone had expected. The evident consequences of ongoing climate change are clearest in high mountains, where both plants and animals have moved upslope in response to warming in several parts of the tropics. Similarly, in the tropical marine realm, coral reefs have suffered unprecedented mass bleaching episodes owing to the rising sea-surface temperature extremes. Nevertheless, few climate change scenarios of the tropics have been published to date.

It is hoped that this book will serve its sole purpose of dissemination of recent insights and information related to climate change occurring over the tropics since the geological past.

Date: January 2023
Place: Lucknow

(Ashok Sahni)

Preface

Climate change is largely affecting every ecosystem worldwide. Because of such an unprecedented shift in climatic conditions, living facing new challenges. The most devastating reflections of global climate change are frequent and intense drought, storms, heat waves, rising sea levels, melting glaciers, warming oceans, etc.

The tropics, impacting nearly half of the world's population witness glaring climate variability. Tropical species are known to have unique life histories and physiologies. Thus, tropical species and communities may show latitudinal responses to climate change. The tropics also witness the El Niño/Southern Oscillation (ENSO) in the Pacific. The challenge lies in understanding the causal link between tropical Pacific sea-surface temperatures and their variability and their response to anthropogenic emissions. We need to trace the fingerprints of climate change in the tropics.

The book *The Role of Tropics in Climate Change—Global Case Studies* attempts to summarize the current state of knowledge on the impacts of climate change in tropical regions and explores and discusses research priorities to better understand the ways in which tropical ecosystems are responding to ongoing climate change.

In the changing climate scenario according to experts as the planet's temperature increases, one of the regions of first land masses in the tropics has to go under severe changes. In the tropics, the effect of global warming on the local heat is dictated very early age. Climate change will be having more adverse impacts on tropical countries. The impact may be severe in Asian and adjoining regions. Generally, the baseline temperatures are increasing at a different rate in various parts of the globe. Sometimes the extreme heat due to global warming is one of the reasons for the increasing population in tropical regions. At times many events due to such extreme weather events are documented as most destructive both economically and socially accounting for more than 80% of people affected in the low latitudinal regions.

The atmospheric concentrations in the form of gases such as a greenhouse gas, carbon dioxide, etc., reflect a balance between emissions from various activities made by humans and animals, sources and sinks, etc. It is proven that due to various human activities the levels of greenhouse gases push the atmosphere on the higher side. These activities are a significant driver of climate change. Forecasting climate, climate variability, and climate change on different scales (space and time), **Rajan and Gautam** have detailed the significant role the tropics play in modulating the climates.

It is a known fact that Southeast Asia, being home to major river systems such as Ayeyarwady, Mekong, and Chao Praya, is of utmost importance in terms of world agriculture. 10% of Gross Domestic Product (GDP) in Southeast Asia is comprised of agriculture and the region's major export goods. Impacts of climate change are widely observed around the world, and Southeast Asia shows high vulnerability to such impacts mainly observed in the agricultural sector. Hence, it is important to understand the likely impacts of climate change on the sector and what could be done to use the climate information to reduce or avoid adverse effects. **Srinivasan** explored the likely impact of climate change based on case studies in Myanmar, Thailand, and Cambodia. Severe climate changes were observed in the recent past in the region, with increases up to 1°C per decade between 1951–2000. Future climate projections indicate that higher temperatures and more hot days are to be expected. Rainfall would likely increase by 5%–12% and 12%–20% for a temperature increase of 1.5 and 2°C, respectively. Such changes could impose direct and indirect negative impacts on agriculture, with increased water demand for crops, reduced growth and yield, and impact on livestock. Shifts in seasons also would result in damage to crops. Requirement for climate resilience is seen in the sector, and therefore different activities that could be implemented to build resilience are discussed. Improving the quality and validity of climate information, enhancing its application by stakeholders, institutional design to ensure the usage of information and developing capacity among stakeholders are among the proposed action.

Similarly, migration in the context of climate change is a justice issue owing to the fact that those countries that have contributed most to climate change have a responsibility for the vulnerable communities forced to migrate due to climate change. Climate justice is an issue of many countries especially coastal nations. Determination and the public discourse on climate change migration is lagging in raising questions of justice. Therefore, **Ashrafuzzaman** has ably addressed these vital issues in his chapter by adopting the quantitative and qualitative method and analyzes through a framework for climate justice. His chapter analyzes the migration behaviors, triggers of migration, different migration period, diverse migration related to climate change, types of migration, reason of migration, migration, and its adaptation options of southwestern coastal region of Bangladesh, which is one of the most climate-stressed zones of the planet. He further proposed in his chapter that migration is a multicausal event; even in situations where the climate is a transcendent driver of movement it is generally compounded by social, economic, political, and different variables. As a result all these environmental factors may turn into the climate injustice that leads the migration of the affected people. Finally, it is important that the climate migration must be provided with human rights and justice.

While **Towfiqul Islam et al.** made climate change projection: Precipitation concentration changes over Bangladesh. Their work intends to assess the variations in extreme event—derived precipitation and drought-probability characteristics over Bangladesh by employing 10 GCMs under the two scenarios including low RCP4.5 and high RCP8.5 scenarios from 2010 to 2099. PCI was used for the characterization of droughts and floods in terms of the frequency, intensity, and the number of precipitation days divided into different categories by each number interval. It was found that the bias-corrected GCM outcomes revealed suitable reliability with the observed dataset and were capable to replicate the actual condition of the drought-flood setting compared to the historical GCMs dataset.

They found that PCI value will increase in both station-wise values of precipitation and in upcoming decades also will increase based on 10 GCMs. Long-term change of PCI value showed an increasing trend in median, minimum, maximum, and second quartile values. Spatial distribution of PCI value showed the decadal change of PCI using percent change and also showed an increasing trend toward foreseeing period. The forthcoming changes in precipitation tended to exhibit a rise under both scenarios; the increasing rate was greater under the RCP8.5 than in the RCP4.5 scenario. The forecasted variation in climatic frequencies of drought-flood frequency based on severity under RCP4.5 and RCP8.5 exhibited a reliable magnitude for any future period, thus indicating the little impact of emission scenarios on the forecasted changes in the extreme drought events. Overall, the drought-flood frequency will probably reduce with time under changing climate about the rise of total precipitation. In particular, the occurrence of severe and moderate droughts and floods will be more frequent at the end of the century, according to our study. The frequency of extreme drought-flood events showed a noteworthy change in the upcoming decade under both scenarios. However, PCI value increase shows no homogeneous pattern for the whole country; some parts of the country will face drought whereas the remaining part of the country will face flood. The increasing value of PCI was not uniform; also in some decades it is showing an observing trend while other decades showed a decreasing trend in terms of the previous decade. Though it is expected that precipitation concentration value will increase significantly in the 2070s (2070–2090), therefore it can be concluded from this study that with increasing PCI value there will be more frequent floods and droughts in the future than in the past. The outcomes obtained in this study represent a vital insight into mitigating the losses in agricultural crop production for drought-flood-prone areas in the forthcoming period. **Priyanthika Jayawardena et al.** on the other hand studied the climate variability, observed climate trends, and future climate projections for Sri Lanka. In their chapter, they found that ENSO and IOD can be considered as one of the potential climatic drivers of interannual rainfall variability in Sri Lanka. El Nino and Positive IOD enhance SIM rainfall with anomalous rising motion while La Nina and Negative IOD diminish SIM rainfall with anomalous sinking motion over Sri Lanka. The significant impact of SWM rainfall was also evident over central parts of the Island with a reduction of seasonal rainfall during El Nino events and enhanced seasonal rainfall during La Nina events. Slight reduction (enhancement) in NEM rainfall over northwestern, northcentral, and central parts is evident during El Nino (La Nina). It is evident that annually averaged mean maximum and minimum temperatures are increasing across most of Sri Lanka. The difference between the maximum and minimum temperatures, diurnal temperature range, is decreasing, indicating that the minimum temperature is increasing faster than the maximum temperature. A significant decrease in the annual occurrence of cold nights and an increase in the annual occurrence of warm nights are also obvious. When compared with temperature changes, a less spatially coherent pattern of change and a lower level of statistical significance were observed in precipitation indices. The annual total precipitation has indicated a significant increase from 1980 to 2015. The trends in extreme precipitation events such as maximum 1-day precipitation, maximum 5-day precipitation, and total precipitation on extreme rainfall days are increasing at most locations, indicating that the intensity of the rainfall is increasing. The increase in precipitation extreme trends indicates that the occurrence of extreme rainfall events notably influences total annual precipitation in Sri Lanka. Therefore, the increases in total rainfall observed in many locations may be due in part to an increase in extreme rainfall events.

Undoubtedly, artefacts, vertebrate remains, and volcanic ash, preserved in the alluvial sediments and limestone caves of Peninsular Malaysia, have been getting attention for a long time to better understand human and animal existences as well as the source area of the distal pyroclastic material. **Singh et al.** made a thorough review to comprehend the history of humans, paleoecology, paleoclimate, and amount of the impact of tephra on the ecosystem. Their chapter suggests that humans were present in the Peninsula before the fresh air fall of Toba ash and stayed after the eruption as well as had genetic correlations with the contemporary African population. It also indicates that hominids were cave dwellers and hunting gatherers. The investigation of vertebrate remains reveals the existence of carnivores, omnivores, and herbivores animal communities that indicates the persistence of grasslands vegetation, swamp environment, and wet climate in the country during the Quaternary period. Geochemically and geochronological, the ash is very well correlatable with the Youngest Toba Tuff (YTT, 75 ka) eruption of Toba Caldera, Sumatra Indonesia, Indian Ocean, South China Sea, and Indian Subcontinent. The same had a modest impact on the living media, although had shifted C_3 forests to C_4 grasslands.

On the contrary, **Aryal et al**. in their chapter explore how climate change is affecting the Kingdom of Bhutan despite being only a net carbon-negative county. Their chapter exclusively assesses the impact of climate change on three specific areas—hydropower, agriculture, and glaciers—and argues for a shift from macro geopolitical analysis to discourses which focus on the microsector-specific aspects. These situated perspectives can help us deal effectively with challenges of climate change.

Similarly, **Shalishe et al.** aimed in their chapter to look at the effects of climate change on the South American continent. South America has experienced climate variability. They found that under three RCPs, both temperature and precipitation will rise in the future. The annual average minimum temperature is expected to rise by 1.25 and 1.6°C for the periods 2031−2050, respectively, compared to the baseline period, while the annual average maximum temperature is expected to rise by 1.25 and 1.85°C for the periods 2031−2050 under RCP4.5 and RCP8.5. Similarly, projected annual average precipitation has been expected to increase. Changes in an increased temperature and geographical and temporal rainfall variability have been evident. Droughts and famines have become more frequent as a result of climate change. Further, flooding, desertification, the loss of wetlands, biodiversity loss, and a drop in agricultural production and productivity have also been observed in the recent past. South America had made numerous measures in various sections of the country to mitigate climate change. Conservation, agriculture, home gardens, traditional agroforestry systems, harvesting non-timber forest products, protected area systems, reforestation and afforestation programs, renewable energy sources, and livestock sales and production are some of the mechanisms in practice for mitigating and adapting to climate change in South America.

In parallel, **Nayak and Takemi** investigate the climate change pattern and its impact on the extreme events over Japan regions from 140 ensemble climate simulation results conducted over Japan for the past 50 years (1961−2010) and the future 60 years (2051−2110). They noticed slightly positive trends in the precipitation amounts and the frequencies and a significant positive trend in temperature over the entire Japan. The climate over Japan is expected to be drier in the future, but the intensity of the extreme precipitation is likely to be much stronger over the entire Japan. The frequencies of precipitation intensities are also expected to increase over some regions of Japan. The variations in the anomalies of the annual precipitation amounts in the future warming climate are likely to be almost the same over the entire Japan as seen in the present climate. The climate in the future is likely to be warmer with stronger hot days with increased frequencies of temperature extremes compared to the present climate. It is noticed that the intensity of the precipitation extremes (averaged of top 10% precipitation days' intensity) in the future climate is likely to increase by up to 5 mm/d over NP and NS, 10 mm/d over ES and EP, and 15 mm/d over WP, WS, and OK. The temperature extremes (mean of top 10% hot days) are also expected to be much hotter in entire Japan. The mean temperature of the top 10% of hot days over NP and NS is likely to increase up to 5°C in future climate, while that of over OK is expected to up to 3°C and over other regions, it is up to 4°C.

The trends and patterns of climate change over the Taiwan region have been highlighted by **Lone et al.** In their chapter, an attempt is made to understand Taiwan's present and past climate. Climatic data from Central Weather Bureau, Taiwan, is used to understand the contemporary climate scenario, and published paleoclimatic records (mostly lacustrine) are reviewed to understand the past climatic changes on this western pacific island. Climatic indices reveal that northern Taiwan receives rains around the year while southern Taiwan is mostly dry during the winters. Furthermore, southern Taiwan is relatively hotter (1.2°C) than northern Taiwan. A comparison of paleoclimatic records reveals no consistency, which may be attributed to errors in the chronology, and/or typhoon's role in altering the climatic signal in these archives.

The lacustrine sediments have been studied by **Chaturvedi et al.** for understanding the Holocene climatic conditions over Ethiopia. They found overall deposition of sediments in Chamo Lake under low energy conditions. The spatial distribution of sediment characteristics and organic matter contents reflects the modern hydrodynamic conditions and organic production in the lake. They recognized the great potential of the Chamo Lake sediments in unraveling the past climatic mysteries over Ethiopia.

Indubitably, the climate is influenced by the coupled Ocean-Atmosphere system across the globe. Global climate prediction and climate modeling that facilitates resource management needs reliable datasets of past climate variables. IPCC in their recommendations reiterated the need for data from different parts of the globe to strengthen general circulation models. Globally there are still gap areas and regions having meagre data. The Northern Indian Ocean is one among such region where data do not exist or as a complex region that is less understood. The Ayeyarwady (Irrawaddy) River of Myanmar (former Burma) is one of the less known large rivers in the world and the second-largest river of Southeast Asia in water discharge; and influenced by both the southwest and the northeast monsoons, its sediment archives host very well-preserved and interesting, high-resolution paleoclimatic records. Being the least understood regions of the Northern Indian Ocean, **Panchang and Nigam** aptly present an account of intradecadal paleoclimatic records for the past 5 centuries and sea-level history along the coast of Myanmar.

Furthermore, to assess tropical climate shift over the Indian Peninsula, the geological response of climate and tectonics for the Pinjor Formation of the Upper Siwalik, Dehradun Subbasin, Garhwal Himalaya, India, through sediment model have been addressed by **Pathak and Shukla** who noticed about 360-m-thick succession of the Pinjor Formation of the Upper Siwalik Subgroup in Badshahibagh Rao area of Dehradun Subbasin which has been classified into six lithofacies, viz.: matrix to clast supported conglomerate, clast supported conglomerate, planar cross-bedded gravelly sandstone, horizontal bedded gravelly sandstone, low angle planar gravelly sandstone, ripple bedded sandstone, and laminated mudstone. With fining upward sandstone units at the base to the multistorey conglomerate units at the top, the Pinjor Formation shows an overall coarsening upward succession formed by prograding small- to moderate-sized alluvial fans in the piedmont setting of the Himalayan Foreland Basin. Paleocurrent data measured from the imbricated clast and planar cross-bedded sandstone denote paleoflow along southwesterly slopes. Pebbles to cobble size clast of the Pinjor are generally quartzite with a minor amount of sandstone, siltstone, and mudstone denoting provenance located in the north comprising the Lesser Himalaya and uplifted parts of the Siwalik rocks. The framework grain composition revealed from sandstone petrography inferred their derivation from metamorphic, igneous, and sedimentary source rocks.

Diatoms (siliceous microscopic algae) are one of the main biotic elements used in assessing the ecological state of all aquatic environments and play a vital role in understanding the ecological conditions in freshwater and marine environments. Being siliceous diatoms are often well preserved in stratigraphic deposits, where they can offer great potential to trace past environmental conditions. The sensitivity of diatoms to a large variety of environmental variables makes them the most widely used paleoecological proxies, providing many research opportunities in paleo sciences ranging from reconstructing present to past environmental conditions. Considering our limited knowledge of the diatom taxonomy and ecology from the tropics, database, including the list of taxa reported till today, their ecological preferences are very scarce. To fill up this gap **Thacker and Balasubramanian** have made an exhaustive review in their chapter from tropical Asia focusing on diatom-based paleoecological reconstructions covering climatology, hydrology, and paleo-water quality assessment using the fossil diatom taxa and their ecological indications. Further, this work provides the first comprehensive checklist of the fossil diatoms from tropical fresh and marine water from Asia, comprising 266 species from 64 genera. This list also provides each taxon's spatial and temporal extent for its meaningful application in environmental reconstruction. This taxa list will be a valuable resource for researchers across the tropics for paleoenvironmental reconstruction.

Similarly, the state of Uttarakhand experienced very heavy rainfall during the period of June 15−17, 2013, causing severe damage to life and property in Uttarkashi, Tehri, Pauri, Rudraprayag, and Chamoli districts of Garhwal and Bageshwar, Almora, and Pithoragarh districts of Kumaon region. Prominent glacier-fed rivers of Pithoragarh District, mainly, Dhauliganga, Goriganga, and Kali rivers, played havoc causing heavy damage to life and property, particularly in the downstream areas.

The rivers with unprecedented discharge swept away many communication routes and settlements/houses located in the villages on the banks of these rivers and led to the emergence of new slides and reactivated the old slides. The investigations included coverage of a quantitative target of 531 line km, making an inventory of 92 landslide incidences including 48 villages in various parts of the Pithoragarh District. **Akhouri et al.** presented an account of the extremes of devastation, the vulnerability of the Himalayan ecosystem, the perceptible shift in the climatic regime, and better preparedness for the future to combat the forces of nature in their chapter. Their study revealed three predominant mass wasting processes viz.: landslides, river bank erosion by flood, and land subsidence. Out of 92 incidences, 52 nos. are landslides, 32 nos. are river bank erosion due to high floods, and 8 nos. are land subsidence. Out of a total of 60 landslides/subsidences, 40 nos. are debris slides, 19 nos. are rock-cum-debris slides, and 1 number was a rock slide. Out of 92 incidences, 21 nos. are new landslides and 39 nos. are old reactivated landslides and all 32 nos. of river bank erosions were new as an aftermath of this event. Approximately, 22 km of the road stretch was severely affected and 21 km of the road needed realignment. In the aftermath of incessant and torrential rains that lasted for nearly 72 hours, the trunk streams like Bhagirathi, Alaknanda, Mandakini, Dhauliganga, Goriganga, Pindar, and Kali swelled enormously. The rivers with unprecedented discharge swept away many settlements/houses on their banks and other infrastructure located in the villages and townships on the banks of the rivers. There has been a lot of speculation and discussion regarding the rainfall data in different parts of the Pithoragarh District. During June 2013 the actual rainfall has been recorded as 400.5 mm which is 34% more as compared to the average normal range, i.e., 299 mm in the last 30 years. During the period of devastation, i.e., between June 15 and 18, 2013, the minimum and maximum rainfall recorded in Pithoragarh and Munsyari areas were 25 and 117 mm, respectively. There is a characteristic deflection from the normal in almost all the districts of Uttarakhand with Dehradun (1436%) recording the maximum departure in rainfall and Pithoragarh District possibly recording the minimum deflection (238%) from normal. This abnormal precipitation cannot be delinked to a climatic aberration and thus could be a testimony to short-term climate change and a precursor to a change that will be irreversible. The lack of preparedness of people in the hills locating themselves on the river-borne terraces of major rivers found themselves sitting on a catastrophe in the wake of the event. The fragility of the mountain chain with high relief, the adverse inclination of the discontinuity planes, intense rain, flash flood, and avalanches made the Himalayan terrain vulnerable to an event like this. Arguably, excessive human interference with a lack of awareness and preparedness contributed to the degradation of the Himalayan ecosystem.

On the contrary, **Nigam** in his chapter identified the importance of the geological/paleontological applications in marine archaeology, which is an upcoming Branch of Science which needs a multidisciplinary approach for more acceptable conclusions. In this, the tools used by geologists and paleontologists can play significant roles as exhibited by the few examples from The Indian waters. India has more than a 7000-km-long coastline, and a large number of researchers are required with inclinations to apply their expertise in the field of marine archaeology. This need requires more training programs and starting of specialized courses in universities and research organizations. Marine archaeological sites especially places like Lothal (oldest dockyard) and Dholavira (oldest evidence of tsunami protection measures and water conservation knowledge) also provide an excellent opportunity to enhance international tourism and thus increase national earning and job opportunities. Whereas, **Ramanan et al.** provided an overview of the National Innovations in Climate Resilient Agriculture (NICRA) programme of India. This is the flagship programme of the Indian Council of Agricultural Research (ICAR) aimed at supporting the Indian farming community against tropical climate vulnerability. Owing to the difference between climate-resilient agriculture and climate-smart agriculture, introspection of the developments of NICRA will help in contemplating the contribution of Indian Agricultural Research to the global climate mandate. For this, we evaluated the scientific contribution of NICRA based on a comprehensive meta-analysis of the publications listed in the Web of Science (WoS). The scientific output under NICRA is commendable as there were 424 research articles and 7 proceeding papers published between 2012 and 2020 by 1336 authors. Most of the research articles were multi-authored documents with 3.09 authors per document. A few certain publications could register more than 100 citations. Also, 47 publications were from researchers affiliated with more than 20 institutions outside the ICAR system indicating the nationwide capacity-building initiative under NICRA. While shortlisting the top 10 authors among the total authorship, most authors were from animal science and crop science backgrounds followed by the natural resource management (NRM) and the fisheries sciences. Agricultural research mostly tends to address local problems; therefore, quite a few research papers might have been published in journals not listed in WoS. Yet, it can contemplate the veracity of research under the NICRA initiative and conclude that scientifically, climate change research in India through NICRA has advanced our overall understanding of the impact of climate variability. The programme further facilitated the development of technology and enabled their interventions to minimize the impact of climate variability.

Interestingly, it is a known fact that the Earth's climate is a function of the Earth's radiative budget which is dependent on the Sun's incoming energy and the energy escaping the Earth's atmosphere. While the Sun is the primary source of heat to the Earth, the structure and composition of the Earth's atmosphere modulates the amount of heat that the Earth receives or reflects. However, the Sun itself is not a constant. The energy that it radiates changes over different time scales due to the Milankovitch cycles and changes in its solar activities, namely, solar flares, geomagnetic storms, coronal mass ejections, solar wind, and emission of solar energetic particles. Such vital causative aspects of the climate change are discussed by **Panchang et al.** covering the historical role of the Sun in controlling the Earth's climate and the various solar cycles observed in the instrumental or paleoclimatic records.

This collection of chapters in this book captures and disseminates some perspectives on climate change from the tropic context. Some of the best-known geologists, meteorologists, archaeologist, and agriculture/climate scientists have put forward their concerns and convictions in this book. It is hoped that this book will serve its sole purpose of dissemination of recent insights and information related to climate change occurring over the tropics since the geological past.

Date: January 2023
Place: New Delhi

(Neloy Khare)

Acknowledgments

It is my great pleasure to express my gratitude and deep appreciation to all the contributing authors from different countries. Without their valuable inputs on various facets of tropical regions in modulating climates and the impact of ongoing climate change observed over various regions around the tropics, this book would not have been possible. Various learned experts who have reviewed different chapters are graciously acknowledged for their timely, constructive and critical reviews.

I sincerely thank Dr. M. Ravichandran, Secretary, Ministry of Earth Sciences, Government of India, New Delhi (India), for his support, and encouragement. Prof. Govardhan Mehta, FRS, has always been a source of inspiration and is acknowledged for his kind support.

The valuable support received from Prof. Avinash Chandra Pandey (IUAC, New Delhi), Prof. Anil Kumar Gupta (IIT Kharagpur), and Dr. K. J. Ramesh (formerly IMD, New Delhi) is gratefully acknowledged. This book has also received significant support from Akshat Khare and Ashmit Khare, who have helped me during the book preparation. Dr. Rajni Khare has unconditionally supported enormously during various stages of this book. Hari Dass Sharma from the Ministry of Earth Sciences, New Delhi (India), has helped immensely in formatting the text and figures of this book and bringing it to its present form. Publishers (Elsevier) have done a commendable job and are sincerely acknowledged.

Date: January 2023
Place: New Delhi

(Neloy Khare)

CHAPTER 1

The importance of tropics in the changing climate

D. Rajan and S. Gautam
Karunya Institute of Technology and Sciences (Deemed to be University), Coimbatore, Tamil Nadu, India

1. Introduction

Advances in Science and Technology can help toward seeking solutions to sustainable development. In this context, it is important that the Global Earth System is monitored/observed efficiently as far as possible and its status can be predicted (Moss et al., 2010) in this changing climate scenario. The issue of climate change is being debated globally and the science of climate change has been at the center of the stage of international concern for the last decade. There is no doubt that there are differences of opinion/schools of thought among the developing and developed countries (IPCC, 2017) about climate change. As we know the effects of climate change are increasing natural disasters, melting glaciers, sea level changes, monsoon variabilities, an increase in temperature, and even problems related to pandemics like COVID, etc. For many years the international level discussion (IPCC, 2017) about the climate change issue is being undertaken as a problem which demands solutions in terms of stopping/limiting burring waste material or pollution/emissions etc and finding a solution through other alternative energy routes and attempting reduction of greenhouse gas emissions. This chapter analyses/documents the issues related to climate change, global warming, and national security. Also, it documents the topic "Climate Change has potential toward bringing insecurity in the geopolitical future of the word".

2. Organization of the chapter

We have organized this chapter with the following items as sections: (1) Crisis of human-caused Climate change, (2) General circulation atmospheric flow pattern, (3) The impact of climate over the tropics, (4) The impact of weather over tropical climate, (5) Implication for the future climate, (6.1) Accordance of Climate change due to various greenhouse gases, (6.2) A Rational view of climate change, (7) Climate change in the Indian contex,t (8) Climate change and national security, (9) Climate change and South Asia, (10) Impact of climate change over Indian agriculture, (11) Tropical teleconnection impacts on the Antarctic climate changes, (12) Projection of precipitations and teleconnection due to climate change, (13) Lesson from pandemic COVID, (14) Adaptation to the heat wave and climate change, (15) Recent observations of greenhouse gas, temperature, and Ocean heat in the warming climate situation, (15.1) Displacement, (15.2) Adaptation, (15.3) Regional climate risk.

Thus by documenting through these sections, our country can gear up for the best climate services through the exchange of future climate projection, current climate information, knowledge, and practices for risk management will ultimately support all the user agents at all levels.

2.1 Crisis of human-caused climate change

The Authors of this chapter experienced the climate change crisis linked with the Heatwave during the year 2022. From March to May heat waves prevailed in India and its adjoining country was about many times more than the normal value that was likely to happen because of 'human-caused climate change. The result of the recent study showed that an unusually long and early onset of Heatwave spells like the one in India and Pakistan (Hindustantimes Daily Newspaper, 2022) have just experienced is very rare, with a chance of occurring only once in 100 years. But this year the human-caused climate change has made it about 30 times more likely to happen. It means that it would have been extraordinarily very rare without the effect of "human-induced climate change" (Hindustantimes Daily Newspaper, 2022). The climate experts document the above statement with the climate data for a long period.

The sixth Intergovernmental Panel on Climate Change Assessment Report has revealed global surface temperature increase from 1850 to 1900 and 2010−201,911 as $0.8°C$ to 44 $1.3°C$, with the best estimate of $1.07°C$ (IPCC, 2021). In this year 2022 during early March most parts of India and Pakistan land region experienced an unusual heat wave condition; which has taken the life of 90 personnel. This India and Pakistan heat wave triggered an extreme Glacial Lake Outburst flood in Northern Pakistan and forest fire in India, particularly in the hill states of Uttarakhand and Himachal Pradesh. On the day of submitting this chapter, Northern India reported an event of a cloudburst. Extreme heat also reduced India's wheat crop yields, causing the government to stop wheat exports. In addition, this heat wave leads to a shortage of coal ultimately implying power outages, etc. This heat wave spell caused serious health impacts and mortality which may not have been recorded in the statistics.

In order to quantify the effect of the climate crisis, the temperature reaches more than $6°C$ above normal in the northern belt of India. The climate modeler forecasted that the return period of such an unusual heat wave spells to be around 100 years in a climate of $1.2°C$ global warming. It is added that the same event would have been about $1°C$ lesser than in a preindustrial climate. The statistics study focuses on the average maximum daily temperatures from March to April 2022 in north-western India and south-eastern Pakistan. The observation result showed that this type of prolonged heat wave is still rare, but mainly the human-caused climate change has made it happen in this current year.

Until overall greenhouse gas emissions are halted, the mean temperature will continue to rise and increased heat wave events will become more often over India. For example, if the global temperature rises $2°C$, a heat wave like this would be expected as often as once every 5−6 years. This means that in a worst-case scenario, a once-in-a-100-year event can return once in 5 years (Grise et al., 2019).

In the Indian scenario, March was the hottest in India in the past 122 years statistics with Pakistan also witnessing record temperatures. March was extremely dry with 62% less rain than normal over Pakistan and 71% below normal over India (Hindustantimes Daily Newspaper, 2022). Many times the temperature was recorded above $49°C$ in the station called Jacobabad in Sindh, and 30% of the country was under the impact of a heat wave. Toward the end of April and May 2020, the heat wave extended into the coastal regions and eastern parts of our country.

Despite setbacks from COVID-19, real-time measurements indicate that global greenhouse gas emissions continued to increase in 2021 (Storymaps.arcgis.com). The current scenario of increasing global temperatures has contributed to more frequent and severe extreme weather events around the world; which include cold and heat waves, floods, droughts, wildfires, cyclones, thunderstorms, etc.

2.2 General atmospheric circulation

It is known that the weather and climate are the consequence of the (a) Rotation of the Earth (b) Earth is heated up by getting energy from the Sun (c) The fundamental laws of physics in particular thermodynamics. The atmospheric circulation can be thought of as a heat engine driven by the main source Sun. And whose energy sink, ultimately, is the blackness of space. In terms of Physics, the work performed by that engine causes the motion of the air masses and in that process; it redistributes the energy absorbed by the Earth's surface near the equator to mid-latitudes. Atmospheric General circulation is the long-term average large-scale movement of air together with ocean circulation. It means that the thermal energy is redistributed on the outermost surface of the Earth. The Earth's atmospheric circulation varies over time, but the large-scale structure of its circulation remains constant. As a schematic diagram, this is the average picture of these circulations that are shown in (Fig. 1.1). In the lesser scale weather systems-mid-latitude depressions, or tropical convective cells—occur chaotically, and long-range weather predictions of those cannot be made beyond 15 days in practical or a month in the research studies.

The planetary-scale atmospheric circulation called "cells" shift toward the pole in warmer periods and comes back due to cold weather. It is true that a property of the Earth's size, rotation rate, heating, and atmospheric depth, all of which change little due to this circulation. Over many millions of years, atectonic uplift can significantly alter their major elements, called the jet stream, and shifting ocean currents. During the peak summer climates, a third desert belt may have existed near

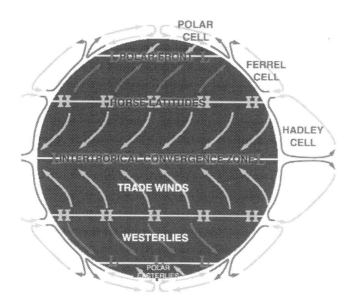

FIGURE 1.1

The schematic diagram shows the regions of high and low pressures. The flow patterns toward the equator from the tropical belt.The weserlies from tropics to mid-lattutdes. Teleconnection-induced climate changes. Schematic depiction of amundsen sea low variability and corresponding impacts on the antarctic climate, including temperature advection, sea ice redistribution, warm water intrusion, and ice sheet melting. *Red arrows [dark gray arrows in print]* and boxes indicate warming processes and impacts, and *blue arrows [light gray arrows in print]* and boxes indicate cooling processes and impacts. The low variability can be linked to tropical teleconnections. *H* indicates the high-pressure center; *L* indicates the low-pressure center.

the equator (Shaw et al., 2019). Thus the general circulation is key and useful information for tropical weather, mid-latitude weather, and climate.

2.3 The impact of climate on the tropical region

The tropics are mainly the warmest regions of the globe, where Hardly circulation prevails during summer. The reader of this chapter may be aware of the general circulation model which is classical textbook information. In this section we have describes that Due to the warming climate, there will be changes over the tropics which include warmer temperatures, rising sea levels, frequent and adverse weather events such as tropical storms, monsoon heavy rainfall, severe flooding, intended drought-like situations, heat waves, and cold waves, etc. A lot of evidence exists that the tropics are expanding during the last 2 decades. Recently it is learned that this activity is continuing. Both the space (Norris et al., 2016) and ground-based observations, and climate model prediction have shown that the edges of the tropics and its associated sub-tropical area are shifting toward higher latitudes due to the warming of the earth.

Here when we explore the reason for expanding the following outcome is listed. From "the General circulation model" it is noted that the mid position of the tropics is shown by rising air and intense convection. This region is popularly called as Inter-Tropical Convergence Zone (ITCZ) when the boundaries of the tropics are categorized by descending air and dry conditions. When we explore the reason for this expansion the pilot study shows that (a) The width of the tropics closely follows the displacement of the oceanic higher meridional temperature gradient, and (b) The mid-latitude ocean experience more surface warming. The enhanced subtropical warming which is independent of natural climate oscillations, such as the Pacific Oscillation leads to poleward advancement (Hu et al., 2020).

In the recent past, scientists find that the dynamic motion drives this phenomenon that has puzzled the climate scientist, however, is still not entirely clear about this expansion. Some more schools of thought are involved in these findings like the rising amount of greenhouse gases, and aerosols also contributing to the tropical expansion. While we talk about the climate modeling results it was documented that when compared to observation and modeling findings are quite simple. The observation from the surface and upper air shows a much more complex evaluation of expanding tropics (Fig. 1.2). Space-based observations help us to better monitor and understand the Earth in a detailed manner (Sikka, 2009). Over each oceanic region, we found that the width of the tropics closely follows the displacement of oceanic mid-latitude meridional temperature gradients (Hu et al., 2020).

It was until a year later that we realized the evidence of a rapid expansion of the tropics, the region that encircles Earth's waist like a green belt. The hearts of the tropic where most lands exist are lush, but the northern and southern edges are comparatively drier. It is noted that these parched borders are growing i.e., expanding into subtropics and pushing them toward the poles. Another major finding of this chapter says that the rate and location of expansion are not uniform. One more argument might involve different forces in the Northern and Southern hemispheres (Heffernan, 2016). South of the equator, the tropical expansion was found to be stronger in the summer (www.nature.com). This leads some scientists to suspect that it is related to the pattern of ozone loss in the stratosphere. The pollutant pushes the ozone molecules above Antarctica in the winter, which triggers circulation changes throughout other parts of the Southern Hemisphere during summer (Polvani et al., 2011).

2.4 The impact of the tropical belt on the climate

The following references support this chapter (a) The expansion commonly referred to as tropical expansion (Archer and Caldeira, 2008) (b) Pole ward shift of the storm tracks (Yin, 2005) (c) Pole ward

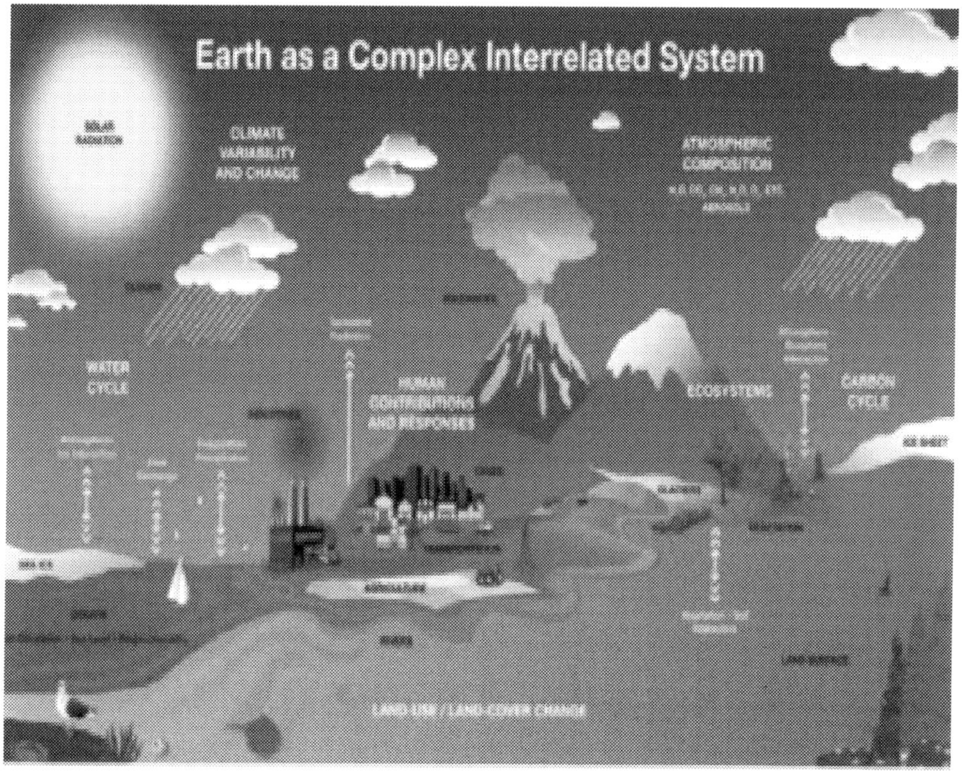

2. Organization of the chapter

FIGURE 1.2

The average annual global greenhouse gas emissions were at their highest levels in human history, but the rate of growth has slowed. Since 2010, there have been sustained decreases of up to 85% in the costs of solar and wind energy, and batteries.

shifts of westerly's (Chen et al., 2008) (d) Shifting of jet streams (Archer and Caldeira, 2008) (e) Change in rainfall patterns (f) Permanent displacement of clouds (Norris et al., 2016) (g) The climate model suggest increasing greenhouse gases (h) Depletion of stratospheric ozone (Polvani et al., 2011) (i) The expansion has an interannual variation (Grise et al., 2019) (j) Observed expanding tropics are due to natural climate oscillations. Even though they are many theoretical studies, the underlying dynamical mechanism driving tropical expansion is still not entirely understood (Shaw, 2019). Many research studies suggest that the atmospheric processes even in the absence of ocean dynamics may lead to tropical expansion.

A very few research study analyzed the width of the tropics in a zonal mean framework, in which the regional characteristics of the tropical area is masked by the zonal average. Many stations lie just outside of the tropics and could soon be smack in the middle of the dry tropical edge. A shift of just a degree of latitude in a particular station near the equator (Heffernan., 2016) in the northern hemisphere—that's enough to have a huge impact on those communities in terms of how much rainfall they will get. This tropical expansion will have an impact on rainfall, evaporation, condensation, etc.

2.5 Implications for the future climate

In recent decades (1979—2010), a significant increase in the global surface temperature trend (>0.17°C/decade) has been observed compared to the 0.07°C/decade trend from 1901 to 2010. The northern/southern hemispheric trends are recorded as 0.08/0.07°C/decade for 1901—2010 and 0.24/0.10°C/decade for 1979—2010 (Morice et al., 2012). Consistency of climate model results with the theory and observations lends strong support to the capacity of global climate models to properly simulate annual mean temperature on the regional scale (Kumar et al., 2020). In this given situation the projected 66% confidence interval for the raise of temperature across all tropical land regions is 20°S ~ 20°N, which is consistent with simulated tropical mean warming 1.4°C or 1.5°C warmer climate. Scenarios of Indian future climate during 2020, 2050, and 2080 simulated at the Indian Institute of Tropical Meteorology have been released in 2009. This highlights that significant improvement is needed in climate models to improve the representation of the boundary layer (Kumar et al., 2020).

2.5.1 Accordance of climate change due to various greenhouse gases

As far as global change is concerned IPCC (2007) indicates that the atmospheric concentration of carbon dioxide and methane has increased substantially. This section of this chapter discusses the physical properties of greenhouse gases and the factors that determine their effectiveness as pollutant gases that can cause global warming. Around the world, national environmental requirements have often steered technological innovation, management improvement, better planning, food security, etc. The earth is a planet in dynamic equilibrium, in that it continually absorbs and emits electromagnetic radiation. In terms of chemistry, Carbine dioxide, and methane currently contribute more the 90% of the total radioactive forcing of greenhouse gases (Fig. 1.2). But it is noted that the control of carbon dioxide levels will not be the complete solution for this warming. Many policymakers are involved in this discussion still it is debatable. Often it is implied by politicians and the media people. Concentration levels of carbon dioxide in the earth's atmosphere are indeed a serious cause of this concern. Many countries are now putting efforts to reduce the activities which emit these gases. Carbon dioxide levels in the atmosphere correlate strongly with the lifestyle of the people. The developed world United State of America (USA), and Europe is taking steps to reduce it. The challenge is to effect guidelines to reduce carbon dioxide concentration considerably without compromising the standard of living of the people and negating all the comfort that engineering technology has brought like Air conditioning, washing machines, etc. Continuously the planet Earth receives ultra-violet rays in the visible range from the Sun; in return, it emits the power gained in the form of energy in the infrared range. The energy balance equation says that "energy in" must equal "energy out" for the temperature of the planet to be constant. This energy balance equation (equality) can be used to calculate the average temperature of the planet will be.

2.5.2 A rational view of climate change

The World Meteorological Organization has pointed out the uncertainties in observations and their interpretation in the annual report 'On the Global Climate. It is not clear here to go deeper into this aspect of uncertainty in global climate change studies and future projections of climate change. From the climate modeling studies, it is learned that the southern hemisphere has been warming at a slower rate compared to the northern hemisphere. Detailed inferences about the impact of global warming on the tropical area (Goswami et al., 2006), however, these findings are drawn only with climate models. The tropical Atlantic sea surface temperature time series represents that of area-weighted mean temperature over the tropical Atlantic belt 20°S ~ 20°N (Fig. 1.3).

India is not mandated for emission cuts as per the popular Kyoto protocol, even then we have the following items that can be considered to reduce emissions and improve energy efficiency (Attri et al., 2009). (a) The setting of Bureau of energy efficiency, (b) Power sector reforms, (c) Options of hydro and renewable energy, (d) Promote Mass Rapid Transport system, (e) Minimize the gas flaring activity, (f) Usage of coal and coal products, (g) Use of cleaner fuel for road and air transport vehicles.

2.6 Climate change in the Indian context

During the British era i.e., 19th and 20th centuries, a lot of emphasis was on the exploitation of India's natural resources. Mainly three areas are identified where the resources are available (Sikka, 2009). They are the biosphere, geosphere, and atmosphere. In those days three scientific departments were established (a) Geological Survey of India which corresponds to minerals (b) Botanical Survey of India which corresponds to forests Zoological Survey of India which corresponds to animals, birds, etc (c) India Meteorological Department which corresponds to the atmosphere. These departments established their networks to monitor the respective environments and provide policy options to the managers of the country. In the Indian context, environmental global warming problems are far too many when compared to other parts of the globe. In the geographical location, India plays a major role as it shares a border with two nations with whom it has already we have fought wars in recent years. If we exclude China, the other neighboring countries Srilanka, Nepal, Pakistan, and Bangladesh are underdeveloped countries. As far as India is concerned the issues concerning climate change are centered around (Sikka, 2009) (A) Adverse impacts of climate change including the costs and capacities of adaption, (B) Position to be taken at global climate change negotiations concerning our emission level, (C) Opportunities on these negotiations might provide to further sustainable development in the country.

2.7 Climate change and national security

Climate change and the security of any nation are interlinked. Many documents on history about this interlink are available depicting climate shifts or extremes of weather triggering debates and even

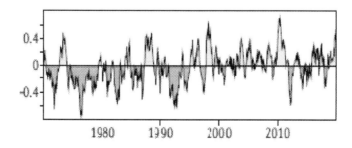

FIGURE 1.3

Distribution of temperature anomalies during the years 1970–2020.

contributing to the rise and fall of civilizations and nations. There is interconnectivity between environmental factors and violent arguments like migration and environmental mismanagement between ethnic conflict and resource debates. During the yesteryears, the issue of conflict was not climate change per se. It concentrated more on the management of resources. The overall research in this field based on realistic situations proves the fact there can be no single real factor to the environment-induced conflicts. Against the backdrop of this, it becomes essential to include environmental issues in debates on national security.

2.8 Climate change and South Asia

Very recently the Asian Development Bank supported financially to handle the climate change situations namely the melting of Himalayan glaciers and other direct threats to potable water, and food security for more than a billion people in South Asia. The South Asian countries Nepal, India, Sri Lanka, Afghanistan, and Bangladesh are particularly vulnerable to falling crop yields caused by glacier retreats, droughts, floods, unseasonal rainfall, and other climate impacts. This chapter concludes that if the current situation persists till the year 2050, the yield of the food grain (Table 1.1) in this South Asian region decreases significantly. The problem of food security will lead to higher hunger and humanity-related issues like starvation, death, etc.

2.9 Impact of climate change on Indian agriculture

Climate change impact will have a disproportional adverse impact on developing countries (IPCC, 2007) like India and Pakistan. Climate change will exacerbate problems related to rapid population growth, increasing poverty, illness, and heavy reliance on agriculture and the environment. There is evidence to show that India has suffered famines throughout its history, and all the more so during the period of British rule. Frequent famines affected many large regions and in some years even the entire India and which brought misery, loss of life, and property. Although famines were primarily caused by a failure of crops due to extremely poor monsoon rains, the difficulties of the populations got compounded by the general apathy of the rulers, lack of relief portions, or inconsiderate measures like increased taxation being introduced at the same time of drought-like situation. During the 19th century, famines kept on affecting different parts of the country, and millions of people died of hunger, and since then had no alternative source of livelihood. In the following year, there was a famine in Madras, and there was a country-wide famine in 1899. The year 1918 is remembered as the year of the great Indian famine. In 1943 also there was a major famine in West Bengal. Even after India got

Table 1.1 India's food grain production against population since the year 1940.

Year	1940	1950	1960	1970	1980	1990	2000	2010	2020	2021
Food grain production (million tonnes)	20.6	50.8	82.1	108.4	129.6	176.4	201.8	241.2	308.1	314.7
Population in millions	240	361	439	548	683	846	1000	1380	1396	1394

independence in 1947, it had to carry on for another 20 years with its battle on the food front. With the partition of the country, it lost some of its most fertile lands and the population (Table 1.1) has been increasing continuously.

Then the Green Revolution and its impact transformed the country into a self-sufficient nation as far as food grain is concerned. The building of huge buffer stocks, a good public distribution system, and an efficient relief and disaster management organization, have all freed the Indian population from the miseries of famine. Today, India has a total of arable land of 180 million hectares. Out of this, the irrigated area is 55 million hectares. Poverty and hunger are projected to persist and some of the most vulnerable and poorest economies will remain in the future.

In recent decades, climate changes have caused impacts on natural and human systems on all continents and across the oceans. The number of hydro-meteorological hazards in particular such as droughts, floods, tropical storms, and forest fires; which were measured on an average of 200 per year during the years 1987−98 increased to 365 per year in 2000−08. The percentage of the irrigated area to the total agricultural land area in India varies from state to state. Although the broad pattern of Indian agriculture is tuned to the annual distribution of rainfall, it is vagaries of the monsoon and the distribution of rainfall across the country. Our country's food grain production in million tonnes against the total population in million is distributed in Table 1.1. The statistics shown in the table speak for themselves.

Even with the modern advanced technologies used in the agricultural sector over the last half a century, climate variability has a large impact on agriculture. Our country's farmers' yield is heavily dependent on seasonal rainfall and temperature. Of the total annual crop losses in world agriculture, many are due to direct weather and climate effects such as floods, drought, untimely rain, frost, hail, and storms. Climate change will exacerbate existing threats to food security and livelihoods which is a combination of increasing frequency of climate hazards, diminishing agricultural production in hard regions, expanding health risks, and heavy demand for water.

2.10 Tropical teleconnection impacts on the Antarctic climate changes

In the modern satellite era, a considerable amount of climate changes have been noticed in the tropical oceans and Antarctic as well. This section summarizes the tropical teleconnections to the southern hemisphere high latitude regions. Here the two important teleconnections are discussed i.e., El Nino-Southern Oscillation (ENSO) and Inter-decadal Pacific Oscillation (IPO). The main focus is given to understanding the physical mechanisms of teleconnections and understanding their long-term climatic impacts (Li, 2021). In the end this section concludes that the tropical-to-polar teleconnections have contributed to the following Antarctic and Southern Ocean changes over the past few decades. The regional rapid surface warming; sea ice expansion and its reduction thereafter, changes in the ocean heat content, accelerated thinning of the Antarctic Ice sheet, etc.

The limited observations and climate model biases restrict our understanding of the relative importance of teleconnections versus those arising from greenhouse gases, ozone recovery, etc. Reducing these uncertainties and improving the understanding requires pan-Antarctic efforts toward sustained, long-term observations and more realistic dynamics and model physics applied within high-resolution climate models. Here we have described the teleconnection patterns triggered by decadal sea surface temperature variability (Fig. 1.4). given here represent the atmospheric teleconnection patterns induced by a positive phase of the Atlantic Multi-decadal Oscillation and a negative phase of the

10   Chapter 1 The importance of tropics in the changing climate

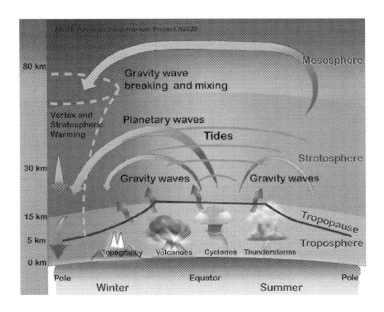

FIGURE 1.4

Dynamical exchange on a large temporal scale demonstrating the interlinking of atmospheric dynamics from the tropical to the polar region.

Obtained from Atmospheric Dynamic Research Infrastructure in Europe program. Image is only for depiction purposes.

Interdecadal Pacific Oscillation. Fig. 1.4 shows a schematic diagram of the different types of atmospheric waves called gravity waves and planetary waves which occur during summer and winter periods.

With the space-based observation, substantial climate changes have been remotely sensed in the Antarctic region, including atmospheric and sea surface temperature, ice sheet thinning, and a general Antarctic-wide expansion of sea ice. Although these changes feature strong zonal asymmetry and increasing greenhouse gas emissions and stratospheric ozone depletion, tropical-polar teleconnections are believed to have a role through Rossby wave dynamics.

During the satellite era, substantial climatic changes have been remotely sensed at the poles like the Antarctic, including all the portion of the atmosphere and oceanic warming, ice sheet thinning, and a general Antarctic-wide expansion of sea ice, followed by rapid loss. Although these changes, featuring strong zonal asymmetry, are partially influenced by increasing greenhouse gas emissions (Figs. 1.2 and 1.4) and depletion of ozone in stratospheric, tropical–polar teleconnections are believed to have a role through the dynamics of atmospheric waves. From the publications, we confirm that understanding of tropical teleconnections to the Southern Hemisphere extra-tropics arising from the El Niño—Southern Oscillation, Inter-decadal Pacific Oscillation, and Atlantic Multi-decadal Oscillation, focusing on the mechanisms and long-term climatic impacts. These teleconnections have contributed to observed Antarctic and Southern Ocean changes, including regional rapid surface warming, sea ice expansion, and its sudden reduction thereafter, changes in ocean heat content, and accelerated thinning of most of the Antarctic ice sheet. However, due to fewer observations and model systematic biases, uncertainties remain in understanding and assessing the role of these teleconnections versus those arising from greenhouse gases, ozone recovery, and internal variability. Sustained pan-Antarctic efforts toward long-term observations, and more advanced model dynamics and modern physics packages employed in the high-resolution climate models, help to reduce these uncertainties.

2.11 Projection/prediction of precipitations and teleconnection due to climate change

From the times of Blanford and Walker until now, forecasts have been searching far and wide, on different spatial as well as temporal scales. The atmospheric teleconnection Fig. 1.4 are mainly three types: (a) Physically explainable processes, (b) Statistically significant relationships, and (c) Physical processes which are also statistically well-correlated (Deser and Wallace, 1987). The climate models are being in our country and other leading numerical weather prediction centers. Here for this section, the model runs were made for 30 years since 2081. The finding of these climate models suggests that significant differences are seen in the projected ENSO teleconnections. In particular, the central and equatorial Pacific precipitations increase and with a relatively small positive ENSO and precipitation are predicted with these models. The above projection is noticed over the Pacific region. In the same manner, the results are seen all other countries as well. When we focus on the land areas, each projection scenario displays a considerable amount of precipitations and changes in the teleconnection patterns. The smallest land area displaying these significant changes is approximately 10%. On the regional scale, the analysis finds significant projected changes in the precipitation pattern. There are several regions that best display the apparent scaling of the teleconnection changes with increasing global warming levels (Kripalani et al., 2007). The teleconnections patterns are seen over the South American monsoon region, South-east South America, Mediterranean region, west-central Asia region, and Indian land. Thus this section concludes that a clear and significant ENSO teleconnection changes in December, January, and February for the period 1981—2100, relative to way back 1950—2014. The remarkable changes were noticed in the eastward shift and intensifications of the atmospheric response to the ENSO event.

2.12 Lesson from pandemic COVID

One of the important lessons learned from the recent Pandemic COVID-19 is "why do we have to promote the study of climate?" The answer is we are only safe as of now as our vulnerable populations. In the case of climate change, the more vulnerable populations and their livelihoods need to be protected first and climate risk for them has to be mitigated on a significant urgent basis. Advocating change in consumption patterns such as further international flights, remote teleconferencing, and other measures to reduce global carbon emissions. Climate goals have to achieve through sustainable growth and conscious mitigation efforts such as energy decarbonization, reforestation, improved and well-planned land use, and so on, besides adequate planning for climate adaptation. It is possible to create better models for sustainable development that balance profit and biodiversity by assigning appropriate monetary value to the ecosystem and the services they provide for the economy of the country (Reference seminar talk Prof Sathese IISC, Bangalore).

2.13 Adaptation to the heat wave and climate change

Adaptation to these kinds of extreme heat and cold can be effective at reducing mortality, however, pointing to India's example of issuing an early heat warning. "Heat Action Plans" that include early warning and early action, awareness raising and behavior-changing messaging, and supportive public services can reduce mortality, and India's rollout of these has been acknowledged. The above statement has been obtained from the Ahmedabad Heat Action Plan, which has been implemented since 2013 and has managed to reduce heat wave-related mortality over the years. The analysis was conducted by the researchers of the World Weather Attribution group about the Heat Action Plan.

In particular, for this year's heat wave it is concluded that High temperatures are common in India but what made this unusual was that it started so early and lasted so long. It is learned that we know this will happen more often due to Climate Change and we need to be prepared for it. The Indian central government organization National Disaster Management Authority is preparing the heat action plans but guidelines for implementation of these plans are being prepared by the state governments. Thus we have documented all the measures to be better prepared for the climate change effects.

2.14 Recent observations of greenhouse gas, temperature, and ocean heat in the warming climate situation

In 2020, the greenhouse gas concentrations reached a new global high when the concentration of carbon dioxide reached 415 parts per million (ppm) globally or in other words 150% of the preindustrial level value. It is observed that in 2021 the rise in the global annual mean temperature was around $1.11 \pm 0.13\ °C$; which is above 1850–1900; less warm than some recent years owing to cooling La Niña conditions at the start and end of the year. The most recent 7 years, 2015–2021, are the seven warmest years on record.

At the surface friction causes anticyclonic winds to diverse, leading to downward movement of air descends, it conserves mass. As the air descends, it adiabatically warms allowing it to hold moisture and thus it increases its boundary cloud cover through increasing relative humidity below the inversion layer (Fig. 1.5). The combination of a warmer, colder, and more humid atmosphere increases downward longwave radiation contributing to surface warming and increasing sea ice melt. The tropical modes of variability including the Asian summer monsoon and other atmospheric oscillations also drive anomalous Antarctic circulation through the stationary Rossby wave train and correspond to stratosphere-troposphere coupling. The circulation anomalies are shown in Fig. 1.5.

Ocean heat reached a high value. The upper 2000 m depth of the ocean continued to warm in 2021 and it is expected that it will continue to warm in the future—a change which is irreversible on centennial to millennial time scales. All data sets agree that ocean warming rates show a particularly strong increase in the past 2 decades. The temperature is penetrating to ever deeper levels. Much of the ocean experienced at least one "strong" marine heatwave at some point in 2021.

2.14.1 Displacement

Humans, livestock, birds, and animals migrate from place to place during peak summer and winter. Evidence birds are migrating from poles also due to extreme weather situations. Extreme weather events and conditions have had major and diverse impacts on population displacement or migration and the vulnerability of people already displaced throughout the year. From Afghanistan to Central

2. Organization of the chapter 13

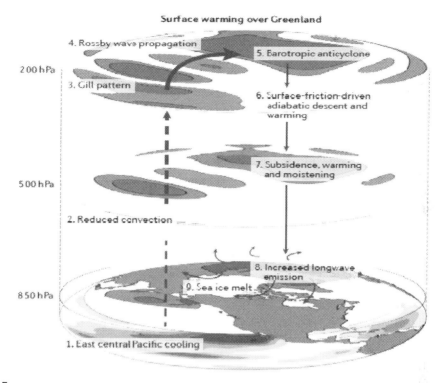

FIGURE 1.5

The schematic diagram vertical distribution of atmospheric waves from surface to upper atmosphere over the central Pacific region.

America, droughts, flooding, and other extreme weather events are hitting those least equipped to recover and adapt.

2.14.2 What can we do?

Fortunately or unfortunately a significant portion of international politicization on climate change issues are being taken place (Ajey, 2009). The fact is science, scientists and book authors have not much to say at the time of making policy on climate change in their respective countries. The climate is changing in unprecedented ways, but there are still many options to alternate the impacts, through both mitigation and adaptation.

2.14.3 Adaptation

As extreme weather events become more frequent and intense, **predictions must go beyond what the weather will be to include what the weather will do.** Early Warning Systems allow people to know hazardous weather is on its way and inform how governments, communities, and individuals can act to minimize the impending impacts. However, **one-third of the world's people,** mainly in the least developed countries and small island developing states, **are still not covered by early warning systems.**

In Africa, it is even worse: 60% of people lack coverage. Ocean Heat Content is a measure of this heat accumulation in the Earth system. It is measured at various ocean depths, up to 2000 m deep. All data sets agree that ocean warming rates show a particularly **strong increase in the past 2 decades.**

2.14.4 Regional climate risk

Through developing a new framework for assessing and explaining regional climate risk using all the available sources of climate information. Some sources are listed here observations, reanalysis, model simulations, better understanding, etc. Ultimately climate information will be made meaningful at the local scale. From the Kyoto Protocol to now the journey toward Copenhagen, the deliberations are mostly at the level of finding a globally accepted solution to the issue (Ajey, 2009).

3. Conclusion and pathways

During this 21st century, various types of threats and risks have become common. In the yesteryears, wars used to cause a large number of causalities but not the case anymore. Certainly, the globe is warming but not at the same rate everywhere. This chapter demonstrates the impact of climate in the tropics and vice versa. Also, it is accounting for the expansion of the tropics. Thus, the tropical region of India has to take this issue seriously.

References

Ajey, L., 2009. Strategic Implications of Climate Change Mausam, pp. 59–64.
Archer, C.L., Calderia, k, 2008. Historical trend of the jet streams. Geophys. Res. Lett. 35, L08803.
Attri, S.D., Mukhopadhyay, B., Bhatnagar, A.K., Singh, V.K., 2009. Emerging trends in environmental Meteorology. Mausam 163–174.
Chen, G., Lu, J., Frierson, D.M., 2008. Phase speed spectra and the latitude of surface westerlies: interannual variability and global warming trend. J. Clim. 21 (22), 5942–5959.
Deser, C., Wallace, J.M., 1987. El Nino events and their relation to the Southern oscillation 1925–1986. J. Geophys. Res. 92, 14189–14196.
Goswami, B.N., 2006. Increasing trend of extreme rain events over India in a warming environment. Science 314, 1442–1444 and co-authors.
Grise, K.M., Davis, S.M., Simpson, I.R., Waugh, D.W., Fu, Q., Allen, R.,J., et al., 2019. Recent tropical expansion: natural variability of forced response? J. Clim. 32 (5), 1551–1571.
Heffernan, O., 2016. The mystery of the expanding tropics. Nature 530, 20–22.
Hindustantimes Daily Newspaper (2022): (Accessed through Internet).
Hu, Y., Lohmann, G., Lu, J., Gowan, E.J., Shi, X., Liu, J., Wang, Q., 2020. Tropical expansion driven by poleward advancing mid-latitude meridional temperature gradients. J. Geophys. Res. Atmos.
Intergovernmental Panel on Climate Change, IPCC Fifth Assessment Report Observed Climate Change Database, 2017. Version 2.01.1. Palisades, NY, Socioecomic Data and Applications Center. NASA, USA https://doi.org/10.7927.
IPCC, 2007. Climate Change (2007): The Physical Science Basis. Cambridge University Press, U.K.
IPCC, 2021. Summary for policymakers. In: Masson-Delmotte, V., Zhai, P., Pirani, A., Conn, S.L. (Eds.), Climate Change (2021): The Physical Science Basis. Contribution of Working Group I to the Sixth Assessment Report of the Intergovernmental Panel on Climate Change, vol. 18, Issues 3–4. IPCC.

Kripalani, R.H., Oh, J.H., Kulkarani, A., Sabade, S.S., Chaudhri, H.S., 2007. South Asian summer monsoon precipitation variability. Coupled climate model simulation and projection under IPCC. Theor. Appl. Climatol. 90, 133–159.

Kumar, R.P., Mukhopadhayay, R., M Krishna, P., Ganai, M., Mahakur, M., Rao, T.N., Kumar, A., Nair, M., Ramakrishna, S.S.V.S., 2020. Assessment of climate models in relation to the low-level clouds over southern Indian Ocean. Q. J. R. Meteorol. Soc. 3306–3325.

Li, X., 2021. Tropical Teleconnections Impacts on Antarctic Climate Changes. Nature Reviews Earth and Environment, 40 co-authors.

Morice, C.P., Kennedy, J.J., Rayner, N.A., Jones, P.D., 2012. Quantifying uncertainties in global and regional temperature change using an ensemble of observational estimates: the HadCRUT4 data set. J. Geophys. Res. Atmos. 117 (8). https://doi.org/10.1029/2011JD017187 n/a n/a.

Moss, R., Edmonds, J., Hibbard, K., et al., 2010. The next generation of scenarios for climate change research and assessment. Nature 463, 747–756. https://doi.org/10.1038/nature08823.

Norris, J.R., Allen, R.J., Evan, A.,T., Zelinka, M.D., OfDell, C.W.,., Klein, S.A., 2016. Evidence for climate change in the satellite cloud record. Nature 536 (7614), 72–75.

Polvani, L.M., Waugh, D.W., Correa, G.J., Son, S.W., 2011. Stratospheric ozone depletion: the main driver of twentieth-century atmospheric circulation changes in the Southern Hemisphere. J. Clim. 24 (3), 795–812.

Shaw, T.A., 2019. Mechanisms of future predicted changes in the zonal mean mid-latitude circulation. Curr. Clim. Change Rep. 5 (4), 345–357.

Sikka, D.R., 2009. Earth system science and its role in India sustainable development in the context of climate change. Mausam 65–90.

Yin, J.H., 2005. A consistent poleward shift of the storm tracks in simulations of the 21st-century climate. Geophys. Res. Lett. 32, L18701.

Further reading

Coping with Global Environment Change, 2011. Disasters and Security. Springer and Science and Business Media, Berlin LLC, p. 61.

Storymaps.arcgis.com. (Accessed through Internet).

The Mystery of the Expanding Tropics, 2016. Nature.

WMO Greenhouse Gas Bulletin (2015): 'The State of Greenhouse Gases in the Atmosphere Based on Global Observations Through' (No. 12|October 24, 2016).

CHAPTER 2

Climate change impacts on water resources and agriculture in Southeast Asia with a focus on Thailand, Myanmar, and Cambodia

G. Srinivasan, Anshul Agarwal and Upeakshika Bandara

Regional Integrated Multi-Hazard Early-Warning System for Africa and Asia (RIMES), Klong Luang, Pathum Thani, Thailand

1. Introduction

Southeast Asia is one of the fastest-growing regions in the world in terms of economy and population, with urbanization expected to continue at a high pace in the coming years (Lorenzo and Kinzig, 2020). Southeast Asia is also highly prone to natural disasters with a high prevalence of floods and cyclones and is among one of the most vulnerable regions to climate change, due to its high population and geographic location (Lee, 2021). The latest IPCC's (Intergovernmental Panel on Climate Change) Working Group 1 assessment states both heat extremes and heavy rainfall will increase over much of Asia in the coming decades and may intensify the existing hydrometeorological disasters (IPCC, 2021). Changes in precipitation patterns may further add stress to the water resources. Sea-level rise and warmer sea surface temperatures (SSTs) are expected to pose additional stress on marine resources that countries in the region depend on.

Southeast Asia is home to major river systems, including the Ayeyarwady, Mekong, Chao Phraya, and the Salween. These rivers are vital for supplying energy, food, and transport in the region. The massive projects for hydropower and irrigation development implemented in the last decade bring great deal of investments, but also significantly impact the river flows and land use. Ensuring that the development guarantees long-term ecological and economic sustainability will require a more system-wide approach that looks beyond country borders and considers large-scale river basins as a whole. Climate change impact assessment on water availability and adaptation planning is vital to sustaining the growth along these major rivers in Southeast Asia. Agriculture is a key sector of Southeast Asia's economy, contributing to 10% of Gross Domestic Product (GDP) and is a vital source of income for over one-third of the population (Zhai and Zhuang, 2012).

Myanmar has a large agricultural sector contributing to about 40% of GDP and providing a living to more than 70% of the population. Out of Myanmar's export earnings, 25%–30% is from agriculture

(Lwin et al., 2017). Rice is one of the major exports from Myanmar along with other cereals, sugar crops, vegetables, and fruits. Ayeyarwady River (drainage area: 372,000 sq. km) is the most important commercial waterway of Myanmar and is among the world's top five rivers with suspended sediments deemed rich for their agricultural values (Rao et al., 2005).

Known as the "Rice Bowl of Asia", Thailand's economy heavily depends on agriculture, providing livelihood to more than 30% of its population. Agricultural export goods from Thailand include rice, canned pineapple, fish, rubber, sugar, and tiger prawns (Lee, 2021). Thailand has 25 major river basins with Chao Phraya and Mun as the two largest rivers. Most water-rich fertile agricultural lands are in the Chao Phraya basin which also hosts the major dams and canal network in Thailand.

Cambodia is an exceptionally endowed country in terms of water resources and had one of the most sophisticated irrigation systems in the world. In Cambodia, agriculture is considered one of the "key growth-enhancing pillars" in the economy and for reducing the poverty in the country as its contribution to GDP is nearly 28% and 85% of the population in the rural area rely on agriculture (Saing et al., 2012). Agriculture is also important for increased foreign exchange earnings. The main agricultural goods in Cambodia are paddy rice, cassava, maize, and soybean. Cambodia is also highly reliant on freshwater in terms of the inland fisheries that support livelihoods and form the main protein source for the Cambodian population, but also due to its tropical climate that brings alternating seasons of shortage and surplus of water. The Mekong River and the Tonle Sap are essential for the Cambodian hydrological system and are the main resource for Cambodian protein supply, which is estimated to be worth 2 billion USD annually (Open Development Cambodia, 2017).

In view of the importance of climate change to lives and livelihoods in Southeast Asia, and the significant contributions made by the region to the world's agriculture, this chapter focuses on the likely impact of climate change on the region's agriculture. It also explores the consequences of climate change on the water sector which is inextricably linked to agriculture. Case studies based in Thailand, Myanmar, and Cambodia are used to illustrate and bring out key points for planning adaptation strategies.

2. Climate change over Southeast Asia

Most of Southeast Asia exhibits a tropical monsoonal climate that is characterized by strong dry and wet seasons. Historical climate data analyses over most of the region show clear increasing trends in temperatures (Eastham et al., 2008). Based on historical temperature records, IPCC's latest assessment report shows an increasing trend in mean surface air temperature in Southeast Asia during the past several decades, with a 0.1–0.3°C increase per decade recorded between 1951 and 2000 (IPCC, 2021). Fig. 2.1 shows the long-term anomalies of annual surface-air temperatures over southeast Asia. Observed rainfall trends show large temporal and spatial variability that masks any historical trends. While there is spatial coherence and confidence of temperature increase over Southeast Asia, the rainfall trends differ with the northern land areas over Myanmar, Thailand, and Cambodia showing a larger increase. A tendency for delay in the onset of the wet season has been also observed in several locations.

Future climate change projections for the region indicate increasing mean temperatures at a rate that is slightly below the global rate of temperature increase. Increased extreme heat events are likely to be experienced with high confidence. Fig. 2.2 shows projected changes in several days with maximum temperatures above 35°C (Tx35) as compared to the baseline climate during 1850–1900

2. Climate change over Southeast Asia 19

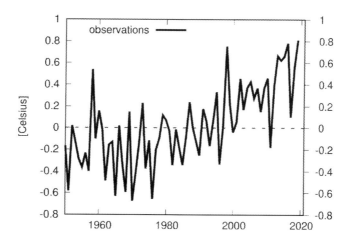

FIGURE 2.1

Observed temperature change from March to June compared to 1980–2000 mean climate over Southeast Asian region 0–35N, 90–120E (land).

Source: CRU TS 4 data.

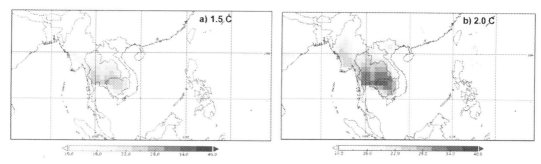

FIGURE 2.2

Change in days with maximum temperatures above 35°C (Tx35) relative to baseline 1850–1900, (A) warming 1.5°C SSP5-8.5 (B) warming 2.0°C SSP5-8.5—annual (CMIP 6–27 GCMs).

Source: Gutiérrez et al. (2021).

based on results from 27 Global Climate Models (GCMs) used in the recent IPCC assessment (IPCC, 2021). Both the global warming levels, 1.5°C and 2.0°C above, presented using the Shared Socioeconomic Pathways, SSP5-8.5 scenario, show an increase in the numbers of Tx35 days over southern Myanmar, most of Thailand and Cambodia. The 2.0°C shows an increase of 28–34 more numbers of Tx35 days. The region is already exposed to "severe" heatwave events during the summer months of April–May and the projected levels of future warming will worsen further. Thirumalai et al. (2017) have attempted to quantify the contribution of long-term warming and the 2015–16 EL Niño to record temperature extremes that occurred over Southeast Asia in April 2016. Dong et al. (2021), based on

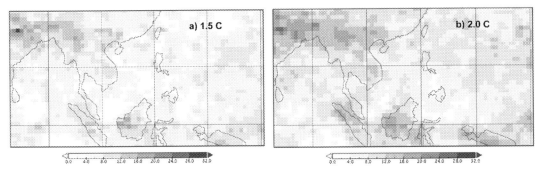

FIGURE 2.3

Percent change in maximum 1-day precipitation (RX1day) relative to baseline 1850–1900, (A) warming 1.5°C SSP5-8.5 (B) warming 2.0°C SSP5-8.5—annual (CMIP 6–33 GCMs).

Source: Gutiérrez et al. (2021).

single model large ensemble simulation results, conclude that frequent heatwaves of longer duration and higher extreme temperatures will occur with rarer events of extreme temperatures becoming common in future over mainland Southeast Asia.

Another important variable showing an increasing tendency over the region is the heavy rainfall events. Fig. 2.3 shows increases in 1-day rainfall amounts by 5%–12% for a 1.5°C and 12%–20% for 2.0°C global warming levels. Such indications are likely to translate into a substantial increase in extreme rainfall events that can cause frequent flooding resulting in damage to both agriculture and livelihoods.

Seas surrounding continental southeast Asia are economic resources for the countries in the region and livelihood providers for people living in the coastal zones. Global warming is projected to increase both SSTs and sea levels in the region. The increase in SSTs will result in marine heatwaves (MHWs) defined as extremely warm episodes of warmer than usual SSTs that impact marine ecosystems (Hobday et al., 2016). Positive sea level trends have been reported over the region during 1950–2009 and are expected to persist causing enhanced impacts in combination with existing decadal fluctuations (Strassburg et al., 2015). Potential climate impacts over southeast Asia based on climate change information from the recent IPCC WG—I report (Ranasinghe et al., 2021) are summarized in Table 2.1 below.

The climate in Southeast Asia is mainly monsoon driven, with Northeast monsoon from November to March and Southwest monsoon from May to October (Loo et al., 2015). During the Southwest monsoon, 80% of rainfall in the region occurs. Most parts of the region have a tropical climate with temperatures above 25°C throughout (Yuen and Kong, 2009). A considerable linkage between the climate and the El Nino-Southern Oscillation (ENSO) is seen in Southeast Asia (Yaya and Vo, 2020). The regional climate change projections need to be considered at the country level for assessing impacts and planning for adaptation strategies.

2.1 Myanmar

Myanmar has a tropical climate with rainfall as high as 5000 mm in the coastal and deltaic area, while the dry parts (Central Myanmar) receive 600 mm of rainfall. The temperature in the country varies from 43°C in Central Myanmar (Central Dry Zone) to 29°C in the East. Myanmar's climate is highly

Table 2.1 Climate Change and its potential impacts on Southeast Asian agriculture and water sector.

Climate indicators	Future changes	Priority impacts—agriculture and water sector
Mean surface temperature	Increase[b]	Impact on crop productivity, increased irrigation requirements, shifts in cropping seasons, and patterns
Extreme heat	Increase[b]	Severe impact on crop yields
Mean rainfall amounts	Increase[a]	Maybe be beneficial in some areas and seasons, and may help meet the increase in crop water requirements due to enhanced evaporative demand
Riverine floods	Increase[a]	Crop losses and decrease in productivity, opportunities to use flooded conditions when possible
Heavy rainfall spells and associated flash flooding	Increase[b]	Severe crop damage and losses
Tropical cyclones	Increase[a]	Threats to agriculture and water infrastructure through high winds, heavy rainfall, and saline water intrusion due to storm surge
Relative sea level	Increase[b]	Impacts on coastal agricultural lands, including losses in arable land due to saline water intrusion
Coastal flooding	Increase[b]	Impacts on coastal agricultural lands, including losses in arable land due to saline water intrusion
Coastal erosion	Increase[b]	Loss of arable land and soil fertility
Marine heatwaves	Increase[b]	Impact on marine fishing
Ocean acidity	Increase[b]	Impact on marine fishing

[a] *medium confidence;*
[b] *high confidence.*

influenced by the Southwest monsoon and ENSO and Indian Ocean Dipole Mode (IOD) (D'Arrigo and Ummenhofer, 2015). There are three dominant seasons in Myanmar; hot season (March—April), wet season (June—October), which coincides with the Southwest monsoon and cool season (November—February). Most of the rainfall in Myanmar comes during the wet season while some rainfall can be observed in Northern and Southern parts during the hot and cold seasons, respectively (Horton et al., 2017).

The overall temperature in Myanmar shows a rising trend, with a varied magnitude over region and season (Nwe et al., 2020). The increase in average annual temperature is expected to range from 0.7 to 1.1°C for moderate and high emission scenarios in the 2030s and from 1.3 to 2.7°C by end of the 2070s. More intense positive change in temperature should be expected during the hot season to wet and cold seasons. The highest increase in temperature is observed in Central Dry Zone and Eastern Hilly region, where increases up to 1.2°C are estimated for the high emission scenario (Horton et al., 2017). Although precipitation too shows an increasing trend in Myanmar, it is not as direct as changes in temperature and spatial and temporal variability in rainfall plays a role in fully understanding the future changes. Hence, there is some uncertainty in the projection of future precipitation. Based on the existing projections, it is suggested that rainfall will be more during the existing wet season from June to October in both the near and far future. Overall rainfall would change by +1% and +11% in overall Myanmar under moderate and high emission scenarios respectively during the 2030s. Towards the end of the century, annual precipitation is expected to increase by 6%–23%. Over the different seasons, rainfall showed high variation, with rainfall likely decreasing during hot and cold seasons in moderate emission scenarios, while high emission scenarios showed positive changes throughout. The wet season showed the highest changes with 2%–12% for recent years and 6%–27% for the far future (Horton et al., 2017).

2.2 Thailand

The climate in Thailand is tropical with temperatures ranging from 19.6 to 30.2°C. Average annual rainfall varies between 1270 and 2000 mm and is concentrated mainly during the rainy season from May to October (Babel et al., 2011). The southern part of Thailand receives relatively higher rainfall. This country is highly influenced by the Southwest and Northeast monsoons of the Indian Ocean. Southwest monsoon occurring from mid-May to mid-October defines the rainy season, with August and September being the wettest months in most parts of Thailand except the East Coast of Southern Thailand. The Northeast monsoon season occurs from mid-October to mid-February. During this time, upper Thailand experiences mild and cold weather while rainfall is seen in Southern parts. The period from mid-February to mid-May is known as summer or pre-monsoon and is the transitional period between monsoons. This period witnessed high temperatures and April is the hottest month where day temperatures exceeding 40°C can occur (Thailand Meteorological Department, 2010). Observed trends over Thailand are consistent with that seen over the region.

In Thailand, extreme weather events are likely to be more intense and more frequent in the future compared to the recent past. Under both the high emission scenarios SSP3-7.0 and SSP5-8.5, the temperature is expected to increase by 3°C toward the end of the 21st century (Fig. 2.4). The climate change projection results are based on multi-model ensembles from the Coupled Model Intercomparison Project 6 (CMIP6) used in the latest IPCC, 2021 WG-I assessment (IPCC, 2021). The higher temperature increase is likely over the Northern inland regions of the country while changes are expected to be smaller in southern coastal areas. Although projections of mean annual rainfall show a consistent increase across models, the amount varies over a wide range. Major changes in seasonality are to be expected, where dry seasons would be dryer and wet seasons wetter, with a higher number of extreme events. In terms of monsoon onset and retreat, the retreat would likely be delayed while onset remains unchanged resulting in longer monsoon seasons (Kiguchi et al., 2020).

Fig. 2.5 illustrates the projected changes in rainfall during different months of the peak rainy season for different time horizons and scenarios. The high emission scenarios show larger increases by the end of the century, but some months like September and October show significant increases even in the near- and mid-term indicated by decades 2031–40 and 2061–70 respectively. Such rainfall increases will impact water resources and agriculture sectors with cascading consequences.

Sea-level rise is important for Thailand as it can cause both physical impacts on the coastline and saline intrusion of the coastal area. Location-specific impacts of sea-level rise depend on modulating local factors such as subsidence or uplift of the ground and erosion.

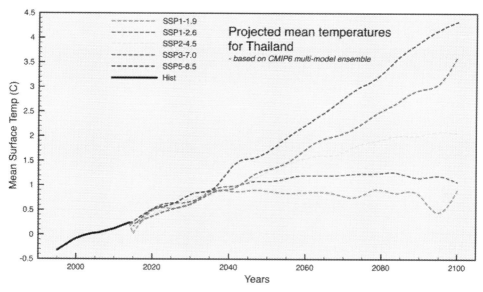

FIGURE 2.4

Projected increase in mean temperatures over Thailand under different shared socioeconomic pathways (SSPs) based on CIMIP6 multi-model ensembles as compared to historical (hist) period 1995–2014 numbers suffixed with SSPs indicate different scenario storylines and linked radiative forcing by 2100.

Source: IPCC (2021).

2.3 Cambodia

Cambodia's monsoonal climate is divided into 6 months of the dry season and 6 months of the wet season. The Southwest monsoon brings most of its rain from mid-May to October while the northeast monsoon brings in cooler air from November to March (Bansok et al., 2011). Rainfall as high as 4000 mm is seen in the Southwest coastal areas of Cambodia, while central and south-central parts receive low rainfall of 1200 and 1400 mm. Other parts of the nation receive rainfall of 1500 mm (RIMES and UNDP, 2020). The country has an average annual temperature of 28°C, where the highest temperature can reach up to 38°C in April while the lowest is seen to be 17°C in January (Thoeun, 2015).

Like Myanmar and Thailand, increasing trends in both temperature and rainfall are observed in Cambodia. Future projections show an increase of temperature up to 3.6°C over the country by the 2090s based on high emission scenarios. The larger increase in temperature is likely to be experienced during the dry season with higher numbers of hot days. The spatial distribution suggests that islands would experience harsher increases in temperature, compared to coastal areas. In terms of rainfall, despite uncertainties in the projections, the consistent indications that emerge are an increase in monsoon season rainfall and more intense extreme events. Spatially, the rainfall projections suggest increases in annual average rainfall during the wet season in the Northwest region, while there is a likelihood for a decrease in the Northeast region during the same period (WBG and ADB, 2021a, b). Possible increases in frequency and intensity of floods that cause severe damage to rice harvests or

24 **Chapter 2** Climate change impacts on water resources

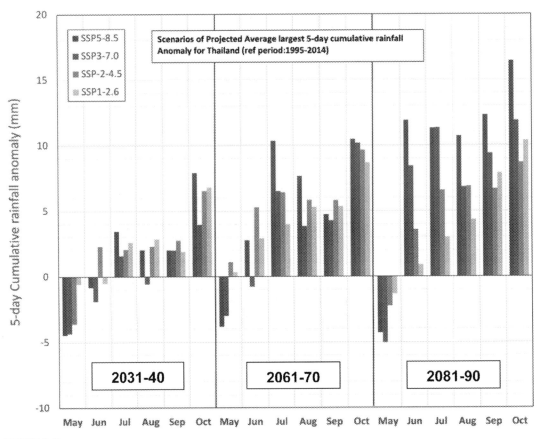

FIGURE 2.5

Projected anomalies of average largest 5-day cumulative rainfall over Thailand under different shared socioeconomic pathways (SSPs) scenarios, time-horizons and monsoon months based on CIMIP6 multi-model ensembles as compared to historical (hist) period 1995–2014.

Source: WBG and ADB (2021a, b).

droughts that result in a significant number of fatalities and economic losses are the major concerns. Sea-level rise will also impact Cambodia's coastal areas like the other two countries discussed in the region.

3. Impacts on agriculture and water resources

With the anticipated changes in future climate across the Southeast Asian region, their impact would likely be immense in every sector. These impacts could either be sudden, such as extreme events and flash floods or could be slow manifesting over the longer term. These impacts are expected to be widespread and varied spatially and temporally and would have a considerable effect on natural, human, and physical systems (Lawrence et al., 2020). Coastal communities are at high risk of being inundated due to sea-level rise. Health systems would be affected due to the frequent occurrence of extreme temperature and rainfall events. Agriculture is highly vulnerable to changes in seasonality and extreme events and changes in the biosphere (Van Meijl et al., 2018).

Agriculture in Southeast Asia is anticipated to be affected by climate change directly, such as agroecological changes and indirectly, in terms of reduced demand, growth, and distribution of income. Extensive impacts on livestock with changes in grasslands and forests (RIMES, 2011a, b). An increase in temperature and reduced availability of water could directly affect crop yield. Water shortages during key stages of paddy cultivation can critically affect growth and yield. The most sensitive periods to water deficit are flowering and the second half of the vegetative period. Climate change will also influence both pre- and post-harvest losses of crops with impacts on crop production. In many parts of the countries in Southeast Asia, the rice yield is likely to decrease under future climate change scenarios as indicated by Chun et al. (2016) (Fig. 2.6).

The water resources sector is closely associated with agriculture and therefore, impacts on water resources also need to be assessed. Proper management of water resources is important for maintaining ecosystems, provision of drinking water, and hydropower production in addition to the agriculture sector. Due to climate change, it is highly unlikely that water resources would remain unaffected (Horton et al., 2017). Higher temperatures would increase loss in evaporation which would decrease the water availability and increase the demand for crop water. Extreme events such as floods and droughts would be more frequent, which would put immense stress on water resources. Changes in seasonality would make water-related decisions hard. Such impacts are already seen in Southeast Asia and are likely to be more intense in the future (RIMES, 2011a, b).

Delgado et al. (2010) using more than 70 years of daily discharge data recorded at four gauging stations downstream of Vientiane, Laos have examined flood trends and variability in the Mekong River that illustrate long-term trends. Analysis of the flood regimes of the Mekong, undertaken by this study shows that there is a decrease in the frequency of the average floods, whereas both very large floods and very small floods are increasing in frequency downstream of Vientiane. The variability in river discharge is increasing, which could be attributable to a variety of causes including climate variability and change, in addition to factors such as changes in land and water use. Planning to deal with climate change needs analysis of both historical data and future projections so that risks are assessed objectively based on existing evidence in combination with science-based projections that have inherent uncertainties. Factors such as land-use changes and similar considerations of changing contexts need to be also considered for future planning.

In the lower Mekong River basin subregion, spread over parts of Myanmar, Thailand, and Cambodia, flood risks from river tributaries escalate during the end of the monsoon season (Sept–Oct) when water from the Mekong River is backed up into the tributaries (Mekong River Commission, 2005). Climate change has already resulted in increased intensity and duration of floods and droughts, as well as rising sea levels in Lower Mekong countries. Lower Mekong countries are among the most vulnerable in the world to the impacts of climate change. Myanmar, Vietnam, and Thailand were among the 10 most-affected countries between 1996 and 2015 on the Global Climate Risk Index, while Cambodia was ranked 13th.

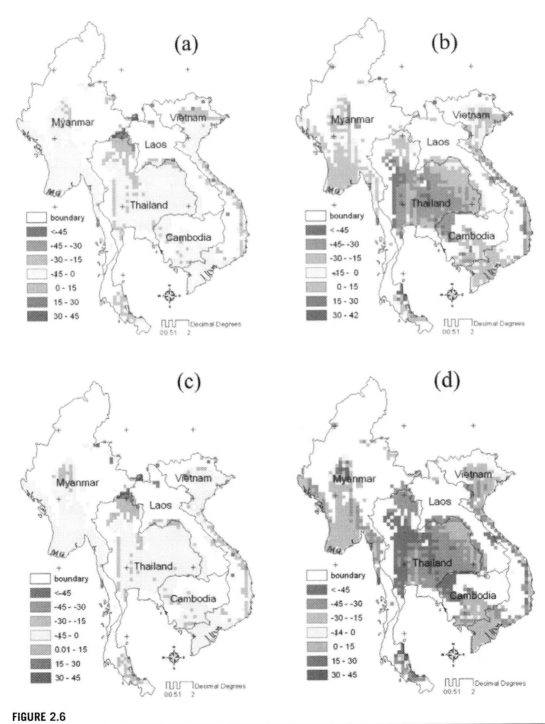

FIGURE 2.6

Projected rice yield change (%) in (A) the 2040s and (B) 2080s under RCP4.5 and (C) 2040s and (D) 2080s under RCP8.5 relative to 1991–2000.

3.1 Myanmar: impacts

Myanmar is highly dependent on agriculture and is likely to be negatively impacted by climate change. Livelihoods, food security, and nutrition would face drastic effects. Crops and livestock show high sensitivity to climate change, especially increase in temperature. Myanmar, being a tropical country, the production yield is expected to reduce by 15%–50% by the end of the century (Swe et al., 2015). Additionally, increased pests, weeds and diseases are likely to increase with climate change, which would threaten food security and result in poverty. Higher amounts of rainfall, heatwaves, and loss of coastal agriculture due to sea level rise would affect the economy (Horton et al., 2017). Coastal townships are especially vulnerable to climate change. They were considered vulnerable to sea-level rise and extreme events such as cyclones (Shrestha et al., 2014).

Seasonal shifts and changes in monsoonal patterns would require readjusting of crop calendars. Farmers are noticing and experiencing these impacts such as a decrease in yield and an increase in pests already, instances of which are likely to further increase in the future. A decrease in yield due to water scarcity was listed as the highest effect of climate change. Other impacts identified by farmers include extreme rainfall-induced crop damage, poor germination of seeds due to high temperature, scarcity of animal feed, and irregular rainfall-enhanced crop damage (Swe et al., 2015).

Eastham et al. (2008) have studied the potential impacts of climate change on runoff over various catchments of the Mekong River basin. These results can be extended to understand the impacts of climate variability and change in other basins as well. Under historical climate conditions, there is strong seasonality in runoff from the basin, with the highest runoff observed in the wet months from May to October when precipitation is greatest. Under the climate projected for the near future, total annual runoff from the basin is likely to increase by about 21%. Although there are uncertainties associated with the climate change projections, median runoff projection results suggest that total basin runoff will increase in all months of the year, with the largest projected increases occurring in the latter half of the wet season.

3.2 Thailand: impacts

Thailand too is likely to face adverse effects in terms of agriculture and water resources from climate change in the future due to its location in the tropics. A decline in rice yield is to be expected by 18% in the 2020s and would increase up to 25% in the 2080s in Northern Thailand (Babel et al., 2011). Higher evaporation and evapotranspiration rates which lead to higher water demand in crops is a factor that would decline rice production and yield. Even if rainfall is expected to be high in the future, the effect of temperature rise would likely be overwhelming (Boonwichai et al., 2018). Similarly, the Sugar Cane yield is also likely to drop by at least 25% in the 2050s due to drops in production. Eastern and central regions are expected to show the least production (Pipitpukdee et al., 2020). It had been identified that impacts would vary spatially, with higher impacts expected in Eastern, South-Central, and Northeastern parts of Thailand, while other areas are likely to suffer less (WBG and ADB, 2021a, b). More intense and more frequent extreme events are the likely scenarios in the future. Thailand's devastating floods in 2011 are a sample of what may be expected more often. Such extreme events would bring chaos to water management practices (Thanasupsin, 2012). An increase in water availability, especially during the dry season would demand the implementation of management structures for floods. Dryer dry months would suggest droughts which would negatively impact the economy, agriculture, and health (Deb et al., 2018).

Rice production in Thailand is highly dependent on monsoon rainfall and supply from canal networks. In recent years, floods and droughts are becoming more frequent and severe as a result of climate change. For example, in 2011 Thailand experienced its worst-ever flood event on record, at US$46 billion for repair and rehabilitation nationally. In 2016, droughts significantly reduced the length of the growing season, as well as agricultural yields. Furthermore, in an economic study focusing on trends in extreme weather conditions along the Chao Phraya river basin, it was projected that in the next 2 decades, extreme droughts could create conditions for dry-season irrigated rice production, where total production levels would be reduced by 30.9% (Anuchitworawong, 2016).

3.3 Cambodia: impacts

Climate-induced extreme events have a drastic effect on the livelihoods of people in Cambodia. Such events inhibit the nation from overcoming poverty and reduce the overall potential for development. Communities in Cambodia are most vulnerable to climate change-induced droughts as it has a high impact on agricultural yield (Chhinh and Poch, 2012). Prolonged drought events are to be expected during the wet and dry seasons based on the projected climate. This would negatively affect not only agricultural activities, but also make water management a daunting process (Oeurng et al., 2019). Loss of yield is inevitable under these conditions. Some of the highest losses in net rice yield in Southeast Asia can be expected in Cambodia, owing to climate change. A recent report estimates rice yields to reduce by 10%–15% by 2040s due to increases in temperature (WBG and ADB, 2021a, b). As agriculture in Cambodia heavily depends on rain instead of irrigation, enhanced effects from reduced rainfall, shifts in seasonality, and intense extreme events make the nation more vulnerable. A drop in labor productivity due to increased temperature has already become an issue and productivity can drop by 20% in the 2050s if the current trends continue.

For Cambodia, particular consideration is the impact on flooding in the Tonle Sap Lake, where the increased seasonal fluctuation and water levels increased both the minimum and maximum area of the lake. This can change fish habitats with negative impacts on agricultural areas, housing, and infrastructure (Eastham et al., 2008).

4. Practices for adaptive actions

As climate change impacts on agriculture and water resources are likely to be exacerbated in the future, it is important to identify actions to reduce these effects. Future climate projections help to make informed decisions and use good practices for resilience in different sectors (Horton et al., 2017).

Climate risks are not new to the agricultural sector in southeast Asia. Even under the current monsoonal climate of the region, farmers and agricultural planners experience the impacts of mid-season dry spells that damage crops in their early stages and late-season floods can cause severe crop loss during harvesting stages. Traditionally, farmers have evolved a range of measures to cope with such climate risks. Chinvanno et al. (2008) have documented rice farmers' experiences to manage climate risks and their perspectives on the potential for applying the same measures to adapt to climate change in Thailand and other countries of the Lower Mekong region. Surprisingly, despite the similarity in climate hazards faced, rice farmers across study areas used significantly different methods to cope with climate risks. Different social-economic, cultural, environmental conditions and national policies influence risk management practices at community levels which underline the need for evolving specific adaptations tuned to user contexts. Adaptation plans, therefore, need to follow a tiered approach with national levels providing broader enabling conditions that can guide and facilitate implementation at context-specific community levels.

Some agricultural practices that can be used are changing harvesting windows and cropping calendars, adjustments in plant density, improving water efficiency, reducing erosion, and choosing plants that have greater tolerance in terms of inundation and extreme weather. For soils with low fertility, Nitrogen application is considered to be effective. Tillage practices and the development of cultivars are also considered good practices to mitigate the risk of climate change (Babel et al., 2011).

As it is likely that wet seasonal rainfall would be more in the future, planning, and management of water resources are required. More dams for flood management, more conservative areas and management of water distribution for consumption, agriculture, and hydropower need to be considered. It is important to involve stakeholders of all sectors in climate-related decision-making so that all aspects are considered, and inclusive decisions can be made. Implementation of such actions would help reduce these impacts to some extent and minimize the toll on the economy of the countries.

5. Priorities for climate-resilient agriculture and water sectors

Historical temperature trends and future projections clearly show increasing temperatures over the Southeast Asian region including the focus countries Myanmar, Thailand, and Cambodia. As the warmer atmosphere holds more moisture, daily rainfall extremes also show increasing trends, both now and in the future. Understanding the combined impact of temperature increase and enhanced daily rainfall extremes, particularly the differentiated impacts across socio-economic contexts is crucial for planning adaptation strategies.

5.1 Enhancing climate information and its use

Climate risk information should be linked to prioritizing structural intervention programs like irrigation facilities in drought-prone provinces, expanding agriculture to regions with lower climate risk and creating climate insurance opportunities in vulnerable communities in climate-risk-prone areas. The use of climate information forums for exchanging and interpreting climate change risk information based on current climate variability and future climate projections among key stakeholders at national, provincial, and community levels should be institutionalized.

Good climate observation networks, including networks for agro-meteorological observations, should be built following standard design criteria for high-density observations in climate zones exhibiting higher variations. This observational system design may be guided by a detailed analysis of existing climate observations in the country. For quality and long-term archival and distribution of national climate data, a suitable climate database system is required. The National Meteorological and Hydrological Services (NMHSs) have already started taking some steps in this direction. This data management system should also enable sharing of data with key agencies involved in agriculture and water sector-related decision-making.

Evaluated climate change scenario information with confidence and uncertainty statistics at a resolution of ~ 20 km should be generated to guide sectors in formulating informed long-term adaptation strategies. Such information is being constantly updated (IPCC, 2021) with endeavors to increase spatial resolutions with the best available science and national research and development institutions must be capable of keeping pace.

5.2 National institutional framework

Both agriculture and the water sectors are complex and cross-cutting with linkage to overarching issues of the Sustainable Development Goals (SDGs) that require the participation of a wide range of institutions right from national to community levels. Given this context, climate information-linked decision support systems in these sectors are necessary to integrate varied information and data generated and modified by different institutions using their specific expertise. This needs to be viewed as a set of processes intended to create the conditions to produce decision-relevant climate change information for use in managing risks in the agricultural and water sector. Hence, institutional structures and linkages are pivotal for the effective translation of climate change information, into decision support services. The chain that starts with the generation of high-quality climate data and future change scenarios has to be translated into useful information that enables "low-risk choices," to build climate resilience.

5.3 Capacities for research and development in the country context

Building capacities of research institutions to take up detailed studies to characterize climate and climate change risks in sectors like animal health, pests and diseases of crops, human health and nutrition, forestry, and coastal zone impacts is an important step toward improving research and development in the country context.

Further, pilot projects should be set up to establish linkages between the science-based information being generated and decision-making at various levels including community-level actions. These projects should also include components for raising community awareness of climate change impacts on food security.

References

Anuchitworawong, C., 2016. Analysis of Water Availability and Water Productivity in Irrigated Agriculture. Thailand Development Research Institute Foundation, Bangkok, Thailand.

Babel, M., Agarwal, A., Swain, D., Herath, S., 2011. Evaluation of climate change impacts and adaptation measures for rice cultivation in Northeast Thailand. Clim. Res. 46 (2), 137–146. https://doi.org/10.3354/cr00978.

Bansok, R., Chhun, C., Phirun, N., 2011. Agricultural Development and Climate Change: The Case of Cambodia. CDRI, Phnom Penh, Cambodia.

Boonwichai, S., et al., 2018. Climate change impacts irrigation water requirement, crop water productivity and rice yield in the Songkhram River Basin, Thailand. J. Clean. Prod. 198 (2018), 1157–1164. https://doi.org/10.1016/j.jclepro.2018.07.146.

Chhinh, N., Poch, B., 2012. Climate change impacts agriculture and vulnerability as expected poverty of Kampong Speu Province, Cambodia. Int. J. Environ. Rural Develop. 3 (2), 28–37.

Chinvanno, S., et al., 2008. Climate risks and rice farming in the lower Mekong river basin. In: Leary, N., et al. (Eds.), Climate Change and Vulnerability. Earthscan, pp. 333–350.

Chun, J., et al., 2016. Assessing rice productivity and adaptation strategies for Southeast Asia under climate change through multi-scale crop modelling. Agric. Syst. 143 (2016), 14–21. https://doi.org/10.1016/j.agsy.2015.12.001.

D'Arrigo, R., Ummenhofer, C., 2015. The climate of Myanmar: evidence for effects of the Pacific Decadal oscillation. Int. J. Climatol. 35 (4), 540–634. https://doi.org/10.1002/joc.3995.

Deb, P., Babel, M., Denis, A., 2018. Multi-GCMs approach for assessing climate change impact on water resources in Thailand. Model. Earth Syst. Environ. 4 (2), 825−839. https://doi.org/10.1007/s40808-018-0428-y.

Delgado, J., Apel, H., Merz, B., 2010. Flood trends and variability in the Mekong river. Hydrol. Earth Syst. Sci. 14 (3), 407−418. https://doi.org/10.5194/hess-14-407-2010.

Dong, Z., et al., 2021. Heatwaves in Southeast Asia and their changes in a warmer world. Earth's Future 9 (7). e2021EF001992. https:doi.org/10.1029/2021EF001992.

Eastham, J., et al., 2008. Mekong River Basin Water Resources Assessment: Impacts of Climate Change. CSIRO: Water for a Healthy Country National Research Flagship.

Gutiérrez, J.M., et al., 2021. Atlas. In: Masson-Delmotte, V., et al. (Eds.), Climate Change 2021: The Physical Science Basis. Contribution of Working Group I to the Sixth Assessment Report of the Intergovernmental Panel on Climate Change. Cambridge University Press.

Hobday, A., et al., 2016. A hierarchical approach to defining marine heatwaves. Prog. Oceanogr. 141, 227−238. https://doi.org/10.1016/j.pocean.2015.12.014.

Horton, R., et al., 2017. Assessing Climate Risk in Myanmar: Technical Report. Center for Climate Systems Research at Columbia University, WWF-US and WWF-Myanmar, New York, NY, USA.

IPCC, 2021. Climate Change 2021: The Physical Science Basis. Contribution of Working Group I to the Sixth Assessment Report of the Intergovernmental Panel on Climate Change. Cambridge University Press. Interactive Atlas Available from: http://interactive-atlas.ipcc.ch/.

Kiguchi, M., et al., 2020. A review of climate-change impact and adaptation studies for the water sector in Thailand. Environ. Res. Lett. 16 (2021).

Lawrence, J., Blackett, P., Cradock-Henry, N., 2020. Cascading climate change impacts and implications. Clim. Risk Manag. 29 (2020), 100234. https://doi.org/10.1016/j.crm.2020.100234.

Lee, S., 2021. In the era of climate change: moving beyond conventional agriculture in Thailand. Asian J. Agricul. Develop. 18 (1362−2021−1176), 1−14.

Loo, Y., Billa, L., Singh, A., 2015. Effect of climate change on seasonal monsoon in Asia and its impact on the variability of monsoon rainfall in Southeast Asia. Geosci. Front. 6 (6), 817−823. https://doi.org/10.1016/j.gsf.2014.02.009.

Lorenzo, T., Kinzig, A., 2020. Double exposures: future water security across urban Southeast Asia. Water 12 (1), 116. https://doi.org/10.3390/w12010116.

Lwin, C., Maung, K., Murakami, M., Hashimoto, S., 2017. Scenarios of phosphorus flow from agriculture and domestic wastewater in Myanmar (2010−2100). Sustainability 9 (8), 1377. https://doi.org/10.3390/su9081377.

Mekong River Commission, 2005. Overview of the Hydrology of the Mekong Basin. Mekong River Commission, Vientiane.

Nwe, T., Zomer, R., Corlett, R., 2020. Projected impacts of climate change on the protected areas of Myanmar. Climate 8 (9), 99. https://doi.org/10.3390/cli8090099.

Oeurng, C., et al., 2019. Assessing climate change impacts on river flows in the Tonle Sap Lake Basin, Cambodia. Water 11 (3), 618. https://doi.org/10.3390/w11030618.

Open Development Cambodia, 2017. Water Resources [Online] Available at: https://opendevelopmentcambodia.net/topics/water-resources#ref-80858-2. (Accessed 15 January 2022).

Pipitpukdee, S., Attavanich, W., Bejranonda, S., 2020. Climate change impacts sugarcane production in Thailand. Atmosphere 11 (4), 408. https://doi.org/10.3390/atmos11040408.

Ranasinghe, R., et al., 2021. Climate change information for regional impact and risk assessment. In: Masson-Delmotte, V., et al. (Eds.), Climate Change 2021: The Physical Science Basis. Contribution of Working Group I to the Sixth Assessment Report of the Intergovernmental Panel on Climate Change. Cambridge University Press, Cambridge.

Rao, P., Ramaswamy, V., Thwin, S., 2005. Sediment texture, distribution and transport on the Ayeyarwady continental shelf, Andaman Sea. Mar. Geol. 216 (4), 239–247. https://doi.org/10.1016/j.margeo.2005.02.016.

RIMES and UNDP, 2020. Proposed Climate Zone of Cambodia, Strengthening Climate Information and Early Warning System in Cambodia, Bangkok: Regional Multi-Hazard Early Warning System for Asia and Africa. RIMES) and United Nations Development Programme (UNDP).

RIMES, 2011a. Managing Climate Change Risk to Food Security in Myanmar. RIMES, Bangkok, Thailand.

RIMES, 2011b. Managing Climate Change Risks for Food Security in Cambodia. RIMES, Bangkok, Thailand.

Saing, C., et al., 2012. Foreign investment in Agriculture in Cambodia. Cambodia Development Resource Institute (CDRI), Food and Agriculture Organization of the United Nations (FAO), Phnom Penh: Rome.

Shrestha, S., Thin, N., Deb, P., 2014. Assessment of climate change impacts on irrigation water requirement and rice yield for Ngamoeyeik Irrigation Project in Myanmar. J. Water Clim. Change 5 (3), 427–442. https://doi.org/10.2166/wcc.2014.114.

Strassburg, M., et al., 2015. Sea level trends in Southeast Asian seas. Clim. Past 11 (5), 743–750. https://doi.org/10.5194/cp-11-743-2015.

Swe, L., Shrestha, R., Ebbers, T., Jourdain, D., 2015. Farmers' perception of and adaptation to climate change impacts in the dry zone of Myanmar. Clim. Dev. 7 (5), 437–453. https://doi.org/10.1080/17565529.2014.989188.

Thailand Meteorological Department, 2010. The Climate of Thailand [Online] Available at: www.tmd.go.th/en/archive/thailand_climate.pdf. (Accessed 16 September 2021).

Thanasupsin, S., 2012. Bangkok, Thailand. Climate Change Impacts on Water Resources: Key Challenges to Thailand CC Adaptation. In 7th THAICID National Symposium.

Thirumalai, K., DiNezio, P., Okumura, Y., Deser, C., 2017. Extreme temperatures in Southeast Asia are caused by El Niño and worsened by global warming. Nat. Commun. 8 (1), 1–8. https://doi.org/10.1038/ncomms15531.

Thoeun, H., 2015. Observed and projected changes in temperature and rainfall in Cambodia. Weather Clim. Extrem. 7 (2015), 61–71. https://doi.org/10.1016/j.wace.2015.02.001.

Van Meijl, H., et al., 2018. Comparing impacts of climate change and mitigation on global agriculture by 2050. Environ. Res. Lett. 13 (6), 064021.

WBG and ADB, 2021a. Climate Risk Country Profile: Thailand. World Bank Group, Washington DC.

WBG and ADB, 2021b. Climate Risk Profile: Cambodia. The World Bank Group and Asian Development Bank, Washington DC.

Yaya, O., Vo, X., 2020. Statistical analysis of rainfall and temperature (1901–2016) in South-East Asian countries. Theor. Appl. Climatol. 142 (1), 287–303. https://doi.org/10.1007/s00704-020-03307-z.

Yuen, B., Kong, L., 2009. Climate change and urban planning in Southeast Asia. SAPI EN. S. Surveys Persp. Integ. Environ. Soc. 15 (2.3).

Zhai, F., Zhuang, J., 2012. Agricultural impact of climate change: a general equilibrium analysis with special reference to Southeast Asia. In: Anbumozhi, V., Breiling, M., Pathmarajah, S., Reddy, V.R. (Eds.), Climate Change in Asia and the Pacific: How Can Countries Adapt. SAGE Publications India Pvt Ltd, New Delhi, pp. 17–35.

CHAPTER 3

Interrelationship between climate justice and migration: A case of south western coastal region of Bangladesh

M. Ashrafuzzaman[1,2,3,4,5]

[1]*Climate Change and Sustainable Development Policies, University of Lisbon & Nova University of Lisbon, Lisbon, Portugal;* [2]*Institute of Social Sciences, University of Lisbon, Lisbon, Portugal;* [3]*Norwich Research Park, University of East Anglia, Norwich, United Kingdom;* [4]*Department of Geography, University of Valencia, Valencia, Spain;* [5]*Department of Anthropology, University of Chittagong, Chattogram, Bangladesh*

1. Introduction

The Intergovernmental Panel on Climate Change (IPCC) identifies carbon dioxide (CO_2) emissions as the primary global warming driver (IPCC, 2022). This leads to adverse effects: higher temperatures, more heatwaves, ice melting, rising sea levels, and extreme weather events (IPCC, 2014). Climate change has disrupted migration patterns, intensifying existing vulnerabilities. It impacts population movements through natural disasters, health and livelihood issues, sea level rise, and resource-based conflicts (IPCC, 2022). People are increasingly displaced by more frequent and severe natural disasters. Elevated temperatures and climate variability have a detrimental impact on health and livelihood, influencing living conditions, public health, food security, and access to clean water. The rising sea levels are making coastal regions uninhabitable, necessitating relocation. Competition for scarce natural resources escalates tensions and can drive conflict-based migration (IPCC, 2014, 2022). Urgent climate change action is needed to address these challenges and reduce injustices faced by affected populations.

Recent academic investigations by Rashid (2013), Martin et al. (2013), Hoffman (2022), and Ashrafuzzaman et al. (2022) have provided valuable insights into the intricate dynamics of climate change impacts. These repercussions are particularly severe in the Global South, affecting countries like the Marshall Islands, Bangladesh, Chile, Kenya, Sudan, and various sub-Saharan nations. These regions grapple with substantial hardship and widespread population movements, occurring both domestically and internationally (Manou and Mihr, 2017). Climate change—induced migration has emerged as a globally significant and contentious issue, garnering the attention of governments and policymakers. This form of migration encompasses diverse categories, including climate refugees, economic migrants, and displaced individuals, with consequences that transcend national borders (Brzoska and Fröhlich, 2016). Factors like conflicts over depleting natural resources, rising sea levels, cyclones, salinity intrusion, and droughts often compel individuals to seek refuge and migrate due to environmental hazards. Although some affected individuals may follow established legal migration routes, the majority undertake these journeys without the benefit of a well-defined legal framework (Islam and Hasan, 2016).

The challenge of climate-induced migration has brought human rights and climate justice to the forefront, addressing the gaps in regional and international legal protection for climate migrants (Rosignoli, 2022). This has prompted global attention to issues of injustice, adaptation, mitigation, and relocation decisions. International bodies like the UN, African Union, European Union, and Organization of American States are actively developing programs to reduce vulnerability among climate-displaced individuals (Cohen and Deng, 2012; Hollifield et al., 2014; Kent, 2018).

The discourse on climate-induced migration, especially in disaster contexts, is prevalent. The IPCC's Sixth Assessment Report (AR6) in 2022 extensively explores climate change's complex links with migration. Bangladesh, highly vulnerable to climate change, faces risks from floods, cyclones, erosion, salinity intrusion, and droughts (Ashrafuzzaman et al., 2022). These challenges result in loss of life, infrastructure damage, social instability, lower living standards for marginalized communities, agricultural losses, food insecurity, and displacement (Mallick et al., 2017).

Climate change projections indicate that approximately 200 million people may be compelled to migrate worldwide by 2050 (Mcdonnell, 2019). In Bangladesh, an annual average of 700,000 individuals face displacement due to climate change, aggravated by events such as Cyclone Sidr (2007), Aila (2009), and Cyclone Amphan (2021), leading to the displacement of millions in coastal regions. Rising sea levels, erosion, salinity intrusion, crop failures, and recurrent flooding are contributing factors to coastal displacement. It is estimated that climate change could displace 13.3 million Bangladeshis by 2050, potentially submerging 17% of coastal land by 2080 (IPCC, 2014). In 2020, approximately 4.4 million coastal residents were displaced as a result of Cyclone Amphan (IDMC, 2020; Sammonds et al., 2021). Addressing challenges like river erosion, inundation, and salinity proves to be a significant hurdle, largely attributed to global and local climate injustices.

Furthermore, there is an increasing focus on studying the relationship between gender, climate change, and migration (Chindarkar, 2012). The interaction between gender, climate change, and migration has led to the concept known as the climate change—gender migration nexus (Kempadoo and Doezema, 2018). Nevertheless, the specific consequences of environmentally induced migration for both men and women have received limited research attention. Women, in particular, encounter substantial challenges during climate-induced displacement, especially in terms of their security and experiences in crisis situations (Brody et al., 2012). However, there is a scarcity of research on the postmigration circumstances of female migrants, encompassing their hydrogeophysical, social, and economic conditions in their new surroundings.

This article concentrates on the Southwest Coastal Region of Bangladesh (SWCRB) to explore the impact of climate change on coastal inhabitants and their resultant migration. It delves into climate change—induced migration and its implications for core human rights, norms, and values. The research specifically focuses on Shyamnagar Upazila in southern Bangladesh. The inquiry seeks to address two central questions: First, what motivates climate change-induced migration to the coastal areas of Bangladesh? Second, what is the relationship between migration and climate justice, both at the local and global levels, in the context of climate migration within the SWCRB? The study's objectives encompass understanding the migration patterns of respondents in the SWCRB, including the factors, duration, motives, types, and variations associated with climate change—induced migration.

Qualitative analysis provides insights into how individuals in the SWCRB region safeguard their livelihoods and adapt their income strategies in response to climate-induced migration.

Additionally, the article delves into the link between climate justice and climate migration, pointing out disparities in migration and justice in the SWCRB. It also offers examples of adaptation strategies aimed at tackling migration-related challenges, especially in the cities of Bangladesh and the SWCRB.

1.1 Theoretical framework: Climate justice and migration

The issue of climate-induced migration encompasses diverse categories of individuals, such as climate refugees, economic migrants, and displaced persons, with significant global implications. However, the lack of a clear legal framework poses a substantial challenge in addressing the migration of these populations (Manou and Mihr, 2017). In response to this challenge, the concepts of human rights and climate justice have gained prominence as vital frameworks. The 1948 Universal Declaration of Human Rights provides a normative framework for addressing migration caused by climate change (Atapattu, 2015; Robinson and Shine, 2018). These frameworks underscore the need to consider both human and environmental aspects, emphasizing the interconnectedness of these issues.

Efficient governance mechanisms necessitate a deep understanding of existing climate and migration policies and their potential to exacerbate inequalities (Scholten, 2020). Although there have been some recent advancements, such as proposals for an international convention to safeguard climate migrants, a comprehensive global framework to tackle these issues is still lacking. The 2015 Paris Agreement highlighted the complexities of incorporating provisions for climate-induced population displacement into international agreements. Subsequent meetings, such as COP22, reinforced climate commitments and stressed the importance of collaboration between nations and stakeholders (Manou and Mihr, 2017).

Climate migration was not a main topic at the 27th Conference of the Parties (COP27) to the UNFCCC, but it resulted in resolutions that renewed pledges to control global temperature rise and urged more emissions reductions and aid for developing countries (Anwar, 2022; UN Climate Press Release, 2022). Identifying climate migrants remains challenging, and the development of comprehensive legal frameworks faces political obstacles. This issue is linked to several human rights, including the right to a good standard of living, self-determination, housing, education, and career opportunities (Manou and Mihr, 2017). The United Nations acknowledges that the overlap between climate-induced migration and the protection of human rights is a significant concern. It poses a risk to both assets and rights, which could worsen gender and ethnic inequalities and disrupt economic and social balance.

1.2 Description of migration

Defining environmental migration presents a challenge due to regional and national variations. In this research, the study uses the International Migration Organization's definition, which defines migration as people moving within a country or across borders without specific criteria for its extent or intent (Brown, 2008). Different angles, like politics, sociology, economics, and anthropology, offer a broad comprehension of migration (Borkert et al., 2006).

The Push and Pull Factor theory explains migration by dividing it into "push" and "pull" elements. Push factors push individuals to leave their homes due to worsening environmental conditions, while pull factors, like job prospects and resource availability, draw people from environmentally stressed areas (Portes, 1995; Piguet, 2010; IMI, 2010). Lee (1966) broadened this concept to include factors such as environmental deterioration, population growth, and perceived shortages as major push factors in environmental migration. Urbanization and income opportunities also influence migration patterns and destinations (Kliot, 2004).

In the context of Bangladesh, migration primarily involves rural-to-urban movement driven by economic, social, and environmental reasons. Natural disasters contribute to severe unemployment in rural areas, leading to migration into urban slums. This displacement exacerbates socio-economic challenges, including land loss and poverty for affected individuals (Poncelet et al., 2010).

1.3 When society and climate experience transformation

The necessity of global collaboration to address a universally significant ecological crisis is underscored by Gare (1995), who argues that humanity is locked in a battle for survival, requiring nations to unite as allies. The concept of the environment has undergone a transformation over time. Initially, it encompassed only the natural world, but its scope gradually expanded to include other species, ecosystems, and, ultimately, humans. For example, the notion that "nature must be managed" (Worster, 1977) is anthropocentric, emphasizing human intervention to manipulate other species within the natural world. Taking a Marxist perspective, environmental economics tackles environmental issues by attributing the global environmental crisis to the primary means of production. Marxism views capitalism as a system that commodifies goods in monetary terms, treating labor as capital, and being dominated by the market. Marx's concept suggests an ongoing interchange between humans and nature, with humans being an integral part of nature (Gooch, 2000). This concept aligns with Marx's term "praxis," indicating an interactive approach involving the natural environment, similar to how the practices of the Van Gujjar transform trees or grass into milk, as related to Marx's concept of praxis (Gooch, 2000).

Geological transformations take millions of years, but shifts in human culture happen within a few hundred to a couple of 1000 years. Academic discussions about today's climate change crisis vary widely. Climate experts contend that humans have become a geological force capable of changing the composition of the atmosphere, leading to higher sea levels, melting polar ice, and changes in climate patterns (Crosby, 1995).

1.4 Study area

The study area, located in the southwestern part of Bangladesh, covers the entire Sundarbans in the Shyamnagar region of the Satkhira district. Geographically, it extends from 21°36′ to 22°24′ north latitudes and 89°00′ to 89°19′ east longitudes, bordering the Sundarbans forest. To the north, it borders Kaliganj (Satkhira) and Assasuni Upazila, while to the south, it is adjacent to the Bay of Bengal and West Bengal, India. The eastern boundary connects with Koyra and Assasuni Upazila, and the western border abuts West Bengal, India. Shyamnagar is the largest upazila in the Satkhira district (Ashrafuzzaman, 2022).

This region, including the Sundarbans, is a component of the world's biggest delta formed by the Ganges-Brahmaputra-Meghna (GBM) rivers, which start in the Himalayas and meet in the Bengal Basin of Bangladesh. It boasts one of the world's largest deltas, with delta sediments stretching around 200 km south from the coastline, creating the GBM Rivers (Ashrafuzzaman, 2022a).

This area is close to a mangrove forest and the Bay of Bengal, making it susceptible to climate change and rising sea levels because it is situated at a low altitude. It was formed by deposits from the Ganges floodplain and has a moderate population density, with an average height above sea level (MASL) between 1 and 5 m. About 45% of this region is at high risk of flooding from a 1-m storm surge (Ashrafuzzaman, 2022b).

According to the 2011 Bangladesh census, Shyamnagar had a population of 318,254, with a fairly equal number of men and women. This Upazila covers an area of 1968.24 square kilometers, including a substantial forested area of 1622.65 square kilometers, and is divided into 12 unions. Shyamnagar Upazila is well-known for its diverse natural climate characteristics (ibid) (Fig. 3.1).

In Shyamnagar, Bangladesh, the local residents have a variety of jobs to support their livelihoods. These jobs include day labor, farming, fish and crab farming, poultry raising, rickshaw and auto-bike taxi driving, mechanics, fishing, selling groceries, and teaching. It is important to highlight that in this community, fish farming is the primary way people make money, and most individuals earn approximately Tk 5000–6000 per month from these main income sources (Ashrafuzzaman et al., 2022).

Nonetheless, some residents who used to depend on the Sundarbans forest for their living, like fishermen and honey collectors, have had to switch to other jobs because of climate-related damage to the forest (Mukhopadhyay, 2016). Additionally, those involved in shrimp farming had to find different ways to earn money as many small farms were destroyed. This pushed both farm owners and workers to seek alternative income options in nearby towns (Afroz et al., 2017).

2. Methodology
2.1 Determining sample size using quantitative methods

For this study, we used a method of selecting samples that accurately reflect the whole population. We aimed for a 95% confidence level and a margin of error of ±5%. Because we had exact counts of beneficiaries, we wanted to make sure that each union had a fair representation in proportion to their population. To do this, we followed a specific sampling technique and statistical formula in our sample design (Table 3.1).

$$n = \frac{z^2 \cdot p \cdot q \cdot N}{z^2 \cdot p \cdot q + (N-1)e^2}$$

Where,

n = Sample size
N = Targeted population size
e = Admissible error in the estimate
p = Proportion of defectiveness or success for the indicator
$q = 1 - p$
z = Standard normal variable at the given level of significance

To ensure each union was fairly represented, we chose the same portion of the sample size for each. We accomplished this by using a method called stratified random sampling.

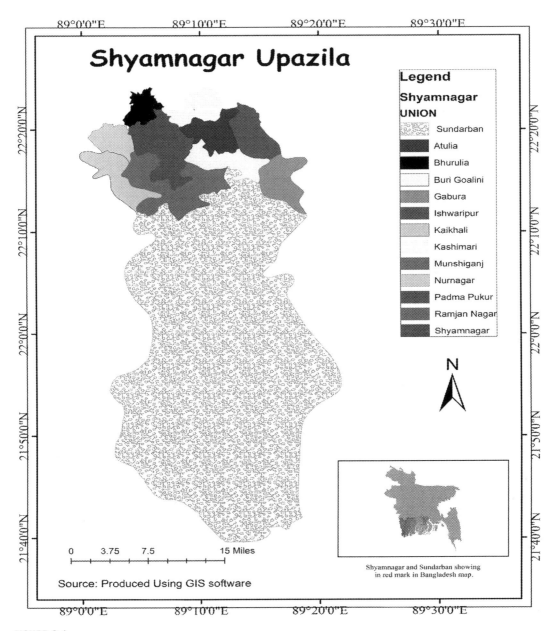

FIGURE 3.1
Map of study areas.

Table 3.1 The detail of quantitative sample size distribution.

Type of respondents	Unions coverage	N = Total population	e = Admissible error in the estimate	Sample size = n	Female	Male	Of them youth
Social vulnerability, justice and adaptation	09	242,392	5.5% admissible error margin	320	98	222	25.5%

2.2 Methods of gathering information

In this study, we collected data by using closed-ended questionnaires in nine union parishads in Shyamnagar Upazila, Bangladesh, from July 2017 to December 2019. We aimed to interview 320 households in each research site. The questionnaire had 128 questions and was carefully designed, tested, and adjusted. To gather data at the union level, we used various sampling techniques like simple random sampling, stratified sampling, and cluster sampling, as suggested by Bryman (2016) and Vehovar et al. (2016). The study focused on the vulnerable coastal area of Bangladesh, which faces various climate-related risks like cyclones, storm surges, erosion, and salinization. This region was selected because it represents a wide range of geographic, socio-economic, and ecological conditions, diverse ecosystems, different land uses, and varying exposure to natural disasters. Each interview took about 50–120 min and primarily involved discussions with the household head or their partner. The survey asked about migration in the Southwestern Coastal Region of Bangladesh, focusing on the primary causes of migration such as sea-level rise, cyclones, floods, salinity, and erosion. After gathering the data, we input all the statistics directly into the SPSS software, following the instructions outlined by Vehovar et al. (2016).

2.3 Qualitative approach

To gain insights for this study, we used a variety of qualitative methods, including Participatory Rural Appraisal (PRA), case studies, observation, Key Informant Interviews (KII), In-Depth Interviews (IDI), focus group discussions (FGD), workshops, and individual interviews. These methods helped us build a comprehensive understanding of the factors behind climate-induced migration in the Southwestern Coastal Region of Bangladesh (SWCRB) (Bryman, 2003; Lune and Berg, 2017).

Participatory Rural Appraisal (PRA) was particularly valuable for revealing migration behaviors, causes, and contributing factors, while FGDs were carried out to explore migration patterns. We conducted a total of 26 FGDs across the 12 unions in the Shyamnagar sub-district. Participants included individuals from various occupations with direct experience of climate-related shocks. Each FGD was led by a panel of 10–12 members, including local government representatives, NGO representatives, doctors, journalists, women representatives, and the researcher (Bryman, 2016).

Additionally, we used various research methods, including case studies, interviews, and questionnaires, to gather information about the factors affecting migration. We developed the interview questions based on existing literature, focusing on migration and climate justice. We held follow-up meetings after each data collection phase, involving participants from different backgrounds, each group led by an experienced farmer or agriculturist knowledgeable about natural hazards and migration.

To collect data, we employed a "multistage sampling" approach, which included random, stratified, and cluster sampling methods. To ensure ethical standards in our qualitative research, we assigned anonymous identifiers to most respondents, such as respondent/participant-1, 2, 3, or A, B, C, etc.

3. Results: Quantitative analysis
3.1 Migration behavior of the respondents

In Fig. 3.2, we inquired about whether people in the Southwestern Coastal Region (SWCR) should consider relocating due to climate change and rising sea levels. Over half of the respondents indicated that SWCR residents should move, either temporarily or permanently, because of these environmental shifts. Only 7.20% of respondents did not have an answer or did not respond to the question.

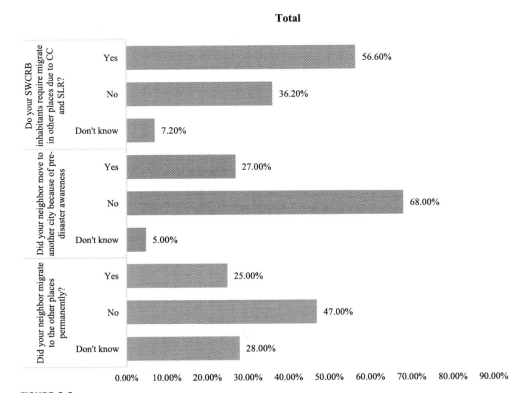

FIGURE 3.2

Migration behavior of respondent.

Field survey, 2017–19, N: 320.

Meanwhile, 36.20% of the respondents believed there was no requirement for them to relocate to other areas, either temporarily or permanently, due to climate change and rising sea levels.

Fig. 3.2 presents survey responses related to the permanent migration of neighbors. Roughly 47% of households responded negatively, indicating that their neighbors had not permanently relocated elsewhere. In contrast, 28% of respondents either claimed not to know or avoided answering this question.

The study also investigated whether the respondents' neighbors had moved to another city as part of predisaster preparedness. Among the respondents, 68% answered "no," indicating that their neighbors had not relocated to another city. Only 27% responded with a "yes," signifying that their neighbors had moved to another city, either temporarily or permanently, while approximately 5% of respondents refrained from providing a response to this specific question.

3.2 Why people initiate migration?

Migration in Bangladesh can happen for various reasons, including climate-related and non-climate-related factors. Climate change events like cyclones, river erosion, floods, droughts, hail, lightning, and salinity are significant drivers of migration in the country. The Chairman of Shyamnagar Upazila, situated in the southwest coast of Bangladesh, notes that the coastal areas in Bangladesh frequently face numerous natural disasters caused by adverse weather conditions. The continuous onslaught of these disasters has made it extremely difficult for people to survive in the region, leading many to leave their ancestral land in search of a safer and more peaceful life.

Respondents also mentioned that people in the region are relocating for two main reasons: to protect their lives by avoiding natural disasters and to safeguard their livelihoods by preventing damage to their homes and crops from these disasters. The table below illustrates that from 1970 to 2020, cyclones caused a loss of life ranging from 0.01% to 6% and damage to homes, crops, and plants ranging from 10% to 75%, 30% to 100%, and 40% to 90%, respectively, which significantly drove migration. In contrast, floods and river erosion in the study areas were reported to cause no loss of life at different times. However, lightning resulted in the deaths of 1476 people in Bangladesh from 2010 to 2016, leading to an increased rate of migration from the affected areas. Table 3.2 summarizes the various natural events that respondents identified as reasons for migration in the research areas.

Table 3.2 Triggers of migration in the studied areas.

Number	Triggers of migration	Year of occurrence	Loss (%)			
			Loss of life	Home damage	Crop damage	Damage to plants
01	Cyclone	1970	6%	70%	80%	85%
02	Cyclone	1988	3%	75%	90%	90%
03	River erosion	1991	—	70%	60%	20%
04	River erosion	2002	—	10%	40%	30%
05	Flood	2000	00	50	60	40
06	River erosion	2004	—	15%	50%	20%
07	Hail	2006	00	2	25	20

Continued

Table 3.2 Triggers of migration in the studied areas.—cont'd

Number	Triggers of migration	Year of occurrence	Loss of life	Loss (%) Home damage	Crop damage	Damage to plants
08	Cyclone Sidr	2007	5	12%	100%	95%
09	Cyclone Akash	2007	0.01	15	50	40
10	Salinity	2007		8	60	80
11	Cyclone Rashmi	2008	0.01	10	30	40
12	Cyclone Bijli	2009	0.03	30	45	45
13	Cyclone Aila	2009	1%	90%	95%	90%
14	Salinity	2009	—	10%	80%	60%
15	Flood	2010	Extensive crop loss and death of cattle due to tidal water and salinity.			
16	Cyclone Mahasen	2013	0.01	30	40	40
17	Cyclone Komen	2015	0.01	25	30	35
18	Cyclone Roanu	2016	0.01	40	50	50
19	Cyclone Mora	2017	0.01	35	40	40
20	Cyclone Fani	2019	0.01	35	45	50
21	Cyclone Bulbul	2019	0.01	30	40	40
22	Cyclone Amphan	2020	0.01	45	60	70
23	River erosion	2007–2020 continue	000	15	30	30
24	Salinity	2000–2020	Health	10	50	50
25	Drought	Every year	Health	00	30	20
26	Lightning/thunderstrom	Every year	Between 1990 and June 2016, Bangladesh witnessed 3086 fatalities, with an average of 114 fatalities annually, resulting in a fatality rate of 0.9 per million (Rahman et al., 2019). In the span of 2010–16, lightning claimed the lives of 1476 individuals in Bangladesh (Rahman et al., 2019). The year 2019 saw 198 fatalities from lightning strikes, while 2020 recorded 255 such fatalities (Ahmed, 2021).			

PRA, FDG, Workshops 2017–19 (Skype, messenger 2020–21).

3.3 Period of migration

According to local residents, individuals in the coastal zone typically engage in seasonal and nonseasonal migration at different times. Many people in the region move to urban areas during disasters and return to their regions when the weather is favorable. However, some people permanently relocate from the region.

In the table below, you can see that the most extended migration period is from September to April, lasting 5–8 months. During this time, migrants go to metropolitan areas to work in brickfields because

Table 3.3 Different migration periods.

Migration period	Sort of undertaking	Duration of residence (months)	Locations
In the months of December–January and July through August	Rice planting, casual work	1–2	Dhaka, Narayanganj, Gazipur, Barisal, Rajsahi, Sylhet, Jessore, Cumilla, Chattogram,
March–May	Paddy reaping	1–3	Faridpur, Barisal, Rajsahi, Rangpur, Netrokona, Sylhet, Jessore, Bagherhat,
September–April	Brick field	5–8	Dhaka, Barisal, Narail, Khulna, Cumilla, Chattogram, (India)
December–March	Lumber processing work (using a portable chainsaw), daily wage labor, etc.	2–4	Bagerhat, Barguna, Gopalgonj, Barisal
January–May	Day labor (e.g., embankment work, pond excavation, etc.)	3–5	Khulna, Bagerhat, Sathkhira Jessore,
Any time of the year, although usually during the monsoon season (June–August)	Casual labor, cycle rickshaw pulling, coastal port/frontier zone, apparel factory, production laborer	1–12	Dhaka, Khulna, Mongla, Jessore, Sylhet, Chittagong
All the year around migration	Not indicated	Not indicated	Maldives, India, Nepal, Bhutan, Middle east, Europe, USA etc.

PRA, FDG, Workshops 2017–19 (Skype, messenger 2020–21).

that is when the brickfields are active. It is worth mentioning that this period aligns with the peak of climate-related disasters.

Migration is a way for people to adapt and survive. They move to cities to find informal jobs like rickshaw pulling, especially during the rainy season, or go to rural areas to work in factories. Some migrants work as day laborers during the paddy harvesting season, staying away from home for 1–3 months. Others leave Bangladesh at various times throughout the year without a set profession or time frame.

Table 3.3 provides information on various migration time periods in the study region.

3.4 Diverse migration related to climate change

The main cause of migration is jobs. About 30.3% of the respondents said that people move because they can find better work, either in cities or other countries. Additionally, 18.8% and 12.5% of all respondents said that land loss, a decrease in economic activities, and lower agricultural production are also important reasons for their migration (see Fig. 3.3).

Climate change leads to river erosion, increased saltiness in groundwater, and a drop in economic activities. The figure shows that after migrating, people take up different jobs like day labor, hawking,

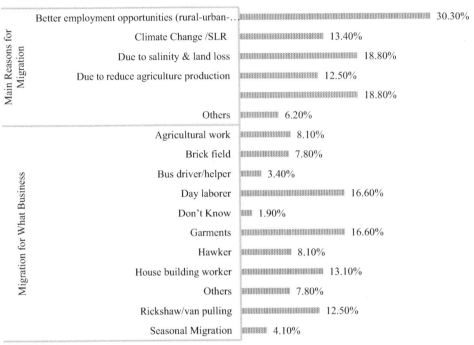

FIGURE 3.3

Climate change diversification migration.

Field survey, 2017–19, N: 320.

driving rickshaws or vans, working as bus drivers, seasonal work, bricklaying, agricultural labor, house construction, and garment work, among others. In all, 16.6% of the respondents stated that people from the southwest coastal region of Bangladesh go from rural to urban areas to find work in daily labor and the garment industry. Additionally, 12.5% and 13.1% of households said that residents also become rickshaw or van pullers and engage in house construction work after migrating. In the coastal area, people choose to move mainly because of natural disasters, which lead to serious job shortages in rural regions, land and property loss, worsening socio-economic situations, and more. As a result, individuals in the study area are forced to relocate to other parts of the country, either for a short time or permanently.

3.5 Types of migration

The presented figure illustrates the primary forms of migration in the region under investigation, categorized by duration. Concerning permanent migration, approximately 30.90% of respondents relocated to various destinations within Bangladesh, while around 14.40% migrated to countries outside of Bangladesh. On the other hand, 37.80% of respondents undertook temporary internal migration within Bangladesh, and 16.90% temporarily moved outside of the country. It is important to highlight that the majority of respondents (54%) stated that individuals in the study area mainly migrate from rural to urban areas (see Fig. 3.4). The coastal region of Bangladesh has been facing climate change effects like higher temperatures, severe floods, more frequent and stronger cyclones, persistent waterlogging, and increased salinity in rivers, groundwater, and soil. Consequently, the region has witnessed agricultural land degradation, diminished food production, riverbank erosion, health issues, unemployment, and economic downturns. These environmental challenges have prompted residents of the southwest coastal region to make migration decisions, often relocating to urban areas. Furthermore, 21.2%, 6.6%, and 4.7% of respondents mentioned that individuals in the study area go to Middle Eastern countries, India, and Malaysia, not only because of climate change but also to seek improved job opportunities and living conditions.

3.6 Reasons for migration

Fig. 3.5 illustrates the primary reasons for climate change−induced migration over the next decade. According to respondents in the surveyed regions, climate change−related disasters such as cyclones (214), salinity (293), floods (198), and waterlogging (173) were the most common triggers for population displacement. Other climatic factors, such as changing rainfall patterns (87) and increasing

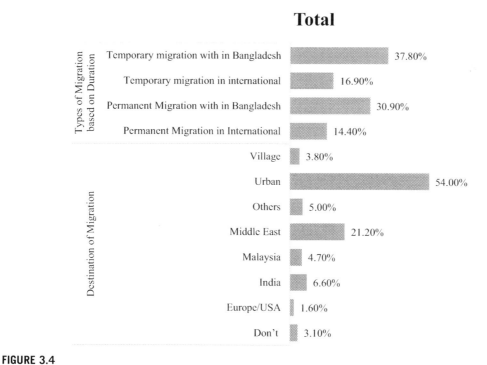

FIGURE 3.4

Types of migration.

Field survey, 2017−19, N: 320.

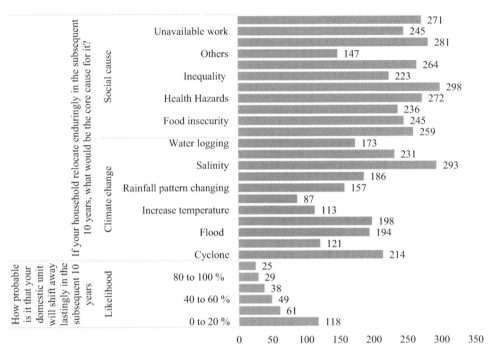

FIGURE 3.5

Reason of migration.

Field Survey, 2017–19; Multiple Response; N: 320.

temperatures (113), had a somewhat lesser influence on people's decisions to move away from their habitats. However, societal factors, including unemployment (271), social inequality (223), food insecurity (245), and health risks (272), also contributed to migration. The majority of respondents (118) stated that they had a 0%–20% likelihood of leaving their current residence in the next 10 years, while 25 or more respondents indicated a high likelihood of doing so, with an 80%–100% chance. Additionally, 38 or more respondents expressed a moderate likelihood of moving from their households, with a 40%–60% chance of doing so in the subsequent decade.

4. Qualitative analysis

4.1 Methods used included participatory rural appraisal, case study, focus group discussions, workshops, interviews, key informant interviews, and in-depth interviews

4.1.1 Initiatives to preserve existence

Participant 201, who serves as the Chairman of the local government in Shyamnagar Upazila, highlighted the rising sea levels in the coastal regions. This increase has led to more frequent cyclones, floods, and waterlogging, all linked to climate change. These destructive events have caused homes to be damaged, fields to be submerged, and crops to be ruined, resulting in the loss of lives and livelihoods. Consequently, numerous individuals have had to depart from their family's long-held residences to ensure their families' safety.

Participant-A, aged 49, hailing from Munshiganj Union, revealed how climate change has driven numerous residents to contemplate changing their living locations, with some even contemplating international migration to countries like India. The destructive force of the rivers has left people homeless and deprived of their belongings, compelling them to seek refuge in areas less susceptible to flooding. These individuals hold onto the hope of finding a better life away from the flood-ridden regions.

Participant-B, aged 45, residing in Burigoalini Union, highlighted the migration of around 3000—4000 villagers from wards one to nine in Burigoalini Union. This mass movement was primarily a result of inadequate cyclone shelters, a lack of safety measures, and insufficient awareness of disaster signals. The consequences of these factors have left these areas desolated.

Participant-C, a 56-year-old resident of Nurnagar Union, described the numerous climate-related dangers that threaten the lives of people in coastal Bangladesh. These dangers include flooding, cyclones, heavy rainfall, typhoons, salinity, drought, riverbank erosion, cloudbursts, and earthquakes. The country's location near the Himalayas makes it susceptible to frequent natural disasters. Every year, Bangladesh faces one or more disasters that disrupt daily life, resulting in the premature loss of many lives, both human and livestock, as well as the destruction of homes, crops, plants, roads, and more. Given this challenging reality and the ongoing impact of climate change, the coastal areas of the country are hit by floods, tornadoes, waterlogging, thunderstorms, droughts, cold waves, heavy fog, and various natural calamities. This relentless series of events has forced thousands of people to leave the coastal regions in search of a safer and more sustainable way of life.

4.1.2 Relocation for income generation

Participant-D, a 38-year-old newspaper reporter from Ramjan Nagar Union, provided valuable insights on how cyclones have severely impacted the region. Over the years, cyclones like Sidr in 2007, Aila in 2009, Mahasen in 2013, Fani in 2019, Bulbul in 2019, and Amphan in 2020 have continuously pounded the coastal areas, causing extensive flooding from seawater. These devastating events, worsened by climate change, have led to ongoing saltwater intrusion, widespread flooding, destruction of shrimp farms, and significant job losses. As a result, people have had to seek new ways of making a living, often forcing them to leave their cherished ancestral lands.

Shrimp farming, once a beacon of hope for many, has proven to be a profitable venture for the affluent, while pushing numerous others into destitution. The practice of crab and shrimp farming thrives in areas like Atulia, Munshiganj, Burigoalini, Gabura, Padmapukur, and various other locations. People living in these areas have adopted comprehensive farming practices that include growing rice, fishing, and cultivating vegetables. Regrettably, residents in coastal communities like Gabura, Padmapukur, Burigoalini, Kaikhali, Ramjannagar, and Sundarbans have faced numerous relocations, mainly due to the relentless effects of climate change.

Participant-E, aged 65, residing in Kaikhali Union, provided insights into the prevailing agricultural practices in the area, with a focus on Aman rice cultivation during the monsoon season (June—November), followed by harvesting in November—December. Nevertheless, unanticipated factors such as river erosion, embankment damage, cyclones, tidal surges, and flooding have resulted in extensive coastal inundation. These events have not only disrupted the livelihoods of farmers but have also inflicted considerable harm to crops. In response to these challenges, many individuals, including Participant-E, have opted to migrate to urban centers. Some have chosen to become rickshaw pullers, while others have sought employment in brick kilns. Additionally, many locals temporarily move to the northern parts of the country for farm work.

Participant-F, the Coordinator of the "Ramjannagar Union Disaster Management Committee" in Shyamnagar Upazila, emphasized the southwest coastal region's vulnerability to global warming and climate change. Many agricultural laborers lost their jobs due to salinity and saltwater shrimp farming.

Participant-G, a 51-year-old resident of Paddapukur Union, recounted his personal struggles during the devastating cyclone Amphan, which resulted in the loss of his farmland, home, livestock, and belongings. Following their stay in a cyclone shelter during Amphan, he and his family made the challenging decision to relocate to an urban area. There, he secured employment in a brick kiln, but this transition marked a period of considerable hardship for his family.

Participant-H, a 55-year-old college instructor hailing from Shyamnagar Union, highlighted the extensive migration of residents from all the unions within the coastal region due to a lack of employment opportunities. Conversations with local residents revealed that in Burigoalini Union No. 9, as many as 80% of the population migrated to urban areas to work as rickshaw pullers and in brick kilns. During the month of Ramadan, approximately 3000 individuals from 4000 families actively seek employment. Similarly, half of Munshiganj Union's residents depend on rickshaw pulling in urban areas for 3–6 months from September to April, supplementing their income by working in brickfieldsIn Atulia Union, about 55% of people have moved to the city, mainly for jobs like rickshaw pulling and working in brick kilns, leading to a significant decline in local male population. Similarly, in Nurnagar Union, approximately 60% of the residents have left for urban areas to work as rickshaw pullers and in brickfields. In the past 15 years, roughly 1250 villages in East Kaikhali, Sayedkhali, Shailkhali, West Kaikhali, Kathamari, Astakhali, Ghoshalpur, Purakhali, Naikathi, Ghagramari, Mendinagar, Shibchandrapur, Nidya, and Jadavpur of Kaikhali Union No. 5 have been devastated by disasters. The middle-class households have faced increasing unemployment, leading to nearly 70% of the union's population moving to cities in search of jobs.

This trend is seen in every union in the upazila, giving a complete picture of climate change-induced migration in the region, as described below:

Kashimari Union: In Kashimari Union, people in villages like Sankarkathi, Kachiharania, Gobindpur, Joynagar, Kashimari, Ghola, and Godara are facing more health problems. Due to limited farming options, around 1500 families lost their jobs and had to move elsewhere.

Burigoalini Union: A 42-year-old named Participant-I from Burigoalini Union highlighted the difficult situation caused by climate change in the union. The absence of a proper drainage system in the area has caused canals to flood during heavy rain, putting residents at risk. Additionally, the lack of fresh water sources has forced families to travel long distances for drinking water. Increased salinity levels have also reduced crop yields and job opportunities, pushing people to seek work in other places.

Gabura Union: Conversations with residents of Gabura Union revealed that conventional farming practices, in place for two decades, have been disrupted. Salinity in the soil, river erosion, and shrimp cultivation, all consequences of climate change, have resulted in substantial job losses. Consequently, inhabitants of this region have relocated as their previous livelihoods, including fishing, leaf picking, honey collection, and employment in shrimp farms, became increasingly challenging for their survival.By using Participatory Rural Appraisal (PRA), Focus Group Discussions (FGD), workshops, and interviews with local residents, we uncovered the reasons people move to Shyamnagar Upazila, potential solutions, and the challenges that hinder nonsolutions (see Table 3.4 for details).

Table 3.4 Reasons for migration and its adaptation options.

Cause of migration	Remedy	Reasons for not remedy
Shortage of farming land and atmosphere apt for agriculture. Illicit shrimp agriculture in coastal regions	To inspire shrimp farmers to cultivate salt lenient paddy and vegetables in an ecologically pleasant method instead of shrimp farming. Conservation of forest ethos and legacy	Water quality, seed supply, irrigation infrastructure, and fishing resources are all negatively impacted by poor planning and management. Threats on ecosystem loss, invasive species, pollution, and climate change, as well as insecure land acquisition within and surrounding coastal regions.
Increase of saltwater in domestic and farming lands.	Effective soil management practices can successfully rid the soil of soluble salts, and this involves two key steps: ensuring proper drainage and utilizing clean and readily available irrigation water. To lower soil salinity levels, the process requires the precise application of defined volumes of water. For instance, applying 6 inches of water can reduce soil salinity by 50%. Doubling the quantity to 12 inches can result in an 80% reduction, and an even more substantial 90% decrease in soil salinity can be achieved by using 24 inches of water.	Lack of better varieties seeds, ineffective seed systems, and insufficient facilities for storing and analysis.
In climate change policy, there's a common problem of gender imbalance that tends to ignore the special viewpoints and	Incorporating gender considerations into local and national policies, disaster risk reduction to ensure gender equality, sustainable reduction of	Absence of gender justice, development policy does not given priorities women or human five basic needs

Continued

Table 3.4 Reasons for migration and its adaptation options.—cont'd

Cause of migration	Remedy	Reasons for not remedy
difficulties faced by women. Women usually have lower income, education, health, mobility, decision-making authority, and are exposed to different forms of gender-based violence. This underscores the necessity for approaches that include gender in dealing with climate change.	existing gender inequalities and vulnerabilities, build on the women capabilities.	
Ban of hunting fish and crabs in the Sundarbans at specific periods.	Construct scientific village to utilizing Sundarbans ecosystem	The overharvesting of commodities from forest, alteration of vegetation patterns, habitat degradation, saltiness, extreme weather events, unrestrained tourism, pollution, and so on.
Consequences of devastating cyclone on river corrosion and austere salinity.	To generate prosperous rehabilitation and substitute occupation instead of hazards.	Stakeholder's initiatives are less than required for rehabilitation.
Advance payments from multiple brick kiln owners for work in other districts	Preparation with the multi-level occupation prospects for surrendering bandits.	Poverty, loss of land, shortage of agriculture activity.

PRA, FDG, Workshops 2017–19 (Skype, messenger 2020–21).

5. Discussions

Migration is a key way to respond to changes in the environment and climate. According to the International Migration Organization (IMO), migration means people moving from one place to another, whether within a country or across international borders (Javed, 2013). In Bangladesh, especially in coastal areas, climate change has created a range of challenges, such as higher temperatures, increased salt in groundwater, regular flooding, rising sea levels, land erosion, and frequent natural disasters, as shown in Table 3.2. These changes in the environment have wide-ranging impacts, influencing agriculture, where people live, public health, biodiversity, animal migration patterns, economic development, and socio-economic conditions, as stated by the IPCC in 2022. It is crucial to recognize that migration results from a complex mix of factors, including social, economic, and political elements, even when climate-related factors are significant. As a result, these environmental factors can lead to situations of climate injustice, ultimately pushing people to migrate in affected communities (Panebianco, 2022).

This study explores how people move in the densely populated SWCRB Upazila due to climate change (see Fig. 3.1). We are particularly interested in understanding the effects of climate change. Migration is a key way people respond to the many impacts of climate change, but there are still many uncertainties about it. When considering the SWCRB area within the broader context of climate change, it is important to mention that the IPCC's fifth assessment report from 2014 highlighted an annual sea-level increase of about 1.5 mm in Bangladesh. Moreover, there looms the ominous prospect of a sea-level surge of one to 5 m should the Greenland, West Antarctic, and Himalayan ice sheets undergo a catastrophic collapse (IPCC, 2014). In the backdrop of Bangladesh, with a population nearing 60 million, a significant portion of who grapple with poverty, diverse climate-related challenges come to the fore. These encompass severe flooding, storm surges, riverbank erosion, salinization, land subsidence, and sea-level escalation. Each of these climate-related facets collectively contributes to the migration dynamics of the coastal populace (Rahman and Salehin, 2013; Nicholls et al., 2016; Hossain et al., 2016).

To encapsulate, the circumstances outlined in this context are a byproduct of anthropogenic climate change, stemming from the cumulative impact of human activities—both past and present. Climate change, arising from global carbon emissions, exerts an immediate influence on human migration and the specter of climate injustice within Bangladesh. Diverse natural perils, particularly those bearing down on coastal communities, operate as potential "push" factors for migration (refer to Table 3.2). Moreover, human actions like turning rice fields into saltwater shrimp farms cause environmental strain and salt buildup, which adds to the reasons people are leaving coastal areas (Falk, 2015; Barai et al., 2019; Bernzen et al., 2019). Climate injustice is not only a global concern but also a matter that implicates local and regional communities. Acknowledging movements precipitated by climate change as displacements are of utmost importance to underscore and legitimize the legal rights of those who are displaced (Bernzen et al., 2019). It is important to understand that the most at-risk groups in society are often those who have not caused much greenhouse gas pollution (Althor et al., 2016), and this makes relocation a clear sign of climate-related unfairness.

When asked about why people in coastal Bangladesh move, participants mentioned a variety of connected reasons, which lead to different migration patterns. These include climate-change-induced natural disasters and economic incentives that either attract people to new places or push them away. These findings match previous research (Warner et al., 2010; Stojanov et al., 2016; Mastrorillo et al., 2016; Hoffmann et al., 2019) and are supported by the data presented in this study (Figs. 3.1, 3.2, and 3.5, as well as Tables 3.2 and 3.3).

People named P, Q, and R from Autulia, Shyamnagar, and Koikhali unions emphasized that many migrants in the research area move temporarily mainly because of economic reasons. Quantitative analysis supports this assertion, with around 37.80% of respondents participating in temporary migration. Furthermore, 30.90% of participants noted that residents migrate from coastal regions to urban areas within Bangladesh, with international migration being influenced by both push and pull factors associated with natural disasters and enhanced employment prospects (Fig. 3.4).

More and more people are moving from rural to urban areas because of both natural disasters and the lure of better economic prospects. In places vulnerable to climate-related problems, often called "hotspots" in Bangladesh, these pressures are likely to push even more rural folks to urban areas, looking for better jobs and a stable life. Around 400,000 people move to Dhaka each year, adding to the number of people living in the city's slums. Many of these migrants had to leave their homes due to climate-related issues like floods and landslides. Unfortunately, they often lack access to essential amenities such as healthcare, education, and food (The Daily Star, 2022). The convergence of natural calamities, marginalized populations, and other societal factors heightens the likelihood of violent conflicts, thereby exacerbating local climate injustice. As municipal authorities grapple with the increasing demands of their populations, addressing the needs of climate migrants and providing essential services becomes a critical exemplar of climate justice (Tegan et al., 2022).

Quantitative analysis reveals that 16.60% of respondents across the southwest coastal region of Bangladesh have indicated that people migrate to urban centers, with a notable focus on employment opportunities in the garment industry, a sector where women are significantly represented (as indicated in Table 3.3 and Fig. 3.3). The flourishing textile industry serves as a major attraction for Bangladeshi women in pursuit of job prospects in urban areas (Gilbert, 2018). However, this migration trend exposes them to urban climate injustice, leading to issues like mental health challenges, food insecurity, and residence in urban slums. The growing urgency of addressing urban environmental health issues within the framework of climate change vulnerability has garnered increasing attention on the global health and climate justice agenda. Climate change dangers like floods, heatwaves, droughts, and food shortages have a more significant impact on already at-risk groups and areas, especially informal settlements. This situation is commonly called "urban climate injustice" (Borg et al., 2021).

People living in urban slums often face hunger because of climate change effects. Issues like lack of nutrients and food insecurity are widespread in these areas due to factors such as high food prices, limited access to healthy meals, reliance on street food, and inadequate health and hygiene services. Climate-related disasters like droughts can disrupt food production and distribution, which affects the urban poor who have moved because of climate issues. This situation violates basic human rights, and those who have migrated due to climate problems are among the most severely affected (Joshi et al., 2019).

Traditional gender roles, particularly women's responsibilities in caring for households and family members, make women more susceptible to the impacts of climate change. For instance, climate-induced water shortages exacerbate women's already substantial responsibilities. During disasters, women frequently take on the care of children and the elderly, exposing them to greater vulnerabilities. Interactions with participants in the SWCRB region revealed that women, during cyclones, encounter challenges like the disruption of natural resource collection, food scarcity, and decreased agricultural productivity, all of which increase the likelihood of displacement. This highlights how female migration due to climate change is closely linked to environmental damage, indicating violations of human rights and gender-based inequalities (Chindarkar, 2012). The former UNFCCC Executive Secretary, Christiana Figueres, stressed in an interview for NDI's "Changing the Face of Politics" podcast series that "Climate change is fundamentally about human rights and justice" (Weber, 2022). This principle extends to social justice, local justice, and global justice. Participants stressed that achieving climate justice requires the active engagement of women in politics and a determined push for gender parity. Women often face resource constraints compared to men, including limited financial means and educational access, alongside exclusion from decision-making processes. Moreover, women's agency and the effectiveness of sustainable management solutions suffer when their demands, objectives, and climate policy expertise are dismissed or underestimated (Tollefson, 2019).

In the southwestern coastal region of Bangladesh, people from Gabura, Padma Pukur, and Burigoyaliny unions said that the main environmental reason for migration is riverbank erosion, as shown in Fig. 3.3. This pervasive problem significantly contributes to displacement issues on the mainland, resulting in substantial insecurity. Riverbank erosion entails the loss of extensive agricultural land, homes, property, casualties, injuries, and disruptions in economic production, education, communication, and sanitation facilities. These stark realities serve as poignant examples of the climate injustices prevalent in the affected coastal regions. Coastal Bangladesh has experienced a noticeable upswing in migration (McLeman, 2017). Field analysis findings suggest that migration is on the rise due to the increasing severity of naturally occurring risks, and this trend is becoming progressively enduring. Migration is a multifaceted phenomenon primarily driven by environmental factors but also influenced by social, economic, and political elements. The complex decision of whether to remain or relocate depends on individual capacities, societal structures, and cultural adaptation to climate change, encompassing both climate-related push and pull factors.

Expanding on our previous conversation, let's discuss the gaps in addressing climate-induced migration and social justice on a national scale in Bangladesh. In 2005, the initial National Adaptation Program of Action (NAPA) was introduced as a framework to tackle climate change and implement adaptation strategies (Ahsan, 2019). However, from the beginning, NAPA saw migration as a negative outcome of climate change, framing internal migration due to climate impacts as something to be prevented rather than proactively managed. While NAPA did acknowledge some links between climate change and migration, it made little progress in creating adaptation programs or policies specifically designed for addressing climate-induced migration (Naser et al., 2019). This has led to policy gaps at both the national level and within local governments and businesses. Climate-displaced migrants often get categorized as part of the urban poor, with no clear guidelines to support their unique needs. Additionally, climate-induced migration is often treated differently in international agreements related to development, disaster management, and climate change.

Let's go back to 2008 when the Bangladesh Climate Change Strategy and Action Plan (BCCSAP) was established, building on the foundation set by the 2005 NAPA initiative (Bhuiyan, 2015). In its updated version, known as the Bangladesh Climate Change Strategy and Action Plan 2009 (BCCSAP-2009), this strategic framework envisions a future where six to eight million people are expected to migrate by 2050, primarily from coastal areas. The driving force behind this mass migration is the diminishing opportunities for livelihoods and a simultaneous drop in agricultural productivity (Ahsan, 2019). It is anticipated that the bustling urban slums in major cities will be the main destinations for these climate-induced migrants. However, the rapid and largely unregulated urbanization in Bangladesh brings a multitude of significant challenges (Naser et al., 2019).

5.1 The relationship between climate justice and climate-induced migration

The idea of environmental justice (EJ) in the United States initially focused on concerns about the location of hazardous waste disposal sites. However, in today's globalized world, the pursuit of justice has become a global issue, attracting attention from activist groups and scholars (Anguelovski and Alier, 2014). "Climate justice" has gained prominence as people recognize that the adverse consequences of climate change and the accountability for those consequences extend beyond national borders. This awareness is supported by the fact that many of the countries most affected by climate change are not the major contributors to global carbon emissions (Walker, 2009). Climate justice, as a complex idea, goes beyond just environmental concerns and covers a wide range of issues, including public health, labor rights, land use, housing, resource distribution, policy development, and community empowerment (Holifield et al., 2009).

Decision-making processes and the creation of policies aimed at addressing climate change-induced migration have proven to be intricate due to the absence of comprehensive legal frameworks on national, regional, and international fronts. Pinpointing the precise number of future migrants and understanding the dynamics of their movement presents a substantial challenge for policymakers, international and regional entities, as well as national and local authorities (Schlosberg, 2017).

Progress in creating new laws and policies to address the negative effects of climate change on migration and human rights violations has been sluggish. Innovative global approaches are needed to develop legal and policy tools that support adaptation and migration for those affected by climate change (Thorp, 2014). Defining environmental or climate migrants (or refugees) is complicated because it is challenging to establish direct links between migration and environmental factors (Mayer, 2016). Nevertheless, climate-induced migration has become a topic of political concern over the past decade in global climate negotiations.

Key milestones include the UNFCCC Cancun Adaptation Framework, which urged countries to cooperate more in dealing with climate change-related displacement, migration, and planned relocation (Manou and Mihr, 2017). A significant milestone was reached when the Task Force on Displacement was established within the Warsaw International Mechanism for Loss and Damage associated with Climate Change Impacts (WIM) during COP21 in Paris. The recommendations of this Task Force for unified approaches to address displacement caused by climate change were approved at COP24 in Katowice (Calliari et al., 2021; Johansson et al., 2022). It is crucial to emphasize the importance of procedural rights, like the right to information and participation, to highlight the significance of respectful migration and adaptation (Cubie, 2018). The connections between climate change, the social impacts of migration, and armed conflicts are clear (Olsson, 2017).

5.2 Example of adaptation in global migration

Some adaptation strategy programs from other climate-vulnerable countries can serve as examples of reducing migration and displacement risks for coastal Bangladesh, mitigating the threats associated with migration (Table 3.5).

Table 3.5 Examples of migration adaptation policies.

Country	Initiatives
Egypt	An important part of the adaptation action plan is to improve the ability of rural communities through the agricultural sector. This involves creating plans, projects, and strategies to make these communities more resilient (Government of Egypt, 2015: 9).
Fiji	The national relocation guidelines (Government of Fiji, 2017: 10) will provide the basis for carrying out a strategy to move coastal communities in response to climate-related risks.
Haiti	A crucial move in addressing climate-related issues is to improve and put into action urban planning and sustainable development plans for urban areas affected by floods, including strategies for relocating the population (Lavenex, et al., 2020).

Table 3.5 Examples of migration adaptation policies.—cont'd

Country	Initiatives
Kiribati	The internal relocation, displacement, and increased urbanization disparities across Kiribati's islands have prompted potential solutions, such as the establishment of domestic policies supporting both government-sponsored and self-supported displaced individuals in their resettlement efforts (Government of Kiribati, 2015).
Sao Tome and Principe	A strategy to decrease the number of people living in at-risk areas is to offer housing in safer locations (Government of Sao Tome and Principe, 2015: 4).
Solomon Islands	One suggested way to deal with the situation is to move the regional headquarters and town from the island to the central region (Government of Solomon Islands, 2015: 12).
Venezuela	Responding to emergencies arising from heavy rains has led to the construction of over 800,000 homes (Government of Venezuela, 2017: 13).
Chad	As a component of the NAPA, the focus is on improving green spaces between communities to help reduce the number of people moving because of climate change (Government of Chad, 2015: 5).
Nigeria	The document on strategies for human settlements and housing clearly shows a desire to bolster rural settlements to decrease the need for relocation (Government of Nigeria, 2015: 21).
Papua New Guinea	The government of Papua New Guinea has officially recognized environmentally triggered relocation as a specific risk (Government of Papua New Guinea, 2015).
Tunisia	Mitigation's impact on sustainable development includes population adjustments and the prevention of regional population decline (Government of Tunisia, 2015: 14).

6. Conclusions

Bangladesh has faced many climate-related challenges over time, like floods, cyclones, waterlogging, saltwater intrusion, and riverbank erosion. These environmental factors have pushed vulnerable coastal communities, especially in the southwest, to migrate as a way to survive. But there has not been much local effort to fully understand and address the social and demographic factors, especially in slow-onset disasters. To bridge this gap, we collected primary data for this study by conducting semi-structured surveys with 320 randomly selected households in Syamnagar Upazila, Satkhira district. This area, situated in the vulnerable southwest coastal region of Bangladesh, is known for its vulnerability to cyclones and saltwater intrusion. Our study found that while the main reason for migration was the pursuit of higher income, it also served as an active strategy to adapt to climate-related challenges. Seasonal migration was a common practice, and the money sent back by migrants was a critical source of income security when local job opportunities were scarce due to climate change pressures. These seasonal migrations did not just happen in nearby areas; they also extended to distant places within the country and even to global destinations, driven by a combination of climate-related factors and economic motivations. The growing recognition that climate change is likely to fuel increased migration, both internally and internationally, has raised concerns about the global framework's readiness to address these issues. Climate-induced relocations can lead to irregular migration surges, place immense strain on refugee systems, and create significant protection gaps for affected migrants. However, it's worth noting that the legal and normative framework for addressing climate-induced migration remains in a nascent stage of development. Climate vulnerabilities have already played a substantial role in ongoing migrations, particularly in coastal areas. This underscores the broader implications of climate change, contributing to global injustices and disparities among nations, thus emphasizing the necessity of adopting a human rights perspective in addressing these challenges. This research aimed to explore how climate change impacts migration in Bangladesh's coastal areas. The study aimed to create a detailed framework that explains climate-related migration patterns, how common they are, and the various implications they have. We employed various approaches to uncover the intricate connections between climate change and migration, underscoring the necessity for well-designed policies to tackle environmental displacement and migration. A thorough understanding of climate change, how people interact with their environment, migration, and how they are all connected is crucial for shaping these policies. We also analyzed existing data to better understand the reasons for climate-induced migration and the effectiveness of adaptation programs. These findings could potentially be used as models to reduce migration risks, not only in coastal Bangladesh but also globally, drawing lessons from countries like Haiti, Fiji, and Sao Tome and Principe.

References

Afroz, S., Cramb, R., Grünbühel, C., 2017. Exclusion and counter-exclusion: the struggle over shrimp farming in a coastal village in Bangladesh. Dev. Change 48 (4), 692–720.

Ahmed, F., 2021. A Case Study for Lightning Scenario and Awareness Building in Bangladesh.

Ahsan, R., 2019. Climate-induced migration: impacts on social structures and justice in Bangladesh. South Asia Res. 39 (2), 184–201.

Althor, G., Watson, J.E., Fuller, R.A., 2016. Global mismatch between greenhouse gas emissions and the burden of climate change. Sci. Rep. 6 (1), 1–6.

Anguelovski, I., Alier, J.M., 2014. The 'Environmentalism of the Poor' revisited: territory and place in disconnected glocal struggles. Ecol. Econ. 102, 167–176.

Anwar, R., 2022. COP27: Time to Address Climate Migration. https://news.cgtn.com/news/2022-11-12/COP27-Time-to-address-climate-migration-1eSmaBdUQ12/index.html. (Accessed 28 November 2022).

Ashrafuzzaman, M., 2022. Climate change driven natural disasters and influence on poverty in the South Western Coastal Region of Bangladesh (SWCRB). SN Soci. Sci. 2 (7), 1–29.

Ashrafuzzaman, M., Santos, F.D., Dias, J.M., Cerdà, A., 2022a. Dynamics and causes of sea level rise in the coastal region of southwest Bangladesh at global, regional, and local levels. J. Mar. Sci. Eng. 10 (6), 779.

Ashrafuzzaman, M., Artemi, C., Santos, F.D., Schmidt, L., 2022b. Current and future salinity intrusion in the south-western coastal region of Bangladesh. Span. J. Soil Sci. 1.

Atapattu, S., 2015. Human Rights Approaches to Climate Change: Challenges and Opportunities. Routledge.

Barai, K.R., Harashina, K., Satta, N., Annaka, T., 2019. Comparative analysis of land-use pattern and socio-economic status between shrimp-and rice-production areas in southwestern coastal Bangladesh: a land-use/cover change analysis over 30 years. J. Coast Conserv. 23 (3), 531–542.

Bernzen, A., Jenkins, J.C., Braun, B., 2019. Climate change-induced migration in coastal Bangladesh? A critical assessment of migration drivers in rural households under economic and environmental stress. Geosciences 9 (1), 51.

Bhuiyan, S., 2015. Adapting to climate change in Bangladesh: good governance barriers. South Asia Res. 35 (3), 349–367.

Borg, F.H., Greibe Andersen, J., Karekezi, C., Yonga, G., Furu, P., Kallestrup, P., Kraef, C., 2021. Climate change and health in urban informal settlements in low-and middle-income countries—a scoping review of health impacts and adaptation strategies. Glob. Health Action 14 (1), 1908064.

References

Borkert, M., Martín Pérez, A., Scott, S., De Tona, C., 2006. Introduction: understanding migration research. Forum Qual. Soc. Res. 7 (3). Art. 3.

Brody, S., Grover, H., Vedlitz, A., 2012. Examining the willingness of Americans to alter behaviour to mitigate climate change. Clim. Pol. 12 (1), 1–22.

Brown, O., 2008. Migration and Climate Change. United Nations.

Bryman, A., 2003. Quantity and Quality in Social Research. Routledge.

Bryman, A., 2016. Social Research Methods. Oxford University Press.

Brzoska, M., Fröhlich, C., 2016. Climate change, migration and violent conflict: vulnerabilities, pathways and adaptation strategies. Migrat. Develop. 5 (2), 190–210.

Calliari, E., Vanhala, L., Nordlander, L., Puig, D., Bakhtiari, F., Hossain, M.F., Huq, S., Rahman, M.F., 2021. Loss and Damage. The Paris Agreement on Climate Change. Edward Elgar Publishing, pp. 200–217.

Chindarkar, N., 2012. Gender and climate change-induced migration: proposing a framework for analysis. Environ. Res. Lett. 7 (2), 025601.

Cohen, R., Deng, F.M., 2012. Masses in Flight: The Global Crisis of Internal Displacement. Brookings Institution Press.

Crosby, A.W., 1995. The past and present of environmental history. Am. Hist. Rev. 100 (4), 1177–1189.

Cubie, D., 2018. Human Rights, Environmental Displacement and Migration. Routledge Handbook of Environmental Displacement and Migration, pp. 329–341.

Falk, G.C., 2015. Land use change in the coastal regions of Bangladesh: a critical discussion of the impact on delta-morphodynamics, ecology, and society. ASIEN 134 (1), 47–71.

Gare, A.E., 1995. Postmodernism and the Environmental Crisis. Routledge, London.

Gilbert, P.R., 2018. Class, complicity, and capitalist ambition in Dhaka's elite enclaves. Focaal 2018 (81), 43–57.

Gooch, P., 2000. At the Tail of the Buffalo: Van Gujjar Pastoralists Between the Forest and the World Arena (India).

Government of Chad, 2015. Intended Nationally Determined Contribution (INDC) for the Republic of Chad. Retrieved 01.10.2021 from: https://www4.unfccc.int/sites/ndcstaging/PublishedDocuments/Chad%20First/INDC%20Chad_Official%20version_English.pdf.

Government of Egypt, 2015. Egyptian Intended Nationally Determined Contribution. Retrieved 01.10.2021 from: https://www4.unfccc.int/sites/ndcstaging/PublishedDocuments/Egypt%20First/Egyptian%20INDC.pdf.

Government of Fiji, 2017. Fiji's National Adaptation Plan Framework. Retrieved 01.10.2021 from. https://cop23.com.fj/wp-content/uploads/2018/03/NAP-Framework-Fiji.pdf.

Government of Kiribati, 2015. Intended Nationally Determined Contribution. Retrieved 01.10.2021 from: https://www4.unfccc.int/sites/ndcstaging/PublishedDocuments/Kiribati%20First/INDC_KIRIBATI.pdf.

Government of Nigeria, 2015. Nigeria's Intended Nationally Determined Contribution. Retrieved 01.10.2021 from: https://www4.unfccc.int/sites/ndcstaging/PublishedDocuments/Nigeria%20First/Approved%20Nigeria%27s%20INDC_271115.pdf.

Government of Papua New Guinea, 2015. Intended Nationally Determined Contribution (INDC) under the United Nations Framework Convention on Climate Change. Retrieved 01.10.2019 from: https://www4.unfccc.int/sites/ndcstaging/PublishedDocuments/Papua%20New%20Guinea%20First/PNG_INDC%20to%20the%20UNFCCC.pdf.

Government of Sao Tome and Principe, 2015. Sao Tome and Principe. Intended Nationally Determined Contribution. Retrieved 01.10.2019 from: https://www4.unfccc.int/sites/ndcstaging/PublishedDocuments/Sao%20Tome%20and%20Principe%20First/STP_INDC%20_Ingles_30.09.pdf.

Government of Solomon Island, 2015. Intended Nationally Determined Contribution. Retrieved 01.10.2021 from: https://www4.unfccc.int/sites/ndcstaging/PublishedDocuments/Solomon%20Islands%20First/SOLOMON%20ISLANDS%20INDC.pdf.

Government of Tunisia, 2015. Intended Nationally Determined Contribution. Retrieved 01.10.2019 from: https://www4.unfccc.int/sites/ndcstaging/PublishedDocuments/Tunisia%20First/INDC-TunisiaEnglish%20Version.pdf.

Government of Venezuela, 2017. Primera Contribución Nacionalmente Determinada de la República Bolivariana de Venezuela para la lucha contra el Cambio Climático y sus efectos. Retrieved 01.10.2021 from: https://www4.unfccc.int/sites/ndcstaging/PublishedDocuments/Venezuela%20First/Primera%20%20NDC%20Venezuela.pdf.

Hoffman, S.M., 2022. Disaster and Climate Change. Cooling Down: Local Responses to Global Climate Change, p. 339.

Hoffmann, E.M., Konerding, V., Nautiyal, S., Buerkert, A., 2019. Is the push-pull paradigm useful to explain rural-urban migration? A case study in Uttarakhand, India. PLoS One 14 (4), e0214511.

Holifield, R., Porter, M., Walker, G., 2009. Spaces of environmental justice: frameworks for critical engagement. Antipode 41 (4), 591–612.

Hollifield, J., Martin, P.L., Orrenius, P. (Eds.), 2014. Controlling Immigration: A Global Perspective. Stanford University Press.

Hossain, M.D., Dearing, J.A., Rahman, M.M., Salehin, M., 2016. Recent changes in ecosystem services and human well-being in the Bangladesh coastal zone. Reg. Environ. Change 16 (2), 429–443.

IDMC, 2020. GRID 2020: Global Report on Internal Displacement. The Internal Displacement Monitoring Centre, Geneva. www.internal-displacement.org/global-report/grid2020/.

International Migration Institute (IMI), 2010. The Environmental Factor in Migration Dynamics—a Review of African Case Studies.

IPCC, 2022. In: Masson-Delmotte, V., Zhai, P., Pirani, A., Connors, S.L., Péan, C., Berger, S., Caud, N., Chen, Y., Goldfarb, L., Gomis, M.I., Huang, M., Leitzell, K., Lonnoy, E., Matthews, J.B.R., Maycock, T.K., Waterfield, T., Yelekçi, O., Yu, R., Zhou, B. (Eds.), Climate Change 2021: The Physical Science Basis. Contribution of Working Group I to the Sixth Assessment Report of the Intergovernmental Panel on Climate Change. Cambridge University Press.

IPCC, 2014. Climate Change: Synthesis Report; Contribution of Working Groups I, II and III to the Fifth Assessment Report of the Intergovernmental Panel on Climate Change (IPCC). Intergovernmental Panel on Climate Change, Geneva, Switzerland.

Islam, M.R., Hasan, M., 2016. Climate-induced human displacement: a case study of Cyclone Aila in the south-west coastal region of Bangladesh. Nat. Hazards 81 (2), 1051–1071.

Javed, M., 2013. Climate Change and Migration Impacts on Bangladesh's Coastal Inhabitants.

Johansson, A., Calliari, E., Walker-Crawford, N., Hartz, F., McQuistan, C., Vanhala, L., 2022. Evaluating progress on loss and damage: an assessment of the executive committee of the Warsaw international mechanism under the UNFCCC. Clim. Pol. 22 (9–10), 1199–1212.

Joshi, A., Arora, A., Amadi-Mgbenka, C., Mittal, N., Sharma, S., Malhotra, B., Grover, A., Misra, A., Loomba, M., 2019. Burden of household food insecurity in urban slum settings. PLoS One 14 (4), e0214461.

Kempadoo, K., Doezema, J., 2018. Women, labor, and migration: the position of trafficked women and strategies for support. In: Global Sex Workers. Routledge, pp. 69–78.

Kent, A. (Ed.), 2018. Climate Refugees. Routledge/Taylor & Francis Group.

Kliot, N., 2004. Environmentally induced population movements: their complex sources and consequences. In: Unruh Maarten, D.S.K., Kliot, N. (Eds.), Environmental Change and its Implications for Population Migration. Kluwer Academic Publishers, Boston, pp. 69–99.

Lavenex, S., Federica, C., Elisa, F., 2020. Environmental Migration Governance at the Regional Level. Environmental Conflicts, Migration and Governance, p. 137.

Lee, E.S., 1966. A The. Migrat., Demograp. 3, 47–57.

Lune, H., Berg, B.L., 2017. Qualitative Research Methods for the Social Sciences. Pearson.

Mallick, B., Ahmed, B., Vogt, J., 2017. Living with the risks of cyclone disasters in the south-western coastal region of Bangladesh. Environments 4 (1), 13.

Manou, D., Mihr, A., 2017. Climate change, migration and human rights. In: Climate Change, Migration and Human Rights. Routledge, pp. 2–8.

Martin, M., Kang, Y.H., Billah, M., Siddiqui, T., Black, R., Kniveton, D., 2013. Policy Analysis: Climate Change and Migration Bangladesh. Refugee and Migratory Movements Research Unit (RMMRU), Dhaka, Bangladesh.

Mastrorillo, M., Licker, R., Bohra-Mishra, P., Fagiolo, G., Estes, L.D., Oppenheimer, M., 2016. The influence of climate variability on internal migration flows in South Africa. Global Environ. Change 39, 155–169.

Mayer, B., 2016. The Concept of Climate Migration: Advocacy and Its Prospects. Edward Elgar Publishing.

Mcdonnell, T., 2019. Climate Change Creates a New Migration Crisis for Bangladesh. https://www.nationalgeographic.com/environment/article/climate-change-drives-migration-crisis-in-bangladesh-from-dhaka-sundabans. (Accessed 11 November 2022).

McLeman, R., 2017. Climate-related migration and its linkages to vulnerability, adaptation, and socioeconomic inequality: evidence from recent examples. In: Research Handbook on Climate Change, Migration and the Law, pp. 29–48.

Mukhopadhyay, A., 2016. Living with Disasters. Cambridge University Press.

Naser, M.M., Swapan, M.S.H., Ahsan, R., Afroz, T., Ahmed, S., 2019. Climate change, migration and human rights in Bangladesh: perspectives on governance. Asia Pac. Viewp. 60 (2), 175–190.

Nicholls, R.J., Hutton, C.W., Lázár, A.N., Allan, A., Adger, W.N., Adams, H., Wolf, J., Rahman, M., Salehin, M., 2016. Integrated assessment of social and environmental sustainability dynamics in the Ganges-Brahmaputra-Meghna delta, Bangladesh. Estuar. Coast Shelf Sci. 183, 370–381.

Olsson, L., 2017. Climate migration and conflicts: a self-fulfilling prophecy?. In: Climate Change, Migration and Human Rights. Routledge, pp. 116–128.

Panebianco, S., 2022. Climate change migration enters the agenda of the wider mediterranean: the long way towards global governance. In: Border Crises and Human Mobility in the Mediterranean Global South. Palgrave Macmillan, Cham, pp. 145–175.

Piguet, E., 2010. Response functions for migration, ethnicity and ageing; land use relationship in rural urban regions. Module 2. http://www.plurel.net/images/D234.pdf. (Accessed 16 April 2022).

Poncelet, A., Gemenne, F., Martiniello, M., Bousetta, H., 2010. A country made for disasters: environmental vulnerability and forced migration in Bangladesh. In: Environment, Forced Migration and Social Vulnerability. Springer, Berlin, Heidelberg, pp. 211–222.

Portes, A. (Ed.), 1995. The Economic Sociology of Immigration: Essays on Networks, Ethnicity, and Entrepreneurship. Russel Sage Foundation, New York.

Rahman, R., Salehin, M., 2013. Flood risks and reduction approaches in Bangladesh. In: Disaster Risk Reduction Approaches in Bangladesh. Springer, Tokyo, pp. 65–90.

Rahman, S.M.M., Hossain, S.M., Jahan, M.U., 2019. Thunderstorms and lightning in Bangladesh. Bangladesh Med. Res. Counc. Bull. 45 (1), 1–2.

Rashid, M.M., 2013. Migration to big cities from coastal villages of Bangladesh: an empirical analysis. Global J. Hum. Soc. Sci. 13 (5), 28–36.

Robinson, M., Shine, T., 2018. Achieving a climate justice pathway to 1.5 C. Nat. Clim. Change 8 (7), 564–569.

Rosignoli, F., 2022. Environmental Justice for Climate Refugees. Routledge.

Sammonds, P., Shamsudduha, M., Ahmed, B., 2021. Climate change driven disaster risks in Bangladesh and its journey towards resilience. J. Br. Acad. 9 (s8), 55–77.

Schlosberg, D., 2017. Reconceiving Environmental Justice: Global Movements and Political Theories. Environmental Justice, p. 29.

Scholten, P., 2020. Mainstreaming versus alienation: conceptualising the role of complexity in migration and diversity policymaking. J. Ethnic Migrat. Stud. 46 (1), 108–126.

Stojanov, R., Kelman, I., Ullah, A.A., Duží, B., Procházka, D., Blahůtová, K.K., 2016. Local expert perceptions of migration as a climate change adaptation in Bangladesh. Sustainability 8 (12), 1223.

Tegan, B., Julia, C., Chris, C., Jessica, K., Rachel, L., 2022. Cimate Change, Migration and the Risk of Conflict in Growing Urban Centers. The United States Institute of Peace. https://www.preventionweb.net/news/climate-change-migration-and-risk-conflict-growing-urban-centers. (Accessed 29 November 2022).

The Daily Star, 2022. Ensuring Climate Justice for the Urban Poor. https://www.thedailystar.net/opinion/views/news/ensuring-climate-justice-the-urban-poor-3110001. (Accessed 28 November 2022).

Thorp, T., 2014. Climate Justice: A Voice for the Future. Springer.

Tollefson, J., 2019. The hard truths of climate change—by the numbers. Nature 573 (7774), 324–328.

UN Climate Press Release, 2022. COP27 Reaches Breakthrough Agreement on New "Loss and Damage" Fund for Vulnerable Countries. https://unfccc.int/news/cop27-reaches-breakthrough-agreement-on-new-loss-and-damage-fund-for-vulnerable-countries. (Accessed 28 November 2022).

Vehovar, V., Toepoel, V., Steinmetz, S., 2016. Non-probability sampling. Sage Handb. Surv. Meth. 329–345.

Walker, G., 2009. Beyond distribution and proximity: exploring the multiple spatialities of environmental justice. Antipode 41 (4), 614–636.

Warner, K., Hamza, M., Oliver-Smith, A., Renaud, F., Julca, A., 2010. Climate change, environmental degradation and migration. Nat. Hazards 55 (3), 689–715.

Weber, H., 2022. Podcasting the climate crisis: what role can podcast hosts play in inspiring climate action? URN: urn:nbn:se:hj:diva-57598, ISRN: JU-HLK-MKA-2-20220456, OAI: oai:DiVA.org:hj-57598, DiVA, id: diva2:1675121; Online available at: https://www.diva-portal.org/smash/get/diva2:1675121/FULLTEXT01.pdf. (Accessed 21 June 2022).

Worster, D., 1977. Nature's Economy: A History of Ecological Ideas. Cambridge, p. 348.

CHAPTER 4

Climate variability, observed climate trends, and future climate projections for Sri Lanka

I.M. Shiromani Priyanthika Jayawardena, D.W.T.T. Darshika, H.M.R.C. Herath and H.A.S.U. Hapuarachchi

Department of Meteorology, Colombo, Sri Lanka

1. Introduction

Sri Lanka is an island located in the Indian Ocean (between 05°55′N and 09°51′N latitudes and 79°41′E and 81°53′E longitudes: south-east of the southern tip of the Indian subcontinent). The island covers a surface of roughly 65,610 square kilometers.

The highlands, mostly above 300 m, occupy the south-central part of Sri Lanka with numerous peaks (Pidurutalagala—2524 m, Kirigalpotte—2396 m), high plateaus, and basins and are surrounded by an extensive lowland area (Fig. 4.1).

Located in the tropics as well as in the Asian monsoon region, the climate of Sri Lanka is characterized as tropical maritime and monsoonal (Pant and Kumar, 1997).

Hence, the two monsoons essentially determine the seasonality of Sri Lanka. The seasons are distinguished only utilizing the timing of the two monsoons and the transitional periods separating them, called intermonsoon seasons (Chandrapala, 1996; Pant and Kumar, 1997; Mamgren et al., 2003). The southwest monsoon from May to September contributed 30% of annual average rainfall, and the northeast monsoon from December to February contributed 26% to the annual average rainfall. The intermonsoon periods, namely, the first intermonsoon and second intermonsoon are from March to April and from October to November, respectively, and contributed 14% and 30% to annual average rainfall (Table 4.1).

1. March to April: First intermonsoon (***FIM***)
2. May to September: Southwest monsoon (***SWM***)
3. October—November: Second intermonsoon (***SIM***)
4. December to February: Northeast monsoon (***NEM***)

Orography plays an important role in the rainfall distribution of Sri Lanka. The central highlands with heights of more than 2500 m above the mean sea level (Fig. 4.1) act as an orographic barrier across the path of the monsoonal winds strongly affecting the spatial patterns of rainfall, temperature, relative humidity, and wind. The windward in one monsoon season becomes leeward side during the other monsoon season and vice versa. Relatively uniform climatic conditions are seen over the larger part of the island mainly over the lowlands due to flat topography and moderate influences of maritime climate (Pant and Kumar, 1997).

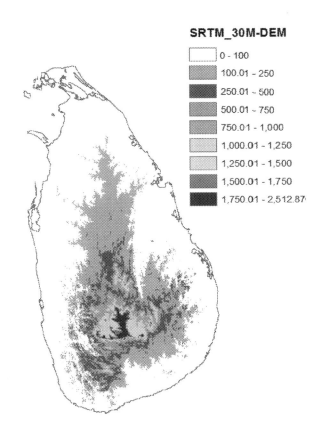

FIGURE 4.1

Topography map of Sri Lanka.

Table 4.1 Seasonal average rainfall (1981–2010) distribution (mm) and contribution to the annual average rainfall during each season (%) (lower).

Season	Period	Average rainfall received	Contribution to national annual average rainfall
FIM	March–April	268 mm	14%
SWM	May–September	556 mm	30%
SIM	October–November	558 mm	30%
NEM	December–February	479 mm	26%
Total annual average rainfall		1861 mm	

1. Introduction 63

1.1 Spatial and temporal variations of rainfall

The annual rainfall shows (Fig. 4.2) remarkable spatial variation, ranging from less than 1000 mm in the driest parts to more than 5000 mm in the wettest parts. The driest regions of the island are situated diametrically opposite to one another in two peripheral regions of Sri Lanka (Pant and Kumar, 1997), the Southeast and Northwest; both regions receive an annual average rainfall of between 800 and 1250 mm (Fig. 4.2). The spatial patterns of the rainfall in different seasons indicate that the southwestern sector of the island receives the highest rainfall in all the seasons except for the NEM season (Fig. 4.3), making it the zone of maximum rainfall in the annual rainfall pattern for the island (Fig. 4.2). The southwestern sector, along with the central highlands, stands out clearly as the wettest part of the island, with a mean annual rainfall exceeding 2000 mm. Within the southwestern sector, the

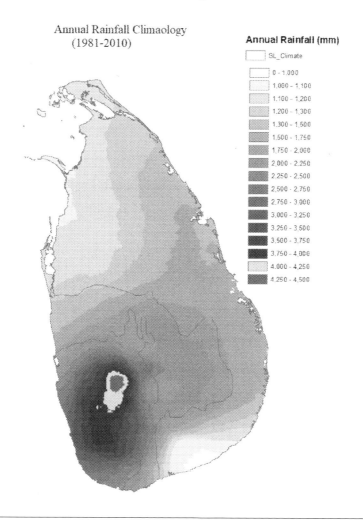

FIGURE 4.2

Annual rainfall distribution (mm).

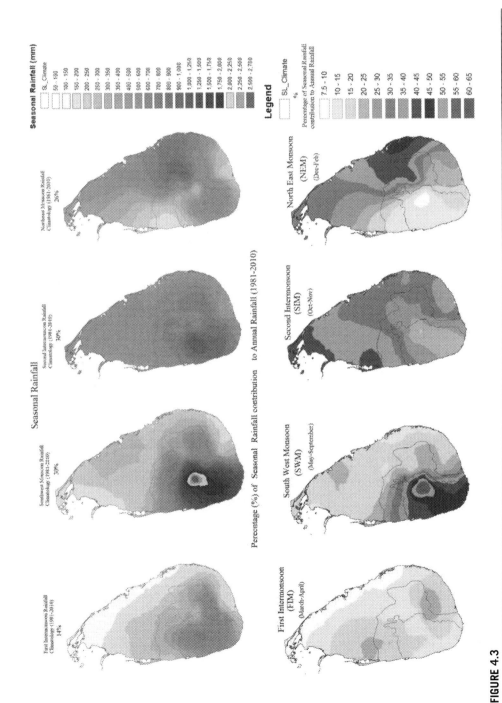

FIGURE 4.3
Spatial maps of seasonal average rainfall (1981–2010) distribution (mm) (upper) and contribution to the annual average rainfall during each season (%) (lower).

maximum rainfall is in the lower to middle altitudes of the western slopes (300—1000 m), and above 1000m the rainfall decreases again. Thus, Sri Lanka has been generalized into three spatial climatic zones: the wet zone in the southwestern region including the central hills of the country; the dry zone covering predominantly the northern and eastern parts of the country; and the intermediate zone skirting the central hills except in the south and the west (Fig. 4.2) (Thambyahpillay, 1954).

SWM season is the lengthiest season extending from May to September. During SWM, the western slopes of central highlands receive over 2000 mm (50%—60% of annual rainfall) seasonal mean rainfall due to the orographic effect, while the leeward side acts as a rain shadow area with less than 250 mm seasonal mean rainfall contributing only 10%—20% to annual rainfall. Southwestern coastal belt above 1000 mm is about 40%—50% of annual rainfall (Fig. 4.3).

NEM season from December to February brings 40%—50% of the annual rainfall to the eastern parts of the country. Moist wind blowing from the Bay of Bengal produces seasonal rainfall of above 600 mm in the eastern parts of the country (Fig. 4.3). The highest rainfall amounts are confined to the north-eastern slopes of the hill country and eastern slopes of the Knuckles/Rangala range. Westward propagating high-frequency equatorial waves such as easterly waves and convectively coupled equatorial Rossby waves play an important role in the enhancement of rain activity over Sri Lanka during NEM (IMD, 1973). On the other hand, cool dry weather can be experienced over many parts of the country, when northerly to northeasterly winds blow from the high-pressure centers located over the Indian landmass (www.meteo.gov.lk).

During the intermonsoon seasons (the transition period of two monsoons), most convective activity is associated with the formation of mesoscale circulations due to differential heating caused by horizontal variations in land surface characteristics.

As there is no predominant wind flow over Sri Lanka, variable light winds, together with orographic lifting, and convergence of sea breeze provided favorable conditions for the formation of convective showers.

During the FIM season (March to April) (Fig. 4.3), the southwestern quarter receives over 300 mm of seasonal mean rainfall with a peak of about 600 mm over the western slope of the central highlands. Most other parts received seasonal mean rainfall of around 100—250 mm.

The SIM season (October to November) shows the uniform distribution of rainfall over SRI LANKA. Almost the entire island receives more than 350 mm of seasonal mean rain during this season, with the southwestern slopes receiving high seasonal mean rainfall in the range of 600—950 mm (Fig. 4.3). Northwestern and southeastern coastal areas, the driest parts of the country, receive more than 35%—45% of annual rainfall during SIM (Fig. 4.3). In addition to the convective activity associated with differential land surface heating, other factors can also contribute to an increase in rainfall and lead to floods and landslides. The overhead position of the ITCZ creates a favorable environment for the formation of convective storms (Suppiah, 1997). Low pressure areas and depressions in the Bay of Bengal can also contribute to an increase in rainfall and lead to flooding and landslides (Malmgren et al., 2003) (Figs. 4.4 and 4.5).

The annual cycle of rainfall over Sri Lanka displays a bimodal pattern with one peak in April to May and the other in October to November (Fig. 4.6) associated with the northward and southward propagation of the ITCZ across the latitudinal belt across Sri Lanka (Suppiah, 1997). Weather disturbances that originate within the ITCZ enhance the rainfall activity forming two peaks in the annual cycle of rainfall (Kane, 1998).

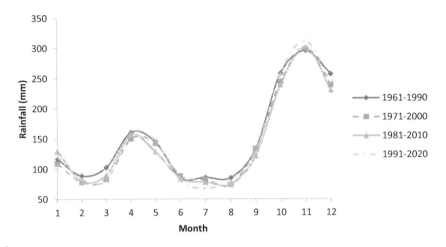

FIGURE 4.4

Annual cycle of country average rainfall.

The annual cycle of rainfall shows spatial variation from the southwest coast to the northeast coast. The bimodal pattern that appeared over the southwest quarter changes gradually into a unimodal pattern with a peak in November–December in northern and north-eastern coastal areas (Fig. 4.7). Central mountains obstruct the south-westerly flow bringing warm dry Kachchan winds (Thambyahpillay, 1958) acting as a rain show over the lee side eastern part of Sri Lanka during FIM and SWM seasons. Northeast monsoon winds bring rainfall to the eastern part of Sri Lanka forming a single peak in the annual cycle of rainfall.

1.2 Temperature variations

Due to the proximity to the equator and maritime influence, the temperature hardly shows significant variation throughout the year. The annual mean surface air temperature is largely uniform (27–29°C) over the lowlands and decreases with altitude experiencing the coolest temperatures over the central highlands (Fig. 4.7). The annual mean surface air temperature is about 15°C at Nuwara Eliya located 1900m above mean sea level in the central highlands (Fig. 4.7). The annual range of maximum temperature is small, and the month-to-month variation is almost barely visible (Fig. 4.8). Compressional warming of dry southwesterly Kachchan wind (Thambyahpillay, 1958) over leeside offset the sea breeze circulation increasing maximum temperature over northeastern coastal areas during SWM (Fig. 4.8) and also creating a 4–5°C range of annual temperature. The mean annual range of temperature change over Sri Lanka varies from around 4–5°C along with eastern and northeastern coastal areas due to 1–2° along with southwestern coastal areas and low-lying areas of southwest parts. The annual range of temperature change is relatively small in the southwestern sector of the country, including the highlands.

Diurnal variation of temperature range from 10–12°C over inland areas to 5–6°C, as typically inherent to low latitude climates (Pant and Kumar, 1997).

1. Introduction   67

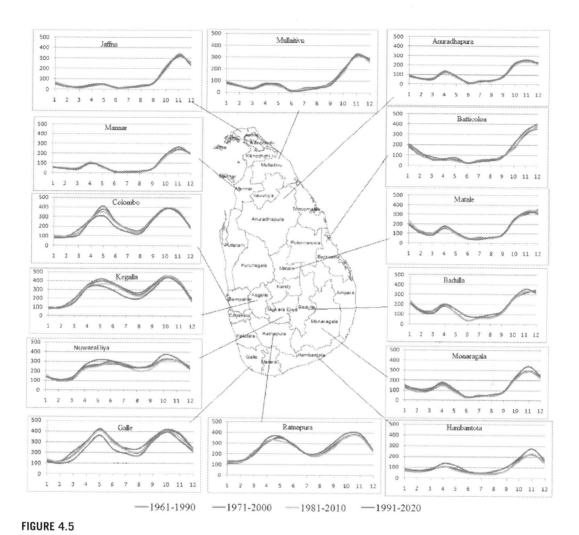

—— 1961-1990   —— 1971-2000   —— 1981-2010   —— 1991-2020

FIGURE 4.5

Annual cycle of rainfall at different locations.

1.3 Spatial and temporal variations of lightning

Sri Lanka is one of the major lightning-prone countries in South Asia (Jayaratne and Gomes, 2012; Jayawardena and Mäkelä, 2021; Gomes et al., 2011). Annual average lightning density (LD) shows remarkable spatial variations. The strongest LD with 150 strokes km^{-2} year^{-1} (st km^{-2} yr^{-1} hereafter) is apparent in the inland areas of western parts of the country, while the least LD less than 20 km^{-2} yr^{-1} were found along with the coastal areas except for western and southwestern coastal areas, Jaffna Peninsular, and high elevation areas of central hills (Jayawardena and Mäkelä, 2021). Lightning hot pots can be seen along the western foothills of the central hills, over the north–central province, and in along the south-eastern foothills.

Chapter 4 Future climate projections for Sri Lanka

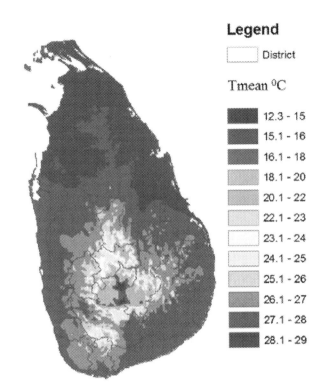

FIGURE 4.6
Spatial variation of annual mean surface air temperature.

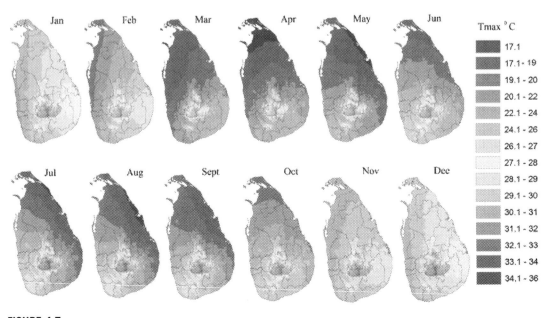

FIGURE 4.7

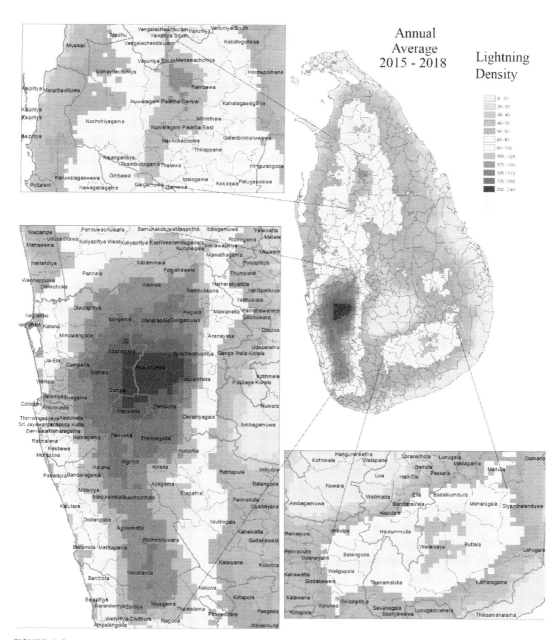

FIGURE 4.8

Spatial distribution of annual average (2015–18) lightning density (LD) (st km^{-2} yr^{-1}) (Jayawardena and Mäkelä, 2021).

Enhanced LFDs along the western and southwestern foothills suggested a link to the sea breeze and orographic lifting. The highest LD of more than 200 st km^{-2} yr^{-1} was observed in a small area of the western part of the island, Ruwanwella divisional secretariat division in Kegalle districts. It is worthy to mention that higher lightning activity in the region depicted by A is a major paddy cultivation region where people often work outside involved in agriculture activities during the daytime and the possibility of becoming victims of lightning fatality is high (Figs. 4.9 and 4.10).

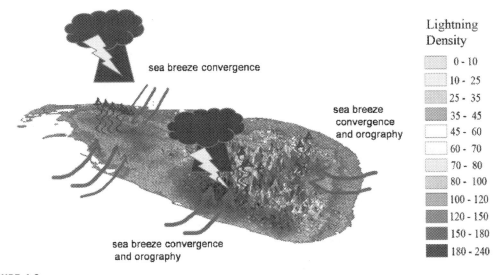

FIGURE 4.9

Spatial distribution of annual average (2015–18) LD (st km^{-2} yr^{-1}) overlay in SRTM DEM (Jayawardena and Mäkelä, 2021).

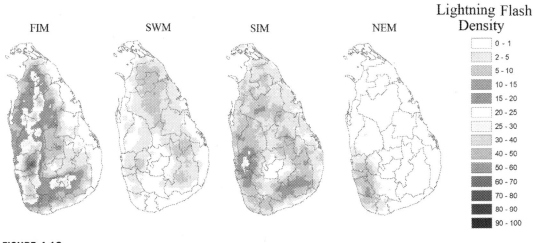

FIGURE 4.10

Spatial distribution of seasonal (st km^{-2} month^{-1}) (Jayawardena and Mäkelä, 2021).

Annual lighting occurrence is dominated by FIM season contributing 50% and then followed by 25% from SIM to the annual lightning occurrence. The last lightning activity is found in NEM (Fig. 4.11). The transition period of two monsoons, the intermonsoon period, as there is no predominant wind flow over Sri Lanka, prevailing variable light winds and convective activity associated with the formation of mesoscale circulations due to differential heating caused by horizontal variations in land surface characteristics, together with orographic lifting, contributed to account for ¾ of annual lightning activity in Sri Lanka(Jayawardena and Mäkelä, 2021). Lightning activity in the two monsoons is confined predominantly to the leeward side but some lightning activity is detected on the windward side during SWM (Fig. 4.11) (Jayawardena and Mäkelä, 2021).

The mean monthly lightning counts show a peak in April followed by May, October, March, and September (Fig. 4.12). In general, lightning counts in April only contributed more than ¼ of the annual lightning counts. Around 10%—15% contribution is coming from May, October, and March, while 5%—10% contribution comes from September and November. The rest of the months plays a very small role in annual lighting counts (Fig. 4.12) (Jayawardena and Mäkelä, 2021). A widespread lightning activity observed during May and October agrees with the propagation of the intertropical convergence zone (ITCZ) twice a year around these 2 months over Sri Lanka (Suppiah, 1997).

Diurnal variation of lightning shows a peak between 1530 and 1730 LST (Fig. 4.10) (Nag et al., 2017; Jayawardena and Mäkelä, 2021 and Weerasekera et al., 2001). Lightning maxima around late afternoon is a common global feature. Lightning maxima in late afternoon indicate that most convective activity in Sri Lanka is associated with land surface heating (Figs. 4.13 and 4.14).

2. Interannual and intraseasonal rainfall variability in Sri Lanka

Seasonal to interannual variability of rainfall, both in amount and distribution, often impact agricultural production and food security as well as other climate-sensitive sectors with adverse socio-economic consequences for society. Production of rice, which is one of the principal crops in Sri Lanka, is highly susceptible to rainfall variability; both deficient and excess rainfall conditions were found to have significantly contributed to the reduction of rice yield (Yoshino and Suppiah, 1983).

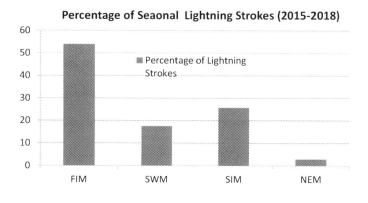

FIGURE 4.11

Temporal distribution of the percentage of seasonal lightning strokes.

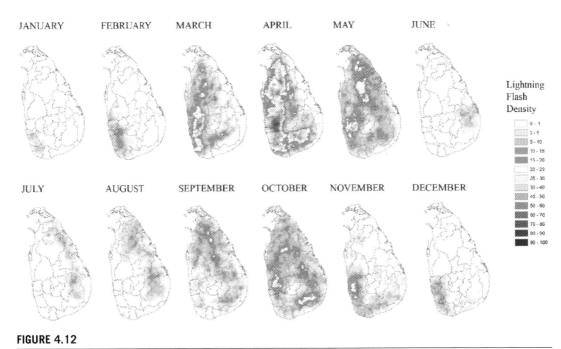

FIGURE 4.12

Spatial distribution of monthly lightning density (LD) (st km^{-2} month^{-1}) (Jayawardena and Mäkelä, 2021).

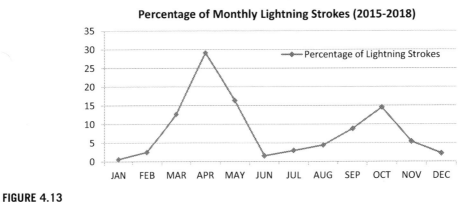

FIGURE 4.13

Temporal distribution of the percentage of monthly lightning strokes.

2.1 Impact of El Nino and La Nina on interannual variability of seasonal rainfall in Sri Lanka

El Nino Southern Oscillation (ENSO) is the strongest interannually varying phenomenon in the tropical coupled ocean-atmosphere system (McPhaden, 2002) modulating the Walker circulation to trigger major shifts in tropical rainfall patterns and deep convection across the globe (McPhaden, 2002; Philander, 1990; Trenberth, 1997).

2. Interannual and intraseasonal rainfall variability in Sri Lanka

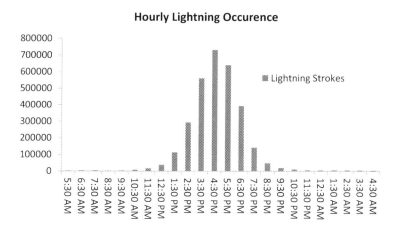

FIGURE 4.14

Hourly lightning counts from 2015 to 2015 over Sri Lanka (Jayawardena and Mäkelä, 2021).

El Nino Southern Oscillation (ENSO) can be considered one of the potential climatic drivers of interannual rainfall variability in Sri Lanka. The relationship between ENSO extremes and rainfall variability in Sri Lanka has been well studied (Rasmussen and Carpenter, 1983; Ropelewski and Halpert, 1987, 1989; Suppiah, 1997, 1996; Kane, 1998; Sumathipala and Punyadeva, 1998; Punyawardena and Cherry, 1999; Malmgren et al., 2003; Zubair and Repolewski, 2006, Hapuarachchi and Jayawardena, 2015). During the El Nino phase, the rainfall is enhanced from October–December (Rasmussen and Carpenter, 1983; Suppiah and Yoshino, 1986) and diminished from January–February (Chandimala and Zubair, 2007). Investigating the year-round rainfall of Sri Lanka, Zubair et al. (2008) identified that El Nino typically leads to wetter conditions from October to December and drier conditions from January to March and July to August on average.

According to Suppiah (1997), positive and negative rainfall anomalies are associated with La Nina and El Nino events during SWM season, but negative and positive rainfall anomalies are associated with La Nina and El Nino events during SIM season. Rainfall anomalies during FIM and NEM seasons do not show clear contrasting patterns as in other seasons.

Using rainfall data from 90 stations and ONI index from 1950 to 2011, Hapuarachchi and Jayawardena (2015) identified that the strongest influence of ENSO extremes is evident during SIM with contrasting spatial patterns are evident in most parts of Sri Lanka, which experience an excess of seasonal rainfall during El Nino years and a deficit of seasonal rainfall in La Nina years (Fig. 4.15). The impact of ENSO extremes on SIM is consistent with findings of previous studies (Rasmussen and Carpenter, 1983; Ropelewski and Halpert, 1987, 1989; Suppiah, 1996, 1997; Kane, 1998; Sumathipala and Punyadeva, 1998; Punyawardena and Cherry, 1999 and Malmgren et al., 2003).

The significant impact of SWM rainfall was also evident over central parts of the island with a reduction of seasonal rainfall during El Nino events and enhanced seasonal rainfall during La Nina events (Fig. 4.15). Sri Lanka SWM rainfall negative deviations during El Nino events (Kane, 1998; Suppiah, 1997).

The significant impact of NEM rainfall is also profound over Northwestern, northcentral, and central parts of the island with a reduction of seasonal rainfall during El Nino events and enhanced

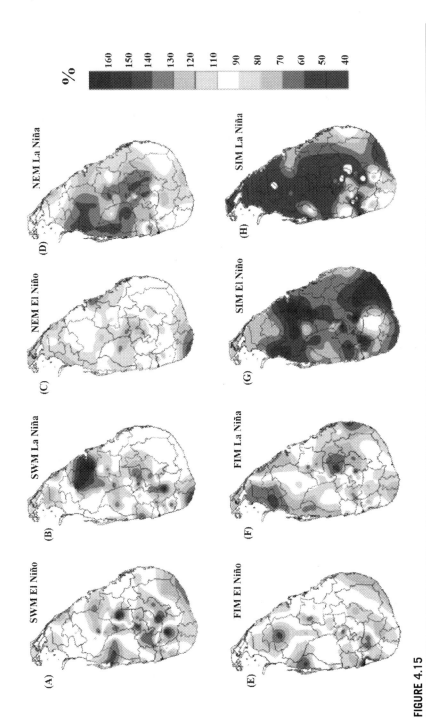

FIGURE 4.15

Composites of seasonal rainfall probabilities (shading) for Southwest Monsoon (SWM) season during El Nino (A) and La Nina (B); for NEM season during El Nino (C) and La Nina (D); for FIM season during El Nino (E) and La Nina (F); and for SIM season during El Nino (G) and La Nina (H). Rainfall probabilities refer to the chance of seasonal rainfall exceeding the median, expressed as a ratio with the mean probability (nominally 50%).

Modified from Hapuarachchi and Jayawardena (2015).

seasonal rainfall during La Nina events (Fig. 4.15). Most of the previous studies found that rainfall anomalies during the NEM seasons do not show a clear contrast in their temporal patterns (Punyawardena and Cherry, 1999; Suppiah, 1997).

Anomalous rising motion with negative-pressure vertical velocity over Sri Lanka enhanced convection in El Nino years, while anomalous sinking motion with positive pressure vertical velocity suppressed convection in La Nina years over Sri Lanka during October and November (Fig. 4.16). Comparatively weak sinking (rising) motion with positive (negative)-pressure vertical velocity over Sri Lanka suppressed (enhanced) rainfall activity in El Nino (La Nina) years during January and February (Fig. 4.16).

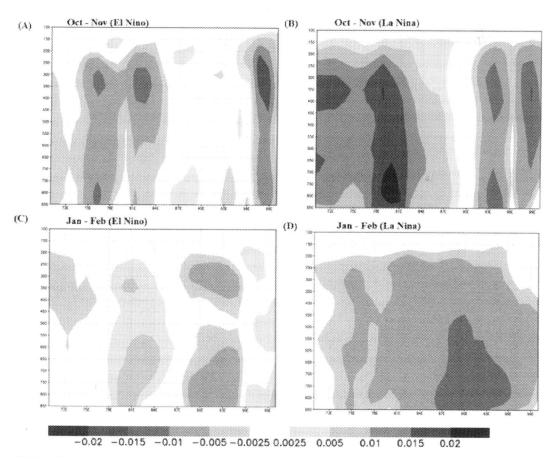

FIGURE 4.16

East-west vertical cross-section of pressure vertical velocity average across Sri Lanka (05°N–10°N) from 70°E to 100°E for October to November during El Nino (A) and La Nina (B) and for January to February during El Nino (C) and La Nina (D).

The weakest influence of ENSO extremes on seasonal rainfall is evident during FIM with no clear contrast in the temporal pattern (Fig. 4.15). Punyawardena and Cherry (1999) also found that the link between FIM rain with both El Nino and La Nina events was less clear (Figs. 4.17—4.20).

2.2 Impact of Indian Ocean Dipole on interannual variability of monthly rainfall from June to December in Sri Lanka

Indian Ocean Dipole (IOD, Saji et al., 1999) is the dominant mode of interannual variability in the tropical Indian Ocean. The positive (negative) IOD is characterized by the anomalously cold (warm) SST in the east and warm (cold) SST in the west Indian Ocean. Anomalous easterly (westerly) equatorial winds couple with anomalous SST and blowing from east (west) to west (east) toward warmer waters in the positive (negative) (Australian Bureau of Meteorology, http://www.bom.gov.au).

2.3 Impact of the Madden-Julian oscillation (MJO) on intra-seasonal rainfall variability in Sri Lanka

The MJO is a global, tropical, eastward-propagating, quasi-regular circulation anomaly with a 30—60 day period. Convection is strongly coupled to the MJO in the Indian Ocean and the West Pacific (Madden and Julian, 1971, 1994). In all seasons, the signal of MJO active phases typically originated

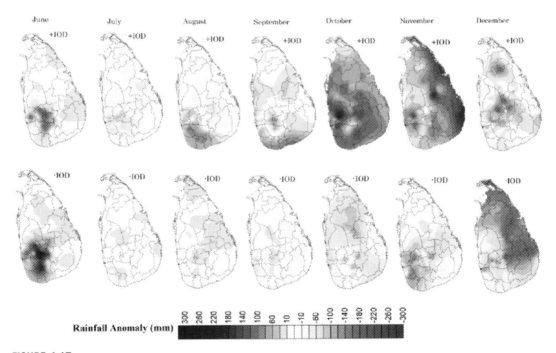

FIGURE 4.17

Composites of monthly rainfall anomalies (shading) for June to December during positive Indian Ocean Dipole (IOD) (upper) and negative IOD (lower).

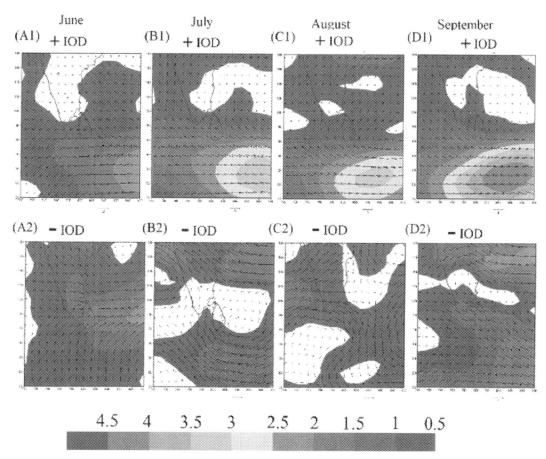

FIGURE 4.18

Composite of 850-hPa wind anomalies during June, July, August and September for positive Indian Ocean Dipole (IOD) (A1, B1, and D1) and negative IOD (A2, B2, C2, and D2) (region: 18N–02S, 70E-90E).

in the western Indian Ocean enhanced convection then moves east along the equator to the Maritime Continent and into the Pacific (Madden, 1986).

The MJO interacts with, and influences, a wide range of weather and climate phenomena and represents an important source of predictability at the sub-seasonal time scale. Weather events under the influence of the MJO include precipitation, surface temperature, tropical cyclones, tornados, floods, wildfires, and lightning (Zhang, 2013).

The MJO impact on extreme rainfall leading to severe floods were studied by Zhu et al. (2003), Tangang et al. (2008), Aldrian (2008), and Jayawardena et al. (2017). A simultaneous combination of local phenomena of terrain effects, cold surges at low levels, MJO wet phase, and a vortex to southeast Sri Lanka cause heavy rainfall that leads to flooding phenomena over Eastern and Northeastern parts of Sri Lanka in December 2014 (Jayawardena et al., 2017).

78 Chapter 4 Future climate projections for Sri Lanka

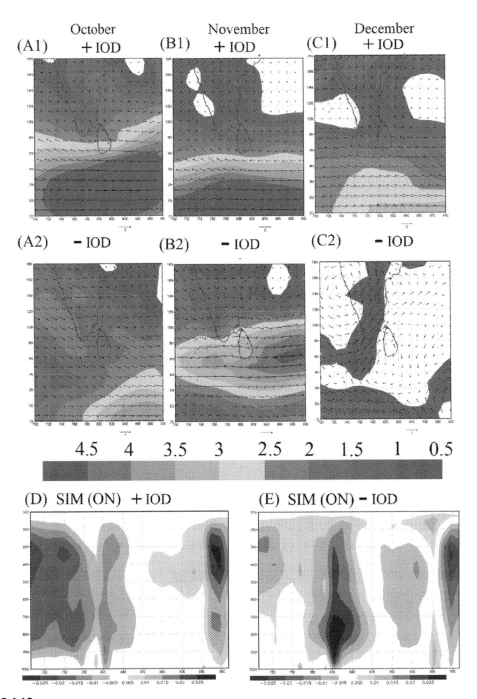

FIGURE 4.19

Composite of 850-hPa wind anomalies Indian Ocean Dipole (IOD) during October, November, and December for positive IOD (A1, B1, and C1) and for negative IOD (A2, B2, and C12) (region: 18N–02S, 70E-90E) and east-west vertical cross-section of pressure vertical velocity composite average across Sri Lanka

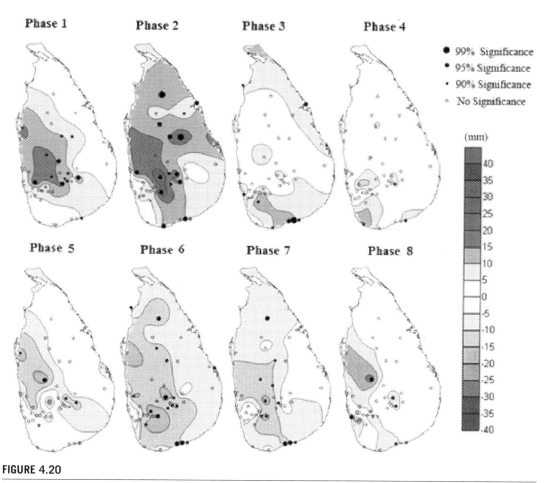

FIGURE 4.20

Madden-Julian oscillation (MJO) composites of daily rainfall anomalies (mm) (shading) for FIM season for phases 1–8 (Jayawardena et al., 2020).

The real-time multivariate MJO (RMM) index developed by Wheeler and Hendon (2004) provides both real-time MJO information (position and strength) and a historical database. Using this index, much work has been done to study the effect of the MJO on rainfall variability over many parts of the world (Wheeler et al., 2009; Zhang et al., 2009; Jia et al., 2010; Martin and Schumacher, 2011; Pai et al., 2011; Matthews et al., 2013; and Peatman et al., 2014).

According to Jayawardena et al. (2020), the greatest impact of the MJO on rainfall over Sri Lanka occurs in the SIM (Fig. 4.22) and SWM (Fig. 4.21) seasons. Enhanced rainfall generally occurs over Sri Lanka during RMM phases 2 and 3 when the MJO convective envelop is in the Indian Ocean and conversely suppressed rainfall in phases 6 and 7. This rainfall impact is due to the direct influence of the MJO's tropical convective anomalies and associated low-level circulations in the Bay of Bengal. In contrast, the MJO influence during the NEM season (Fig. 4.23) is slightly less than during the SWM and SIM seasons as a result of the southward shift of the MJO convective envelop during boreal winter.

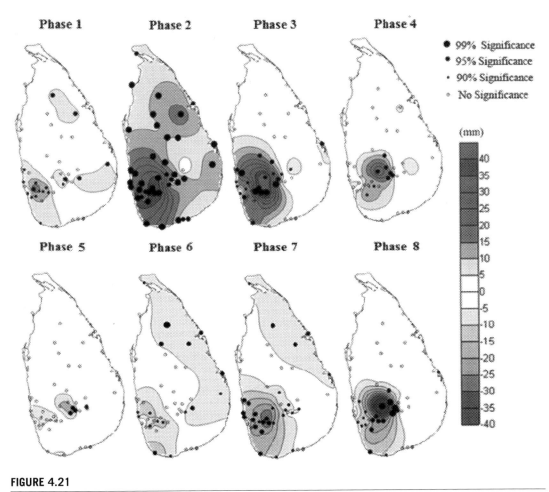

FIGURE 4.21

Madden-Julian oscillation (MJO) composites of daily rainfall anomalies (mm) (shading) for Southwest Monsoon (SWM) season for phases 1–8 (Jayawardena et al., 2020).

2.4 Impact of the boreal summer intra-seasonal oscillation (BSISO) on intra-seasonal SWM rainfall variability in Sri Lanka

Not all of the monsoon intraseasonal variance is associated with the MJO. MJOs strength undergoes a strong seasonal cycle, which is strongest in boreal winter and weakest in boreal summer (Madden and Julian, 1994; Wheeler and Hendon, 2004). Furthermore, the MJO shows a complex character with prominent northward propagation and variability extending much further from the equator during boreal summer (Madden and Julian, 1994). Some studies indicate that the MJO cannot capture the northward propagating monsoon intraseasonal oscillation completely (Lee et al., 2013) during the summer monsoon season.

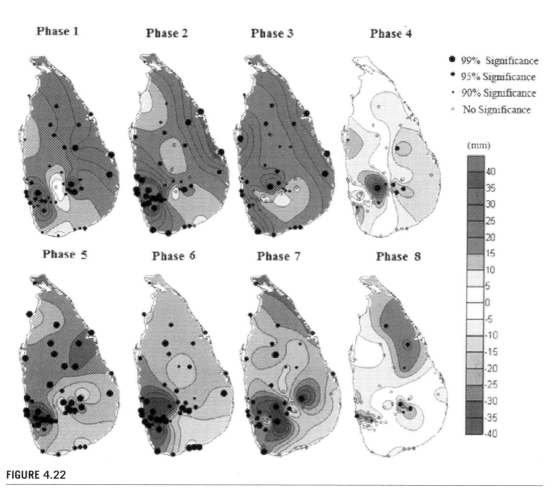

FIGURE 4.22

Madden-Julian oscillation (MJO) composites of daily rainfall anomalies (mm) (shading) for SIM season for phases 1—8 (Jayawardena et al., 2020).

The impact of the BSISO on SWM seasonal rainfall in Sri Lanka is investigated by Jayawardena et al. (2016), using BISISO indices (Lee et al., 2013) and the observational rainfall data from 1981 to 2010. BSISO one represents the canonical northward propagating variability that often occurs in conjunction with the eastward MJO with quasi-oscillating periods of 30—60 days and BSISO two represents the northward and northwestward propagating variability with periods of 10—30 days existing primarily during the premonsoon and monsoon onset season (Lee et al., 2013).

According to Jayawardena et al. (2016), widespread positive anomalies are evident in phases 1 to 3 with the strengthening of positive anomalies in phase 2, while widespread dry conditions can be seen from phases 5 to 8 with the strengthening of negative anomalies apparent over central hills in phase 7 and 8 for BSISO 1. In phase 4, slightly above normal rainfall is evident over the southwest quarter, while slightly below normal rainfall is obvious elsewhere (Fig. 4.24).

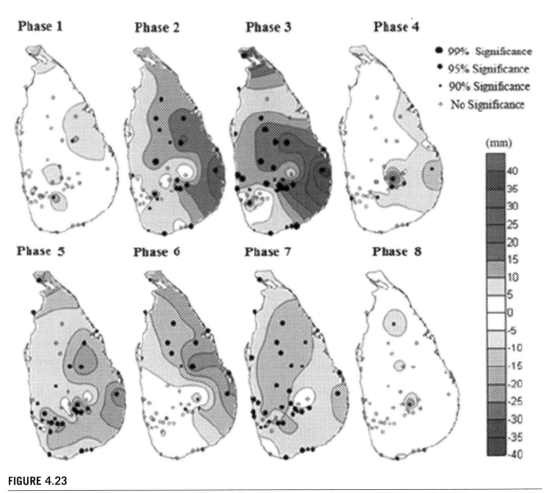

FIGURE 4.23

Madden-Julian oscillation (MJO) composites of daily rainfall anomalies (mm) (shading) for NEM season for phases 1—8 (Jayawardena et al., 2020).

For BSISO2, widespread positive rainfall anomalies are evident only in phase 1. During phase 2, the country was nearly divided into two halves with positive anomalies over the southwestern part and negative anomalies elsewhere. Enhancement of positive rainfall anomalies over western slopes of central hills is also evident in phase 2. Phases 3 and 4 are similar to phase 2 with the weakening of wet conditions over the southwest quarter and strengthening of dry conditions elsewhere. Enhancement of negative rainfall anomalies over the southwest quarter is also evident in phases 6 and 7 (Fig. 4.25) (Jayawardena et al., 2016).

3. Observed local trends

There are several studies have been conducted relevant to climate change and climate variability based on trend analysis of climatic variables such as precipitation and temperature in Sri Lanka. An

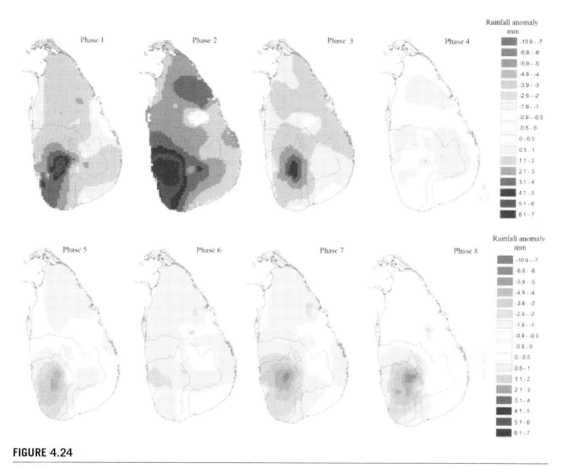

FIGURE 4.24

BSISO1 composites of daily rainfall anomalies (mm) (shading) for Southwest Monsoon (SWM) season for phases 1—8 (Jayawardena et al., 2016).

increasing trend in air temperature is evident in most parts of the country (Sheikh et al., 2015; Basnayake, 2002; Eriyagama et al., 2010). It has been reported that both mean maximum and mean minimum air temperatures have increased and an increase in minimum air temperature contributes more to an average increase in annual temperature than daytime maximum air temperature (Basnayake, 2002).

Previous studies indicated that no clear pattern or trend has been observed in precipitation. Some studies identified that average rainfall is showing a decreasing trend (Sheikh et al., 2015; Basnayake, 2002), while Premalal (2009) observed that heavy rainfall events have become more frequent in central highlands during the recent period. Analyzing fluctuations in rainfall associated with the four climatic seasons using rainfall data for nearly 130 years (1870—2000), 15 rainfall stations, Malmgren et al. (2003) identified a decrease in rainfall in higher elevation areas and an increase in rainfall in lowlands in the southwestern sector of Sri Lanka during SWM season. None of the stations shows any significant

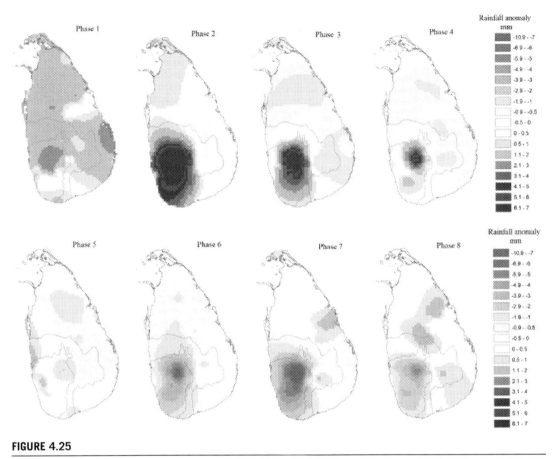

FIGURE 4.25

BSISO2 composites of daily rainfall anomalies (mm) (shading) for Southwest Monsoon (SWM) season for phases 1–8 (Jayawardena et al., 2016).

change in NEM precipitation through time. While analyzing rainfall data for more than 100 years (1895–1996), Jayawardena et al. (2005) found that no coherent increase or decrease in rainfall was observed in stations in the wet or dry zones. Nisansala et al. (2020) indicated that an increasing rainfall trend is evident in eastern, southeastern, north, and north–central regions, while the rainfall trend is decreasing in the western and central parts of the country for 31 years from 1987 to 2017.

A significant increase in annual rainfall from 1989 to 2019 in all climatic zones including wet, dry, intermediate, and semi-arid of Sri Lanka is identified with a maximum increasing trend in the wet zone and the minimum increasing trend in the semi-arid zone by Alahakoon and Edirisinghe (2021).

3.1 Recent observed trends in climatic extremes in Sri Lanka (1981–2015)

Precipitation and temperature indices developed by the World Meteorological Organization–Expert Team on Climate Change Detection and Indices (WMO-ETCCDI) (Easterling et al., 2003; Alexander et al., 2006) were calculated using daily precipitation, daily maximum temperature, and daily minimum temperature data, from 19 surface weather stations of the Department of meteorology, Sri Lanka, covering the period from 1980 to 2015. The long-term trends for these indices in Sri Lanka during the period from 1980 to 2015 were then examined, using quality-controlled daily station data. A 5% level (large triangles) and 10% level (small triangles) of statistical significance were also taken into consideration.

An increasing trend is apparent in maximum temperature as well as in maximum temperatures in most of the stations. Diurnal temperature range (DTR), the difference between daytime maximum temperature and nighttime minimum temperature, shows a significant decreasing trend in most of the stations (Fig. 4.26). The change in DTR is regional rather than to be global. Many studies have revealed that there is a decrease in DTR since the mid-20th century (Easterling et al., 1997). It is shown clearly that DTR is decreasing over most stations except Hambantota and Trincomalee (Fig. 4.26). This decreasing trend is due to the rapidly rising trend of minimum temperatures.

Over 60% of the stations show significant increasing trends in the percentage of warm nights, and 70% of the stations show significant decreasing trends in the percentage of cold nights (Fig. 4.27). This

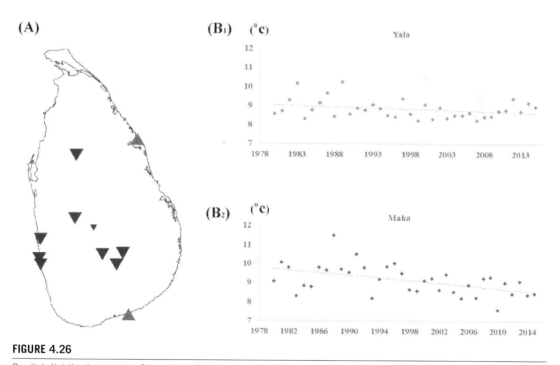

FIGURE 4.26

Spatial distribution maps of trends in diurnal temperature range (A) and time-series for diurnal temperature range for Yala (B_1) and Maha (B_2) seasons at Anuradhapura. The upward-pointing *red* [*gray in print*] *triangles* show increasing trends, while the downward-pointing *blue* [*dark gray in print*] *triangles* indicate decreasing trends. Significant changes at the 5% level are indicated by large triangles, and the 10% level is indicated by small triangles.

From Jayawardena et al. (2018).

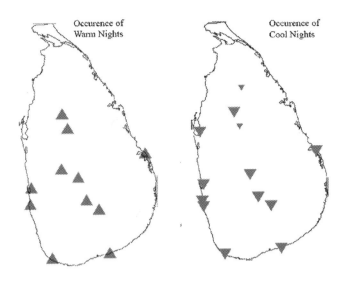

FIGURE 4.27

Spatial distribution maps of trends in occurrence of warm nights (left) and occurrence of cold nights (right).

From Jayawardena et al. (2018).

means that many stations have seen fewer cold nights and more warm nights during the past three-four decades. This result is consistent with previous studies conducted by Alexander et al. (2006) who found that over 70% of the land area showed a significant increase in the annual occurrence of warm nights, while the occurrence of cold nights showed a similar proportion of significant decrease. Over nearly all parts of the globe, both tails of the minimum temperature distribution have warmed at a similar rate (Alexander et al., 2006).

An increasing trend was observed for the annual total rainfall (Fig. 4.28) as well as several days above 10, 20, and 30 mm rainfall (Fig. 4.30) in Sri Lanka for the period used in this report, that is, 1980–2015. 65% of stations show a significant increasing trend for annual precipitation at a 5% or 10% level (Fig. 4.30). Comparing with decadal changes (Table 4.2), it is found that increase in maximal 1-day precipitation and maximal 5-day precipitation during the period from 2010 to 2015 in stations is located in all three climatic zones. A 60% and 50% increase in maximal 1-day and 5-day precipitation amounts, respectively, at Anuradhapura from 2010 to 2015. Similarly, 110% and 60% increase in maximal 1-day and 5-day precipitation amount can be seen in Batticoloa for the same period compared to the 30-year average. 90% of stations show a nonsignificant increasing trend in total precipitation on extreme rainfall days (R95p and R99p), but only 20%–25% of the station trends are significant at the 5%–10% level (Fig. 4.29).

Consecutive wet days (CWD) and consecutive dry days (CDD) show mixed trends. There is an increasing trend in CWD in Katugastota and decreasing trend in CWD in Nuwaraeliya and Ratnapura. Less consecutive dry day (CDD) is observed in stations located in western coastal areas such as Colombo, Katunayake, Ratmalana, and Puttalam.

When compared with temperature changes, a less spatially coherent pattern of change and a lower level of statistical significance were observed in precipitation indices. The trends in extreme

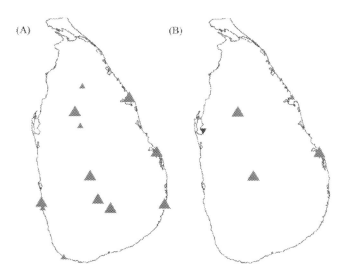

FIGURE 4.28

Spatial distribution maps of trends for annual total precipitation (A) and simple daily intensity of rainfall (B).

From Jayawardena et al. (2018).

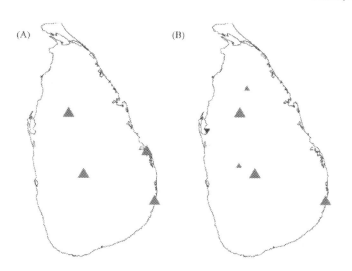

FIGURE 4.29

Spatial distribution maps of trends for very wet days (A) and extremely wet days (B).

From Jayawardena et al. (2018).

precipitation events such as maximal 1-day precipitation, maximal 5-day precipitation, and total precipitation on extreme rainfall days (R95p and R99p) are increasing at most locations, indicating that the intensity of the rainfall is increasing. The increase in precipitation extreme trends indicates that the occurrence of extreme rainfall events notably influences total annual precipitation in Sri Lanka. Therefore, the observed increases in total rainfall observed in many locations may be due in part to an increase in extreme rainfall events.

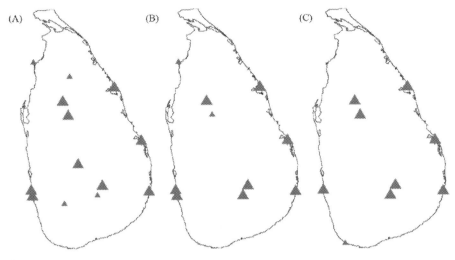

FIGURE 4.30

Spatial distribution maps of trends for some days, above 10 mm rainfall (A), above 20 mm rainfall (B), and above 30 mm rainfall (C).

Jayawardena et al. (2018).

4. Climate change—future climate change projections

The Intergovernmental Panel on Climate Change (IPCC) in its Sixth Assessment Report in 2021 indicated the intensification of hydro-meteorological hazards in South Asia with changing climate. The vulnerability has increased due to high population densities and encroachment in urban areas imposing a higher threat to the region.

Climate change projections also referred to as climate scenarios are widely used for assessments of the potential impacts of climate change on natural processes and human activities, including assessments conducted at the local/regional scale.

The standard set of scenarios used in the Fifth Assessment Report of the Intergovernmental Panel on Climate Change (IPCC AR5) is called representative concentration pathways (RCP) (Van Vuuren et al., 2011). The RCPs describe four different 21st-century pathways of GHG emissions and atmospheric concentrations, air pollutant emissions, and land use. The RCP 2.6 represents a low emission scenario; RCP 4.5 and RCP six represent a moderate emission scenario, while RCP 8.5 represents a high emission scenario. The fifth phase of the Coupled Model Intercomparison Project (CMIP5) (Tayler, 2012) produced a multimodel dataset in support of the AR5.

4.1 Future climate projections of seasonal rainfall and annual rainfall for Sri Lanka

NASA Earth Exchange Global Daily Downscaled Projections (NEX-GDDP) dataset (Nemani et al., 2015) is comprised of downscaled climate scenarios for the globe that are derived from the general

Table 4.2 Decadal changes in extreme rainfall events defines as maximum 5-day rainfall amount (mm) and maximum 1-day rainfall amount (mm).

Extreme rainfall event defined as maximum five-day rainfall (5Day_Max) and maximum one day rainfall (1Day_Max)

	Dry zone			Intermediate zone			Wet zone		
5Day_Max	Anuradhapura	Puttatum	Batticoloa	Kurunegala	Badulla	Bandarawela	Katunayake	Katugastota	NuwraEliya
Average 30	175	193	251	221	192	159	142	93	85
Decade 80	166	213	238	229	237	177	123	95	91
Decade 90	176	189	225	214	168	144	162	94	94
Decade 2000	184	180	286	219	171	160	133	90	70
2010–15	281	159	528	301	257	192	146	127	95
1Day_Max	Anuradhapura	Puttatum	Batticoloa	Kurunegala	Badulla	Bandarawela	Katunayake	Katugastota	NuwraEliya
Average 30	89	113	126	123	95	76	257	177	188
Decade 80	73	133	130	123	110	78	243	178	217
Decade 90	99	114	113	115	94	75	265	179	197
Decade 2000	94	95	135	131	81	75	252	173	150
2010–15	135	79	200	172	124	91	263	249	198

circulation model (GCM) runs conducted under CMIP5 (Taylor et al., 2012) and across two of the four greenhouse gas emissions scenarios RCP 4.5 and RCP 8.5. NEX-GDDP data for 6 GCM climate models (25-kilometer (km) grid resolution) given in Table 4.3 were used to develop figure climate projections. The representative concentration pathways (RCP) RCP 8.5 and 4.5 scenarios from the IPCC AR5 2013, representing futures under high emission and moderate emission, respectively, were adopted, with three time periods—2030, 2050s, and 2080s. To address this source of uncertainty, single climate projections from climate models mentioned in Table 4.3 are used to generate a multimodel ensemble. Multimodel ensemble projections have higher reliability and consistency when several independent models are combined (Doblas-Reyes et al., 2003; Yun et al., 2003).

SWM rainfall anomaly is positive and increasing in both moderate (RCP 4.5) and high (RCP 8.5) emission scenarios with the increasing anomaly being significant in the wet zone (Fig. 4.31).

Northeast monsoon rainfall anomaly is negative for short-term, medium-term, and long-term projections, and a negative trend is observed under moderate emission scenario RCP 4.5. Northeast monsoon rainfall anomaly slightly positive in short-term projection 2020–2040, and negative thereafter for medium-term and long-term projections under high emission scenario. A negative trend is observed for the high emission scenario RCP 8.5. Decreasing anomaly is significant over the dry zone (Fig. 4.32).

FIM rainfall anomaly is negative in 2020–2040, slightly negative in 2040–2060, and positive except in northeastern parts under moderate emission scenario RCP 4.5 (Fig. 4.33 (upper)). FIM rainfall anomaly is negative in all 3-time frames with no significant trend under high emission scenario RCP 8.5 (Fig. 4.3 (lower)).

SIM rainfall anomaly is negative in the dry zone and positive in wet zone parts in 2020–2040 (Fig. 4.34 (upper)). Seasonal rainfall anomaly is increasing after that under RCP 4.5 (Fig. 4.34 (upper)). SIM rainfall anomaly is positive and increasing under RCP 8.5 scenarios with a significant increase of positive rainfall anomaly over the southwestern and southeastern parts (Fig. 4.34 (lower)).

Multimodel ensemble projections indicated that the annual rainfall anomaly is negative in northeastern parts and positive in southwestern parts in 2020–2040, while the annual rainfall anomaly is positive and increasing thereafter under moderate emission scenario RCP 4.5. Annual rainfall anomaly

Table 4.3 CMIP5 models used to develop climate projections for Sri Lanka.	
CanESM2	The Second Generation Coupled Global Climate Model Canadian Center for Climate Modeling and Analysis (2.8*2.8)
CNRM-CM5	National Centre for Meteorological Research/Meteo-France (1.4*1.4)
CSIRO-MK3-6−0	Commonwealth Scientific and Industrial Research Organisation (CSIRO) and the Queensland Climate Change Centre of Excellence (QCCCE). (1.895*1.875)
GFDL-CM3	Geophysical Fluid Dynamic Laboratory NOAA, USA Coupled Climate Model (2*2.5)
MRI-CGCM3	Global Climate Model of the Meteorological Research Institute, Japan (1.132*1.125)
NCAR-CCSM4	National Center for Atmospheric Research, USA Coupled Climate Model (0.942*1.25)

4. Climate change—future climate change projections

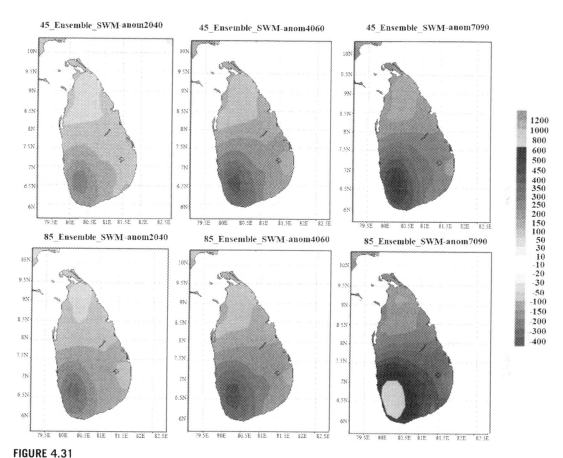

FIGURE 4.31

Multimodel ensemble of change in Southwest Monsoon (SWM) rainfall, relative to 1975–2005 for moderate emission scenario (representative concentration pathways [RCP] 4.5) (upper) and high emission scenario (RCP 8.5) for time periods (2020–2040) (left), (2040–2060) (middle), and (2070–2090) (right).

is positive and increasing under high-emission scenario RCP 8.5 with the increasing anomaly being significant in the wet zone (Fig. 4.35).

Increases in the frequency and severity of extreme rainfall events observed during the recent past and projected in future indicated that the occurrence of extreme rainfall events notably influences total annual precipitation in Sri Lanka. Therefore, the observed and projected increase in total rainfall may be due in part to an increase in extreme rainfall events. Patterns of change in precipitation extremes are more heavily influencing the climate variability by aggravating the variability with the occurrence of more floods, landslides, and droughts.

Changes in annual as well as SWM seasonal rainfall compare to the baseline climatology, indicate that positive rainfall anomaly in the wet zone will rise with the time under high as well as moderate emission scenarios. The western slopes of the central hills are prone to natural disasters like landslides that will impact agriculture by land degradation and soil erosion.

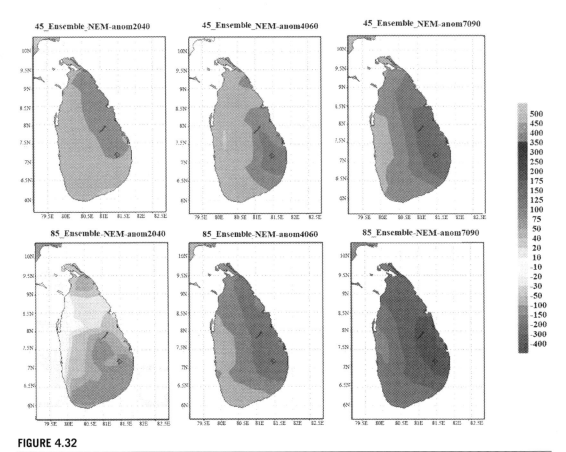

FIGURE 4.32

Multimodel ensemble of change in NEM rainfall, relative to 1975−2005 for moderate emission scenario (representative concentration pathways [RCP] 4.5) (upper) and high emission scenario (RCP 8.5) for time periods (2020−2040) (left), (2040−2060) (middle), and (2070−2090) (right).

Changes in NEM seasonal rainfall compare to the baseline climatology clearly indicate that negative rainfall anomalies especially in the dry zone will befall with the time under both high as well moderate emission scenarios. Reduction in NEM rainfall (Dec−February) may increase the vulnerability of the agriculture sector as nearly 70% of the paddy cultivate in the Maha season (September to March) in the dry zone of Sri Lanka. More frequent droughts can be expected in the dry and intermediate zones.

4.2 Future climate projections of annual temperature for Sri Lanka and projected heatwave occurrence in Colombo

Multimodel ensemble prediction indicates an increase in maximum temperatures as well as minimum temperatures for all three time periods in 2020−2040, 2040−2060, and 2070−2090 for both moderate-

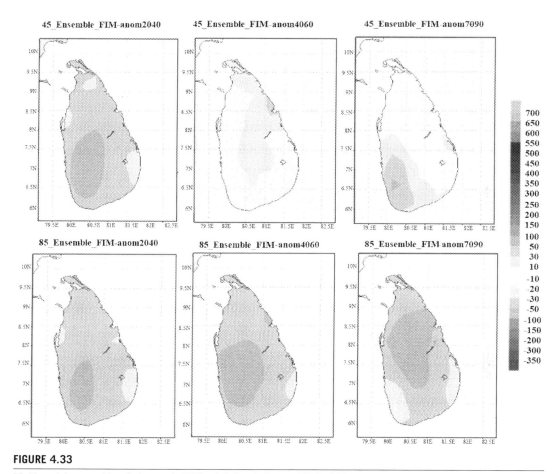

FIGURE 4.33

Multimodel ensemble of change in FIM rainfall, relative to 1975—2005 for moderate emission scenario (representative concentration pathways [RCP] 4.5) (upper) and high emission scenario (RCP 8.5) for time periods (2020—2040) (left), (2040—2060) (middle), and (2070—2090) (right).

emission (RCP 4.5) and high-emission scenarios. For the moderate-emission scenario, multimodel ensemble prediction indicated that an increase in minimum and maximum temperature in 0.7—1.2°C, 1.0—1.6°C, and 1.5—2.3°C can be expected during 2020—2040, 2040—2060, and 2070—2090, respectively (Figs. 4.36 and 4.37). For the high emission scenario, multimodel ensemble prediction indicated an increase of minimum temperature in 1.1—1.5°C, 1.6—2.5°C, and 2.4—3.5°C, while an increase of maximum temperature in 1.0—1.5°C, 1.4—2.3°C, and 2.2—3.2°C can be expected in 2020—2040, 2040—2060, and 2070—2090, respectively, (Figs. 4.36 and 4.37).

Characteristics of the future heat waves (HW hereafter) under the RCP 4.5 moderate emission scenario and RCP 8.5 high emissions scenario from 2020 to 2099 over the Colombo area were investigated using the Global Climate Model of the Meteorological Research Institute, Japan MRI-CGCM3.

94 **Chapter 4** Future climate projections for Sri Lanka

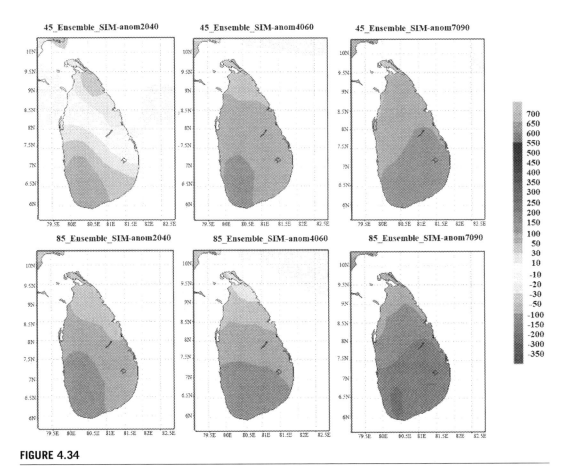

FIGURE 4.34

Multimodel ensemble of change in SIM rainfall, relative to 1975—2005 for moderate emission scenario (representative concentration pathways [RCP] 4.5) (upper) and high emission scenario (RCP 8.5) for time periods (2020—2040) (left), (2040—2060) (middle), and (2070—2090) (right).

MRI-CGCM3 model simulations indicated an increasing trend in HW frequency and the number of days that contribute to heatwaves under both emission scenarios (Fig. 4.38). The increasing trend in the high emission scenario is higher than in the moderate emission scenario. HW duration, the length of the longest HW, increases exponentially under high emission scenarios than in moderate emission scenarios over Colombo (Fig. 4.39). Fig. 4.40 represents the HW number defined as several individual HWs, which show an increasing trend under both emission scenarios. An increasing trend in HW frequency, duration, and number are prominent after 2030 (Figs. 4.38—4.40).

The increasing trend of HW in Colombo, a densely populated city, in the future climate will make the population more vulnerable to the impact of HWs. The consequence of future HWs might be severe; therefore, there is an urgent need to prepare a strategy to deal with its likelihood consequences.

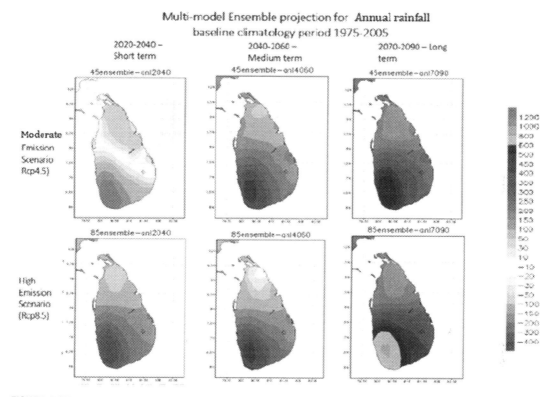

FIGURE 4.35

Multimodel ensemble of change in annual rainfall, relative to 1975–2005 for moderate emission scenario (representative concentration pathways [RCP] 4.5) (upper) and high emission scenario (RCP 8.5) for periods (2020–2040) (left), (2040–2060) (middle), and (2070–2090) (right).

Further, high humidity compounds the effects of the temperatures being felt by human beings. Extreme heat can lead to dangerous, even deadly, consequences, including heat stress and heatstroke.

While the effects of heat may be exacerbated in cities like Colombo, due to the urban heat island effect, the livelihoods, and social well-being can also be severely disrupted during and after periods of unusually hot weather. Urban surfaces such as roads and roofs absorb, hold, and re-radiate heat, raising the temperature in our urban areas. This effect is often worsened by development activity when green spaces are replaced with more hard surfaces that absorb heat and human activities such as traffic, industry, and electricity usage generate heat that adds to the urban heat island effect.

Improving the urban tree canopy provides cooling shade, preventing dark surfaces from absorbing and releasing heat from the sun. Water released from the tree leaves through transpiration also has a slight cooling effect. Over a large enough area, urban parks, and water bodies can significantly cool a city (Ban-Weiss et al., 2011).

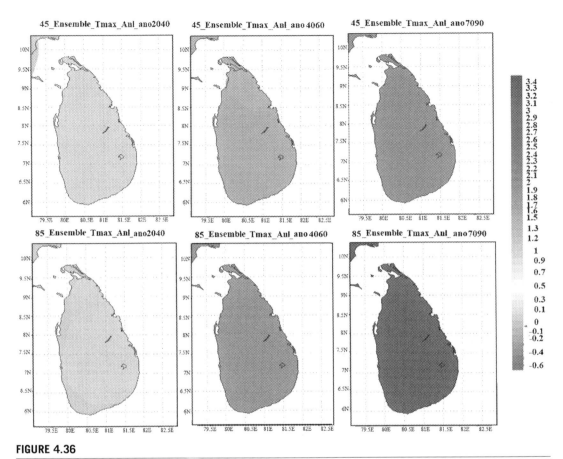

FIGURE 4.36

Multimodel ensemble of change in maximum temperature relative to 1975–2005 for moderate-emission scenario (representative concentration pathways [RCP] 4.5) (upper) and high-emission scenario (RCP 8.5) (lower) for periods (2020–2040), (2040–2060), and (2070–2090).

5. Conclusions

ENSO and IOD can be considered as one of the potential climatic drivers of interannual rainfall variability in Sri Lanka. El Nino and positive IOD enhance SIM rainfall with anomalous rising motion, while La Nina and negative IOD diminish SIM rainfall with anomalous sinking motion over Sri Lanka. The significant impact of SWM rainfall was also evident over central parts of the island with a reduction of seasonal rainfall during El Nino events and enhanced seasonal rainfall during La Nina events. Slight reduction (enhancement) in NEM rainfall over northwestern, northcentral, and central parts is evident during El Nino (La Nina).

MJO can be considered as one of the potential climatic drivers of intraseasonal rainfall variability in Sri Lanka. Considering all four seasons, Sri Lanka rainfall appears to be directly influenced by the

5. Conclusions

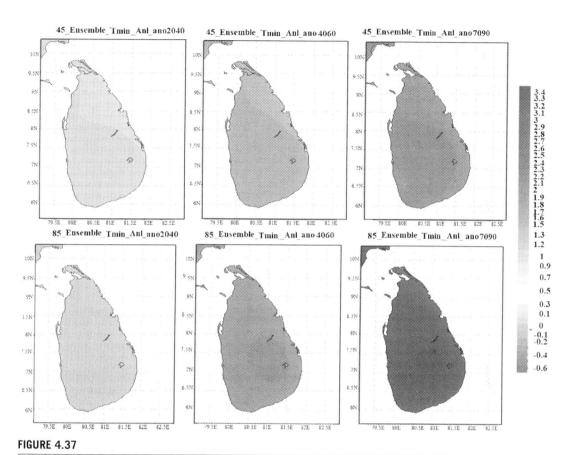

FIGURE 4.37

Multimodel ensemble of change in minimum temperature, relative to 1975–2005 for moderate-emission scenario (representative concentration pathways [RCP] 4.5) (upper) and high-emission scenario (RCP 8.5) (lower) for periods (2020–2040), (2040–2060), and (2070–2090).

MJO's tropical convective signal with the largest positive anomalies in phases 2 and 3 when the MJO convective envelope is located over the Indian Ocean and largest negative anomalies in phases 6 and 7 when the MJO convective envelope is located over the western Pacific. Of the four seasons, the greatest impact occurs during SIM with well-marked wet signals in phases 1 to 3 and dry signals in phases 5 to 7, respectively. The MJO's impact on Sri Lanka rainfall during SWM is also significant with wet signals in phases 2 to 4 and dry signals in phases 6 to 8. A relatively smaller MJO influence is found in the other two seasons (FIM and NEM) with wet signals in phases 2 and 3 and dry signals in phases 5, 6, and 7.

BSISO also modulate the intraseasonal rainfall variability during SWM with widespread positive anomalies at phase 1 to 3, while widespread dry conditions can be seen from phase 5 to 8 for BSISO 1. For BSISO2, widespread positive rainfall anomalies is evident only in phase 1 and negative anomalies

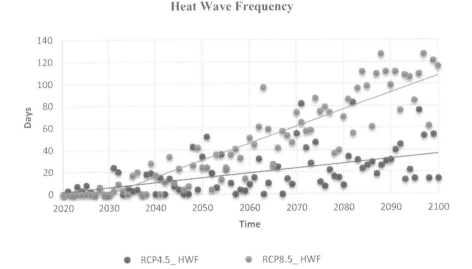

FIGURE 4.38

Time series of heat wave (HW) frequency (days) from 2020 to 2100 at Colombo for moderate-emission scenario (representative concentration pathways [RCP] 4.5) (*blue* [*dark gray in print*] *circles*), and high-emission scenario (RCP 8.5) (*red* [*gray in print*] *circles*).

in phase 5. Rainfall enhancement (diminution) is apparent over the western slopes of the central hill in phases 2 (6), 3(7), and 4(8).

It is evident that annually averaged mean maximum and minimum temperatures are increasing across most of Sri Lanka. The difference between the maximum and minimum temperatures and diurnal temperature range is decreasing, indicating that the minimum temperature is increasing faster than the maximum temperature. A significant decrease in the annual occurrence of cold nights and an increase in the annual occurrence of warm nights are also obvious.

When compared with temperature changes, a less spatially coherent pattern of change and a lower level of statistical significance were observed in precipitation indices. The annual total precipitation has indicated a significant increase from 1980 to 2015. The trends in extreme precipitation events such as maximum 1-day precipitation, maximum 5-day precipitation, and total precipitation on extreme rainfall days are increasing at most locations, indicating that the intensity of the rainfall is increasing. The increase in precipitation extreme trends indicates that the occurrence of extreme rainfall events notably influences total annual precipitation in Sri Lanka. Therefore, the observed increases in total rainfall observed in many locations may be due in part to an increase in extreme rainfall events.

Multimodel ensemble projections show enhancement in annual as well SWM rainfall in both moderate- (RCP 4.5) and high (RCP 8.5)-emission scenarios with the increasing anomaly being significant in the wet zone.

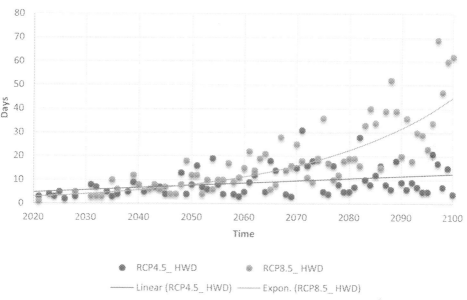

FIGURE 4.39

Time series of heat wave (HW) duration (days) from 2020 to 2100 at Colombo for moderate-emission scenario (representative concentration pathways [RCP] 4.5) (*blue [dark gray in print] circles*), and high-emission scenario (RCP 8.5) (*red [gray in print] circles*).

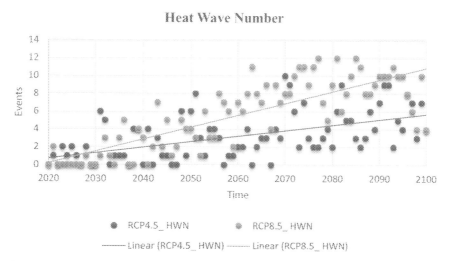

FIGURE 4.40

Time series of heat wave (HW) number (events) from 2020 to 2100 at Colombo for moderate-emission scenario (representative concentration pathways [RCP] 4.5) (*blue [dark gray in print] circles*), and high-emission scenario (RCP 8.5) (*red [gray in print] circles*).

A negative rainfall anomaly is apparent in NEM under both emission scenarios in the medium-term and long-term projections under the high-emission scenario. A negative trend is observed for the high-emission scenario RCP 8.5. Decreasing anomaly is significant over the dry zone.

FIM rainfall anomaly is negative in all 3-time frames with no significant trend under high emission scenario RCP 8.5.

SIM rainfall anomaly is positive and increasing under RCP 8.5 scenarios with a significant increase of positive rainfall anomaly over the southwestern and southeastern parts.

Multimodel ensemble prediction indicates an increase in maximum temperatures as well as minimum temperatures under both emission scenarios.

References

Alahacoon, N., Edirisinghe, M., 2021. Spatial variability of rainfall trends in Sri Lanka from 1989 to 2019 as an indication of climate change. ISPRS Int. J. Geo-Inf. 10 (2), 84.

Aldrian, E., 2008. Dominant factors of Jakarta's three largest floods. J. Hidrosfir Indonesia 3, 105–112.

Alexander, L.V., Zhang, X., Peterson, T.C., Caesar, J., Gleason, B., Klein Tank, A.M.G, Haylock, M., Collins, D., Trewin, B., Rahimzadeh, F., Tagipour, A., 2006. Global observed changes in daily climate extremes of temperature and precipitation. J. Geophys. Res. Atmos. 111 (D5).

Ban-Weiss, G.A., et al., 2011. Climate forcing and response to idealized changes in surface latent and sensible heat. Environ. Res. Lett. 6.

Basnayake, B.R.S.B., Fernando, T.K., Vithanage, J.C., 2002. Variation of air temperature and rainfall during Yala and Maha agricultural seasons. In: Proceedings of the 58th Annual Session of Sri Lanka Association for the Advancement of Science (SLASS). Section E1, p. 212.

Chandimala, J., Zubair, L., 2007. Predictability of streamflow and rainfall based on ENSO for water resources management in Sri Lanka. J. Hydrol. 335 (3–4), 303–312.

Chandrapala, L., 1996. Long term trends of rainfall and temperature in Sri Lanka. In: Abrol, Y.P., Gadgil, S., Pant, G.B. (Eds.), Climate Variability and Agriculture. Narosa Publishing House, New Delhi, pp. 150–152.

Doblas-Reyes, F.J., Pavan, V., Stephenson, D.B., 2003. The skill of multimodel seasonal forecasts of the wintertime North Atlantic Oscillation. Clim. Dynam. 21, 501–514. https://doi.org/10.1007/s00382-003-0350-4.

Easterling, D.R., Horton, B., Jones, P.D., Peterson, T.C., Karl, T.R., Parker, D.E., Salinger, M.J., Razuvayev, V., Plummer, N., Jamason, P., Folland, C.K., 1997. Maximum and minimum temperature trends for the globe. Science 277 (5324), 364–367.

Easterling, D.R., Alexander, L.V., Mokssit, A., Detemmerman, V., 2003. CCl/CLIVAR workshop to develop priority climate indices. Bull. Am. Meteorol. Soc. 84 (10), 1403–1407.

Eriyagama, N., Smakhtin, V., Chandrapala, L., Fernando, K., 2010. Impacts of Climate Change on Water Resources and Agriculture in Sri Lanka: A Review and Preliminary Vulnerability Mapping. International Water Management Institute, Colombo, Sri Lanka. https://doi.org/10.5337/2010.211, 51pp. (IWMI Research Report 135).

Gomes, C., Kadir, M.Z.A.A., 2011. A theoretical approach to estimate the annual lightning hazards on human beings. Atmos. Res. 101, 719–725. https://doi.org/10.1016/j.atmosres.2011.04.020.

Hapuarachchi, H.A.S.U., Jayawardena, I.M.S.P., 2015. Modulation of seasonal rainfall in Sri Lanka by ENSO extremes. Sri Lanka J. Meteorol. 1, 3–11.

India Meteorological Department, 1973. Northeast Monsoon. FMU Report No. IV-18.4.

Jayaratne, K.C., Gomes, C., September 2012. Public Perceptions and Lightning Safety Education in Sri Lanka. In2012 International Conference on Lightning Protection (ICLP). IEEE, pp. 1–7.

Jayawardena, I.M.S.P., Mäkelä, A., 2021. Spatial and temporal variability of lightning activity in Sri Lanka. In: Multi-Hazard Early Warning and Disaster Risks. Springer, Cham, pp. 573—586.

Jayawardena, I.M.S.P., Sumathipala, W.L., Basnayake, B.R.S.B., 2016. Impact of the intra-seasonal oscillations (ISO) on rainfall variability during Southwest Monsoon in Sri Lanka. In: 16th Conference of the Science Council Asia: Science for People: Mobilizing Modern Technologies for Sustainable Development in Asia, Abstracts and Proceedings 31 May-02 June 2016, Colombo, Sri Lanka, pp. 104—109.

Jayawardena, I.M.S.P., Sumathipala, W.L., Basnayake, B.R.S.B., 2017. Impact of Madden Julian oscillation (MJO) and other meteorological phenomena on the heavy rainfall event from 19th— 28th December 2014 over Sri Lanka. J. Natl. Sci. Found. Sri Lanka 45 (2).

Jayawardena, I.M.S.P., Darshika, D.T., Herath, H.R.C., 2018. Recent trends in climate extreme indices over Sri Lanka. Am. J. Clim. Change 7 (4), 586—599.

Jayawardena, I.M.S.P., Wheeler, M.C., Sumathipala, W.L., Basnayake, B., 2020. Impacts of the Madden-Julian oscillation (MJO) on rainfall in Sri Lanka. Mausam 71 (3), 405—422.

Jia, X., Chen, L., Ren, F., Li, C., 2010. Impacts of the MJO on winter rainfall and circulation in China. Adv. Atmos. Sci. 28, 521—533.

Kane, R.P., June 30, 1998. ENSO relationship to the rainfall of Sri Lanka. Int. J. Climatol. 18 (8), 859—871.

Lee, J.-Y., Wang, B., Wheeler, M.C., Fu, X., Waliser, D.E., Kang, I.-S., 2013. Real-time multivariate indices for the boreal summer intra- seasonal oscillation over the Asian summer monsoon region. Clim. Dynam. 40, 493—509.

Madden, R.A., 1986. Seasonal variations of the 40-50 day oscillation in the tropics. J. Atmos. Sci. 43 (24), 3138—3158.

Madden, R.A., Julian, P.R., 1971. Detection of a 40-50 day oscillation in the zonal wind in the tropical Pacific. J. Atmos. Sci. 28, 702—708.

Madden, R.A., Julian, P.R., 1994. Observations of the 40-50-day tropical oscillation: a review. Mon. Weather Rev. 122, 814—837.

Malmgren, B.A., Hulugalla, R., Hayashi, Y., Mikami, T., 2003. Precipitation trends in Sri Lanka since the 1870s and relationships to El Niño—southern oscillation. Int. J. Climatol. 23 (10), 1235—1252.

Martin, E.R., Schumacher, C., 2011. Modulation of Caribbean precipitation by the Madden-Julian oscillation. J. Clim. 24 (3), 813—824.

Matthews, A.J., Pickup, G., Peatman, S.C., Clews, P., Martin, J., 2013. The effect of the Madden-Julian Oscillation on station rainfall and river level in the Fly River system, Papua New Guinea. J. Geophys. Res. Atmos. 118, 10926—10935. https://doi.org/10.1002/jgrd.50865.

McPhaden, M.J., 2002. El Niño and La Niña: Causes and Global consequence Encyclopedia of Global Environmental Change. Anonymous John Wiley and Sons, LTD, pp. 353—370.

Nag, A., Holle, R.L., Murphy, M.J., January 2017. Cloud-to-ground lightning over the Indian subcontinent. In: Postprints of the 8th Conference on the Meteorological Applications of Lightning Data. American Meteorological Society, Seattle/Washington, pp. 22—26.

Nemani, R.R., Thrasher, B.L., Wang, W., Lee, T.J., Melton, F.S., Dungan, J.L., Michaelis, A., December 2015. NASA Earth Exchange (NEX) Supporting Analyses for National Climate Assessments. In: AGU Fall Meeting Abstracts.

Nisansala, W.D.S., Abeysingha, N.S., Islam, A., Bandara, A.M.K.R., 2020. Recent rainfall trend over Sri Lanka (1987—2017). Int. J. Climatol. 40 (7), 3417—3435.

Pai, D.S., Bhate, J., Sreejith, O.P., Hatwar, H.R., 2011. Impact of MJO on the intraseasonal variation of summer monsoon rainfall over India. Clim. Dynam. 36, 41—55. https://doi.org/10.1007/s00382-009-0634-4.

Pant, G.B., Kumar, K.R., 1997. Climates of South Asia. John Wiley &Sons.Ltd., Baffins Lane, Chichester, pp. 219—224.

Peatman, S.C., et al., 2014. Propagation of the Madden–Julian Oscillation through the Maritime Continent and scale interaction with the diurnal cycle of precipitation. Quarter. J. Royal Meteorol. Soc. 140, 814–825. https://doi.org/10.1002/qj.2161.

Philander, S.G.H., 1990. El Nino, La Nina, and the southern oscillation. Int. Geophys. Ser. 46.

Premalal, K.H.M.S., April 2009. Weather and climate trends, climate controls and risks in Sri Lanka. In: Presentation Made at the Sri Lanka Monsoon Forum. Department of Meteorology, Sri Lanka.

Punyawardena, B.V.R., Cherry, N.J., 1999. Assessment of the predictability of the seasonal rainfall in Ratnapura using southern oscillation and its two extremes. J. Natl. Sci. Counc. Sri Lanka 27 (3), 187–195.

Rasmusson, E.M., Carpenter, T.H., 1983. The relationship between eastern equatorial Pacific sea surface temperature and rainfall over India and Sri Lanka. Mon. Weather Rev. 110, 354–383.

Ropelewski, C.F., Halpert, M.S., 1987. Global and regional scale precipitation patterns associated with the El Nino/southern oscillation. Mon. Weather Rev. 115, 1606–1626.

Ropelewski, C.F., Halpert, M.S., 1989. Precipitation patterns associated with the high index phase of the Southern Oscillation. J. Climate 268–284.

Saji, N.H., Goswami, B.N., Vinayachandran, P.N., Yamagata, T., 1999. A dipole mode in the tropical Indian Ocean. Nature 401, 360–363.

Sheikh, M.M., Manzoor, N., Ashraf, J., Adnan, M., Collins, D., Hameed, S., Manton, M.J., Ahmed, A.U., Baidya, S.K., Borgaonkar, H.P., Islam, N., 2015. Trends in extreme daily rainfall and temperature indices over South Asia. Int. J. Climatol. 35 (7), 1625–1637.

Sumathipala, W.L., Punyadeva, N.B.P., 1998. Variation of the rainfall of Sri Lanka in relation to El Nino. In: Proceedings of the Annual Sessions of the Institute of Physics (Sri Lanka, Colombo).

Suppiah, R., 1996. Spatial and temporal variations in the relationships between the Southern Oscillation phenomenon and the rainfall of Sri Lanka. Int. J. Climatol. 16 (12), 1391–1407.

Suppiah, R., 1997. Extremes of the southern oscillation and the rainfall of Sri Lanka. Int. J. Climatol. 17 (1), 87–101.

Suppiah, R., Yoshino, M.M., 1986. Some agroclimatological aspects of rice production in Sri Lanka. Geogr. Rev. Jpn. B 59 (2), 137–153.

Tangang, F.T., Juneng, L., Salimun, E., Vinayachandran, P.N., Seng, Y.K., Reason, C.J., Behera, S.K., Yasunari, T., 2008. On the roles of the northeast cold surge, the Borneo vortex, the Madden-Julian Oscillation, and the Indian Ocean Dipole during the extreme 2006/2007 flood in southern Peninsular Malaysia. Geophys. Res. Lett. 35, L14S07. https://doi.org/10.1029/2008GL033429.

Taylor, K.E., Stouffer, R.J., Meehl, G.A., 2012. An overview of CMIP5 and the experiment design. Bull. Am. Meteorol. Soc. 93 (4), 485–498.

Thambyahpillay, G., 1954. The rainfall rhythm in Ceylon. Univ. Ceylon Rev. 12 (4), 223–274.

Thambyahpillay, G., 1958. The kâcchchān—a föuhn wind in Ceylon. Weather 13 (4), 107–114.

Trenberth, K.E., 1997. The definition of El Niño. Bull. Am. Meteorol. Soc. 78, 2771–2777.

Van Vuuren, D.P., Edmonds, J., Kainuma, M., Riahi, K., Thomson, A., Hibbard, K., Hurtt, G.C., Kram, T., Krey, V., Lamarque, J.F., Masui, T., 2011. The representative concentration pathways: an overview. Climatic Change 109 (1–2), 5.

Weerasekera, A.B., Sonnadara, D.U.J., Fernando, I.M.K., Liyanage, J.P., Lelwala, R., Ariyaratne, T.R., 2001. The activity of cloud-to-ground lightning was observed in Sri Lanka and the surrounding area of the Indian Ocean. Sri Lankan J. Phys. 2.

Wheeler, M.C., Hendon, H.H., 2004. An all-season real-time multivariate MJO index: development of an index for monitoring and prediction. Mon. Weather Rev. 132, 1917–1932.

Wheeler, M.C., Hendon, H., Cleland, S., Donald, A., Meinke, H., 2009. Impacts of the Madden–Julian oscillation on Australian rainfall and circulation. J. Clim. 22, 1482–1498.

Yoshino, M.M., Suppiah, R., 1983. Climate and paddy production: a study on selective districts in Sri Lanka. In: Climate, Water and Agriculture in Sri Lanka.

Yun, W.T., Stefanova, L., Krishnamurti, T.N., 2003. Improvement of the multimodel ensemble technique for seasonal forecasts. J. Clim. 16, 3834–3840.

Zhang, C., 2013. Madden–Julian oscillation: bridging weather and climate. Bull. Am. Meteorol. Soc. 94, 1849–1870.

Zhang, L.N., Wang, B.Z., Zeng, Q.C., 2009. Impacts of the Madden-Julian oscillation on summer rainfall in southeast China. J. Clim. 22, 201–216.

Zhu, C., et al., 2003. The 30–60 day intraseasonal oscillation over the western North Pacific Ocean and its impacts on summer flooding in China during 1998. Geophys. Res. Lett. 30 (18).

Zubair, L., Ropelewski, C.F., 2006. The strengthening relationship between ENSO and Northeast Monsoon rainfall over Sri Lanka and southern India. J. Clim. 19, 1567–1575.

Zubair, L., Siriwardhana, M., Chandimala, J., Yahiya, Z., January 1, 2008. Predictability of Sri Lankan rainfall based on ENSO. Int. J. Climatol. 28 (1), 91–101.

Further reading

Australian Bureau of Meteorology. http://www.bom.gov.au/.

Jayawardene, H.K.W.I., Sonnadara, D.U.J., Jayewardene, D.R., 2005. Trends of rainfall in Sri Lanka over the last century. Sri Lankan Journal of Physics 6.

Meteo. www.meteo.gov.lk.

Webster, P.J., Moore, A.W., Loschnigg, J.P., Leben, R.R., 1999. Coupled ocean-atmosphere dynamics in the Indian Ocean during 1997–98. Nature 401, 356–360.

Zubair, L., Rao, S.A., Yamagata, T., 2003. Modulation of Sri Lankan Maha rainfall by the Indian Ocean Dipole. Geophys. Res. Lett. 30 (2).

CHAPTER 5

A review study on the artifacts, vertebrate remains, and volcanic ash of peninsular Malaysia and their significance to reconstruct paleoclimate

Ajab Singh[1] and Neloy Khare[2]

[1]*Department of Geology, Faculty of Science, University of Malaya, Kuala Lumpur, Malaysia;* [2]*Ministry of Earth Sciences, New Delhi, India*

1. Introduction

The Malaysian Peninsula, located at the southernmost extremity of the Asian mainland, has long piqued the interest of scientists due to its significance in reconstructing human and animal prehistories in Southeast Asia (Andrews, 1905; Tweedie, 1953; Peacock and Dunn, 1968; Taha, 1987; Saidin, 2006; Ibrahim et al., 2012, 2013; Tshen, 2013; Goh et al., 2020) and interpreting source area of widespread distal ash (Screvenor, 1930; Stauffer, 1973; Nishimura and Stauffer, 1981; Storey et al., 2012; Gatti et al., 2013; Singh et al., 2021b). It is believed that the Peninsula was likely used as a land bridge by early humans and animals to traverse the Asian mainland southwardly during the Pleistocene period and submersion of Sundaland in the sea (Tweedie, 1953; Saidin, 2012). Various archeological explorations and subsequent typological connections with the contemporary African artifact industries in the country have been carried out at various locations to classify them as Paleolithic sites (Sieveking and Ann de, 1958; Majid, 1997; Goh et al., 2020). According to Dahlbom et al. (2001), artifacts are materially objectified ancient man-made objects that are crafted for sustaining livelihood and are being/have been employed as techniques to comprehend origin, and survival, migration, and anatomic changes. Based on the morphology and carvings of the artifacts, the source of their producers, either *Homo sapiens* or *Neanderthal*, can be accessed precisely (Dahlbom et al., 2001). The *Homo sapiens*, also known as hominids of the Middle Stone Age of Africa, are modern humans thought to have emerged around 200,000 ka years ago in the African continent and then, expanded over southeast Asia (Nitecki and Nitecki, 1994; Stringer and Mckie, 1996; Clark and Willermet, 1997). Whereas, the *Neanderthals* were archaic men who lived between 250,00,00−40,000 ka in Europe and Western Asia (Schwalbe, 1906; Klaatsch, 1909; Hoffecker, 2009). As per Johnson (2001), the *H. sapiens* and *Neanderthals* have been differentiated based on anatomical characteristics; however, artifacts produced by both human races reveal substantial similarities, that is, (1) comparable stone tool features with less bone, antler, and ivory indicators and (2) artifacts of both hominids were employed to hunt down less dangerous animals. Several researchers studied artifacts from the Pleistocene to Holocene periods at the Kota Tampan site in Lenggong valley area, Perak State, Malaysia, and its environs to correlate the industries produced either by *H. sapiens* or Neanderthals, concluding that they were produced by the modern humans (Saidin, 2012; Goh and Saidin, 2018; Goh et al., 2020).

Studies of faunal remains preserved in alluvial sediments and caves have been carried out for a long time to deduce the diversity and timeframe of animal existence in the Peninsula (Andrews, 1905; Sartono, 1973; Ibrahim et al., 2012; Tshen, 2013; Muhammad et al., 2021). Andrews (1905) first reported the faunal remains from the Salak area near Kuala Lumpur, and later on, many researchers have explored remnants of animals from differennt areas (Peacock and Dunn, 1968; Sartono, 1973; Ibrahim et al., 2012; Tshen, 2013; Muhammad et al., 2021). Based on the examinations of vertebrate remains, Sartono (1973) proposed that the animal population likely arrived in Malaysia from the adjacent continents and countries during the Pleistocene period and stayed until they migrated to other parts of Southeast Asia.

The light gray, fine-grained pyroclastic material acts as a horizon marker in the Quaternary deposits, is widely dispersed in alluvial deposits of various river basins, and is relatively less documented for recreating paleoclimate of the region. Based on the field studies, geochemistry and geochronology, it has been identified as the remains of the YTT eruption (Screvinor, 1930; Stauffer, 1973; Storey et al., 2012; Gatti et al., 2013; Pearce et al., 2019; Singh et al., 2021b). The YTT eruption occurred at the Toba Caldera, Sumatra, Indonesia, around 75 ka BP from a 100 km long and 30 km wide caldera and produced about 2800 km^3 dense rock equivalent (DRE) (Ninkovich et al., 1978; Chesner et al., 1991; Mason et al., 2004). Before the YTT eruption, the same caldera witnessed three more volcanic events as Haranggaol Dacite Tuff (HDT, 1.2 Ma) (Nishimura et al., 1977), Oldest Toba Tuff (OTT, 804 ka) (Chesner et al., 1991), and Middle Toba Tuff (MTT, 540 ka) (Diehl et al., 1987). Among these eruptions, the YTT is most debated among the scientific community because of its high magnitude and impact on climate and global dispersion (Ninkovich et al., 1978; Rose and Chesner, 1987). The literature review vividly demonstrates that the investigations on artifacts, fossil remains, and volcanic ash have been carried out in the past to better understand human history, paleoecology, and paleoclimate in the Peninsula; however, aerially and no significant information has been consolidated. Thus, a systematic review is required to assemble material in a single attempt. Therefore, the present attempt is being framed to review the previous studies made on these three aforementioned aspects to produce substantial knowledge regarding human existence, paleoecology, and paleoclimate in the country during the Quaternary Period. It will also offer information on animal diversity, vegetation scenes, and the impact of the YTT ash on the environment.

2. History of human occupation in the Peninsula

Investigations on human occupation in Malaysia have been going on for a long time to discover prehistoric hominids' traces and their origins. Fig. 5.1 reveals the locations of human sites in different states and districts of the country. Earl (1863) discovered evidence of human occupation in the river terraces of the Perak river and Wray (1905), in the caves of Perak and Kedah States. Collings (1938) carried out the first archeological surveys at the Kota Tampan locality, located on the western fringe of the Perak river basin in the Lenggong Valley, and recognized as the country's Paleolithic site

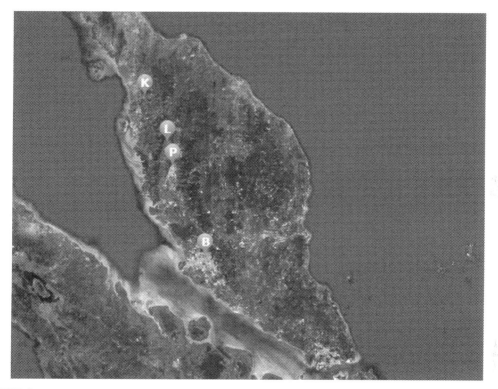

FIGURE 5.1

Map revealing the preservation of artifacts and vertebrate remains at various sites in different states of Malaysia. *B*, Batu caves; *K*, Kedah; *L*, Lenggong; *P*, Perak.

(Sieveking, 1954). The chronology of the tools at the aforesaid site is a matter of concern among the scientific community as Collings (1938) dated the same using comparative typology, a technique extensively used in the 90s, and regarded them as Lower Paleolithic industry akin to the Patijan culture of Java. Using stratigraphic linkage proposed by Walker and Sieveking (1962) assigned the age of the same industry as the Late Early or Early Middle Pleistocene. From these attempts onward, Majid and Tjia (1988) excavated the area and found volcanic ash preservation between and over the incomplete artifacts entrapped in the load mix of boulders and gravels in the alluvial deposits, correlating the same with the YTT eruption of Toba Caldera, Indonesia, based on the petrographic and chemical signatures. They claimed that the presence of ash between and above the artifacts suggests contemporaneity in the event of volcanic eruption and the last occupation of humans in the area as indicated by the presence of incomplete tools that reflect the abandonment of the site, most likely due to the impact of the eruption on the mass population. Goh et al. (2020) investigated the same site and found large stone tool assemblages dating back 70 ka, theorizing that these stone tools are the products of Anatomically Modern Humans who were technologically advanced enough to produce tools with advanced carvings and arts (Goh et al., 2020). The same authors reinvestigated 69 stones in a technological context and noticed that assemblages are predominantly made up of small flakes, bipolar artifacts, and a small number of core stones with freehand and bipolar percussions being the two most common stone flake technologies (Goh et al., 2020). Goh et al. (2020) identified multiplatform reduction, single-platform, and bidirectional-platform reduction technologies, concluding that the multiplatform reduction technology was primarily employed by hominids to produce artifact industry. Based on this study, they also stated that the stone factory in the Kota Tampan area including entire Peninsular Malaysia is attributable to products of Anatomically Modern Humans that unequivocally resembled 40,000 year-old artifacts reported from the Bukit Bunuh site, a few hundred meters away from the Kota Tampan area (Saidin, 2006). Apart from the Kota Tampan locality, other tools bearing localities in the Lenggong valley area have also been reported as Bukit Jawa (Majid, 1997), Kampung Temelong (Saidin, 1997a), Lawin (Saidin, 1997b), and Bukit Bunu (Saidin, 2005) with stone tools anvils, cores, hammerstones, pebble tools (chopper, handaxe, etc), flake tools, and debitages (Saidin, 2005).

3. Origin, migration, and settlement of hominids in Malaysia

Based on the regional, intercontinental, environmental, and volcanic temporal markers, several alternative models have predicted that modern humans evolved in the Levant, Malawi, and South Africa localities in the African continent (Aiello, 1993; Lahr, 1996; Harpending et al., 1993; Sharry et al., 1994; Rogers and Jorde, 1995; Ambrose, 1998, 2019). Regional continuity (multiregional evolution) (Wolpoff, 1989) and hybridization (Africo-European *sapiens* and mixing) (Bräuer, 1984) models have supplied considerable knowledge of human history based on fossil data interpretations (Ambrose, 1998). According to Wolpoff (1989), the regional continuity model posits that the *Homo erectus* population was dispersed from its native Africa to the rest of the old world 100,000 years ago and eventually, evolved into modern humans as *Homo sapiens*. They further went on to say that regional gene flow precluded the emergence of new species, however, allowed all local populations to create a group of modern humans (Ambrose, 1998). Several researchers have attributed dates to the *H. erectus* fossils found in Java, China, and Georgia, implying that hominids dispersed around 180,000 years ago (Swisher et al., 1994; Huang et al., 1998; Gabunia and Veuka, 1995). According to Bräuer (1984), the hybridization model aids in understanding the significant role of African populations in the evolutions of modern humans through hybridization techniques as well as distinguishing them in Europe and western Asia. Nevertheless, these models are not much in the scientific domain than the Weak Garden of Eden Model proposed by Harpending et al. (1993) based on evidence from fossils, archeology, paleogeography, paleoclimatology, and genetic structures of dead humans. The model emphasizes the origin, differentiation, and dispersal of modern humans in Africa and other parts of the world as well as their bottleneck and subsequent population growth during the Late Pleistocene period. The growth in hominid masses is hypothesized based on the Upper Paleolithic or Later Stone Age industries, a new technology found in a different part of the globe (Ambrose, 1998). Apart from the preceding model, Harpending et al. (1993) proposed another model "strong Garden of Eden" alleging the origin and dispersal of modern humans across the world as well as the evolution of *H. sapiens* in a sub-population of *Homo erectus*. The recent archeological excavations in southeast Asia including Peninsular Malaysia have revealed that around 100,000 years ago, the population of Africa began to migrate to other continents from Africa via a southern route connecting Asia, Australia, and Arabia (Amrbose, 1998, 2019; Oppenheimer, 2003, 2009; Goh et al., 2020). The dates of human's arrival in different parts of the world are disputed and contradictory, but, it is posited that the hominids first settled in the African continent and then spread to other regions (Oppenheimer, 2003, 2009). Oppenheimer (2009) drew a diagram revealing the estimated period of contemporary hominids' inhabitation in various continents and Peninsulas (Fig. 5.2). It demonstrates that hominids were grown in the African continent around 150,000 ka years ago, traveled to Egypt and Israel around 1,20,0000 ka ago, and died about 90,000 years ago. It also suggests that a group of humans moved from the southern Arabian Peninsula to India ca. 85,000 years ago, however, according to Oppenheimer (2003), the ancestors of this group were all non-Africans. The sketch again reveals modern humans first arrived in Europe roughly 40,000–50,000 years ago. Ca. 40,000 years ago, hominids began to migrate north from Pakistan to the Indus River and Central Asia, and at the same time, they started to move from the East Asia coast to the west via following the silk road. According to Oppenheimer (2009), the human population in Central Asia trekked to Europe and Beringia between 20,000 and 30,000 years ago, then from Timor to Australia around 65,000 years ago. The population crossed the Bering land bridge connecting Siberia and Alaska approximately 22,000–25,000 years ago then went to America and Chile around 15–19,000 and 12,500 years ago (Oppenheimer, 2009). It implies that the ages of human occupation in the different continents are provisional and vary, however, demonstrates constant population mobility incontinent after continent. The ages reported for the artifacts at the Kota Tampan

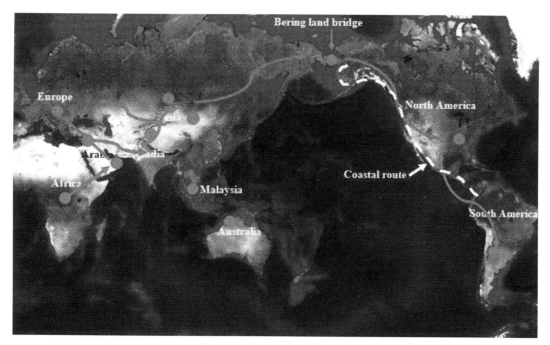

FIGURE 5.2

Map demonstrating dispersal of the humans in different parts of the old world from Africa to South America via following both land and coastal routes.

Modified and updated from Oppenheimer (2009).

site are between 40,000 and 70,000 years old, indicating that this site was occupied by modern humans. However, the authenticity of data and period of human occupation in the country raise questions as hominids reached Australia around 65,000 years ago through Malaysia, implying that the age for the human occupation in the former should be ideally older than in Australia. Thus, we suggest reasons behind the discrepancies in the ages could be (1) use of unreliable techniques and (2) wrong typological correlations of the artifacts.

4. Evolution and migration of quaternary faunas and their detailed studies

The Malaysian Peninsula, restricted within the constraints of Sundaland, forms a tropical East Asia with Borneo, Sumatra, Java, and Bali islands and experiences mixed hot and wet climates that are conducive to the growth of evergreen rainforests (Corbet and Hill, 1992; Hock, 2007; Ratnayeke et al., 2018). Because of its unique geographical position in the equatorial region and diverse ecology, it manifests great interest in studying the Quaternary faunas that are poorly compiled to comprehend the biogeographic evolution and origins of animals in the country (Sartono, 1973). Based on the volume of paleontological and geological research on the Pleistocene vertebrates, Sartono (1973) opined that the animals in Peninsular Malaysia and its adjacent regions evolved within the Asian continent and migrated to other countries via following the archipelago route. He/she further suggested that during the Upper Pleistocene period, these animals would have likely followed two distinct routes to disperse in other old regions, that is, (1) a western route connecting the Sundaland area and (2) an eastern route through Taiwan and the Philippines linking Sundaland and Sulawesi area, indicating that both the routes were connected with the Sundaland. The studies on the vertebrate remains of the Quaternary period have been made exceptionally from the alluvial deposits and caves to reconstruct palaeontology, paleoclimate, and paleoecology of the region (Ingham and Bradford, 1960; Hooijer, 1962).

4.1 Animal remains from the alluvial deposits

The first study of the vertebrate remains of Malaysia was carried out under colonial rule by Andrews (1905); however, several subsequent researchers attempted similar investigations (Roe, 1953; Savage, 1937; Ingham and Bradford, 1960; Jones, 1966; Peacock and Dunn, 1968; Tshen et al., 2013) and encountered difficulties in detailing due to dense rainforests. Notwithstanding dense vegetation, Andrews (1905) reported isolated teeth of *Elephus namadicus* from a tin-mine site excavated in the alluvium of Perak river exposed in the Salak area, Kuala Kangsar; however, Peacock and Dunn (1968) questioned his identification of animal remains due to lack of profundity in qualifying the criteria to include the same in the *Elephas* class. Roe (1953) discovered the third upper molar of *Elephas maximus*, an Indian elephant species, in a 6 m thick tin-bearing deposit of Batu Valley, exposed near the Batu Caves, Selangor area; however, no morphological details of the same were provided. In subsequent years, Jones et al. (1966) reported vertebrate bones in the form of teeth from fluvial deposits uncovered in the Bukit Kaki area of Perlis State and identified them as upper molar of *Elephas maximus*. Peacock and Dunn (1968) found skeletons of the same in the Lukut area of Negeri Sembilan state and noted that there can be a possibility of exploring more vertebrate remains; however, disrupted alluvium exposures and the existence of dense vegetation in the region limit further description

(Fig. 5.1). Based on the morphological correlations, Tshen et al. (2013) reviewed the *Elephas* skeletons reported from fluvial environments and proposed that the ancient elephants of the Malaysian Peninsula could be direct descendants of Southeast Asian species.

4.2 Animal remains from the cave deposits

Falconer and Cautley (1846), reporting *Elephas namadicus,* were the first to investigate animal bones, preserved in the several limestone caves primarily in the Lenggong valley and Tumban areas of Perak State, Batu area of Petaling Jaya near Kuala Lumpur, and other parts of the country (Fig. 5.2). Following that, Andrews (1905) examined the bones of Perak caves and found remnants of *Rhinoceros, Hexaprotodon,* and *Duboisia,* all of which were Middle Pleistocene in age (Cranbrook and Piper, 2013). Hooijer (1962) retrieved large faunal remnants of *antelope* and *Hippopotamus* from the Tambun area of Perak State, implying the existence of grasslands and swampy bodies in the region. Davison (1994) documented remains of tiger, wild dog, bats, primates, civet, bamboo rat, mouse deer, porcupine, tapir, seladang barking deer, and sambar deer in the form of bones and teeth from the Gua Gunung Runtuh and Gua Kelawar areas, suggesting that these remains were likely parts of the animals used for food by ancient humans. Muhammad and Yeap (2000) conducted extensive research in the Gua Badak cave situated in the Lenggong valley and reported petrified teeth and bones of mammals, *Bos Gaurusm Rus unicolor, Muntiacus muntjak,* and *Paradoxurus.* According to Ibrahim et al. (2012), the Malaysian Nature Society 1992 recovered large preservation of animal bones preserved in the cave's ceiling of the Gua Naga Mas, Gunung, and Lanno areas of Ipoh district of Perak State, though, witnessed a lack of attention (Ibrahim et al., 2012). Nevertheless, Tjia (2000) sought to identify the same remains and concluded that these bones belonged to a small tiger, bear, or wild dog. Muhammad et al. (2020) recently reported a cheek tooth of *Stegodon,* an extinct genus of Proboscidea, from the embedded infillings of a cave situated in the Gopeng area of Perak State, concluding that the evolution, migration, and existence of animals in this species in the Peninsula has yet to be attempted.

Apart from the caves described above, other limestone caves have also been studied for the same aspect. The Batu cave, a tourist site in Petaling Jaya, has been researched to interpret the prehistory of the region. In this regard, several investigations have been made in this area to better understand the ecology and occupancy of living media (Ridley, 1899; Ibrahim et al., 2012; Sahak et al., 2019; Muhammad et al., 2020). Ridley (1899) reported remnants of large wild animals viz., tiger, bear, wild ox, pig, muntjac, deer, and elephant from the same cave. As per Ibrahim et al. (2012), the recent exploration is not able to be traced the preservation of vertebrate fauna in the cave due to the deforestation and developmental work in and around the aforementioned site. Nevertheless, Ibrahim et al. (2012) attempted to investigate the same cave and found numerous mammalian fossils and identified them as *Sus scrofa* (common wild pig), *Sus barbatus* (bearded pig), Ursidae-gen. et sp. indet (bear), *Macaca* sp. (macaque), nonhuman hominoid (primates), *Panthera tigris* (tiger), *Capricornis sumatraensis* (southern serow), *Muntiacus muntjak* (red muntjac/barking deer), *Rusa unicolor* (sambar deer), *Tapirus indicus* (Asian tapir), gen. et sp. Indet. (rhinoceros), and bat teeth. They further went on to say that the skeletons are mainly of carnivores, omnivores, and herbivores animal communities that indicate variations in the vegetational environment. Ibrahim et al. (2013) discovered Pleistocene *orangutan* (Pongo sp.) fossil remains from the same area, claiming that they would have crossed the biogeographic divide between mainland Southeast Asia and the Sunda sub-region around 500 ka to enter Malaysia, implying the existence of an evergreen, tall rainforest environment. In the course of the study, Sahak et al. (2019) carried out a comprehensive study and reported 39 teeth and jaw fragments of Murinae fossils from the breccia deposits of the Late Pleistocene age. They further identified the same as *Chiropodomys gliroides*, *Leopoldamys sabanus*, L. minutus, *Maxomys whiteheadi*, *M. rajah*, and *Rattus rattu* and asserted that their presence indicates the existence of a low vegetated environment.

5. Occurrence and source of volcanic ash

The light-gray colored, powered, semi-consolidated to consolidated, silt-sized grained pyroclastic material is widely distributed in the Quaternary strata of numerous Malaysian river basins and lakes with thicknesses ranging from 1 to 4 m (Screvenor, 1930; Stauffer, 1973; Stauffer and Batchelor, 1978; Nishimura and Stauffer, 1981; Debaveye et al., 1986; Gatti et al., 2013; Singh et al., 2021b). The same material has been examined for various interpretations including deposition (Gatti et al., 2013; Singh et al., 2021b), geochemistry (Westgate et al., 1998, 2013; Pearce et al., 2014, 2019) and geochronology (Storey et al., 2012) to interpret paleoenvironment, source area, and age. Based on its physical appearance, preservational conditions, and sedimentological attributes, it has been classified as primary and reworked/secondary tephra units with varying thicknesses (Singh et al., 2021b). Gatti et al. (2013) excavated ash columns with 04 m thickness uncovered at the Kampung Bukit Sapi and Kampung Luat localities lying along the Perak river basin and noticed two distinct units with varying hues, thicknesses, and physical characteristics, dividing them into reworked columns. According to them, the reworked unit is mostly yellowish-brown in hue, hard to compact, comparatively coarse-grained and characterized by abundant low-energy generated sedimentary structures, that is, laminations, cross-beddings and soft sediment deformational features. Based on the particle size and magnetic susceptibility studies, Gatti et al. (2013) posited that the reworked tephra was transported from other locations and deposited in the low energy environment, primarily a swamp and lake (Gatti et al., 2013). Singh et al. (2021b) recently carried out extensive field investigations of several ash-bearing sites exposed in the Padang Terap basin and Lenggong valley area of the Perak river and observed deposition of the same aforementioned units in a complete thick ash column, however, also noticing a new type of horizon within the same as compact reworked tephra. According to them, the compact reworked tephra, occupying the top of the column and at places, alternated primary and reworked, exhibits a yellowish-brown color, very hard to compact and laminated nature. Singh et al. (2021b) asserted that the field and laboratory observations are in good accordance with Gatti et al. (2013), nevertheless, suggested that primary tephra was likely deposited straightly from the atmosphere, whereas the reworked was likely transported little distance from the other locations in the low current energy set-up. In the case of compact reworked tephra, it would have the part of the reworked unit, however, turned hard to compact during a dry climate in the region for a brief length of time.

Several geochemical attempts have been made in the past to fix the composition of the ash to correlate it with the source area of Toba Caldera (Ninkovich et al., 1978; Jasin, 1987; Debaveye et al., 1986; Westgate et al., 1998; Gatti et al., 2013; Pearce et al., 2019). Ninkovich et al. (1978) collected the samples from the Kota Tampan area for geochemical fingerprinting and reported the silicic nature of ash based on the high content of silica and alkalis. Subsequently, they established a link between it and the YTT, claiming that the tephra distributed throughout Southeast Asia came from the same source. Jasin (1987) explored a thick and elongated bed revealed along the road-cut at the Kampung Temong

Hilir locality lying along the Perak river basin to determine its source using microscopic and geochemical characteristics. He proposed that the pyroclastic material consists principally of transparent and bubble ball-shaped glass shards and is geochemically rhyolite in nature, making it comparable to the YTT. After subsequent years, more studies on the same aspect have been constructed for its intracontinental correlations (Westgate et al., 1998; Gatti et al., 2013; Pearce et al., 2019). Westgate et al. (1998) collected ash samples from both proximal areas of Toba caldera and distal sites in the Indian Ocean, Malaysia, and river basins of India and carried out geochemical fingerprinting of the glass shards. Based on the elemental compositions, they concluded that the distal tephra shows unequivocal geochemical comparability with the proximal deposits. Similar geochemical fingerprinting as of Westgate et al. (1998), Gatti et al. (2013) and Pearce et al. (2019) for the ash deposits of Lenggong valley and Padang Terap river, respectively. However, Pearce et al. (2019) asserted that the minor fluctuations in elemental compositions are most likely due to magma ejection from different strata of the chamber during the eruption, although, the ash in Southeast Asia belongs to the YTT. More investigations have been done to determine the source area of the same material exposed in other river basins of Malaysia (Debaveye et al., 1986). Debaveye et al. (1986) thoroughly investigated the tephra uncovered in the Padang Terap river basin to interpret its source area and examined the geochemistry of the same. Based on the elemental concentrations, they concluded that the ash is rhyolitic and comparable with the YTT.

The geochronology of tephra exposed at several localities has been conducted to assess the age of the same for correlation with the source area, however, inconsistent even though using various techniques. Stauffer (1980) analyzed fission tracks of zircon, concentrated from the ash of Kota Tampan, Ampang lake of Kuala Lumpur, and ash layer of Toba Caldera, and assigned their ages as 31,000, 30,000, and 30,000, respectively. Similarly, Nishimura and Stauffer (1981) collected the ash from interstratified fluvio-lacustrine sediments of the Serdang area, Kuala Lumpur, and isolated 10 zircons for fission-track analyses. They reported ages ranging from 23,000 to 30,000 on each grain, however, considered an average of 30,000. Based on these numbers, the same authors suggested that the ash in Peninsular Malaysia is from a single eruption of Toba Caldera, however, did not specify the name of the eruption as the Toba Caldera had produced 03 more volcanic eruptions instead of YTT, that is, HDT, OTT, and MTT. Storey et al. (2012) sampled ash from the Kota Tampan using the bore drilling technique to isolate biotite for dating and intimidated the same with the $^{40}Ar/^{39}Ar$ method. They assigned the age to be 73.5 ka, which is consistent with the well-established date of the YTT. The study suggests that the ash in Peninsular Malaysia, technologically, geochemically, and geochronologically, is related to the YTT eruption.

6. Controversies over the impact of YTT ash on the ecosystem

Based on the premise of high magnitude and widespread fine-grained ejecta of the YTT eruption, earlier researchers have projected that the said eruption would have wreaked havoc on the vegetation and the lives of humans as well as their migration from Africa to other continents including Asia (Ambrose, 1998; Williams et al., 2009). The ejecta was principally composed of powdered pyroclastic debris and sulfur gases, in which, the former had formed volcanic clouds in the atmosphere, while the latter interacted with the prevalent H in the stratosphere and produced H_2SO_4. The volcanic clouds prevented sun radiation from reaching the surface, causing cooling for about 06 years, while, H_2SO_4

was dropped in the form of acidic rains (Rampino and Self, 1992). Zielinski et al. (1996) reported YTT sulfate maxima from the ice cores of Greenland and suggested that the winter was persistent globally for less than a decade after the eruption. As per Ambrose (1998), these two calamities, that is, cooling and acidic rains would have obliterated the living media including anatomically modern humans and their migration from continent to continent. Several investigations have revealed preservations of human industries in the pre- andpost-YTT tephra successions of stratified alluvial deposits in Southeast Asia including Malaysia (Goh et al., 2020) and India (Petraglia et al., 2007, 2012; Haslam et al., 2011; Clarkson et al., 2020). It suggests that humans existed in Southeast Asia and Malaysia long before and after the YTT eruption (Goh et al., 2020). Based on the paleoclimatological and volcanological data, Jones (2010) claimed that the impact of the eruption was partial. Similarly, based on the Earth Simulation Model (ESM), Timmereck et al. (2012) documented that the regions of Southeast Asia overpowered by humans were likely experiencing a chronically warm environment before the eruption and further stated that the living media, however, could have easily endured the temperature. They went on to explain that the shift in the vegetation, mostly from forest to savannah, would have posed a challenge for humans but, only for a short time (Timmereck et al., 2012). Based on the isotopic studies of calcium carbonates nodules from the ash-bearing successions, Williams et al. (2009) and Singh et al. (2021) have already established the alteration in the vegetation in Southeast Asia after the eruption. Williams et al. (2009) anticipated a transformation of C_3 forests to C_4 grasslands to woodlands across the Indian Subcontinent using isotopic study and also asserted that the YTT eruption was succeeded by cooling and protracted warm and dry environment based on the pollen study from the marine cores of Bengal basin. The same authors concluded that the Toba eruption caused long-term cooling and a warm, dry environment in Southeast Asia, and they questioned the hypotheses of minimal impact of the eruption. Based on the isotopic study of carbonate nodules assorted from the ash uncovered in the Padang Terap river basin, Singh et al. (2021) recently reported the existence of C4 vegetation in the Malaysian Peninsula after the said eruption.

7. Conclusions

To reconstruct the prehistory of humans, animals, and the source area of the pyroclastic material in the country, a complete review study on the aspects of artifacts, vertebrate remains, and volcanic ash preserved in the alluvial sediments and limestone caves of Peninsular Malaysia have been compiled. The main conclusions of the study are as follows;

1. The artifacts, preserved mainly in the alluvial deposits and caves, are produced by anatomically modern humans who were cave dwellers and hunting gathers and genetically similar to the African hominids.
2. The study suggests that the population arrived and stayed in the country much before and after the YTT eruption as evidenced by two explorations, (1) the OSL dating of artifact-bearing sediments revealing the existence of humans before 70 ka and (2) typological correlation suggesting humans survival in the Peninsula between 40,000 and 11,000.
3. Vertebrate remains principally of *Elephas, Hippopotamus*, antilope, *Rhinoceros, Hexaprotodon, Duboisia, Tiger, Wild dog, Dear, Paradoxurus,* and *Stegodon* are preserved mainly in the alluvial sediments and caves, indicating the presence of carnivores, omnivores, and herbivores animal communities as well as suggest the existence of grasslands, swampy environment, and wet environment.

4. Volcanic ash is exposed in three distinct columns with varying thicknesses and hues as primary, reworked and compact reworked tephra and has been deposited in a tranquil aquatic environment. Geochemically and geochronological, it is closely related to the ejecta of the YTT eruption of Toba Caldera, Indonesia.
5. Evidence of both C_3 and C_4 vegetations, specifically in the YTT ash-bearing river basins of Southeast Asia including Malaysia, have been established based on the isotopic studies of calcrete nodules. Hence, it suggests the transformation of forests into grasslands during the said eruption.

Acknowledgments

We authors are thankful to Dr. Ng Tham Fatt, Head, Department of Geology, Faculty of Science, the University of Malaya for providing infrastructure facilities during this study. We also acknowledge Dr. Ros Fatihah Muhammad, Senior Lecture of the same aforementioned department for her moral support.

References

Aiello, L.C., 1993. The fossil evidence for modern human origins in Africa: a revised view. Am. Anthropol. 95, 73–96.

Ambrose, S.H., 1998. Late Pleistocene human population, bottlenecks, volcanic winter, and differentiation of modern humans. J. Hum. Evol. 34, 623–651.

Ambrose, S.H., 2019. Chronological calibration of Late Pleistocene modern human dispersals, climate change and archaeology with geochemical isochrons. In: Reyes-Centeno, H., Sahle, Y., Bentz, C. (Eds.), Modern Human Origins and Dispersal. DFG Center for Advanced Studies. University of Tübingen, Germany. Kerns Verlag, Tübingen, pp. 171–213.

Andrews, C.W., 1905. Fossil tooth of *Elephas namadicus* from Perak. J. Feder. Malay States Mus. 1, 81–82.

Brauer, G., 1984. The afro-European sapiens hypothesis and hominid evolution in East Asia during the late middle and upper Pleistocene. In: edu, Andrews, P., Franzen, J.L. (Eds.), The Early Evolution of Man with Special Emphasis on Southeast Asia and Africa, vol. 69. Courier Forschungsinstitut Senckenberg, pp. 145–165.

Chesner, C.A., Rose, W.I., Deino, A., Drake, R., Westgate, J.A., 1991. Eruptive history of the Earth's largest Quaternary caldera (Toba, Indonesia) is clarified. Geology 19 (3), 200–203.

Clark, G.A., Willermet, C.M., 1997. Conceptual Issues in Modern Human Origins Research. Aldine de Gruyter, New York.

Clarkson, C., Harris, C., Li, B., Neudorf, C., Roberts, M., Richard, G., Lane, C., Norman, K., Pal, J.N., Jones, S., Shipton, C., Koshy, J., Gupta, M.C., Mishra, D.P., Dubey, A.K., Boivin, N., Petraglia, M., 2020. Human occupation of northern India spans the Toba super-eruption ~74,000 years ago. Nature Commun. 11, 961.

Collings, H.D., 1938. Pleistocene site in the Malay Peninsula. Nature 142, 575–576.

Corbert, G.B., Hill, J.E., 1992. The Mammals of the Indomalayan Region: A Systematic Review. Oxford University Press, Oxford, p. 488.

Cranbrook, E.O., Piper, P.J., 2013. Palaeontology to policy: the Quaternary history of Southeast Asian tapirs (Tapiridae) in relation to large mammal species turnover, with a proposal for conservation of Malayan tapir by reintroduction to Borneo. Integr. Zool. 8, 95–120.

Dahlbom, B., Beckman, S., Nilsson, G.B., 2001. The nature of artefacts. In: Artefacts and Artificial Science. Almqvist and Wiksell International, Stockholm.

Davison, G.W.H., 1994. Some remarks on the vertebrate remain from the excavation at Gua Gunung Runtuh, Perak. In: Zuraina, M. (Ed.), The Excavation of Gua Gunung Runtuh and the Discovery of the Perak Man in Malaysia. Department of Museums and Antiquity Malaysia, Kuala Lumpur, pp. 141–148.

Debaveye, J., Dapper, M.D., Paepe, P.D., Gijbels, R., 1986. Quaternary tephra deposits in the Padang Terap district, Kedah, peninsular Malaysia. Geosea V Proceed. I, Geol. Soc. Malaysia, Bulle. 19, 533–549.

Diehl, J.F., Onstott, T.C., Chesner, C.A., Knight, M.D., 1987. No short reversals of the Brunhes age were recorded in the Toba Tuffs, north Sumatra, Indonesia. Geophys. Res. Lett. 14 (7), 753–756. https://doi.org/10.1029/gl014i007p00753.

Earl, G.W., 1863. On the shell-mounds of province Wellesley in the Malay Peninsula. Trans. Ethnol. Soc. 2, 119–129.

Falconer, H., Cautley, P.T., 1846. Fauna Antiqua Sivalensis, Is the Fossil Zoology of the Siwalik Hills, in the north of India. Smith, Elder, London (plates only).

Gabunia, L., Vekua, A., 1995. A plio-pleistocene hominid from Dmanisi, East Georgia, Caucasus. Nature 373, 509–512.

Gatti, E., Saidin, M., Talib, K., Rashidi, N., Gibbard, P., Oppenheimer, C., 2013. Depositional processes of reworked tephra from the late Pleistocene Youngest Toba Tuff deposits in the Lenggong Valley, Malaysia. Quat. Res. 79 (2), 228–241. https://doi.org/10.1016/j.yqres.2012.11.006.

Goh, H.M., Bakry, N., Saidin, M., Shahidan, S.I., Curnoe, D., Saw, C.Y., bin Ariffin, Z., Kiew, Y.M., 2020. The palaeolithic stone assemblage of Kota Tampan, west Malaysia. Antiquity 94.

Goh, H.M., Saidin, H.M., 2018. The prehistoric human presence in Gua Kajang: ancient lifeways in the Malay Peninsula. J. Malays. Branch R. Asiatic Soc. 91, 1–18.

Harpending, H.C., Sherry, S.T., Rogers, A.R., Stoneking, M., 1993. The genetic structure of ancient human populations. Curr. Anthropol. 34, 483–496.

Haslam, M., Harris, C., Clarkson, C., Pal, J.N., Shipton, C., Crowthera, A., Koshy, J., Bora, J., Ditchfield, P., Rame, H.P., Price, K., Dubey, A.K., Petraglia, M., 2011. Dhaba: an initial report on an acheulean, middle palaeolithic and microlithic locality in the middle son valley, North-Central India. Quat. Int. 258 (2012), 191–199.

Hock, S.S., 2007. The Population of Peninsular Malaysia. Institute of Southeast Asian Studies. ISEAS—Yusof Ishak Institute.

Hoffecker, J.F., 2009. The spread of modern humans in Europe. Proc. Natl. Acad. Sci. USA 106 (38), 16040–16045.

Hooijer, D.A., 1962. Report upon a collection of Pleistocene mammals from tin-bearing deposits in a limestone cave near Ipoh, Kinta Valley, Perak. Feder. Malaya Memoir 7, 1–5.

Huang, W., Fu, Y.-X., Benny, H.-J., Chang, X.G., Jorde, L.B., Li, W.-H., 1998. Sequence variation in ZFX introns in human populations. Mol. Biol. Evol. 15, 138–142.

Ibrahim, Y.K., Peng, L.C., Cranbrook, E., Tshen, L.T., 2012. Preliminary report on vertebrate fossils from Cistern and swamp caves at Batu caves near Kuala Lumpur. Bull. Geol. Soc. Malays. 58, 1–8. December 2012.

Ibrahim, Y.K., Tshen, L.T., Westaway, K.E., Cranbrook, E.O., Humphrey, L.T., Muhammad, R.F., Zhao, J., Peng, L.C., 2013. First discovery of Pleistocene orangutan (Pongo sp.) fossils in Peninsular Malaysia: biogeographic and paleoenvironmental implications. J. Hum. Evol. 65 (6), 770–797.

Ingham, F.T., Bradford, E.F., 1960. The geology and mineral resources of the Kinta valley, Perak. District Memoir No. 9. Geological Survey, Federation of Malaya 347.

Jasin, B., Jantan, A., Abdullah, I., Said, U., 1987. Volcanic ash beds at Kampung Temong, Kuala Kangsar, Perak. Warta Geologi 13 (5), 205–2011.

Johanson, D., 2001. Origins of Modern Humans: Multiregional or Out of Africa? Action Bioscience. American Institute of Biological Sciences, Washington, DC.

Jones, C.R., Gobbett, D., Kobayashi, J.T., 1966. Summary of the fossil record in Malaya and Singapore 1900–1965. In: Jones, C.R., Gobbett, D.J., Kobayashi, T. (Eds.), Geology and Palaeontology of Southeast Asia, vol. 2. University of Tokyo Press (Japan), pp. 309–355.

Jones, S.C., 2010. Palaeoenvironment response to the ~74 Toba ash-fall in the Jurreru and Middle Son valleys in southern and north-central India. Quat. Res. 73, 336–350.

Klaatsch, H., 1909. Evidence that *Homo Mousteriensis Hauseri* belongs to the Neanderthal type. L'Homme Préhistorique. 7, 10–16.

Lahr, M., 1996. The Evolution of Modern Human Diversity. Cambridge University Press, Cambridge.

Majid, Z., 1997. The discovery of Bukit Jawa, Gelok, a middle-late palaeolithic site in Perak, Malaysia. J. Malays. Branch R. Asiatic Soc. 70 (273), 49–52, 2.

Majid, Z., Tjia, H.D., 1988. Kota Tampan, Perak: the geological and archaeological evidence for a late Pleistocene site. J. Malays. Branch R. Asiatic Soc. 61, 123–134.

Mason, B.G., Pyle, D.M., Oppenheimer, C., 2004. The size and frequency of the largest explosive eruptions on Earth. Bull. Volcanol. 66 (8), 735–748.

Muhammad, R.F., Tshe, L.T., Ibrahim, N., Rajak, M.A.A., Razif, F.M., Kem, Z., Tat, C.B., Yuen, T.C., Lee, N.Y., Chiew, C.J., Fai, C.J., Rahman, M.N.A., Hussein, S.R., 2020. First discovery of stegodon (Proboscidea) in Malaysia. Warta Geologi 46 (3), 196–1998.

Muhammad, R.F., Lim, T.T., Ibrahim, N., Abdul Razak, M.A., Mohd Razif, F., Kem, Z., Ching, B.T., 2021. The first discovery of Stegodon (Proboscidea) in Malaysia. Warta Geologi 46 (3), 196–198.

Muhammad, R., Yeap, E.B., 2000. Proposed conservation of Badak Cave C, Lenggong as vertebrate fossil site extraordinary. In: Proceedings of the Geological Society of Malaysia Annual Geological Conference, pp. 189–196.

Ninkovich, D., Shackleton, N.J., Abdel-Monem, A.A., Obradovich, J.D., Izett, G., 1978. K–Ar age of the late Pleistocene eruption of Toba, North Sumatra. Nature 276 (5688), 574–577.

Nishimura, S., Stauffer, P.H., 1981. Fission-track of zircons from the Serdang tephra, Peninsular Malaysia. Warta Geolog 7 (2), 39–41.

Nishimura, S., Yojoyama, T., Wirasantosa, S., Dharma, A., 1977. Danau Toba-the outline of lake Toba, North Sumatra, Indonesia. Paleolimnol. Lake Biwa. Japan Pleistocene 5, 313–332.

Nitecki, M.H., Nitecki, D.V., 1994. Origins of Anatomically Modern Humans. Plenum Press, New York.

Oppenheimer, C., 2003. Climatic, environmental and human consequences of the largest known historic eruption: Tambora volcano (Indonesia) 1815. Prog. Phys. Geogr. Earth Environ. 27 (2), 230–259.

Oppenheimer, S., 2009. The great arc of dispersal of modern humans: Africa to Australia. Quat. Int. 202, 2–13.

Peacock, B.A.V., Dunn, F.L., 1968. Recent archaeological discoveries in Malaysia, 1967 (West Malaysia). J. Malays. Branch R. Asiat. Soc. 41, 171–179.

Pearce, N.J.G., Westgate, J.A., Gatti, E., Pattan, J.N., Parthiban, G., Achyuthan, H., 2014. Individual glass shard trace element analyses confirm that all known Toba tephra reported from India is from the c. 75-ka Youngest Toba eruption. J. Quat. Sci. 29 (8), 729–734.

Pearce, N.J., Westgate, J.A., Gualda, G.A., Gatti, E., Muhammad, R.F., 2019. Tephra glass chemistry provides storage and discharges details of five magma reservoirs that fed the 75 ka Youngest Toba Tuff eruption, northern Sumatra. J. Quat. Sci. 35 (1–2), 256–271. https://doi.org/10.1002/jqs.3149.

Petraglia, M., Korisettar, R., Boivin, N., Clarkson, C., Ditchfield, P., Jones, S., Koshy, J., Lahr, M.M., Oppenheimer, C., Pyle, D., Roberts, R., Schwenninger, J.-L., Arnold, L., White, K., 2007. Middle Paleolithic assemblages from the Indian subcontinent before and after the Toba super-eruption. Science 317 (5834), 114–116.

Petraglia, M., Ditchfield, P., Jones, S., Korisetter, R., Pal, J.N., 2012. The Toba volcanic super-eruption, environment change, and hominin occupation history in India over the last 140,000 years. Quat. Int. 258, 119–134.

Rampino, M.R., Self, S., 1992. Volcanic winter and accelerated glaciation following the Toba supereruption. Nature 359, 50−52.

Ratnayake, S., van Manen, F.T., Clements, G.R., Kulaimi, N.A., Sharp, S.P., 2018. Carnivore hotspots in Peninsular Malaysia and their landscape attributes. PLoS One 13 (4), 0194217. https://doi.org/10.1371/journal.pone.0194217.

Ridley, H.N., 1899. Report: appendix to "caves of the Malay peninsula-report of the Committee consisting of sir W. H. Flower (Chairman), Mr H. N. Ridley (secretary), Dr R. Hanitsch, Mr Clement Reid, and Mr A Russell Wallace, appointed to explore certain caves in the Malay Peninsula, and to collect their living and extinct fauna. In: Report of the Sixty-Eighth Meeting of the British Association for the Advancement of Science Held at Bristol in September 1898. John Murray, London, pp. 572−582.

Roe, F.W., 1953. The geology and resources of the neighbourhood of Kuala selangor and Rasa, selangor, federation of Malaya, with an account of the geology of the Batu arang coalfield. Geol. Surv. Depart. Fed. Malaya, Memoir 7, 163.

Rogers, A.R., Jorde, L.B., 1995. Genetic evidence on modern human origins. Hum. Biol. 67, 1−36.

Rose, W.I., Chesner, C.A., 1987. Dispersal of ash in the great Toba eruption, 75 ka. Geology 15 (10), 913.

Sahak, I.H., Lim, T.T., Muhammad, R.F., Nur, S.I.A.T., Mohammad, A.A.A., 2019. First systematic study of late Pleistocene rat fossils from Batu caves: new record of extinct species and Biogeography implications. Sains Malays. 48 (12), 2613−2622.

Saidin, M., 1997a. Palaeolithic Culture in Malaysia: The Contribution of Sites Lawin, Perak and Tingkayu, Sabah, Unpublished PhD Thesis. Universiti Sains Malaysia, Penang.

Saidin, M., 1997b. Comparative study between palaeolithic sites of Kampung Temelong and Kota Tampan, and its contribution to the Southeast Asia late Pleistocene culture. Malaysia Mus. J. 34.

Saidin, M., 2005. Cave formations of sites with skeletal remains in Lenggong, Perak. In: Zuraina, M. (Ed.), The Perak Man and Other Prehistoric Skeletons of Malaysia. Penerbit University Sains Malaysia, Pulau Pinang, p. 119e126.

Saidin, M., 2006. Bukit Bunuh, lenggong, Malaysia: new evidence of late pleistoceneculture in Malaysia and Southeast Asia. In: Bacus, A.E., Glover, I.C., Pigott, V.C. (Eds.), Uncovering Southeast Asia's Past: 60−64. NUS Press, Singapore.

Saidin, M., 2012. From Stone Age to Early Civilization in Malaysia. Universiti Sains Malaysia Press, Penang.

Sartono, S., 1973. On Pleistocene migration routes of vertebrate fauna in Southeast Asia. Bull. Geol. Soc. Malays. 6, 273−286.

Savage, H.E.F., 1937. The Geology of the neighbourhood of Sungei Siput, Perak, Federated Malay States, with an Account of the Mineral Deposits. Memoir No. 1 (New Series). Geological Survey Department, Federated Malay States, p. 46.

Schwalbe, G., 1906. In: Nägele, E. (Ed.), Studies on the Pre-History of Man (in German). Stuttgart. https://doi.org/10.5962/bhl.title.61918 hdl:2027/uc1.b4298459.

Scrivenor, J.B., 1930. A recent rhyolite-ash with sponge-spicules and Diatoms in Malaya. Geol. Mag. 67 (9), 385−393. https://doi.org/10.1017/s0016756800100445.

Sherry, S.T., Rogers, A.R., Harpending, H.C., Soodyall, H., Jenkins, T., Stoneking, M., 1994. Mismatch distributions of mtDNA reveal recent human population expansions. Hum. Biol. 66, 761−775.

Sieveking, Ann de G., 1958. The Palaeolitic industry of Kota Tampan, Perak, Northwestern Malaya. Asian Perspective 2.

Sieveking, G. de G., 1954. Gua Cha and the Malayan stone age. Malaysian History J. 1, 111−125.

Singh, A., Muhammad, R.F.B.H., Taib, N.I., Jha, D., Srivastava, A.K., 2021a. Surface texture, mineralogy and stable isotope studies of nodular calcretes preserved in the YTT ash of Padang Terap river basin and lenggong valley, peninsular Malaysia: implications in its origin and paleoclimatic reconstruction. Rhizosphere 19, 100380.

Singh, A., Muhammad, R.F.B.H., Taib, N.I., Srivastava, A.K., 2021b. Field data, grain size and petrography of Youngest Toba Tuff (YTT ca. 75 ka) ash from Padang Terap river basin and Lenggong Valley, Peninsular Malaysia: emphasis on transportation history and depositional configuration. J. Earth Syst. Sci. (in press).

Stauffer, P.H., 1973. The late Pleistocene age indicated volcanic ash in west Malaysia. Geol. Soc. Malays. Newsl. 40, 1−4.

Stauffer, P.H., Batchelor, B., 1978. Quaternary tephra and associated sediments at Serdang, selangor. Warta Geologi 4 (1), 7−10.

Stauffer, P.H., Nishimura, S., Batchelor, B.C., 1980. Volcanic ash in Malaya from a catastrophic eruption of Toba, Sumatra, 30,000 years ago. In: Nishimura, S (Edu): Physical Geology of Indonesian Island Arc. Kyoto, Japan, University of Kyoto, pp. 156−164.

Storey, M., Roberts, R.G., Saidin, M., 2012. Astronomically calibrated $^{40}Ar/^{39}Ar$ for the Toba supereruption and global synchronization of late Quaternary records. Proc. Natl. Acad. Sci. U. S. A 109 (46), 18684−18685.

Stringer, C., McKie, R., 1996. African Exodus: The Origins of Modern Humanity. Henry Holt, New York.

Swisher, C.C., Curtis, G.H., Jacob, T., Getty, A.G., Suprijo, A., Widiasmoro, 1994. Age of the earliest known hominids in Java, Indonesia. Science 263, 1118−1121.

Taha, A.J., 1987. Archaeology in peninsular Malaysia: past, present and future. J. Southeast Asian Stud. V. XVII (2), 1−2.

Timmereck, C., Graf, H.-F., Zanchettin, D., Hagemann, S., Kleinen, T., Krüger, K., 2012. Climate response to the Toba super-eruption: regional changes. Quat. Int. 258, 30−44.

Tjia, H.D., 2000. A large vertebrate fossil at Naga Mas cave, Bukit Lanno, Kinta valley. Warisan Geologi Malaysia (Geol. Herit. Malaysia) 3, 209−218.

Tshen, L.T., 2013. The Quaternary Elephas fossils from Peninsular Malaysia: historical overview and new material. Raffles Bull. Zool. 29, 139−153.

Tweedie, M.W.F., 1953. The stone age in Malaysia. J. Malayan Branch Royal Asiatic Soc. (JMBRAS) 26 (2).

Walker, D., Sieveking, A. G. de, 1962. The palaeolithic industry of Kota Tampan, Perak, Malaysia. Proc. Prehist. Soc. 6, 103−139.

Westgate, J.A., Pearce, N.J.G., Perkins, W.T., Preece, S.J., Chesner, C.A., Muhammad, R.F., 2013. Tephrochronology of the Toba tuffs: four primary glass populations define the 75-ka Youngest Toba Tuff, northern Sumatra, Indonesia. Journal of Quaternary Science. https://doi.org/10.1002/jqs.2672.

Westgate, J.A., Shane, P.A.R., Pearce, N.J.G., Perkins, W.T., Korisetter, R., Chesner, C.A., Williams, M.A.J., Acharyya, S.K., 1998. All Toba tephra occurrences across peninsular India belong to the 75,000 yr B.P. eruption. Quaternary Research 50, 107−112.

Williams, M.A.J., Ambrose, S.H., Van der Kaars, S., Ruehlemann, C., Chattopadhyaya, U., Pal, J., Chauhan, P.R., 2009. Environmental impact of the 73 ka Toba super-eruption in South Asia. Palaeogeogr. Palaeoclimatol. Palaeoecol. 284, 295−314.

Wolpoff, M.H., 1989. Multiregional evolution: the fossil alternative to Eden. In: Mellars, P., Stringer, C.B. (Eds.), The Human Revolution: Behavioural and Biological Perspectives on the Origin of Modern Humans. Edinburgh University Press, Edinburgh, pp. 62−108.

Wray, L., 1905. Further Notes on cave dwellers of Perak. J. Feder. Malay States Museum 1 (1), 13−15.

Zielinski, G.A., Mayewski, P.A., Meeker, L.D., Whitlow, S., Twickler, M.S., Taylor, K., 1996. Potential atmospheric impact of the Toba Mega-Eruption ∼71,000 years ago. Geophys. Res. Lett. 23 (8), 837−840. https://doi.org/10.1029/96gl00706.

CHAPTER 6

Bhutan and the geography of climate change

Jeetendra Prakash Aryal[1], Medha Bisht[2] and Dil Bahadur Rahut[3]

[1]*International Center for Biosaline Agriculture, Dubai, United Arab Emirates;* [2]*South Asian University, New Delhi, India;* [3]*Asian Development Bank Institute, Tokyo, Japan*

1. Introduction

A landlocked country, Bhutan's geographical location is unique. While its geo-strategic location makes it central to strategic discourses in Eastern Himalayas, its biophysical profile with contrasting physiographic zones makes it central for studying the impact of climate change. The altitude along Bhutan's Southern Belt, bordering India, ranges between 200 and 2000 m. While its central mid-Himalayas comprises of fast-flowing river systems with altitudes extending between 2000 and 4000 m, its Greater Himalayas in the North are characterized by snow-capped mountains and alpine meadows higher than 4000 m. Thus, while captivatingly, the distance between North and South is 175 km, while its altitude rises swiftly from 150 m in the South to higher than 7000 m in the North (Bisht, 2013). Notably, of the total land area of 38,294 sq km, 71% is covered with forest, 7% is covered with snow, only 3% of the land is cultivable and meadows, 4% is barren or under pastures (12th 5-year plan). Though Bhutan is a net carbon-negative, it is highly vulnerable to climate change due to its rugged and mountainous terrain. As climate change escalates further due to an increase in global greenhouse gas (GHG) emissions, it threatens the livelihood, food, and nutritional security of poor and vulnerable families worldwide.

Therefore, this chapter explores how climate change adversely affects the Kingdom of Bhutan despite being only a net carbon-negative county. This paper confines to assess the impact of climate change on three specific areas—hydropower, agriculture, and glaciers. The rationale for choosing these sectors is their salient role in Bhutan's economic, ecological, and social profile. For instance, hydropower has been central to Bhutan's economic growth, and its exports are concentrated heavily in the hydropower sector. While it was expected that by 2020, Bhutan would produce around 10,000 MW of power, the current electricity generation stands at 2326 MW (Dorji, 2018). Bhutan has been able to achieve 100% access to electricity. However, the sector has been posing distinct ecological and financial challenges, particularly with regard to Bhutan's external debt situation.

Agriculture provides livelihood to a significant portion of the population and contributes to the food and nutritional security of the nation, besides contributing to the GDP, agro-based industrial development, poverty alleviation, and foreign exchange income through export. However, agriculture depends on weather conditions; hence, it is the most vulnerable sector. Because agriculture contributes to a quarter of GHG emissions and is the most vulnerable sector, investment in GHG mitigation and adaptation to climate change is crucial for ensuring food security and reducing vulnerability to poverty in developing countries.

Bhutan has several glacial lakes across the country. With the increase in temperature, the glacial lake is expected to melt down, resulting in the Glacial Lake outburst flood (GLOF). Such GLOF is anticipated to wash away land, property, and life downstream in Bhutan and India. Although the Government of Bhutan has initiated several programs to reduce the lake burst and its impact by shifting the village, the effort is inadequate given the number of glacial lakes and the anticipated impact. Support from international and bilateral agencies to manage the GLOF is crucial.

2. The discourse on climate change in Bhutan

While the 11th Five-year Plan prioritized climate change mitigation strategies, the 12 Five Year Plan (2018−23) has integrated climate change as the sixth National Key Result Area (NKRA). In fact, all 17 NKRA are synchronized with the Sustainable Development Goals (SDGs) and related target indicators. Significantly, Bhutan adopted its climate change strategy in 2020 with a vision of, "a prosperous, resilient and carbon neutral Bhutan where the pursuit of gross national happiness for the present and future generations is secure under a changing climate" (National Environmental Commission, 2020). Focusing on four broad policy objectives—mitigation, adaptation, support in the form of finance, technology, capacity building, research and awareness, and effective engagement among stakeholders through effective coordination and collaboration, pursuing a carbon neutral development and adopting adaptive capacity and resilience, has been identified as some of the primary goals in Bhutan's Climate Change Policy. Indeed as the second Nationally Determined Contribution Report notes that, "the national institutions for coordination of climate change actions across key agencies and stakeholder groups have been revitalised with the Climate Change Coordination Committee (C4) from the erstwhile Multisectoral Technical Committee on Climate Change. In addition, a climate change 'one-stop platform' is being set up to help coordinate multi-stakeholder dialogue to develop and implement climate-related work in Bhutan, with the aim to improve coordination between the different climate-sensitive sectors, enhance knowledge management and improve reporting and monitoring of all climate actions in Bhutan" (NDC, p. 4). Bhutan also endorsed the Kigali Amendments and the Montreal Protocol on Ozone Depleting Substances in 2019 and activated the system for licensing the import and export of HFCs (Royal Government of Bhutan, 2021).

While much has been written, and research has been proliferating regarding climate change issues in Bhutan, there is little analysis on how and in what ways this discourse could shape and inform the emerging "geographies of climate change," particularly in Bhutan and Eastern Himalayas in general (Chaturvedi and Timothy, 2015). However, before one addresses some of these issues, it would be appropriate to introduce the sector-specific impact of climate change in Bhutan.

3. Hydropower in Bhutan in context to climate change

Bhutan has five major and minor river systems, which hinge on the glacial and snow melt and rainfalls. Five major river systems are Drangme Chhu (Manas River), Wang Chhu (Raidak River), Punatsang Chhu (Sankosh River), Ammo Chhu (Torsa River), and Mangde Chhu (Tongsa River) (Dorji, 2016;

cited in Bisht, 2019). India—Bhutan joint hydro projects are mostly built on these five major river systems. Jaldakha in Samtse, Aiechhu in Sarpang, Nyera Amari in Samdrup Jongkhar, and Jomori in Samdrup Jongkhar/Trashigang and are the minor river systems. According to some estimates, 74 dams have been envisaged across these river basins (Dorji, 2018).

Hydropower and agriculture have been identified as the climate-sensitive sectors in Bhutan. The power sector has been the primary driver of growth through direct export earnings (Indian rupee earning) and has a consequential impact on the construction sector and power-intensive industries. While the agriculture sector contributed significantly in the initial stages of economic development, the share of GDP to agriculture has declined sharply over the last few decades. The share of agriculture, forestry, and fishing to GDP declined from 39.8% in 1980% to 24.3% in 2000% and 19.2% in 2020. Despite this, the agriculture sector remains the primary employer, with nearly 57% of the labor force currently dependent on it (Gross National Happiness Commission, 2019).

With increasing reliance on hydropower in recent years, hydropower has been termed as a "strategic resource," identified as important for ensuring energy security (Ministry of Economic Affairs, 2021). Bhutan's journey toward hydropower development can be neatly divided into two phases—pre-2007 and post-2009. In the first phase, Bhutan witnessed the completion of its three signpost projects—Chuka, Kurichu, and Tala—a marker of adopting an integrated approach to developing bilateral hydropower cooperation. Marked by a unique model of partnership and collaboration, the cooperation had positive consequences for the improved socio-economic indicators in Bhutan. The discourse on climate change and the environmental consequences of the dam was not an issue of concern, given that most of the hydel dams were run of river projects. The second phase started post 2009, resulting from an MoU signed between India and Bhutan in Dec 2009, with a commitment by India to purchase 10,000 MW from Bhutan by 2020 (Bisht, 2011). While the political, economic, and social aspects of hydropower cooperation between the two countries have been dwelled upon by most scholars (Saklani, 2019; Ranjan, 2018; Bisht, 2013), analysis on the impact of climate change on hydropower has not received enough attention.

Bhutan is among the countries with the high per capita water availability, around 94,500 cubic meters, the highest in South Asia (UNDP, 2020). Although 99.5% of Bhutanese have access to improved water sources, only 63% have 24-hour access to drinking water (UNDP, 2020). Predictions by climate scientists and government and international agencies note that temperatures in Northern Bhutan will continue to rise. Rising temperatures due to the accelerating rate of snowmelt and temperature increases can peak river discharge sooner than later, resulting in flooding and GLOFs, adversely affecting families and infrastructure near rivers (World Bank, 2021). Further, these climatic changes could affect water availability and river flow, affecting Bhutan's hydropower generation and economy. Due to global warming and climatic changes, prolonged drought, erratic rainfall, and changes in rainfall patterns, agricultural productivity declines, thereby threatening food and nutritional security.

Precipitation on the southern foothills is anticipated to increase during the monsoon leading to flash floods. Lugye Tsho burst in 1994, resulting in flooding and a catastrophic impact, was the first documented evidence of the climate-associated glacial lake outburst (Lotay, 2015; Acharya, 2015 cited in Bisht, 2019), and it followed in 2000, 2004, 2009, 2013, and 2016. Likely, upsurges in the days with heavy rainfall might intensify flooding risk, erosion, and rates of river discharge (World Bank, 2021, p. 17). This will not only lead to the disruption of average flow, affecting the generation of hydropower but also increase the uncertainty in the magnitude of flow, impacting hydropower generation. There is also an increased likelihood of damaging the infrastructure, transmission pipelines, and power distribution where the increase in sediments also adversely affects the performance of hydropower generation (National Environment Commission, Thimphu cited in Asian Water Cycle initiative; Dorji, 2018).

4. Climate change, global warming, and glaciers in Bhutan

With nearly 3% of the total surface area under glaciers and being located in the eastern Himalayan region, Bhutan has been one of the major hotspots of climate change and associated global warming (Bajracharya et al., 2014). Climate variability and climate extremes have thus increased substantially in the country (Gurung et al., 2017; Mahagaonkar et al., 2017). These glaciers are responsible for feeding the majority of the rivers in the country and major sources of freshwater downstream (Hagg et al., 2021; Molden et al., 2022; Rinzin et al., 2021; Rupper et al., 2012). Though the exact number of glaciers in the Bhutan Himalayas varies across the studies, it contains about 1583 glaciers (Nagai et al., 2017). In recent years, temperature variability—both at spatial and temporal levels—has been experienced across the country (Hoy et al., 2016). Such variability in temperature and precipitation patterns, primarily caused by global climate change, has increased the intensity of glacial melting in the Himalayan region of Bhutan (Williams et al., 2018). Glaciers and melt-water largely affect the hydrological system in Bhutan, which is not only crucial to ensure the perennial flow of water in the region but also to sustain the entire economy owing to its significant contribution to agriculture, drinking water supplies, and the generation of hydropower (Mahagaonkar et al., 2017; Mahanta et al., 2018; Tariq et al., 2021). Increasing glacier melt will severely affect Bhutanese people's livelihood as it affects the overall environment, ecosystems, and the economy.

Global warming caused by climate change affects the glacial retreat and the formation of glacial lakes in the Himalayan region. Thus, global warming has already greatly impacted glacier stability by intensifying its melting (Ding et al., 2019). More interestingly, some glaciers will continue to shrink irrespective of future temperature rise, thereby increasing the number of surging glaciers (Ding et al., 2019). On the global scale, it is predicted that almost one-fourth of the total mountain glacier mass may disappear due to global warming by the year 2050, and this loss may reach 50% by the end of 21st century if the present trend of global warming continues (IPCC, 1996; Johannes, 1994). Shrinkage and intensified retreat of glaciers, expansion of existing glacial lakes, and formation of new ones increase the likelihood of glacial lake outburst floods (GLOFs) (Gurung et al., 2017; Hagg et al., 2021). The steady retreat of glaciers over a period of the last 3 decades has led to the expansion of glacial lakes by almost four to eight-folds in Bhutan (Mahagaonkar et al., 2017). With an extensive study of the topographic map and satellite images from 66 glaciers in Bhutan, Karma et al. (2003) showed that the rate of glaciers retreat is slightly more than 8% in the region. According to Karma et al. (2003), the glacier area declined to 134.94 sq km in 1993 from 146.87 sq km in 1963. Their study found that several small glaciers, which used to exist during 1963, have completely disappeared in 1993. Data on the status of glaciers in Bhutan for a period of 3 decades, starting from 1980, showed that, though the number of glaciers are increasing, the area under it is decreasing (Bajracharya et al., 2014).

Rapid accumulation of glacial-melt water in these lakes often causes GLOF that discharges a huge amount of water and debris, leading to catastrophic effects in downstream (Bajracharya et al., 2014; Veettil et al., 2016; Veh et al., 2020). Hagg et al. (2021), in their study in the Mo Chu River Basin, showed that GLOF could destroy many public infrastructures as well as agricultural land. They reported that GLOF could severely affect the hydropower system located as far as 120 km downstream. The loss and damage due to GLOF downstream can be extensive as it destroys multiple infrastructures, including roads, bridges, villages, agricultural lands, and human lives. In Bhutan, where most of the rivers are glacier-fed, GLOFs can have huge consequences, primarily because more than 70% of human settlements are along the river valleys (Wangchuk et al., 2019). The GLOF also has socio-economic impacts in some cases as when agricultural lands are swept away and filled with debris, resulting in severe negative impacts on human livelihoods. Rapidly melting glaciers and other climate extremes are estimated to cause nearly 6% loss per year in the gross domestic product of Bhutan (Asian Development Bank, 2014).

5. Agriculture in the context of climate change

Bhutan has mountainous topography, the arable land is relatively small in comparision to the total land area in Bhutan (World Bank, 2021). Despite limited arable land area, almost 57% of the Bhutanese people secure their living from agriculture (Chhogyel and Kumar, 2018), and in 2019, the contribution of GDP to the agriculture sector is about 15.8% of the GDP. Though the Royal Government of Bhutan initiated several policies and programs to promote commercialization in the agriculture sector, subsistence farming is predominant, primarily due to small land size and low farm mechanization levels (Ministry of Agriculture and Forests, 2020).

Majority of the farmers in Bhutan are smallholders. Resource constraints smallholder farmers in the country tend to disengage from farming as it is challenging to generate sufficient income to sustain their livelihood, and climate change has been increasing the risk and uncertainties in the agriculture sector. The major cereals that farmers commonly grow are paddy, maize, wheat, barley, millet, and buckwheat. Dairy cattle and poultry have gradually become more common livestock (Ministry of Agriculture and Forests, 2020). Nevertheless, crop and livestock types vary largely across the agro-ecological zones of the country (Katwal et al., 2015). Almost 53% of the total areas of Bhutan belong to mountainous terrain with alpine and temperate zone, which makes the country more vulnerable to climatic variability (Katwal and Bazile, 2020).

Climate change is pervasive and affects all but poor countries with limited resilient capacity, and the agriculture sector is most vulnerable to climate risk as it is dependent on climatic conditions. Like other countries in South Asia, agriculture in Bhutan suffers from the adverse impacts of climate extremes and associated rainfall and temperature variability (Aryal et al., 2020). As Bhutanese agriculture entirely depends on the monsoon rain, even a small variation in the onset and retreat of the monsoon considerably affects agricultural production (National Environment Commission, 2006). The agriculture sector in Bhutan suffers from multiple climatic risks, including unusual outbreaks of crop pests and diseases, unpredictable rainfalls, hailstorms, and floods (Chhogyel et al., 2020; Chhogyel and Kumar, 2018).

Farming in Bhutan is constrained by its mountainous terrain in addition to the less fertile soil and frequent weather variability and extreme events. In mountain communities, landslides and soil erosion are common during the rainy season. The probability of climate-related disasters may increase with changes in precipitation patterns and rise in temperature across the globe. These changes will lead to a higher risk of GLOF in many parts of Bhutan, causing a loss of fertile land and damage to other property (Veh et al., 2020). With increasing climate change, mountain communities are more likely to face several climate risks, and hence, it is imperative to have a deeper understanding of its impacts and also more cautious adaptation planning. The agriculture sectors in Bhutan, particularly in the upper highland, are vulnerable to climate risk, which is likely to impose great challenges to livelihood and food security in the country (Chhogyel and Kumar, 2018). With climate change, Bhutan's agriculture sector would be affected through water stress, soil erosion, and pest and disease outbreak. Under RCP 8.5 and to some extent under 4.5, it was found that potatoes will not be suitable in lower altitudes of southern Bhutan (less than 1000 mts), while there are opportunities for expansion in mid altitudes (1000−3000 mts) (Parker et al., 2017).

Given the importance of agriculture to the national economy, adaptation to climate change is crucial to secure food and nutritional security. Improving adaptive capacity requires more investment in climate-smart agriculture, crop insurance provision, modern technology, including digital agriculture and better extension support. Holistic farmers-centric rural development policy would not only help in improving the yield, income, and livelihood of the rural farm household in Bhutan but also contribute to food and nutritional security, reverse migration, provide inputs to the agro-based industries for value addition, gross domestic product, and foreign exchange earning through the export of the raw and value-added agricultural product.

6. Geographies of climate change

These emerging challenges draw our attention to distinct geographies of climate change, which are scalar in nature. They also offer insights on potential areas of cooperation, which can act as vantage points to revisit cooperation in the Eastern Himalayas in new ways. This is important because borders and passes in South Asia have animated the geopolitics in the region in distinct ways, where borders in South Asia have often been negotiated through shifts in population and politics (Walcott, 2010). Both colonial and postcolonial discourse regarding control over passes and headwaters has transformed the South Asian space into a strategic construct, thereby strengthening the discourse on macro geopolitical changes. What needs to be reckoned with here are the nonlinear, micro impact of climate change on South Asian countries, particularly, and what implications this could have for potential geopolitical changes. Thus, while questions related to the impact such changes can have on lower-riparian countries such as India (particularly Assam) in the South and Bhutan's relationship with China in the North are important, the emergent impact of climate change also calls for deepening ones analysis and calls for a more holistic, alternative approach to address consequences of climate change.

For instance, flash floods from Eastern Bhutan have been a point of tension between the people inhabiting the bordering district. However, they have also been entry points for initiating alternative ways of cooperation across borders, which in many ways enhance the resilience of these communities (Bisht, 2019). However, such cases of cooperation in South Asia are sparse and need to be strengthened, which can only happen when we move our gaze from the macro analysis to a more microanalysis. Some of these concerns are real as frequently the Assam government has highlighted the issue of flash floods, soil erosion, and inundation due to dams being built in Bhutan, Nagaland, and Arunachal Pradesh.

A less-researched area is how the climate change discourse can impact Bhutan−China relations. It needs to be highlighted that Bhutan is fed by rivers from China (Iwata et al., 2004). While the northern basin has only 59 glaciers, the area occupied by the glaciers in this basin is the largest (Mool, 2001). Thus, the northern glacial boundary between Bhutan and China (Gasa district) has the potential to create further misperceptions and confusion on border demarcation. The Northern part of the Gasa district is already claimed by China and is under negotiations between both countries. Climate change has already been affecting the Sino–Bhutan dispute. Until a few years ago, regular human activity or movement along the China–Bhutan border was not possible due to the extreme weather conditions. Bisht notes that, "National Assembly members in Bhutan have been raising the issue of easy entrance for Chinese nationals into Bhutan as a result of melting glaciers along the border. In 2005, National Assembly members pointed out that more passes had opened up in Laya and Lunana, located in the extreme north and sharing borders with Tibet, thus enabling easier access for Tibetans entering Bhutan" (Bisht, 2011).

Research from satellite images suggests that many large and dangerous glacial lakes exist in the headwater areas of Kuri Chhu in China. Kuri Chhu flows from China to Bhutan, and if a GLOF from these lakes occurs, areas in the lower reaches of Kuri Chhu in Bhutan could witness heavy losses. Notably, in the recent past, flash floods from Kurichhu Dam have been a constant factor of tensions between Bhutan and India. While the Bhutanese Government is concerned about this danger, it cannot take preventative action because the lakes are located in China (Iwata et al., 2004). These facts hint toward a more cooperative rather than confrontational approach to issues emerging out of the discourse on climate change. It also demands that we shift our gaze from macrogeopolitical analysis to discourses that focus on the microsector-specific aspects, which focus on preparedness and mitigation. For Bhutan, which is a landlocked country, resilience to climate change can only happen when cooperative rather than confrontational geopolitics replaces the region. This demands that we consciously work toward transforming the geo-climatic narrative, as geographies of climate science connect the upstream with the downstream countries in a rather unique way.

References

Aryal, J.P., Sapkota, T.B., Khurana, R., Khatri-Chhetri, A., Rahut, D.B., Jat, M.L., 2020. Climate change and agriculture in South Asia: adaptation options in smallholder production systems. Environ. Dev. Sustain. 22, 5045–5075. https://doi.org/10.1007/s10668-019-00414-4.

Asian Development Bank, 2014. Melting Glaciers, Climate Extremes Threaten Bhutan's Future - Report [WWW Document]. URL https://www.adb.org/news/melting-glaciers-climate-extremes-threaten-bhutans-future-report. (Accessed 23 December 2021).

Bajracharya, S.R., Maharjan, S.B., Shrestha, F., 2014. The status and decadal change of glaciers in Bhutan from the 1980s to 2010 based on satellite data. Ann. Glaciol. 55, 159–166. https://doi.org/10.3189/2014AoG66A125.

Bisht, M., 2011. India-Bhutan Power Cooperation: Between Policy Overtures and Local Debates. IDSA Issue Brief, New Delhi, pp. 1–13. October 7, 2011.

Bisht Medha, 2013. Bhutan and climate change: identifying strategic implications. Contemp. S. Asia 21 (4), 398–412.

Bisht Medha, 2019. From the edges of borders: reflections on water diplomacy in South Asia. Water Pol. 21 (6), 1123–1138. https://doi.org/10.2166/wp.2019.124, 1 December 2019.

Chaturvedi, Timothy, 2015. Climate Terror: A Critical Geopolitics of Climate Change. Palgrave Macmillan, New York.

Chhogyel, N., Kumar, L., 2018. Climate change and potential impacts on agriculture in Bhutan: a discussion of pertinent issues. Agric. Food Secur. 7 (79). https://doi.org/10.1186/s40066-018-0229-6.

Chhogyel, N., Kumar, L., Bajgai, Y., Hasan, M.K., 2020. Perception of farmers on climate change and its impacts on agriculture across various altitudinal zones of Bhutan Himalayas. Int. J. Environ. Sci. Technol. 17, 3607−3620. https://doi.org/10.1007/s13762-020-02662-8.

Climate Risk Country Profile: Bhutan, 2021. The World Bank Group and the Asian Development Bank.

Ding, Y., Zhang, S., Zhao, L., Li, Z., Kang, S., 2019. Global warming weakening the inherent stability of glaciers and permafrost. Sci. Bull. 64, 245−253. https://doi.org/10.1016/j.scib.2018.12.028.

Dorji, T., 2018. Long Term Climate Impact on Hydropower in Bhutan. ICIMOD. https://www.irena.org/-/media/Files/IRENA/Agency/Events/2018/Dec/Bhutan/5-Tashi-Dorji-ICIMOD−Long-Term-Climate-Impacts-on-Hydro-Power-in-Bhutan.pdf?la=en&hash=6F3169F85E00BDB261C1AED12CA5A24D03A55912.

Gross National Happiness Commission, 2019. Twelfth five year plan 2018−2023. Main Document 1.

Gurung, D.R., Khanal, N.R., Bajracharya, S.R., Tsering, K., Joshi, S., Tshering, P., Chhetri, L.K., Lotay, Y., Penjor, T., 2017. Lemthang Tsho glacial Lake outburst flood (GLOF) in Bhutan: cause and impact. Geoenviron. Dis. 4, 17. https://doi.org/10.1186/s40677-017-0080-2.

Hagg, W., Ram, S., Klaus, A., Aschauer, S., Babernits, S., Brand, D., Guggemoos, P., Pappas, T., 2021. Hazard assessment for a glacier Lake Outburst flood in the Mo Chu River Basin. Bhutan. Appl. Sci. https://doi.org/10.3390/app11209463.

Hoy, A., Katel, O., Thapa, P., Dendup, N., Matschullat, J., 2016. Climatic changes and their impact on socio-economic sectors in the Bhutan Himalayas: an implementation strategy. Reg. Environ. Change 16, 1401−1415. https://doi.org/10.1007/s10113-015-0868-0.

IPCC, 1996. Climate Change 1995: Impacts, Adaptation and Mitigtion of Climate Change, Scientific and Technical Analyses.

Iwata, S., Karma, K., Ageta, Y., Sakai, A., Narama, C., Naito, N., 2004. Glacial goemorphology in the Lunana area in the Bhutan Himalaya: moraine stages, glacial lakes, and rock glaciers. Himal. J. Sci. 2, 10.3126/hjs.v2i4.853.

Johannes, O., 1994. Quantifying global warming from the retreat of glaciers. Science 264 (80), 243−245. https://doi.org/10.1126/science.264.5156.243.

Karma, T., Ageta, Y., Naito, N., Iwata, S., Yabuki, H., 2003. Glacier distribution in the Himalayas and glacier shrinkage from 1963 to 1993 in the Bhutan Himalayas. Bull. Glaciol. Res. 20, 29−40.

Katwal, T.B., Bazile, D., 2020. First adaptation of quinoa in the Bhutanese mountain agriculture systems. PLoS One 15, e0219804.

Katwal, T.B., Dorji, S., Dorji, R., Tshering, L., Ghimiray, M., Chhetri, G.B., Dorji, T.Y., Tamang, A.M., 2015. Community perspectives on the on-farm diversity of six major cereals and climate change in Bhutan. Agric. For. https://doi.org/10.3390/agriculture5010002.

Lotay, Y., 2015. Bhutan Disaster Management. Country Report, Asian Disaster Reduction Center.

Mahagaonkar, A., Wangchuk, S., Ramanathan, A.L., Tshering, D., Mahanta, C., 2017. Glacier environment and climate change in Bhutan—an overview. J. Clim. Change 3, 1−10. https://doi.org/10.3233/JCC-170010.

Mahanta, C., Mahagaonkar, A., Choudhury, R., 2018. Climate change and hydrological perspective of Bhutan. In: Mukherjee, A. (Ed.), Groundwater of South Asia. Springer Singapore, Singapore, pp. 569−582. https://doi.org/10.1007/978-981-10-3889-1_33.

Ministry of Agriculture and Forests (2020). Agriculture Statistics, 2020. Royal Government of Bhutan.

Ministry of External Affairs, 2021. Launch of Sustainable Hydro Development Policy. April 21, at. https://www.moea.gov.bt/?p=10582.

Molden, D.J., Shrestha, A.B., Immerzeel, W.W., Maharjan, A., Rasul, G., Wester, P., Wagle, N., Pradhananga, S., Nepal, S., 2022. The great glacier and snow-dependent rivers of Asia and climate change: heading for troubled waters. In: Biswas, A.K., Tortajada, C. (Eds.), Water Security under Climate Change. Springer Singapore, Singapore, pp. 223–250. https://doi.org/10.1007/978-981-16-5493-0_12.

Mool etal, 2001. Inventory of glaciers, glacial lakes and Glacial Lake Outburst floods monitoring and early warning systems in the hindu Kush-himalayan region Bhutan. ICIMOD.

Nagai, H., Ukita, J., Narama, C., Fujita, K., Sakai, A., Tadono, T., Yamanokuchi, T., Tomiyama, N., 2017. Evaluating the scale and potential of GLOF in the Bhutan Himalayas using a satellite-based integral glacier–glacial lake inventory. Geoscience. https://doi.org/10.3390/geosciences7030077.

National Environment Commission, 2006. National Adaptation Programme of Action. National Environment Commission, Royal Government of Bhutan (Thimphu, Bhutan).

National Environmental Commission, 2020. Climate Change Policy of the Kingdom of Bhutan. Royal Government of Bhutan.

Parker, L., Nora, G., Than Thi, N., Chimi, R., Dawa, T., Dorji, W., Yadunath, B., et al., 2017. Climate Change Impacts in Bhutan: Challenges and Opportunities for the Agricultural Sector, 2017.

Ranjan, A., 2018. India-Bhutan hydropower projects: cooperation and concerns. In: ISAS Working Paper, p. 309.

Rinzin, S., Zhang, G., Wangchuk, S., 2021. Glacial Lake area change and potential outburst flood hazard assessment in the Bhutan himalaya. Front. Earth Sci. 9, 1136. https://doi.org/10.3389/feart.2021.775195.

Royal Government of Bhutan, 2021. Secondly Nationally Determined Contribution. June 5. Thimphu.

Rupper, S., Schaefer, J.M., Burgener, L.K., Koenig, L.S., Tsering, K., Cook, E.R., 2012. Sensitivity and response of Bhutanese glaciers to atmospheric warming. Geophys. Res. Lett. 39. https://doi.org/10.1029/2012GL053010.

Saklani, U., Tortajada, C., 2019. India's development cooperation in Bhutan's hydropower sector: concerns and public perceptions. Water Altern. (WaA) 12 (2), 734–759.

Tariq, M.A., Wangchuk, K., Muttil, N., 2021. A critical review of water resources and their management in Bhutan. Hydrology. https://doi.org/10.3390/hydrology8010031.

UNDP, 2020. Water and Climate Change, March 2020 at. https://www.bt.undp.org/content/bhutan/en/home/stories/water-and-climate-change.html.

Veettil, B.K., Bianchini, N., de Andrade, A.M., Bremer, U.F., Simões, J.C., de Souza Junior, E., 2016. Glacier changes and related glacial lake expansion in the Bhutan Himalaya, 1990–2010. Reg. Environ. Change 16, 1267–1278. https://doi.org/10.1007/s10113-015-0853-7.

Veh, G., Korup, O., Walz, A., 2020. Hazard from Himalayan glacier lake outburst floods. Proc. Natl. Acad. Sci. USA 117, 907–912. https://doi.org/10.1073/pnas.1914898117.

Walcott, S., 2010. Bordering the Eastern Himalaya: boundaries, passes, power contestations. Geopolitics 15, 62–81.

Wangchuk, S., Bolch, T., Zawadzki, J., 2019. Towards automated mapping and monitoring of potentially dangerous glacial lakes in Bhutan Himalaya using Sentinel-1 Synthetic Aperture Radar data. Int. J. Rem. Sens. 40, 4642–4667. https://doi.org/10.1080/01431161.2019.1569789.

Williams, P.A., Crespo, O., Abu, M., Simpson, N.P., 2018. A systematic review of how vulnerability of smallholder agricultural systems to changing climate is assessed in Africa. Environ. Res. Lett. 13, 103004. https://doi.org/10.1088/1748-9326/aae026.

Further reading

Acharya, G., 2017. Flash floods the dangerous new normal in Bhutan. Thirdpole Net. July 17. Available at. https://www.thethirdpole.net/en/2017/07/18/flash-floods-the-dangerous-new-normal-in-bhutan/. (Accessed 6 April 2019).

Katwal, T., 2013. Multiple cropping in Bhutanese agriculture: present status and opportunities. In: Regional Consultative Meeting on Popularizing Multiple Cropping Innovations as a Means to Raise Productivity and Farm Income in SAARC Countries. Peradeniya, Kandy Srilanka).

Komori, J., 2008. Recent expansions of glacial lakes in the Bhutan Himalayas. Quat. Bar Int. 184, 177–186. https://doi.org/10.1016/j.quaint.2007.09.012.

United Nations Children's Fund, 2021. The Climate Crisis is a Child Rights Crisis: Introducing the Children's Climate Risk Index. United Nations Children's Fund (UNICEF), New York.

CHAPTER 7

Study of spatio-temporal climate variability and changes in South America using climate data tools

Amba Shalishe[1], Anirudh Bhowmick[1], Subodh Kumar Chaturvedi[2] and Jai Ram Ojha[3]

[1]*Faculty of Meteorology and Hydrology, Arba Minch Water Technology Institute, Arba Minch University, Arba Minch, Ethiopia;* [2]*Institute of Hydrocarbon, Energy and Geo-Resources, ONGC Center for Advanced Studies, University of Lucknow, Lucknow, Uttar Pradesh, India;* [3]*Department of Geology, College of Natural Sciences, Arba Minch University, Arba Minch, Ethiopia*

1. Introduction

Climate change is a major global environmental hazard that has a significant impact on agricultural productivity and affects humanity in a variety of ways (Tamiru and Fekadu 2019). Both, rich and developing countries continue to face significant challenges as a result of global climate change and the associated weather extremes. Food shortages and chronic diseases caused by climate change have impacted billions of people in underdeveloped countries. Central and South America (CA/SA) have unique ecosystems, the highest biodiversity in the world, and a diverse range of eco-climatic gradients. Unfortunately, as agricultural boundaries advance as a result of rapidly increasing agricultural and cattle production, these natural riches are jeopardized (Grau and Aide, 2008).

Because South America is a less developed region of the world, the adaption of human practices to climate change will be furthermore difficult than in more developed places such as North America, Europe, or Australia. Climate changes have been seen, such as variations in precipitation and rising temperatures. More extreme weather events, species extinctions, water stress, lower rice and soybean yields, and negative consequences on beaches from increasing sea levels are all expected results.

Climate variability and catastrophic weather events have wreaked havoc on Latin America in recent years. Intense rains, flooding, and drought in the Amazon, as well as Hurricane Catarina in 2004 (not to be mistaken with Hurricane Katrina in 2005, the first known hurricane-sized tropical cyclone to occur in the South Atlantic. Temperature and precipitation patterns have shifted during the last few decades. Rainfall in southeast Brazil, Paraguay, Uruguay, and parts of Argentina has increased. Rainfall in southern Chile, southwest Argentina, and southern Peru has decreased. Land degradation (including deforestation) has occurred as a result of land-use changes, which may have influenced weather patterns. Smoke from burning trees has been shown to affect regional temperatures and rainfall in southern Amazonia, as well as harm human health through air pollution.

Climate variability can be seen across Latin America on a wide range of time scales, from intraseasonal to long term. This climate variability is generally connected with phenomena that currently have significant social and environmental effects in several sub-regions of Latin America, which could be worsened by global warming and accompanying climate change (Basso et al., 2001). Some of the investigations conducted by researchers in the region have revealed signals that can be linked to variability and/or change in climate conditions for South America, particularly for streamflow, precipitation, temperature, glacier oscillations, general circulation, and extreme occurrences. Climate change scenario studies for several sub-regions of South America were used to estimate potential future climate conditions (Basso et al., 2001).

2. Data and methodology

From 2000 to 2020, total precipitation data were available from the Global Surface Summary of the Day (GSOD), which was provided and hosted by the National Climatic Data Center (NCDC) of the National Oceanic and Atmospheric Administration (NOAA) for South America (Fig. 7.1).

The data quality was evaluated, and stations with greater than 20% missing data were removed from consideration. Daily rainfall records for each meteorological station were used to construct decadal, monthly, and seasonal statistics (Berhane et al., 2020). To improve accuracy and reliability, the majority of satellite rainfall products use data from ground-based meteorological stations. As a result, in the study location, evaluation with an independent dataset is critical to determine the satellite product that best reproduces the observed data (Bayissa et al., 2017). According to an overview of operational weather station data from Bolivia, Chile, Colombia, Ecuador, Peru, Venezuela, and Argentina, South America has a total of 14,595 stations (meteorological and hydrological) (Cndom et al., 2020).

However, most of them are blended in the majority of satellite rainfall outputs. The independent weather stations were chosen based on their relative position in various agro-climatic zones and the availability of good data during the study period. The grid values of the satellite image used to estimate rainfall are compared to ground-based observations, regardless of where the weather stations are placed inside the grid box. This means that point-to-grid comparison is used in this study, regardless of the different grid widths of each satellite product. The Climate Data Tools (CDT) and the data utilized as an initial and surface boundary condition are described in the following subsections. CDT is an R package that is free and open-source. It was developed as part of the Enhancing National Climate Services (ENACTS) project at Columbia University's Earth Institute, which is part of the International Research Institute for Climate and Society (IRI).

3. Results and discussions
3.1 Precipitation variability

In some places, precipitation patterns show marked shifting, and temperatures are increasing, and it shows that the frequency and severity of weather extreme conditions for example like heavy rains are changing. The repercussions are manifold, ranging from melting Andean glaciers to devastating floods and droughts. The Pacific and Atlantic Oceans, which flank the continent, are warming and getting more acidic, while the sea level rises. Unfortunately, as the atmosphere and oceans continue to quickly

3. Results and discussions

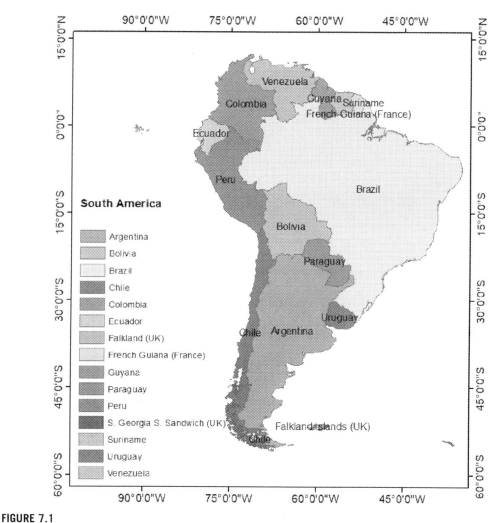

FIGURE 7.1

Map of South America.

change, the region will be hit even more. There will be a disruption in food and water supplies. Towns and cities, as well as the infrastructure that supports them, will become increasingly vulnerable. Human health and welfare, as well as natural ecosystems, will be harmed. Climate change and extreme disasters have wreaked havoc on Latin America. Between 2000 and 2013, the Intergovernmental Panel on Climatic Change reported 613 extreme climate and hydro-meteorological events. Hydro-meteorological events include hurricanes and typhoons, hailstorms, thunderstorms, blizzards, tornadoes, heavy snowfall, avalanches, floods, and coastal storm surges, including flash floods, cold spells and heatwaves, and drought. People have been displaced, there have been several fatalities, and there have been enormous economic losses as a result of this. Tropical storms from both the Atlantic and

Pacific oceans have wreaked havoc in Mexico, Central America, and the Caribbean. Aside from the destruction caused by the storms in coastal areas, their torrential rains inland have inflicted far more harm. According to the IPCC, Hurricane Mitch alone impacted 600,000 people in 1998, primarily due to floods and landslides caused by severe rains.

The snow depth in the cordillera is significantly connected with precipitation in Santiago, Chile. From the late 1800s through the mid-1970s, recorded precipitation showed a downward trend, but this pattern has since reversed. A similar pattern has been seen in the region's streamflow (Minetti and Sierra, 1989; Carril et al., 1997; Compagnucci and Vargas, 1998; Compagnucci, 2000). A negative trend in precipitation and streamflow has been seen in southern Chile and the Andean cordillera (Quintela et al., 1993). The propensity for increased winter precipitation in north-western Mexico has resulted in positive trends in river water levels. Interannual climate variability has increased as a result of more heavy winter precipitation (Magaa and Conde, 2000). On the other hand, depending on the catchment's orientation, some portions of southern Mexico and Central America see positive or negative rainfall patterns (Aparicio, 1993; IPCC, 1996; Jáuregui, 1997; IPCC, 2022). Rainfall research for Nicaragua from 1961 to 1995 revealed negative trends in the country's north and northwest. On the Caribbean coast, there was a systematic increase, while there was essentially little fluctuation along the central and Pacific coasts (MARENA, 2000).

For the period 1990–2020, rainfall trends at a regional level showing variation of 0 mm. to approximately 600 mm for the South American continent. Maximum rainfall is shown in Columbia. Parts of Peru and the Andes Mountainous area show scarcity of rainfall. Such conditions are supported by (Mesa et al., 1997) who said the rainy seasons have arrived earlier in central Colombia in recent years than they did 25 years ago. Colombian river stream-flow trends are variable, but the main river catchments, such as the Cauca and Magdalena Rivers, are on the decline. River outflows have been decreasing in recent years, which could be due to deforestation (Poveda and Mesa, 1997). The monthly distribution of rainfall (Fig. 7.2) from 1990 to 2020 shows maximum rainfall is on the western coast of Columbia. And minimum rainfall was observed in the area of leeward slopes of the Andes Mountains as well as the northeastern seashore. Fig. 7.3, shows the average distribution of minimum temperature

FIGURE 7.2

Spatial distribution of monthly precipitation from 1990 to 2020 in South America.

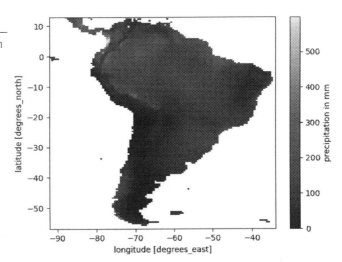

3. Results and discussions 135

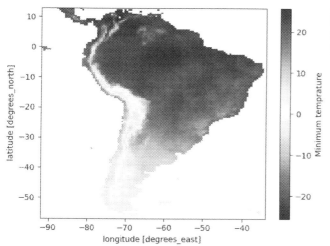

FIGURE 7.3

Spatial distribution of minimum temperature from 1990 to 2020.

in the continent. West-central part of the continent with higher topographic elevation area show at higher latitude the temperature reduces below −20°C. Fig. 7.4, shows the spatial distribution of average high temperature which shows the part of the tropics and Brazil is mostly dominated by high temperature. Also, in comparison to Climate Research Unit (CRU) in Northern Brazil, the simulations are colder throughout this season. It is worth noting that both simulations reproduce the lowest temperatures across the eastern sector of South-Eastern Brazil in both summer and winter.

Climate modeling has shown to be incredibly beneficial in developing climatic change projections and future climate scenarios under various forcing situations. General circulation models have proved their ability to realistically reproduce large-scale characteristics of observed climate; as a result, they are frequently employed to analyze the effects of increased atmospheric loading of greenhouse and

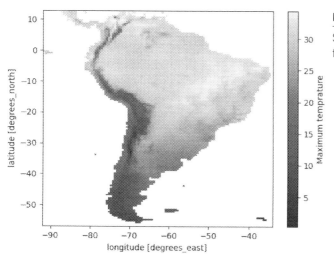

FIGURE 7.4

Spatial distribution of maximum temperature from 1990 to 2020 in South America.

other gases on the climate system. Although there are variances in how different models portray climate system dynamics, they all produce equivalent results on a global scale. However, they have a hard time reproducing regional climatic trends, and there are significant differences among models. Local effects of topography and other thermal differences commonly influence the distributions of surface variables such as temperature and rainfall in different parts of the planet, and GCMs' coarse spatial resolution cannot resolve these effects. As a result, large-scale GCM scenarios should not be directly employed for impact analyses, particularly at the regional and local levels; downscaling procedures are necessary.

The rates of mean annual temperature changes in the parts of the Latin American region for the next century are anticipated to be 0.2–2°C under the (B1) low-emissions scenario developed as part of the IPCC Special Report on Emission Scenarios. For the higher emissions case, the warming rate might vary between 2 and 6°C (A2). On a global scale, most GCMs generate similar estimates for temperature increases; nevertheless, anticipated precipitation changes remain very uncertain.

It is critical to have a projection of concurrent changes in temperature and precipitation at the regional scale for impact analyses. Various climate change scenarios for Latin America have been proposed based on GCM projections under the IS92a scenario. The majority of these regional scenarios are based on GCM simulations that have been downscaled using statistical methods. Climate change projections for Mexico indicate that the country's climate will become drier and warmer. Reduced precipitation and rising temperatures are putting a strain on certain hydrological zones of Mexico .

To estimate changes in surface temperature and precipitation, several climate change scenarios for different portions of Latin America use linear interpolation of GCM output (Carril et al., 1997; MARENA, 2000). Under the IS92a scenario for the year 2100, the results for Costa Rica suggest a small increase in precipitation for the Southeastern Caribbean region and a significant decrease—close to 25%—in the Northwestern Pacific region. As a result of El Nino and increased demand from infrastructure for tourists and irrigation, this latter region is already experiencing water shortages. The average temperature of Costa Rica is anticipated to climb by more than 3°C by 2100 under the same climate scenario, while trends in actual climate data (1957–97) show a 0.4°C increase per 10 years for the more continental Central Valley locations. This last estimate could be influenced by factors other than climate change.

Rainfall patterns have varied throughout time and space, making it difficult to identify trends. Annual rainfall has increased in the northern and central areas of South America since the 1960s, whereas the southern portions have gotten drier on average. Extreme precipitation occurrences have also become more common. In South America, annual rainfall has increased by at least 0.2–0.4% year over year. Rainfall forecasts for South America are largely speculative. Rainfall is projected to increase in general, with the greatest increase occurring during the short rainy season, which runs from October to December. Heavy rain occurrences will likely become more frequent .

3.2 Temperature variability

3.2.1 Average temperature

Fig. 7.3 depicts the expected temperature change from 2011 to 2040. The results suggest a modest increase in the north of the domain, primarily over the Amazon region, which can experience temperature increases of more than 4°C (about longitude 50W), while a slight drop is expected in the south. Other studies have found this warming across the Amazon, but the expected increase in temperature in this region is larger here.

The temperature rises throughout a greater region from 2041 to 2070, with the greatest increase occurring over the Amazon (Fig. 7.4). Warming is extended across the entire continental domain for the period 2071—2100, with the biggest increase in the Amazon region on the order of 60°C (Fig. 7.5). The strengthening of the greenhouse effect of water vapor is the primary cause of this rise in temperature minimum temperature.

The rise in downward longwave radiation at the surface, which is higher over tropical regions, demonstrates the severity of the greenhouse effect (Fig. 7.6). At the same point, the highest increase in downward longwave radiation, larger than 60 Wm^2, occurs. Despite the maximum temperature and the fact that the anthropogenic greenhouse gas trend was not taken into account in the OLAM regional climate projections, the greenhouse effect was amplified by the feedback of water vapor created by ocean warming.

3.2.1.1 Future climate change and variability prediction for 30 years

The future pattern of temperature, rainfall, and estimated evapotranspiration was analyzed under RCP4.5 and RCP8.5 scenarios for the study area. The evaluation was made in 30 years of dataset ranging from 2021 to 2050 and 2051—80 represented for analysis. The future change was calculated for each RCP scenario from the baseline (1986—2005) period for both future time frames. As a result, maximum and minimum temperatures were calculated separately for a better understanding of rainfall, as seen below. Precipitation change from January to December on the Intergovernmental Science-Policy Platform on Biodiversity and Ecosystem Services (IPBES) South America platform, compared to the whole CMIP5 ensemble from 1986 to 2005. On the left, one line per model is presented for each scenario, along with the multimodel mean; on the right, percentiles of the entire dataset: the box spans from 25% to 75%, the whiskers from 5% to 95%, and the horizontal line signifies the median (50%) seen in Fig. 7.5.

The average daily rainfall in South America for the baseline period (198—2005) is 4 mm. However, the average daily rainfall in South America is projected to decrease in the future under RCP 4.5 and RCP 8.5 scenarios for the period 2031—50 (Fig. 7.5). The rate of decrease is higher (0.1 mm/day) in the 2031—2050 time period under the RCP 4.5 scenario in the same as RCP 8.5. However, the difference in model daily rainfall is expected to vary in the future. Previous studies also report that future rainfall may increase or decrease in many parts of Thailand and Nepal (Shrestha and Roachanakanan, 2021; Boonwichai et al., 2018; Shrestha et al., 2018; Arunrat and Pumijumnong, 2015; Babel et al., 2011).

Maximum and minimum temperatures change IPBES South America January—December with respect to 1986—2005 full CMIP5 ensemble (Figs. 7.6 and 7.7). On the left, one line per model is displayed for each scenario, along with the multimodel mean, and on the right, percentiles for the entire dataset: The box covers a range of 25%—75%, the whiskers a range of 5%—95%, and the horizontal line represents the median (50%).

According to Figs. 7.6 and 7.7, the projected annual average maximum and minimum temperature show an increase in South America in the period 2031—50 for RCP 4.5 and RCP 8.5. Increments in temperature for both RCP scenarios are comparable; however, at this time, RCP 8.5 has a projected maximum increase in both maximum and minimum temperature. In the period 2031—50, RCP 4.5 and 8.5 have projected increase in maximum temperature from 1°C to 1.5 °C and 1.5 °C—2.0 °C,

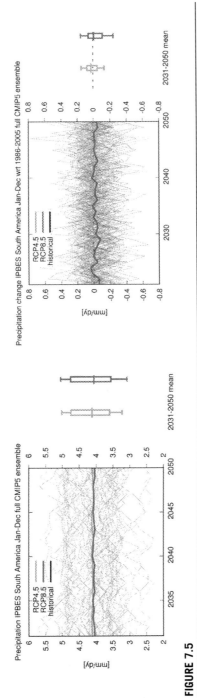

FIGURE 7.5

Projected future daily average precipitation with respect to 1986–2005.

3. Results and discussions

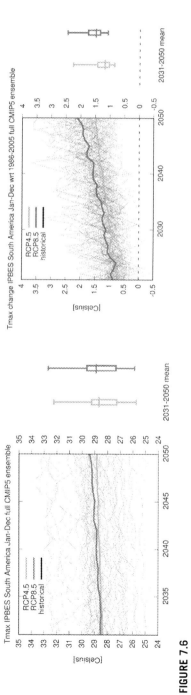

FIGURE 7.6

Projected future average annual maximum temperature.

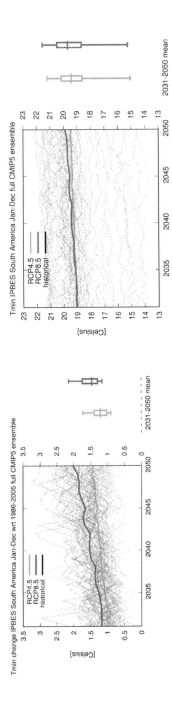

FIGURE 7.7

Projected future average annual minimum temperature.

respectively, and increase in minimum temperature from 1°C to 1.5 °C for RCP 4.5 and 1.0 °C−2.2 °C 8.5 for RCP scenarios. In general, the projected increase in minimum temperature is slightly higher than the maximum temperature for the RCP 8.5 scenario.

4. Conclusions

This study has presented changes in rainfall and temperature and relationships with climate and weather extremes in South America. The findings of the study show that there are significant intraregional differences in rainfall amount, variability, and trend. The results showed there is an increase in temperature and spatiotemporal rainfall variation increasing in South America. In comparison to the baseline period, this study found that temperature and precipitation in South America will rise in the future. The forecast band between RCP 4.5 and RCP 8.5 has offered a basic range of this increase, which will be defined in the future by socioeconomic and emission routes. The change in climatic conditions will have serious implications in various hydrology, agriculture, and the environment in the country, which is already facing such disasters even without the effect of climate change. The types of animals and plants (including crops) that may thrive in specific places are also determined by annual and seasonal temperature and precipitation patterns. Temperature and precipitation variations can disturb a wide range of ecological processes, especially if they happen faster than plants and animals can adapt. Suitable mitigation and adaptation measures should be designed in South America like construction of multipurpose reservoirs and rainwater harvesting systems, utilizing water-efficient irrigation technologies, and adapting growing seasons to the shift in the rainy season and dry season.

References

Aparicio, R., 1993. Meteorological and oceanographic conditions along the southern coastal boundary of the Caribbean Sea, 1951−1986. In: Maul, G.A. (Ed.), Climate Change in the Intra-americas Sea. United Nations Environment Program, IOC, Arnold, London, United Kingdom, pp. 100−114.

Arunrat, N., Pumijumnong, N., 2015. The preliminary study of climate change impact on rice production and economic in Thailand. Asian Soc. Sci. 11, 275−294. https://doi.org/10.5539/ass.v11n15p275.

Babel, S.M., Agarwal, A., Swain, D.K., Srikantha, H., 2011. Evaluation of climate change impacts and adaptation measures for rice cultivation in Northeast Thailand. Clim. Res. 46, 137. https://doi.org/10.3354/cr00978.

Basso, E., Compagnucci, R., Fearnside, P., Magrin, G., Marengo, J., Moreno, A.R., Suárez, A., Solman, S., Villamizar, A., Villers, L., Argenal, F., Artigas, C., Cabido, M., Codignotto, J., Confalonieri, U., Magaña, V., Morales-Arnao, B., Oropeza, O., Pabón, J.D., ..Vargas, W., Nobre, C. (Eds.), 2001. Latin America. In: Climate Change 2001: Impacts, Adaptation, and Vulnerability. Cambridge University Press.

Bayissa, Y., Tadesse, T., Demisse, G., Shiferaw, A., 2017. Evaluation of satellite-based rainfall estimates and application to monitor meteorological drought for the upper blue nile basin, Ethiopia. Rem. Sens. 9 (7), 669.

Berhane, A., Hadgu, G., Worku, W., Worku, W., Berhanu, A., 2020. Trends in extreme temperature and rainfall indices in the semi-arid areas of Western Tigray, Ethiopia. Environ. Syst. Res. 9, 3.

Boonwichai, S., Shrestha, S., Babel, M., Weesakul, S., Datta, A., 2018. Climate change impacts irrigation water requirement, crop water productivity and rice yield in the Songkhram River Basin, Thailand. J. Clean. Prod. 198. https://doi.org/10.1016/j.jclepro.2018.07.146.

Carril, A., Doyle, M., Barros, V., Nuñez, M., 1997. Impacts of climate change on the oases of the Argentinean Cordillera. Clim. Res. 9, 121–129. https://doi.org/10.3354/cr009121.

Compagnucci, R.H., 2000. ENSO events' impact on hydrological systems in the Cordillera de Los Andes during the last 450 years. In: Volkheimer, W., Smolka, P. (Eds.), Southern Hemisphere Paleo-And Neo-Climates: Key Sites, Methods, Data and Models. Springer Verlag, Berlin, Germany, pp. 175–185.

Compagnucci, R.H., Vargas, W.M., 1998. Inter-annual variability of Cuyo river's streamflows in the Argentinean Andean mountains and ENSO events. Int. J. Climatol. 18, 1593–1609.

Condom, T., Martínez, R., Pabón, J.D., Costa, F., Pineda, L., Nieto, J.J., López, F., Villacis, M., 2020. Climatological and hydrological observations for the SouthSouth American Andes: in situ stations, satellite, and reanalysis data sets. Front. Earth Sci. 8.

Grau, H.R., Aide, M., 2008. Globalization and land-use transitions in Latin America. Ecol. Soc. 13 (2), 16. URL: http://www.ecologyandsociety.org/vol13/iss2/art16/.

IPCC, 1996. In: Watson, R.T., Zinyowera, M.C., Moss, R.H. (Eds.), Climate Change 1995: Impacts, Adaptations and Mitigation of Climate Change: Scientific-Technical Analyses. Contribution of Working Group II to the Second Assessment Report of the Intergovernmental Panel on Climate Change. Cambridge University Press, Cambridge, United Kingdom and New York, NY, USA, p. 880.

IPCC, 2022. In: Carter, T.R., La Rovere, E.L., Jones, R.N., Leemans, R., Mearns, L.O., Nakicenovic, N., Pittock, A.B., Semenov, S.M., Skea, J., Gromov, S., Jordan, A.J., Khan, S.R., Koukhta, A., Lorenzoni, I., Posch, M., Tsyban, A.V., Velichko, A., Zeng, N., Gupta, S., Hulme, M. (Eds.), Impacts, Adaptation and Vulnerability, pp. 145–190.

Jáuregui, E., 1997. Climate changes in Mexico during the historical and instrumented periods. Quat. Int. 43/44, 7–17.

Magaña, V., Conde, C., 2000. Climate and freshwater resources in northern Mexico, Sonora: a case study. Environ. Monit. Assess. 61, 167–185.

MARENA, 2000. Climatic and Socioeconomic Scenarios for Nicaragua for the 21st Century. (Publication PNUD-NIC/98/G31). Ministry of the Atmosphere and Natural Resources of Nicaragua, Nicaragua.

Mesa, O.J., Poveda, G., Carvajal, L.F., 1997. Indroducción al Clima de Colombia (Introduction to the Climate of Colombia). National University Press, Bogota, Colombia.

Minetti, J.L., Sierra, E.M., 1989. The influence of general circulation patterns on humid and dry years in the Cuyo Andean region of Argentina. Int. J. Climatol. 9 (1), 55–68.

Poveda, G., Mesa, O.J., 1997. Feedbacks between hydrological processes in tropical South America and large scale oceanic-atmospheric phenomena. J. Clim. 10, 2690–2702.

Quintela, R.M., Broqua, R.J., Scarpati, O.E., 1993. Posible impacto del cambio global en los recursos hídricos del Comahue (Argentina). In: Proceedings of X Simpósio Brasileiro de Recursos Hídricos and I Simpósio de Recursos Hídricos do Cone Sul, 7–12 November, 1993, Brasil, vol 3. Anais, Brazil, pp. 320–329.

Shrestha, S., Roachanakanan, R., 2021. Extreme climate projections under representative concentration pathways in the Lower Songkhram River Basin, Thailand. Heliyon 7 (2), E06146.

Shrestha, U.B., Sharma, K.P., Devkota, A., Siwakoti, M., Shrestha, B.B., 2018. Potential impact of climate change on the distribution of six invasive alien plants in Nepal. Ecol. Indicat. 95 (1), 99–107. https://doi.org/10.1016/j.ecolind.2018.07.009. ISSN 1470-160X.

Tamiru, L., Fekadu, H., 2019. Effects of climate change variability on agricultural productivity. Int. J. Environ. Sci. Natural Res. 17 (1), 14–20.

Further reading

Canziani, P.O., Gerardo Carbajal Benitez, 2012. Climate impacts of deforestation/land-use changes in central South America in the PRECIS regional climate model: mean precipitation and temperature response to present and future deforestation scenarios. Sci. World J. 20.

Haylocka, M.R., Petersonb,T.C., Alvesc, L.M., Ambrizzid, T., Anunciaçãoe, Y.M.T., Baezf, J., Barrosg, V.R., Berlatoh, M.A., Bidegaini, M., Coronelj, G., Corradik, V., Garcial, V.J., Grimmm, A.M., Karolyn, D., Marengoc, J.A., Marinoo, M.B., Moncunillp, D.F., Nechetq, D., Quintanar, J., Rebelloe, E., Rusticuccig, M., Santoss, J.L., Trebejot, I., Vincentu, L.A., J. Clim. vol 19: 8, pp. 1490–1512.

Nuñez, E., Cabeza, J., Escudero, J.C., 1989. Relación entre la biomasa de jarales y su rendimiento energético por pirolisis. In: Bellot, J. (Ed.), Jornadas sobre las bases ecológicas para la gestión en ecosistemas terrestres. CIHEAM, Zaragoza, pp. 345–350.

South America: Climate Change Impacts, 2022. Climate change. In: Context (Encyclopedia.com).

Trends in Total and Extreme South American Rainfall in 1960–2000 and Links with Sea Surface Temperature, 2006.

Climate change scenario over Japan: trends and impacts

CHAPTER 8

Sridhara Nayak[1,2] and Tetsuya Takemi[2]

[1]*Research and Development Center, Japan Meteorogical Corporation, Osaka, Japan;* [2]*Disaster Prevention Research Institute, Kyoto University, Kyoto, Japan*

1. Introduction

Japan is a country of several thousands of islands with various topographies and climates. This country is located in the Pacific Ocean off the East Asian coast and lies between the parallels of approximately 20−45N and 123−154E. It has five main islands—Hokkaido in the North, Honshu in the Central, Shikoku in the West, Kyushu and Okinawa in the South, and more than other 6800 small islands (Bureau, 2021). The topography of Japan is characterized by intensely undulating mountain regions, which are roughly 73% of the total area, various lowlands and plains such as Kanto, Nobi and Osaka plain, etc., several short and steep rivers and freshwater lakes. The mountain areas of Japan are at high elevations and largely extended from north to south. As a result, the main islands of Japan are separated into two main regions—the Sea of Japan side and the Pacific Oceanside. Because of these topographical and physical differences, Japan has varied climates. According to Koppen's climate classification (Oliver and Wilson, 1987), the climate of Japan consists of three major climatic zones such as mild-summer (cool humid) continental climate over Hokkaido, hot-summer (warm humid) continental climate over northeast Honshu, and subtropical (warm tropical rainforest) climate over central and southern Japan.

Since the country lies between the Sea of Japan and the Pacific Ocean, the climate of Japan is largely influenced and regulated by the factors such as water vapors, winds, ocean currents, etc., those flow on the two giant bodies of water toward the Japan land. The winter monsoon carries heavy precipitation to the Sea of Japan side regions particularly Honshu in comparison to the Pacific Ocean side. Therefore, the climate in the regions of the Sea of Japan side remains humid in winter, while that in the regions of the Pacific Ocean side remains dry. On the other hand, the summer monsoon carries high humidity to the regions of the Pacific Ocean side and low precipitation to the regions of the Sea of Japan side (Bureau, 2021). The temporal and spatial patterns of various climatic variables over Japan are reported in many empirical as well as modeling studies (Xu et al., 2002; Yue and Hashino, 2007; Kanada et al., 2010; Nayak et al., 2018; IPCC, 2001, 2007; IPCC et al., 2014, 2021). In a study, Xu et al. (2002) documented that the annual mean precipitation over Japan corresponds to 1634.44 mm, and the annual mean temperature over Japan corresponds to 12.87°C during the 20th century.

The climate signals during the last century show remarkable changes in the patterns and trends in the climatic variables over Japan (Case and Tidwell, 2007; IPCC, 2021). The mean annual temperature increased by 1.06°C per century in Japanese rural areas over the 20th century, 1.77°C per century in the urban cities, and 2.70°C per century in the metropolitan cities (Higashino and Stefan, 2014, 2020). The mean annual precipitation amount in Japan over the 20th century decreased by 60 mm per century with a change of 20 ± 11 mm per century in rural to metropolitan areas. Several studies have also discussed the possible changes in the climate patterns and trends over Japan in future scenarios (Fujita et al., 2019; Nosaka et al., 2020; IPCC, 2012, 2014, 2021). Winter temperature at higher latitudes over Japan such as Hokkaido is expected to increase by 3°C or more by the end of the 21st century based on the SRES A1B scenario, while it is expected by 2–3°C in eastern to western Japan regions. The snowfall is likely to decrease with an exception over Hokkaido and highly elevated regions (Ohba and Sugimoto, 2020). Under the SRES A2B scenario, the precipitation amount is expected to increase in June through September by the end of the 21st century, while it is likely to decrease slightly or remain the same in other months (Kurihara et al., 2005). Similar changes in the climate patterns over Japan are also expected in the RCP scenarios (IPCC, 2014, 2021) and other future warming climates at 1.5, 2, and 4K (Nosaka et al., 2020).

The recent research and the Intergovernmental Panel on Climate Change (IPCC) Assessment Reports revealed that the frequency of warm days (>35°C) and intense precipitation events (>50 mm per hour) are increased in recent years (JMA, 2021) and projected to increase in future climate climates (Fuzita et al., 2019; Nayak et al., 2018). This indicates that the trends and patterns in the temperatures and precipitation amounts over Japan, particularly the extremes, are expected to change severely toward the end of the 21st century due to the anticipated changes in climate. As a consequence, it may have an impact on the socio-economic development of Japanese regions (O'Gorman et al., 2009; IPCC, 2007, 2012, 2021; Takemi et al., 2016c). Therefore, understanding the extreme temperature and precipitation events and their patterns over Japanese regions in the present and future climates is important to resist future climate change issues. This study attempts to investigate the climate change pattern and its impact on the extreme precipitation events over Japanese regions by analyzing 140 ensemble climate simulation results conducted over Japan for the past 50 years (1961–2010) and the future 60 years (2051–2110).

2. Present climate over Japan

In this section, the present climate in terms of precipitation and temperature climatologies for the past 50 years (1961–2010) is explored over Japan and its seven regions. The seven regions of Japan are classified based on the two sides (the Sea of Japan side and the Pacific Ocean side) and previous studies (Iizumi et al., 2011; Tsunematsu et al., 2013). The seven regions are shown in Fig. 8.1 and named as the Pacific Ocean side of Northern Japan (NP), Eastern Japan (EP), Western Japan (WP), the Sea of Japan side of Northern Japan (NS), Eastern Japan (ES), Western Japan (WS), and Okinawa (OK).

The datasets used for this analysis include the 0.25 degrees grid daily observed precipitation and temperature from the Asian Precipitation—Highly—Resolved Observational Data Integration Toward Evaluation (APHRODITE, Yatagai et al., 2012) the 20 km grid hourly precipitation and temperature datasets from and 50 ensemble climate simulation results from the database for Policy Decision making for Future climate change (d4PDF, Mizuta et al., 2017). The d4PDF datasets were prepared at

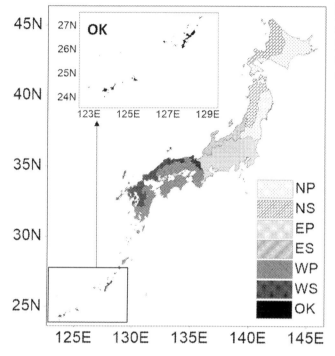

FIGURE 8.1

Seven regions of Japan. The inset figure highlights Okinawa. The regions are classified based on previous studies (Iizumi et al., 2011; Tsunematsu et al., 2013).

Meteorological Research Institute (MRI) Japan by using MRI Atmospheric General Circulation Model (AGCM) and available at 20 km grid resolution over Japan for the past 60 years (1951–2010) with 50 ensemble climate simulations and the future 60 years (2051–2110) under 4K warming with ensemble climate simulations. The model descriptions and simulation specifics are detailed in Mizuta et al. (2017).

2.1 Precipitation

Fig. 8.2 illustrates the climatology, annual variations, and extremes of precipitation over Japan derived from the d4PDF ensemble experiment results and that of APHRODITE. The annual mean precipitation corresponds to a minimum of 10 mm d-1 precipitation over many areas of southern to eastern regions of Japan, while it accounts for a maximum of 2 mm d-1 precipitation over some areas of eastern to northern Japan (Fig. 8.2A–B). The annual total precipitation amounts over Japan and its seven regions are shown in Table 8.1. The d4PDF ensemble mean indicates that the annual total precipitation over Japan accounted for 1856 mm and varies in the range of 1411–2520 mm within the seven regions, while the APHRODITE observation shows 1756 mm over Japan and varies within the seven regions in the range between 1227–2246. These results are slightly higher than the previous research (e.g., Xu et al., 2002), which finds 1634.44 mm of annual precipitation over Japan during the 20th century. This may be due to the use of different periods. Xu et al. (2002) analyzed the data for the period 1886–2000, while we analyzed for the period 1961–2010. Thus, the comparison indicated increased

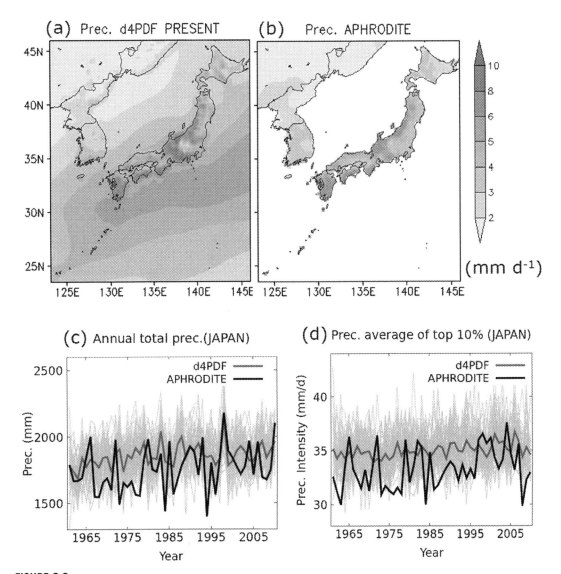

FIGURE 8.2

Precipitation climatology from (A) the d4PDF ensemble mean and (B) APHRODITE for the period 1961–2000. (C) Annual total precipitation over Japan land. (D) Annual average of top 10% precipitation intensities over Japan during 1961–2010. The *blue thin lines [light gray lines in print]* in (C) and (D) correspond to the results from the ensemble member.

Table 8.1 Averaged annual total precipitation (mm) during 1961−2010.

	d4PDF	APHRODITE
NP	1488.05	1227.03
NS	1411.19	2246.99
EP	2520.23	1691.01
ES	2141.15	1865.37
WP	1786.61	1902.25
WS	2034.31	1903.17
OK	1653.26	1756.17
Japan	1856.87	

precipitation in recent periods. The EP receives the highest precipitation amounts, while NP receives the lowest annual precipitation. The variation of annual total precipitation over Japan is noticed in the range of ∼1400−2300 mm (Fig. 8.2C). It indicates very small positive trends of 23.83 and 34.48 mm per decade in the annual precipitation in d4PDF and APHRODITE, respectively. The average of top 10% precipitation days (i.e., the mean of the precipitation intensities exceeding the 90th percentile) over Japan corresponds to 30−37 mm in observation and 30−40 mm in d4PDF ensemble experiments on a daily scale (Fig. 8.2D).

The precipitation anomalies during 1961−2010 indicate a variation of about ±500 mm precipitation from the mean over entire Japan with few exception years in some regions where it shows higher magnitudes (Fig. 8.3). The frequency of the extremes with precipitation intensity exceeding 30 mm/d accounted for up to ∼10 days each year over NP and NS and ∼30 days in each year over other regions of Japan (Fig. 8.4). The frequency of such days averaged over entire Japan shows about 10 days in each year with an uncertainty range of 3−20 days in each year. Overall, precipitation amounts and the frequency show a slightly positive trend in the years between 1961−2010, but no significant changes in the patterns of the annual total precipitation and the frequencies of the extremes are noticed.

2.2 Temperature

The climatology, annual variations, and extremes of temperature over Japan are shown in Fig. 8.5. The annual mean temperature is noted as 20°C or higher over southern regions, about 15−19°C in many regions of western regions, roughly 10−14°C in eastern regions and 10°C or lower over northern regions (Fig. 8.5A−B). The d4PDF ensemble mean indicates that the annual mean temperature over Japan corresponds to 13.33°C, which is again slightly higher than the annual temperature documented in previous studies (12.87°C in Xu et al., 2002). As we mentioned earlier, Xu et al. (2002) analyzed the entire 20th century, while our results are computed for the period 1961−2010. Thus, the comparison indicated an increased temperature in recent periods. The annual mean temperatures over NP and NS show 7.76 and 7.48°C, respectively, while that over EP and ES indicate 11.25 and 12.73°C, respectively. It shows 14.91 and 15.33°C respectively over WP and WS and 24°C over OK. These results closely follow the annual mean temperatures over Japanese regions seen in the APHRODITE observation (Table 8.2). The annual variation of mean temperature over Japan during 1961−2010 shows a decreasing pattern of mean temperature until roughly 1985 and an increasing pattern

150 Chapter 8 Climate change scenario over Japan: trends and impacts

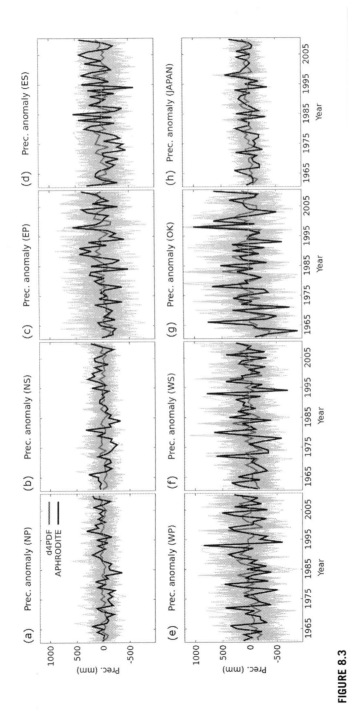

FIGURE 8.3

Precipitation anomalies over (A–G) seven regions of Japan and (H) whole of Japan. The results correspond to the deviation of annual total precipitation from the climatology during 1961–2010. The blue thin lines (light gray lines in print) correspond to the results from the ensemble members.

2. Present climate over Japan 151

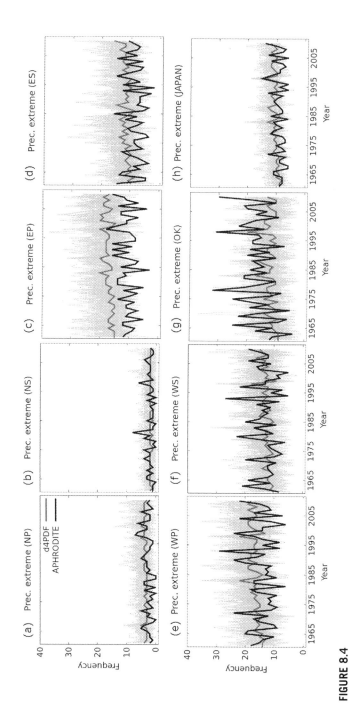

FIGURE 8.4

Annual frequency of precipitation days exceeding 30 mm/d during 1961–2010 over (A–G) seven regions of Japan and (H) whole of Japan. The *blue thin lines* [*light gray lines in print*] correspond to the results from the ensemble members.

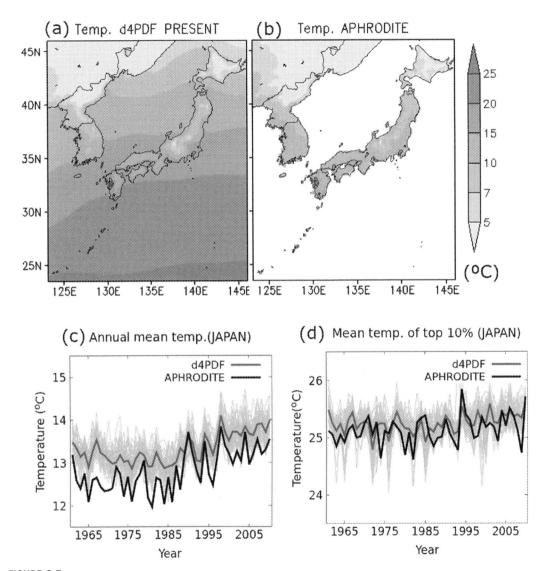

FIGURE 8.5

Temperature climatology from (A) the d4PDF ensemble mean and (B) APHRODITE for the period 1961–2000. (C) Annual mean temperature over Japan land. (D) Annual mean of top 10% temperature days over Japan during 1961–2010. The *blue thin lines [light gray lines in print]* in (C) and (D) correspond to the results from the ensemble member.

Table 8.2 Averaged annual mean temperature (°C) during 1961–2010.

	d4PDF	APHRODITE
NP	7.62	7.60
NS	7.48	7.33
EP	11.25	11.09
ES	12.73	12.09
WP	14.91	14.33
WS	15.33	14.21
OK	24.00 13.33	23.05
Japan		12.81

afterward and corresponds to a range within 12–14°C in observation and within 12–15°C in ensemble experiment results (Fig. 8.5C). The rate of increase of annual mean temperature over Japan is found as 0.15°C and 0.20°C per decade in d4PDF and APHRODITE, respectively. The mean of top 10% temperature days (i.e., above 90th percentile) over Japan shows nearly 25°C during 1961–2010 (Fig. 8.5D).

The temperature anomalies over seven subregions of Japan indicate a similar increasing pattern as seen in cases of the whole of Japan in recent years during 1961–2010 (Fig. 8.6). The frequency of the temperature extremes exceeding 25°C accounted for up to ∼10 days over NP and NS, up to ∼50 days over EP and ES, up to ∼80 days over WP and WS, and more than 100 days and up to ∼250 over OK (Fig. 8.7). The frequency of such extreme temperature days averaged over entire Japan shows about 50 days each year. Overall, a significant positive trend in temperature over all the seven regions of Japan is noticed, but no substantial change in the frequencies of extreme temperature days is found over Japanese regions except over OK.

3. Projected climate change over Japan

Future scenarios suggest that the global temperature will continue increasing and significantly changing the patterns of precipitation (Fujita et al., 2019; Nayak and Takemi, 2022; IPCC, 2014, 2021). In this section, the precipitation and temperature climatologies for the future 60 years (2051–2110) are explored over Japanese regions. The annual variation and the extremes of the projected precipitation and temperature are also discussed. For this analysis, the 90 ensemble future climate simulation results of d4PDF datasets at 20 km resolution are utilized. As mentioned in Section 2, these simulations were conducted under 4K warming relative to preindustrial levels, similar to the RCP8.5 scenario.

3.1 Precipitation

Fig. 8.8 presents the annual mean precipitation climatology for the period 2051–2110 and its expected future change from the period 1961–2010. The distribution of the annual mean precipitation in future climate is likely to be the same as it was seen in the present climate but with a lighter intensity. In a study, Murata et al. (2015) also reported that annual precipitation over many regions of Japan does not exhibit significant changes under the RCP8.5 scenario. An increase in the annual

154 Chapter 8 Climate change scenario over Japan: trends and impacts

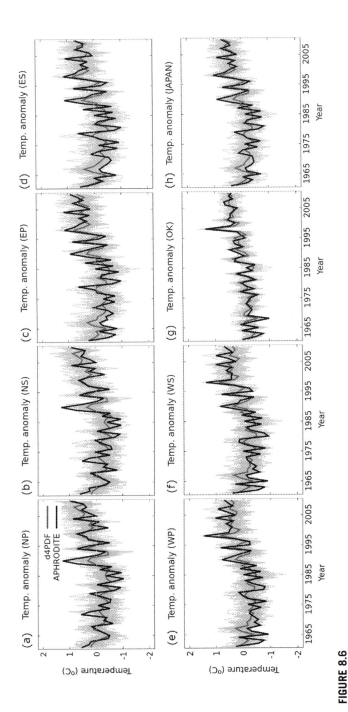

FIGURE 8.6

Temperature anomalies over (A–G) seven regions of Japan and (H) whole Japan. The results correspond to the deviation of annual mean temperature from the climatology during 1961–2010. The *blue thin lines [light gray lines in print]* correspond to the results from the ensemble members.

3. Projected climate change over Japan 155

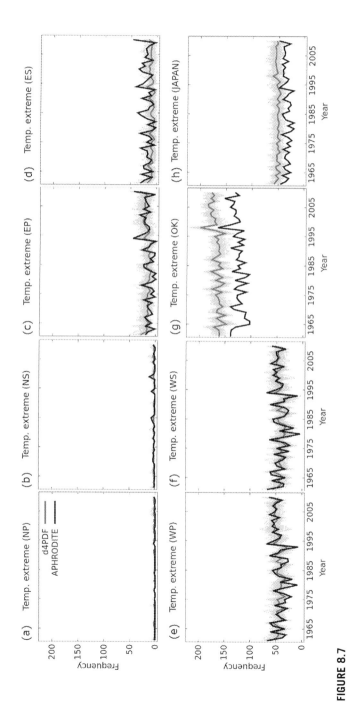

FIGURE 8.7

Annual frequency of temperature days exceeding 25°C during 1961–2010 over (A–G) seven regions of Japan and (H) whole Japan. The *blue thin lines (light gray lines in print)* correspond to the results from the ensemble members.

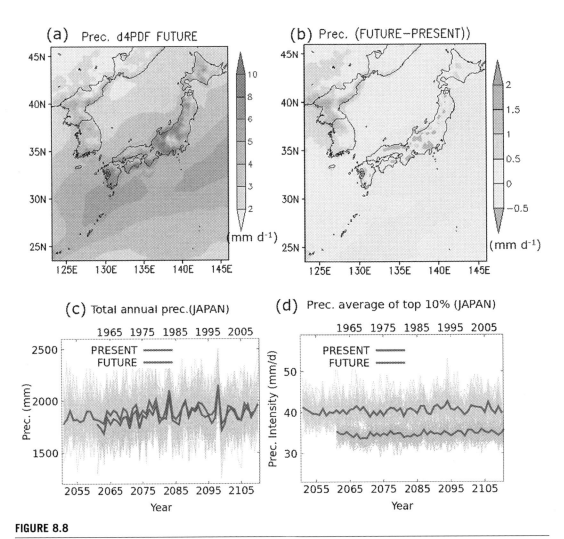

FIGURE 8.8

(A) Precipitation climatology for the period 2051–2110 from the d4PDF ensemble mean. (B) Future changes in the precipitation climatology. (C) Annual total precipitation over Japan land from present and future climate. (D) Annual average of top 10% precipitation intensities over Japan land in present and future climate periods. The *thin lines* in (C) and (D) correspond to the results from the ensemble members.

mean precipitation intensity is expected over the northern and southern regions of Japan (Fig. 8.8A–B). It is found that the total annual precipitation amount is expected to increase by 41–48.62 mm over NP and NS, 51.91 mm over WP, 18.75 mm over WS and 104.34 mm over OK, while it is likely to decrease by 3.17 mm over ES and 37.25 mm over EP. Overall Japan shows an increase in total annual precipitation by 33.26 mm (Table 8.3). Even though the annual precipitation amount in future climate averaged over EP is likely to decrease, some areas over EP are expected to

Table 8.3 **Projected change in averaged annual total precipitation (mm).**

	Future	Change
NP	1536.67	48.62
NS	1452.19	41.00
EP	2482.98	−37.25
ES	2137.98	−3.17
WP	1838.52	51.91
WS	2053.06	18.75
OK	1757.60	104.34
Japan	1890.13	33.26

receive relatively more precipitation compared to other regions of Japan. The comparison between the annual variations of precipitation amounts in the present climate and its next 100 years does not indicate any significant difference (Fig. 8.8C). However, the comparison between the averages of the top 10% precipitation days in the two climates showed a significant increase in future climate (Fig. 8.8D). The intensity of the extreme precipitation in a future warming climate is likely to increase by ∼5 mm/d over Japan.

The regional variation of the anomalies of the annual precipitation amounts in both climates is shown in Fig. 8.9. It indicated that the precipitation anomalies in future climates are likely to vary almost in the same tendencies as noticed in the present climate. The frequency of the precipitation extremes with precipitation intensity exceeding 30 mm/d in future climates shows some higher values compared to that in the present climate (Fig. 8.10). There are no considerable changes in the number of extreme precipitation days expected over NP, NS, and EP in future climate, however, the number of extreme precipitation days over other regions show an increase of up to 10 days each year. Recent IPCC reports (e.g., IPCC, 2021) also documented an increase in extreme precipitation days over Japan under various future scenarios. Overall, the climate over Japan is expected to be drier in the future warming climate, but the intensity of the extreme precipitation is likely to be much stronger over entire of Japan (this is discussed in Section 4.1). The frequencies of precipitation intensities are also expected to increase over some regions of Japan. The variations in the anomalies of the annual precipitation amounts in the future warming climate are likely to be almost the same over the entire Japan as seen in the present climate.

3.2 Temperature

The spatial distribution of temperature climatology in the future climate indicated about 12°C in northern regions, ∼16−17°C in eastern regions, roughly 19°C in western regions, and ∼27°C in southern regions of Japan (Fig. 8.11A). Entire Japan is expected to be warmer by 3−5°C by the end of the 20th century (Fig. 8.11B). Considering regional averages, NP and NS are likely to be warmer up to 5.11°C; EP, ES, and WP are expected to get warmer up to 4.71°C, and OK is likely to be warmer up to 3.26°C (Table 8.4). The recent IPCC reports highlighted that the annual mean temperature over Japan will increase up to 1.7°C under the RCP2.6 scenario and up to 5.4°C under the RCP8.5 scenario (IPCC, 2021). The magnitude of the temperature extremes over the entire of Japan is expected to be much warmer in the future compared to the present climate (this is discussed in Section 4.1). The

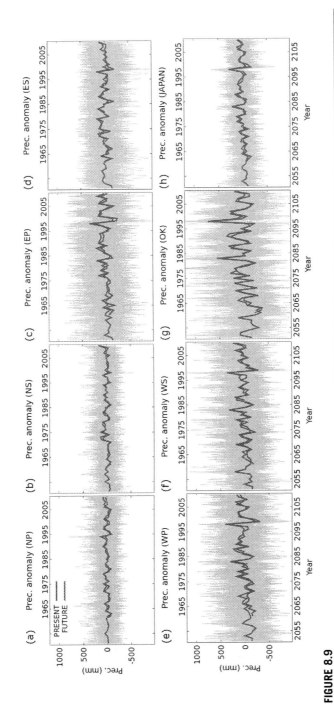

FIGURE 8.9

Precipitation anomalies in present and future climates over (A–G) seven regions of Japan and (H) whole of Japan. The results correspond to the deviation of annual total precipitation from the climatology of the corresponding period. The *thin lines* correspond to the results from the ensemble members.

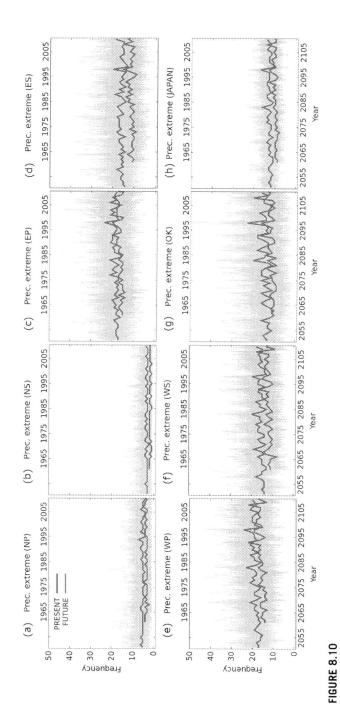

FIGURE 8.10

Annual frequency of precipitation days exceeding 30 mm/d in present and future climates over (A–G) seven regions of Japan and (H) whole of Japan. The *thin lines* correspond to the results from the ensemble members.

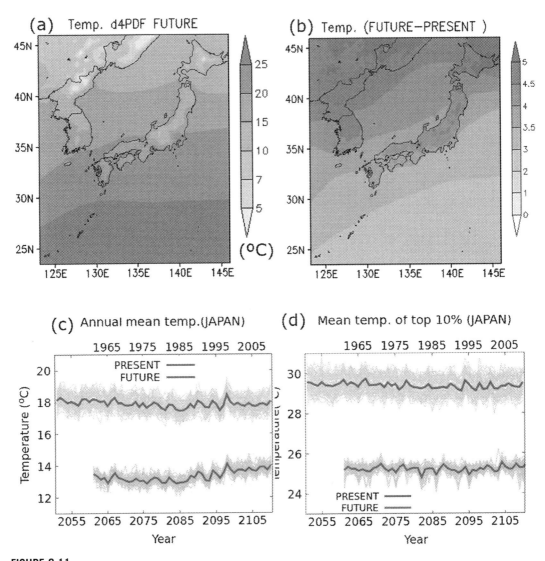

FIGURE 8.11

(A) Temperature climatology for the period 2051–2110 from the d4PDF ensemble mean. (B) Future changes in the temperature climatology. (C) Annual mean temperature over Japan land from present and future climate. (D) Annual mean of top 10% temperature days over Japan land in present and future climate periods. The *thin lines* in (C) and (D) correspond to the results from the ensemble members.

variations of the annual mean temperature in future climate showed about 4°C warmer in each year compared to the corresponding past 100 years in present climate (Fig. 8.11C). The mean temperature of the top 10% hot days in each year is also expected to increase by ~4°C in the corresponding next 100 years in future climate (Fig. 8.11D).

Table 8.4 Projected change in averaged annual mean temperature (°C).

	Future	Change
NP	12.65	5.02
NS	12.59	5.11
EP	15.96	4.71
ES	17.36	4.63
WP	19.32	4.41
WS	19.66	4.34
OK	27.26	3.26
Japan	17.88	4.55

The temperature anomalies over Japanese regions in the two climates indicated similar patterns; however, the years after 1998, the corresponding next 100 years, that is, the years after 2098 showed lower values indicating a possible less warming rate at the end of the 20th century 267 as compared to the end of the 19th century (Fig. 8.12). The comparison between the frequencies of the temperature extremes exceeding 25°C in present and future climates indicated significant increases over entire Japan (Fig. 8.13). The number of extreme temperature days in future climate is expected to increase up to ~50 days over NP and NS, and up to ~80 days over other regions. Several studies conducted with different temperature thresholds (e.g., 30°C in IPCC, 2021) also highlighted an increase in the number of hot days exceeding that threshold in future climate (Nosaka et al., 2020; IPCC, 2021). Overall, the climate in the future is likely to be warmer with stronger hot days with increased frequencies of temperature extremes compared to the present climate.

4. Impacts

Climate change causes various impacts such on agriculture, ecosystem, natural disasters, etc., over Japan. In this section, the climate change impacts on Japan are discussed specifically on the extreme events such as precipitation and temperature extremes and typhoons.

The impact of climate change on the precipitation and temperature extremes over seven regions of Japan is presented in Fig. 8.14 and Fig. 8.15, respectively. It is noticed that the intensity of the precipitation extremes (averaged of top 10% precipitation days' intensity) in the future climate is likely to increase by up to 5 mm/d over NP and NS, 10 mm/d over ES and EP, and 15 mm/d over WP, WS, and OK (Fig. 8.14). The temperature extremes (mean of top 10% hot days) are also expected to be much hotter in the entire of Japan (Fig. 8.15). The mean temperature of the top 10% of hot days over NP and NS is likely to increase up to 5°C in future climate, while that of over OK is expected to up to 3°C and over other regions, it is up to 4°C. The recent IPCC reports and a multitude of other studies also highlighted the projected temperature over the Japanese region in similar magnitudes (IPCC, 2012, 2014, 2021; Murata et al., 2012; Ishii and Mori, 2020; Nayak and Takemi, 2020a). The extremely high daily minimum and maximum temperatures are also likely to increase in many regions of Japan (Murata et al., 2012).

As we discussed earlier, the frequency of the precipitation and temperature extremes are also expected to increase in most of the Japanese regions under a future warming climate. The frequency of

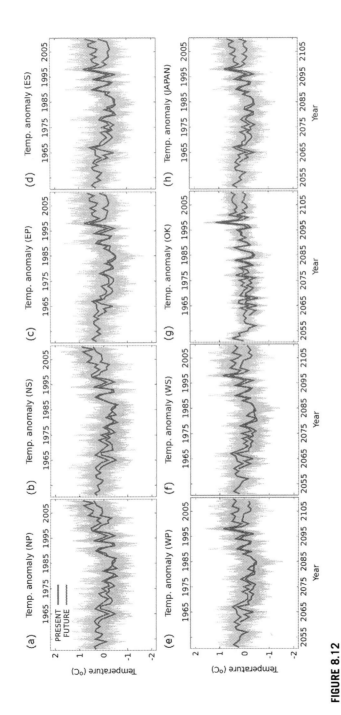

FIGURE 8.12

Temperature anomalies in present and future climates over (A–G) seven regions of Japan and (H) whole Japan. The results correspond to the deviation of annual mean temperature from the climatology of the corresponding period. The *thin lines* correspond to the results from the ensemble members.

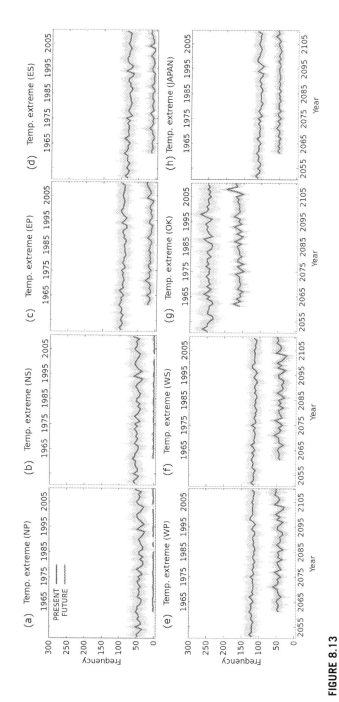

FIGURE 8.13

Same as Fig. 8.10, but with temperature days exceeding 25°C.

164 Chapter 8 Climate change scenario over Japan: trends and impacts

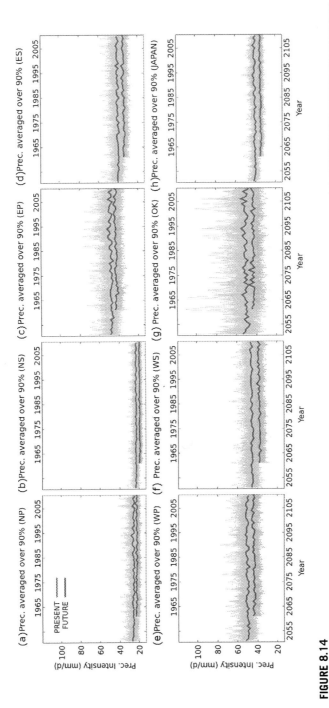

FIGURE 8.14

Annual average of top 10% precipitation intensities over (A–G) seven regions and (H) whole Japan land in present and future climate periods. The *thin lines* in (C) and (D) correspond to the results from the ensemble members.

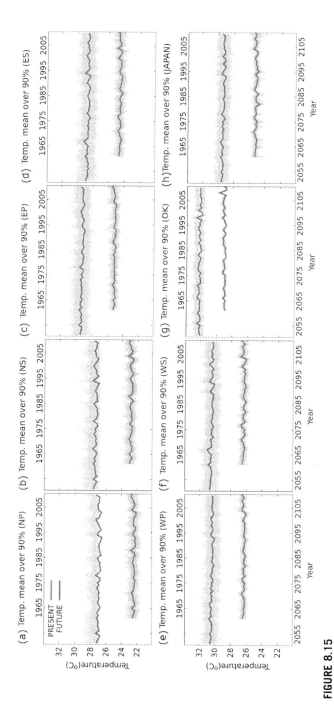

FIGURE 8.15

Annual average of top 10% temperature days over (A–G) seven regions and (H) whole Japan land in present and future climate periods. The *thin lines* in (C) and (D) correspond to the results from the ensemble members.

extreme precipitation events during the Baiu season (it is the rainy period of early summer in Japan) are likely to increase in future climate (Osakada and Nakakita, 2018). The extreme precipitation intensities over Japan would be more intense in a future warming climate (Nayak and Dairaku, 2016; Nayak et al., 2018). Likewise, heavy snowfall is likely to increase under future warming climate over the Sea of Japan side regions Japan (Kawase et al., 2016; Ohba and Sugimoto, 2020). The typhoons in a future warming climate are also expected to intensify and bring strong winds and excess precipitation to the landfall regions of Japan (Takayabu et al., 2015; Takemi et al. 2016a,b,c; Ito et al., 2016; Nayak and Takemi, 2019a, 2019b, 2020b; Takemi et al., 2019; Takemi and Unuma, 2019; Takemi and Unuma, 2020). Nayak and Takemi (2020b) examined the impact of climate change on various individual typhoons that made landfall over Northerner Japan and reported that the maximum wind speeds of those typhoons would increase up to ~ 10 m/s and the minimum central pressure would be decreased up to ~ 20 hPa. All these impacts discussed above alarm the severity of the damages due to climate change.

5. Summary

This study investigates the climate change pattern and its impact on the extreme events over Japanese regions from 140 ensemble climate simulation results conducted over Japan for the past 50 years (1961–2010) and the future 60 years (2051–2110). We noticed slightly positive trends in the precipitation amounts and the frequencies and a significant positive trend in temperature over the entire of Japan. The climate over Japan is expected to be drier in the future, but the intensity of the extreme precipitation is likely to be much stronger over the entire of Japan. The frequencies of precipitation intensities are also expected to increase over some regions of Japan. The variations in the anomalies of the annual precipitation amounts in the future warming climate are likely to be almost the same over the entire Japan as seen in the present climate. The climate in the future is likely to be warmer with stronger hot days with increased frequencies of temperature extremes compared to the present climate. It is noticed that the intensity of the precipitation extremes (averaged of top 10% precipitation days' intensity) in the future climate is likely to increase by up to 5 mm/d over NP and NS, 10 mm/d over ES and EP, and 15 mm/d over WP, WS, and OK. The temperature extremes (mean of top 10% hot days) are also expected to be much hotter in entire Japan. The mean temperature of the top 10% of hot days over NP and NS is likely to increase up to 5°C in future climate, while that of over OK is expected to up to 3°C and over other regions, it is up to 4°C.

6. Declarations
6.1 Availability of data and material

The d4PDF data (the database for Policy Decision making for Future climate change) dataset supporting the conclusions of this article is obtained from the Data Integration and Analysis System (DIAS, http://search.diasjp.net/en/dataset). The precipitation and temperature observation dataset supporting the validation of this study is obtained from the Asian Precipitation-Highly-Resolved Observational Data Integration Toward Evaluation (APHRODITE's Water Resources, https://www.chikyu.ac.jp/precip/english/products.html).

Funding

This study was supported by the TOUGOU program grant number JPMXD0717935498 and the SENTAN program grant number JPMXD0722678534 funded by the Ministry of Education, Culture, Sports, Science, and Technology, Government of Japan.

Authors' contributions

SN proposed the topic, designed the study, analyzed the data, and drafted the manuscript. TT helped in the interpretation and the construction of the manuscript. All authors read and approved the final manuscript.

Acknowledgments

This study was supported by the TOUGOU program Grant Number JPMXD0717935498 and the SENTAN program grant number JPMXD0722678534 and funded by the Ministry of Education, Culture, Sports, Science, and Technology, Government of Japan. The Japan Meteorological Agency (JMA) is acknowledged for providing the radar/rain gauge-analyzed precipitation product.

References

Bureau, S., 2021. Ministry of internal affairs and communications. . Statistical Handbook of Japan. Available from: https://www.stat.go.jp/english/data/handbook/index.html (Accessed on 17 November 2021).

Case, M., Tidwell, A., 2007. Nippon Changes: Climate Impacts Threatening Japan Today and Tomorrow. WWF International: Gland, Switzerland. Available from. https://www.wwf.or.jp/activities/lib/pdf_climate/environment/WWF_NipponChanges_lores.pdf. (Accessed 17 November 2021).

Fujita, M., Mizuta, R., Ishii, M., Endo, H., Sato, T., Okada, Y., Watanabe, S., et al., 2019. Precipitation changes in a climate with 2-K surface warming from large ensemble simulations using 60-km global and 20-km regional atmospheric models. Geophys. Res. Lett. 46 (1), 435–442.

Higashino, M., Stefan, H.G., 2014. Hydro-climatic change in Japan (1906–2005): impacts of global warming and urbanization. Air Soil. Water Res. 7. ASWR-S13632.

Higashino, M., Stefan, H.G., 2020. Trends and correlations in recent air temperature and precipitation observations across Japan (1906–2005). Theor. Appl. Climatol. 140 (1), 517–531.

Iizumi, T., Nishimori, M., Dairaku, K., Adachi, S.A., Yokozawa, M., 2011. Evaluation and intercomparison of downscaled daily precipitation indices over Japan in present-day climate: strengths and weaknesses of dynamical and bias correction-type statistical downscaling methods. J. Geophys. Res. Atmos. 116 (D1).

IPCC, 2001. In: Climate Change 2001: Synthesis Report. A Contribution of Working Groups I, II, and III to the Third Assessment Report of the Intergovernmental Panel on Climate Change [Watson, R.T. and the Core Writing Team. Cambridge University Press, Cambridge, United Kingdom, and 383 New York, NY, USA, p. 398.

IPCC, 2007. Summary for policymakers. In: Solomon, S., Qin, D., Manning, M., Chen, Z., Marquis, M., Averyt, K.B., Tignor, M., Miller, H.L. (Eds.), Climate Change 2007: The Physical Science Basis. Contribution of Working Group I to the Fourth Assessment Report of the Intergovernmental Panel on Climate Change. Cambridge University Press, Cambridge, United Kingdom and New York, NY, USA.

IPCC, 2012. In: Field, C.B., Barros, V., Stocker, T.F., Qin, D., Dokken, D.J., Ebi, K.L., Mastrandrea, M.D., Mach, K.J., Plattner, G.-K., Allen, S.K., Tignor, M., Midgley, P.M. (Eds.), Managing the Risks of Extreme Events and Disasters to Advance Climate Change Adaptation. A Special Report of Working Groups I and II of the Intergovernmental Panel on Climate Change. Cambridge University Press, Cambridge, UK, and New York, NY, USA, p. 582.

IPCC, 2014. In: Field, C.B., Barros, V.R., Dokken, D.J., Mach, K.J., Mastrandrea, M.D., Bilir, T.E., Chatterjee, M., Ebi, K.L., Estrada, Y.O., Genova, R.C., Girma, B., Kissel, E.S., Levy, A.N., MacCracken, S., Mastrandrea, P.R., White, L.L. (Eds.), Climate Change 2014: Impacts, Adaptation, and Vulnerability. Part A: Global and Sectoral Aspects. Contribution of Working Group II to the Fifth Assessment Report of the Intergovernmental Panel on Climate Change. Cambridge University Press, Cambridge, UK and New York, USA, p. 1132pp.

IPCC, 2021. Summary for policymakers. In: Masson-Delmotte, V., Zhai, P., Pirani, A., Connors, S.L., Péan, C., Berger, S., Caud, N., Chen, Y., Goldfarb, L., Gomis, M.I., Huang, M., Leitzell, K., Lonnoy, E., Matthews, J.B.R., Maycock, T.K., Waterfield, T., Yelekçi, O., Yu, R., Zhou, B. (Eds.), Climate Change 2021: The Physical Science Basis. Contribution of Working Group I to the Sixth Assessment Report of the Intergovernmental Panel on Climate Change. Available from: https://www.ipcc.ch/report/ar6/wg1/downloads/report/IPCC_AR6_WGI_SPM_final.pdf. (Accessed 10 January 2022).

Ishii, M., Mori, N., 2020. d4PDF: large-ensemble and high-resolution climate simulations for global warming risk assessment. Prog. Earth Planet. Sci. 7 (1), 1–22.

Ito, R., Takemi, T., Arakawa, O., 2016. A possible reduction in the severity of typhoon wind in the northern part of Japan under global warming: a case study. SOLA 12, 100–105.

JMA, 2021. Climate change in Japan. In: Climate Change Monitoring Report 2020. Available from: https://www.jma.go.jp/jma/en/NMHS/ccmr/ccmr2020.pdf. (Accessed 10 February 2022).

Kanada, S., Nakano, M., Kato, T., 2010. Climatological characteristics of daily precipitation over Japan in the Kakushin regional climate experiments using a non-hydrostatic 5-km-mesh model: comparison with an outer global 20-km-mesh atmospheric climate model. SOLA 6, 117–120.

Kawase, H., Murata, A., Mizuta, R., Sasaki, H., Nosaka, M., Ishii, M., Takayabu, I., 2016. Enhancement of heavy daily snowfall in central Japan due to global warming as projected by a large ensemble of regional climate simulations. Climatic Change 139 (2), 265–278.

Kurihara, K., Ishihara, K., Sasaki, H., Fukuyama, Y., Saitou, H., Takayabu, I., Noda, A., et al., 2005. Projection of climatic change over Japan due to global warming by high-resolution regional climate model in MRI. SOLA 1, 97–100.

Mizuta, R., Murata, A., Ishii, M., Shiogama, H., Hibino, K., Mori, N., Kawase, H., et al., 2017. Over 5,000 years of ensemble future climate simulations by 60-km global and 20-km regional atmospheric models. Bull. Am. Meteorol. Soc. 98 (7), 1383–1398.

Murata, A., Nakano, M., Kanada, S., Kurihara, K., Sasaki, H., 2012. Summertime temperature extremes over Japan in the late 21st century are projected by a high-resolution regional climate model. J. Meteorol. Soc. Jpn. 90, 101–122.

Murata, A., Sasaki, H., Kawase, H., Nosaka, M., Oh'izumi, M., Kato, T., Nagatomo, T., et al., 2015. Projection of future climate change over Japan in ensemble simulations with a high-resolution regional climate model. SOLA 11, 90–94.

Nayak, S., Dairaku, K., 2016. Future changes in extreme precipitation intensities associated with temperature under SRES A1B scenario. Hydrological Research Letters 10 (4), 139–144. https://doi.org/10.3178/hrl.10.139.

Nayak, S., Takemi, T., 2019a. Quantitative estimations of hazards resulting from Typhoon Chanthu (2016) for assessing the impact in current and future climate. Hydrological Research Letters 13 (2), 20–27.

Nayak, S., Takemi, T., 2019b. Dynamical downscaling of Typhoon Lionrock (2016) for assessing the resulting hazards under global warming. J. Meteor. Soc. Japan 97 (1), 69–88. https://doi.org/10.2151/jmsj.2019-003.

Nayak, S., Takemi, T., 2020a. Clausius-Clapeyron scaling of extremely heavy precipitations: case studies of the July 2017 and July 2018 heavy rainfall events over Japan. J. Meteorol. Soc. Jpn. 98 (6), 1147−1162.

Nayak, S., Takemi, T., 2020b. Robust responses of typhoon hazards in northern Japan to global warming climate: cases of landfalling typhoons in 2016. Meteorol. Appl. 27 (5), e1954. https://doi.org/10.1002/met.1954.

Nayak, S., Dairaku, K., Takayabu, I., Suzuki-Parker, A., Ishizaki, N.N., 2018. Extreme precipitation linked to temperature over Japan: current evaluation and projected changes with multi-model ensemble downscaling. Clim. Dynam. 51, 4385−4401.

Nayak, S., Takemi, T., 2022. Assessing the impact of climate change on temperature and precipitation over India. In: Sumi, T., Kantoush, S., Saber, M. (Eds.), Wadi Flash Floods: Challenges and Advanced Approaches for Disaster Risk Reduction. Springer Singapore, pp. 121−142. https://doi.org/10.1007/978-981-16-2904-4 (Chapter 4).

Nosaka, M., Ishii, M., Shiogama, H., Mizuta, R., Murata, A., Kawase, H., Sasaki, H., 2020. Scalability of future climate changes across Japan was examined with large-ensemble simulations at+ 1.5 K, + 2 K, and +4 K global warming levels. Prog. Earth Planet. Sci. 7 (1), 1−13.

O'Gorman, P.A., Schneider, T., 2009. The physical basis for increases in precipitation extremes in simulations of 21st-century climate change. Proc. Natl. Acad. Sci. USA 106 (35), 14773−14777.

Ohba, M., Sugimoto, S., 2020. Impacts of climate change on heavy wet snowfall in Japan. Clim. Dynam. 54 (5), 3151−3164.

Oliver, J.E., Wilson, L., 1987. Climate classification. In: Oliver, J.E., Fairbridge, R.W. (Eds.), The Encyclopedia of Climatology. Van Nostrand Reinhold Company, New York, NY, USA.

Osakada, Y., Nakakita, E., 2018. Future change of occurrence frequency of Baiu heavy rainfall and its linked atmospheric patterns by multiscale analysis. SOLA 14, 79−85.

Takayabu, I., Hibino, K., Sasaki, H., Shiogama, H., Mori, N., Shibutani, Y., Takemi, T., 2015. Climate change effects on the worst-case storm surge: a case study of Typhoon Haiyan. Environ. Res. Lett. 10 (6), 064011.

Takemi, T., 2019. Impacts of global warming on extreme rainfall of a slow-moving typhoon: a case study for Typhoon Talas (2011). SOLA 15, 125−131. https://doi.org/10.2151/sola.2019-023.

Takemi, T., Unuma, T., 2020. Environmental factors for the development of heavy rainfall in the eastern part of Japan during Typhoon Hagibis (2019). SOLA 16, 30−36.

Takemi, T., Ito, R., Arakawa, O., 2016a. Effects of global warming on the impacts of typhoon Mireille (1991) in the Kyushu and Tohoku regions. Hydrol. Res. Lett. 10 (3), 81−87.

Takemi, T., Ito, R., Arakawa, O., 2016b. Robustness and uncertainty of projected changes in the impacts of Typhoon Vera (1959) under global warming. Hydrol. Res. Lett. 10 (3), 88−94.

Takemi, T., Okada, Y., Ito, R., Ishikawa, H., Nakakita, E., 2016c. Assessing the impacts of global warming on meteorological hazards and risks in Japan: Philosophy and achievements of the SOUSEI program. Hydrological Research Letters 10 (4), 119−125.

Takemi, T., Unuma, T., 2019. Diagnosing environmental properties of the July 2018 heavy rainfall event in Japan. SOLA 15A 60−65. https://doi.org/10.2151/sola.15A-011.

Tsunematsu, N., Dairaku, K., Hirano, J., 2013. Future changes in summertime precipitation amounts are associated with topography in the Japanese islands. J. Geophys. Res. Atmos. 118, 4142−4153.

Xu, Z.X., Takeuchi, K., Ishidaira, H., 2002. Long-term trends of annual temperature and precipitation time series in Japan. J. Hydrosci. Hydraul. Eng. 20 (2), 11−26.

Yatagai, A., Kamiguchi, K., Arakawa, O., Hamada, A., Yasutomi, N., Kitoh, A., 2012. APHRODITE: constructing a long-term daily gridded precipitation dataset for Asia based on a dense network of rain gauges. Bull. Am. Meteorol. Soc. 93 (9), 1401−1415.

Yue, S., Hashino, M., 2007. Probability distribution of annual, seasonal and monthly precipitation in Japan. Hydrol. Sci. J. 52 (5), 863−877.

CHAPTER 9

A brief review of climate in Taiwan

Mahjoor Ahmad Lone[1], Kweku Afrifa Yamoah[2] and Tsai-Wen Lin[3,4]

[1]*Department of Geography and Environmental Sciences, Northumbria University, Newcastle upon Tyne, United Kingdom;* [2]*Department of Archaeology, University of York, York, United Kingdom;* [3]*Department of Geosciences, National Taiwan University, Taipei, Taiwan;* [4]*Alfred Wegener Institute of Polar and Marine Research, Bremerhaven, Germany*

1. Introduction: hydroclimate of Taiwan

In Taiwan, climatic variability has not been studied extensively as compared to mainland China and the Indian sub-continent. The prevailing hydroclimatic conditions are influenced by the southwesterly monsoon, which brings moist air from the Pacific Ocean and precipitates over land during summer, whereas the northeasterly monsoon, which brings dry, cool air from the northeast dominates in winter. Precipitation patterns are not homogenous over this western Pacific Island. Central and southern Taiwan receives enormous rainfall only during summers, whereas the north of Taiwan receives rainfall throughout the entire year. Rainfall in Taiwan is also modulated strongly by the Kuroshio Current, a major boundary current that transports a large volume (18–25 Sverdrup = 10^6 m^3 s^{-1}) of warm, salty seawater from the western tropical Pacific (WTP) to the North Pacific (Selvaraj et al., 2007; Wang et al., 2015).

Typhoons, which are tropical storms, also cause high precipitation over Taiwan. It has been estimated that approximately 60% of rainfall in Taiwan is from typhoons. The high rate of typhoon occurrences in Taiwan is likely due to the geographical location between the South China Sea and the Philippian Sea, which are both major sources of tropical storms. The typhoons from the western North Pacific Ocean are more frequent and powerful compared to the ones emanating from the South China Sea. On average, three tropical cyclones arrive annually in Taiwan with the highest storm frequency occurring from June to August. Köppen's classification characterizes Taiwan into four climate types, viz., mild, humid climate (Cfa) in the north, monsoon, and trade-wind coastal climate (Am) in the south, wet-dry tropical climate (Cwa) in the west, and temperate rainy climate with dry winter (Cw) in mountain areas.

2. Contemporary climate in Taiwan

The climate in Taiwan (formerly known as Formosa) is mainly controlled by the monsoon. Due to summer (Southwesterly) and winter (Northeasterly) monsoons, the climate behaves differently in different areas on this western Pacific Island. The summer monsoon brings higher rains in southern

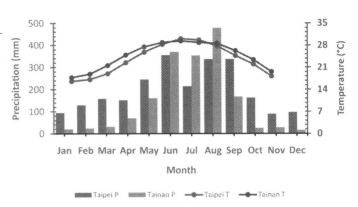

FIGURE 9.1

Average monthly precipitation and temperature (1991–2020) in Northern (Taipei) and Southern (Tainan) Taiwan.

Taiwan with scanty rainfall during the winter resulting in average annual rainfall of 1741 mm. On the other hand, northern Taiwan receives rain all year round with major rainfall during the summer monsoon summing up to 2370 mm on the annual average (Fig. 9.1).

Southern Taiwan is relatively hotter than northern Taiwan with an average annual temperature difference of 1.2°C. However, this difference nearly diminishes during the summer monsoon when it rains all over the island (Fig. 9.1). Taiwan is surrounded by the East China Sea to the north, the Pacific Ocean on the east, the South China Sea on the southeast, the Philippine Sea to the south, and Taiwan Strait to the west. This surrounding water leads to dramatic changes in weather that enhance the atmospheric vapor during the summer causing extremely high humidity and chilly winds during the winter. During typhoons, the weather also gets extremely rainy and stormy.

3. Hydroclimate variability in Taiwan during the late Holocene

Most of the published paleoclimate archives from Taiwan only extend to a few hundreds of years. The scarcity of long-range terrestrial records can be related to Taiwan's rugged topography, steep gradient, and rapid uplift, rendering most lakes short-lived and shallow, thereby limiting their suitability as paleoclimatic archives.

Here, we compiled paleoclimatic records, mainly pollen, emanating from different lakes across Taiwan (Fig. 9.2). Interestingly, comparisons between the reconstructed rainfall dynamics over the same geographical location reveal contrasting climatic scenarios over the same timescales. The hydroclimate reconstruction from Duck Pond (Chen et al., 2009), Site Kiwulan (Lin, 2004), and Sonlou Lake (Liew, 2014), based on pollen assemblages, show a cold wet interval during the little ice age (LIA) although significant biases are observed for the age models. This cold wet pattern is in contrast with cold and dry conditions observed from Cueifong Lake (Wang et al., 2015) during the LIA. Cueifong and Retreat lakes (Wang et al., 2014) show a warm wet interval, while the more southern lakes Chitsai (Liew and Huang, 1994) and Great Ghost (Lou and Chen, 1998) show a warm dry period during the MWP. The rest of the paleoclimate datasets from the different lakes do not show any recognizable climatic differences between LIA and MWP for lakes.

Overall, the inconsistencies between intra and intersite proxy comparisons are complicated by the ambiguity associated with the age models. Detailed limnological changes based on different

3. Hydroclimate variability in Taiwan during the late Holocene

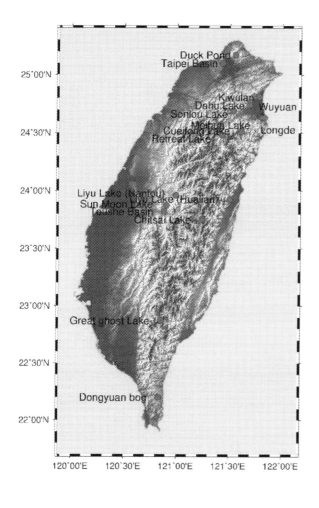

FIGURE 9.2

Map of Taiwan showing topography and the location of paleoclimatic records.

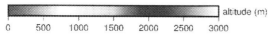

sedimentation rates and processes were often not considered in the climate reconstructions. For instance, the timescale defined for the LIA in Duckpond (Chen et al., 2009) is entirely different from Site Kiwulan (Lin, 2004), and that of Sonlou Lake (Liew, 2014) is also different from Cuifong Lake (Wang et al., 2015). This discrepancy can be seen in the reconstructed rainfall events from all the lakes across Taiwan, thus, making it difficult to detect the different hydroclimatic conditions between the north and south as observed in present-day instrumental records. A detailed compilation of the proxy records from Taiwan is shown in Fig. 9.3.

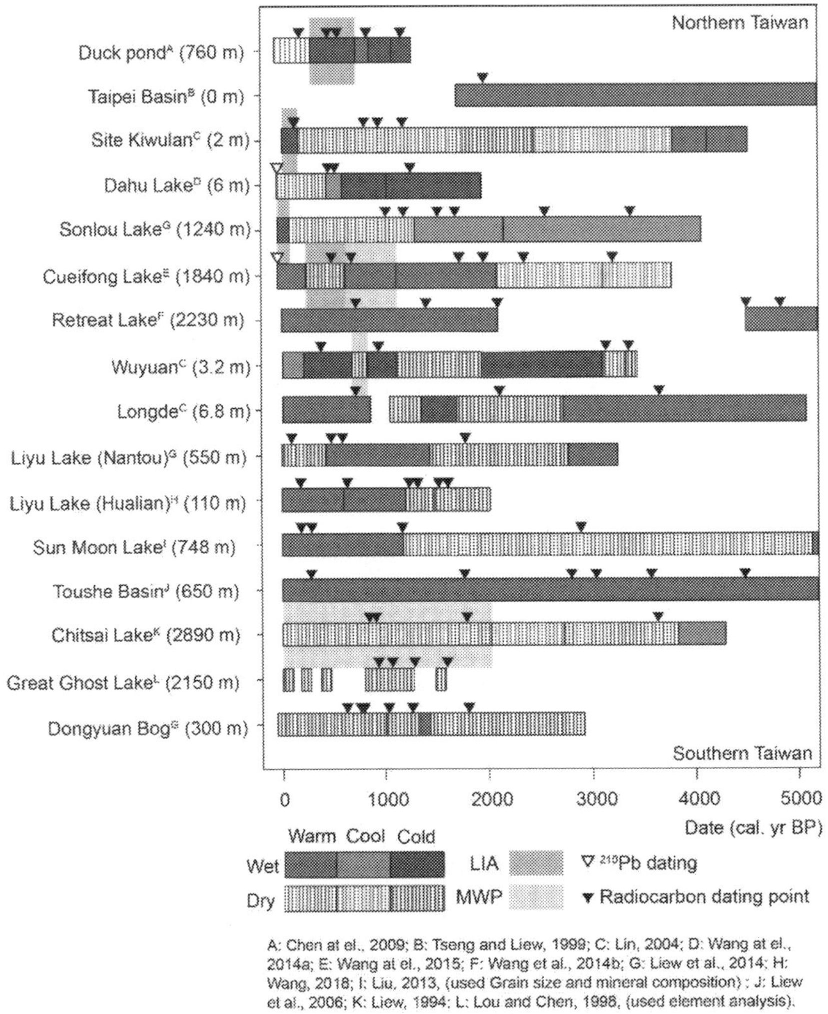

FIGURE 9.3

A compilation of late Holocene paleoclimatic records in Taiwan.

Also, these inconsistencies may be either due to erroneous interpretations of the proxies or the proposition of inaccurate mechanisms to adequately explain the hydrological signal observed. Taiwan's climate is mainly influenced by monsoon and typhoon occurrences. Other variables such as humidity, temperature, moisture source, etc., may influence these variables. The proxies may likely be recording only any or a combination of the signals. The rainfall signal for instance could be either a monsoon or typhoon or both. Therefore, these proxies must be applied in such a way that these variables are delineated to present robust reconstructed paleoclimatic conditions.

In summary, Taiwan's hydroclimate over the late Holocene remains ambiguous due to incoherent and inconsistent proxy records within the same lakes as well as different lakes. Due to the complications associated with Taiwan's lakes, further studies are encouraged in a more concerted and collaborative manner between researchers following a multidisciplinary approach to climate reconstruction.

4. Critical issues remaining

4.1 Dynamical relationship between ENSO and typhoons

Preceded by an increase in atmospheric greenhouse gases, global warming is expected to lead to higher sea surface temperatures—an important driver of El Nino Southern Oscillation (ENSO) dynamics. The variability of ENSO leads to noticeable changes in hydrological patterns not only in the tropical Pacific region but also worldwide through ocean-atmospheric teleconnections (Collins et al., 2010). The intensity and frequency of tropical hurricanes—called typhoons in Asia—have been directly linked to ENSO variability. Thus, unraveling the mechanisms of ENSO control on typhoon frequencies and monsoon dynamics provides a basis to assess the impacts of future climate change on Taiwan. However, our understanding of the dynamic relationship between ENSO and tropical typhoons remains scanty. This raises critical questions that remain to be answered: (1) Are ENSO dynamics and typhoon frequency closely connected, or are there other, independent, factors driving each of these climatic phenomena? (2) Is the dynamical relationship between ENSO and typhoon frequencies only valid during the Holocene, characterized by relatively constant boundary conditions, or also during the last glacial period with very different boundary conditions?

4.2 Unavailability of hydroclimate datasets over the LGM in Taiwan

Robust paleoclimatic data synthesis is fundamental in assessing the regional and global impact of the predicted future climate change in any area. However, this is not the case for Taiwan, where contrasting climatic scenarios over the same timescales have been inferred from sedimentary records either from the same lake or lakes with close proximities. Moreover, long-term climate records covering glacial times are almost nonexistent. The scarcity of longer terrestrial records can be related to Taiwan's rugged topography, steep gradient, and rapid uplift, rendering most lakes short-lived and shallow, thereby limiting their suitability as paleoclimate archives. In such a situation, speleothems (another continental paleoclimatic archive) could have provided glacial-interglacial paleoclimatic records. However, the absence of karstified terrain in Taiwan leaves no scope for carbonate cave formation with a minor exception though. This, therefore, presents a challenge in assessing the mechanisms driving the long-term local rainfall variability of Taiwan over the last glacial period. Critically, it makes it difficult to contextualize the local hydroclimatic variability of Taiwan, which is dominated by typhoons, vis-à-vis the more regional East Asian monsoon.

4.3 Dating uncertainties for proxy records

Precision dating is fundamental to the mechanistic understanding of hydroclimatic conditions beyond its recent limit. Presently, instrumental data records exist but are limited in their spatial coverage and only extend to the last centuries. Therefore, providing a historical perspective of climate dynamics through time and space depends on the accuracy of the age model applied to the record. However, our data compilation (Fig. 9.3) reveals incoherent and inconsistent ages, thus making it difficult to tease out a more general hydroclimate pattern over the Holocene, where most of the records extend.

References

Chen, S.H., Wu, J.T., Yang, T.N., Chuang, P.P., Huang, S.Y., Wang, Y.S., 2009. Late Holocene paleoenvironmental changes in subtropical Taiwan inferred from pollen and diatoms in lake sediments. J. Paleolimnol. 41 (2), 315–327.

Collins, M., An, S.I., Cai, W., Ganachaud, A., Guilyardi, E., Jin, F.F., Jochum, M., Lengaigne, M., Power, S., Timmermann, A., Vecchi, G., 2010. The impact of global warming on the tropical Pacific Ocean and El Niño. Nat. Geosci. 3 (6), 391–397.

Liew, P.M., Huang, S.Y., 1994. A 5000-year pollen record from Chitsai Lake, central Taiwan. Terr. Atmos. Ocean Sci. 5 (3), 411–419.

Liew, P.-M., Wu, M.-H., Lee, C.-Y., Chang, C.-L., Lee, T.-Q., 2014. Recent 4000 years of climatic trends based on pollen records from lakes and a bog in Taiwan. Quat. Int. 349, 105–112.

Lin, S.-F., 2004. Environmental and Climatic Changes in Ilan Plain over the Recent 4200 Years as Revealed by Pollen Data and Their Relationship to Prehistory Colonization (PhD Thesis). National Taiwan University, Taiwan, p. 179.

Lou, J.-Y., Chen, C.-T., 1998. Paleoenvironmental records from the elemental distributions in the sediments of Great Ghost lake in Taiwan. J. Lake Sci. 10 (3), 13–18.

Selvaraj, K., Chen, C.T.A., Lou, J.Y., 2007. Holocene East Asian monsoon variability: links to solar and tropical Pacific forcing. Geophys. Res. Lett. 34 (1).

Wang, L.-C., Behling, H., Chen, Y.-M., Huang, M.-S., Arthur Chen, C.-T., Lou, J.-Y., Chang, Y.-P., Li, H.-C., 2014. Holocene monsoonal climate changes tracked by multiproxy approach from a lacustrine sediment core of the subalpine Retreat Lake in Taiwan. Quat. Int. 333, 69–76.

Wang, L.C., Behling, H., Kao, S.J., Li, H.C., Selvaraj, K., Hsieh, M.L., Chang, Y.P., 2015. Late Holocene environment of subalpine northeastern Taiwan from pollen and diatom analysis of lake sediments. J. Asian Earth Sci. 114, 447–456.

CHAPTER 10

Potential of the lake sediments study of Ethiopia in understanding the Holocene climatic conditions: a case study of Chamo Lake, Southern Nations, Nationalities, and People's Region, Ethiopia

Subodh Kumar Chaturvedi[1], Anirudh Bhowmick[2], Gosaye Berhanu[3], Geremu Gecho[3] and Jai Ram Ojha[3]

[1]Institute of Hydrocarbon, Energy and Geo-Resources, ONGC Center for Advanced Studies, University of Lucknow, Lucknow, Uttar Pradesh, India; [2]Faculty of Meteorology and Hydrology, Arba Minch Water Technology Institute, Arba Minch University, Arba Minch, Ethiopia; [3]Department of Geology, College of Natural Sciences, Arba Minch University, Arba Minch, Ethiopia

1. Introduction

Climate is a changing phenomenon. It governs the culture, tradition, food habits, welfare of people, flora and fauna, nature of the soil of the regions, etc. In the past, climate changes have caused the rise and fall of many civilizations in different parts of the globe. However, it has absorbed many such changes and kept on flourishing as changes were natural and generally took a long time to reach the maximum, thus giving sufficient time for adaptation and sometime even migration (Chaturvedi, 2001; deMenocal, 2001; Nigam, 2003). However, the uncontrolled emission of greenhouse gases in the atmosphere due to rapid industrialization since the 18th century has accelerated climate changes. Of several consequences perceived, accelerated sea-level rise, a change in the rainfall pattern, and increasing intensity and frequency of storms are of major concern worldwide (Pearman, 1988; Warrick et al., 1996; Capobianco et al., 1999; Nigam and Chaturvedi., 2006).

Ethiopia, the land of human origin, is a testimony to the favorable climatic conditions in the recent geological past. The country constitutes one of the most significant environmental and cultural reserves on Earth (Asrat et al., 2012). The varied geography and climate, rich fertile land, and diverse life in a tropical country like Ethiopia has all the potential not only to become self-sufficient in food grains but also to export to other countries. Currently, a certain agricultural product such as coffee of Ethiopia is one of the important sources of foreign revenue generation for the country. Agriculture in Ethiopia is largely dependent on rainfall. The frequent occurrences of drought and famine in a country not only impact agricultural products but also have a greater impact on the country's economy. Under such circumstances, a country often depends on foreign aid for the sustenance of life. Therefore, a long-term climate prediction on a decadal to century basis is the need of the hour for guarding against the furies of nature and to enable the government agencies to plan and make necessary arrangements in advance for its population. A high-resolution paleoclimatic investigation especially for the Holocene Period is the need of the hour to model the future climatic scenario of the country. Since the instrumental records or direct observations for climatic fluctuation are not available beyond the past century, we have to depend on proxies for paleoclimatic reconstructions.

Ethiopia is endowed with several large natural lakes both in the rift valley and high elevated plateau that have stored a repository of climatic evidence. The sediment supply in the lake comes from the transported materials through wind, rainwater, river, and stream water from the lake catchment area and to some extent from the in-lake production. Hence, the lake forms a reservoir for sediment deposits. Further, the increasing population, deforestation, and other associated impact of anthropogenic activities together with fluctuations in climatic conditions must have affected not only the lake water but also the lake sediments and fauna contained within. These continuous deposits of organic and inorganic materials in the lakes serve as an invaluable proxy to infer the changing climate records of geological past, paleo-productivities, hydrodynamic conditions, sediment depositional environment, hydrocarbon deposits of the sedimentary basin, etc (Nigam and Hashimi, 1995; Last, 2001; Mazumder et al., 2013).

Despite its great geological significance, except few researchers (Kassa, 2015, Kebedew et al., 2020, Ojha et al., 2020; Chaturvedi et al., 2021), the lake sediments have not received adequate attention from the researchers in this part of the region due to lack of infrastructure and expertise.

In view of this, an attempt has been made to study the spatial distribution of the sediment's textural parameters and geochemical and organic matter of the Chamo Lake to understand the depositional environment in the lake. The spatial distribution of the lake floor sediments and its various sedimentological and geochemical parameters not only reflect the modern hydrological conditions and organic products of the lake, but study can form baseline data to investigate the fluctuations in lake level and paleoclimate through the study of the sediment core from the Chamo Lake.

2. Description of the study area
2.1 Physiography of Ethiopia

Ethiopia is home to a diverse range of landscapes and landforms. They were formed by a combination of tectonism, upliftment, erosion, and depositional processes working on a variety of rocks. Ethiopia's geological development, with alternating stages of mountain building (orogenesis), lava emplacement, crustal upcoming, faulting, pine plantation, and deep fluvial dissection, has imprinted specific traits on the country's geomorphological landscapes, in places that are unique on Earth.

Ethiopia covers a total area of approximately 1,132,328 km^2. It occupies a considerable chunk of Africa's inner horn, and it is a landlocked country that lacks a border with the Red Sea or the Indian Ocean. The Rift Valley bisects Ethiopia's landmass, dividing it into the eastern and western highlands, each with adjacent lowlands. Based on SRTM 30 m resolution data, the contrast in relief is striking, with land elevations ranging from −214 m at Asal Lake in the Afar valley (Africa's lowest point) to 4529 m at Mt. Ras Dejen of the Simen Mountains (Fig. 10.1). The plateau lies between 1500 and 3000 m above sea level and is dotted by a series of volcanoes that form high mountain ranges, the highest of which are the Simen Mountains in the north and the Bale Mountains in the south (Fig. 10.1).

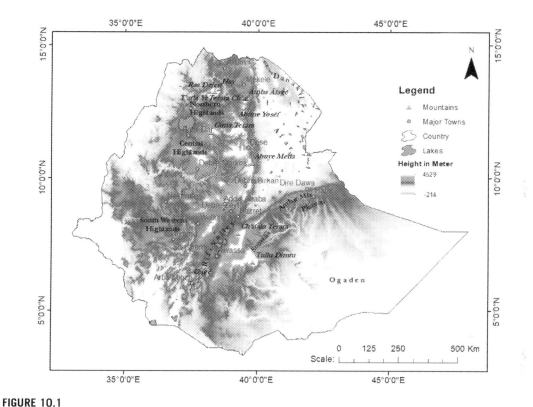

FIGURE 10.1

Physiography of Ethiopia.

The northern, central (the Blue Nile River watershed in downstream of Tana lake), and southern areas of the northwestern highlands are much more wide and mountainous. The southwestern part of the plateau (also known as the Somali Plateau) is undulated rough, but it is slightly lower in elevation (the highest peak Mt. Tullu Dimtu, 4383 m a.s.l.). It is separated into a northern area with the Ahmar and Garamullata Mountains, a center region with the Chercher Mountains, and a southern segment with the Bale and Haranna-Mandebo mountains, as well as the Ogaden. Deep river valleys penetrate through the northwestern and southeastern mountains, some as deep as 1500 m, such as the Blue Nile Gorge (Ayalew and Yamagishi, 2004), and gently descend toward the Sudan lowlands. The Ethiopian part of the Great Rift Valley spans more than 900 km, with an average width of 50–100 km, between the Kenyan and Djibouti borders, reaching heights as low as 500 m above sea level. To the north, the Afar triangle defines the fault system, while to the south, it extends past the Kenyan border and Turkana Lake. The highest peaks on the Rift floor are around the Dubeta Col (1670 m a.s.l.) between Ziway Lake and Koka Reservoir, which separates the Awash River basin from the endorheic drainage systems of the Main Ethiopian Rift (MER). The Rift's bottom descends gradually from Djibouti's Asal Lake, at 155 m above sea level, to Turkana Lake, at 361 m above sea level. The Rift Valley floor is not uniformly flat, since scattered volcanoes or volcanic systems rise more than 1000 m above sea level (e.g.,

Zikwala, 2,989, Boset, 2,447, Fantalé, 2007 m a.s.l.). The Ethiopian Rift's main trunk contains seven major lakes, all closed systems of tectonic or volcano-tectonic origin (Ziway, Langano, Abijata, Shala, Awasa, Abaya—the largest with a surface area of 1162 km^2 and Chamo). Other lakes can be found in the Danakil depression (Africa Lake, 70 km^2) and the northernmost half (Abe Lake, 350 km^2), fed by the Awash River (Wood and Talling, 1988). The northern and central highlands also have large lakes. Tana Lake, with a surface area of 3600 km^2, is Ethiopia's largest, while other highland lakes such as Hayk (23 km^2) and Ashenge (20 km^2) are significantly smaller. The Blue Nile and its tributaries are the most notable river system, the largest of which being the Tekeze/Atbara, which joins the Blue Nile in Sudan territory. Many significant rivers are the tributaries of the Abay/Blue Nile system, which drains a large area of their northern and central parts due to the typical westward slope of the western highlands.

About half of the country's water is discharged through the Blue Nile, the Tekeze, and the Baro rivers. Other large rivers that flow into the Indian Ocean originate from the Somali Plateau (Genale/Juba and Wabe Shebele rivers). Several smaller river drainage basins are closed systems, with the Awash and Omo rivers being the largest (Fig. 10.2).

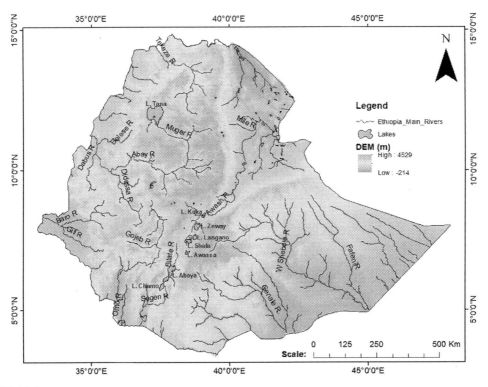

FIGURE 10.2

Main rivers and lakes of Ethiopia.

The Abay River is another name for the Blue Nile River which accounts for around 80% of the Great Nile River's volume. The Abay, which originates in Ethiopia's Lake Tana, and the White Nile, which originates in Lake Victoria, join at Khartoum, Sudan's capital, to form the world's longest river, which empties into the Mediterranean Sea. The Blue Nile Falls are located in Tis-Abay town, 30 km east of Bahir Dar, and are a 20–30 min walk from Tiss Abay. Before reaching Lake Turkana/Rudolf, the Omo River travels its way through a high, impassable gorge before slowing as it approaches the lowlands and meandering over the flat, semi-desert vegetation area. The Baro River area, which may be reached by land or air from Gambela in western Ethiopia, remains a thrilling and challenging destination for rafting. The Awash River is a river that originates in the mountains west of Addis Ababa and travels south. By the time, it reaches its destination; it has transformed into a vast flow of muddy waterfall that cascades into a 100-meter-deep canyon on its route to extinction in the Danakil Depression's lowlands.

2.2 Chamo Lake

The Chamo Lake is situated between the 5 degrees 58′ 17″ N and 5 degrees 42′ 15″ N latitude and between 37 degrees 27′ 17″ E and 37 degrees 38′ 20″ E longitude in the southern part of the Main Ethiopian Rift. The lake is observed to be elongated in shape and oriented in NNE–SSW direction (Fig. 10.3). The total catchment and surface areas of the Chamo Lake area are about 1942.65 km^2 and 328.63 Km^2, respectively.

The maximum water depth in the lake is reported to be 14 m in the central portion of the lake (Awulachew, 2007). While river Gelana, Milate, Gidabo, Hare, Baso, and Amesa are the major rivers that flow into Lake Abaya; the rivers Kulfo, Sile, Argoba, Wezeka, and Sagan flows into the Chamo Lake. Abaya and Chamo Lakes are also fed by several small streams and ephemeral rivers. The Abaya and Chamo Lakes are linked together hydrologically. An overflow from Lake Abaya during the rainy season flows into the Kulfo River and ultimately drains to Lake Chamo. The River Metenafesha flows out from the eastern shore of Lake Chamo and joins the Sermale stream. These tributaries together form part of the Segen river, which flows into the Chew Bahir Basin. Lake levels closely correlate with the prevailing precipitation pattern (Schutt and Thiemann, 2006).

2.3 Geomorphology

Ethiopia's geomorphology is mostly determined by its geological structure, but weathering, erosion, and deposition processes all had a role in sculpting the country's current landscapes and landforms. The crustal evolution, which was defined by a significant swelling within one of the planet's most active extensional areas, produced three primary morphostructural units, as well as sub-regions with distinct geomorphic traits. (1) the northwestern plateau, which is the source of many river systems can be divided into three sections: (i) the northern highlands, which comprises the Adwa volcanic plug belt, (ii) the central highlands, and (iii) the southwestern highlands; (2) the Rift Valley, which is divided into three sections: the northern, central, and southern part, as well as it constitutes the Afar and Danakil depressions; (3) the southern-eastern plateau, which is divided into northern and southern sectors and also includes the Ogaden tableland. The plateaus were formed by the domal uplift of the Arabian-Ethiopian region (Merla et al., 1979) and the accumulation of flood basalts, whose thickness ranges from 1000 to 2000 m in some areas (Kieffer et al., 2004) and even 3000 m in the Chercher

182　Chapter 10 Potential of the lake sediments study of Ethiopia

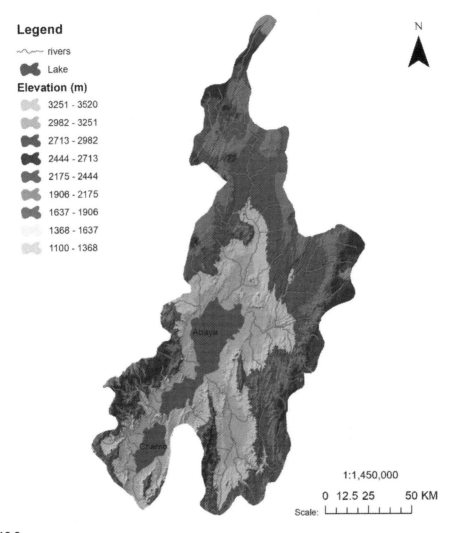

FIGURE 10.3

Triangulated Irregular Network (TIN) map of Abaya Chamo Lake basins.

highlands (Juch, 1975) (Fig. 10.1). The plateaus' highest points are around 3000 m above sea level, but they are occasionally topped by steep volcanic mountains, some of which reach elevations of over 4000 m. Close to the Sudanese border, the main plateaus drop radially and gradually to altitudes of 500–600 m and 200–400 m near the Somalia border. The dominant topographic gradient is greater along the rift margins (about 0.004–0.006), then reduces to 0.001-0.002 in the distal portions, the inselbergs of granitic composition are very common and distinctive geomorphological feature of these areas. The shift from plateau borders to rift was abrupt, with step-like topography and several extended grabens, as well as asymmetrical and unpaired horsts and a bottom, interrupted locally by recent trachyte plugs. In some spots, spectacular fault-generated escarpments can be seen, with long and high-exposed fault planes and alluvial fans at their foot.

2.4 Climate

Ethiopia's land is composed of various topography; therefore, the country has a variety of climates, as well as temperature and precipitation differences between locations. High rainfall and humidity characterize Ethiopia's equatorial rainforests in some pockets of the southern part and the southwest; however, the Afro-Alpine type of climatic condition on the peaks of the Semien and Bale Mountains, as well as in the north-east, east, and south-east plains, experience desert-like conditions. Cooler climates can be found in the country's central and northern highlands. The eastern part of the country experiences extremely desert conditions, with very little rainfall. Seasonal variation of rainfall in the parts of Ethiopia is routinely subjective by the migration of the Inter-Tropical Convergence Zone (ITCZ), with significant interannual fluctuation. Bega, Belg, and Kiremt are Ethiopia's three rainy seasons. Kiremt is the main rainy season, and it lasts from middle June to middle September and contributes to 50%–80% of yearly precipitation. Belg, a periodic secondary wet season that occurs from February to May in parts of northern and central Ethiopia, receives significantly less rainfall than the main wet season. In Ethiopia's southern regions, there are two separate wet seasons: Belg, which lasts from February to May, and the Bega, which lasts from October to December, and is drier and cooler. The southwestern highlands receive about 2000 mm of yearly rainfall, whereas the south-eastern and north-eastern plains receive less than 300 mm. Across Ethiopia, temperatures can range from $-15°C$ in the highlands to over $25°C$ in the lowlands.

2.5 Geology

The geological setup in the Ethiopian region reveals that it has a geological history of around one billion years (Fig. 10.4). The continental drift suggests that the first event was the formation of the Ethiopian basement ranging in age from 880 to 550 Ma and closed the Mozambique Ocean between West and East Gondwana. This bowed and slanted Proterozoic basement was subjected to extensive erosion over a 100 million years, obliterating any Precambrian orogen relief. In Ethiopia, Ordovician to Silurian is recognized by fluviatile sediments, while Late Carboniferous to Early Permian period was marked by glacial deposits, which were deposited above an Early Paleozoic planation surface. Particularly, the Jurassic Period is marked by flooding of most of the part of Horn of Africa, triggered by a marine incursion from the Paleotethys sea and the embryonic Indian/Madagascar Ocean at the start of Gondwana's separation. The Ethiopian region was exposed land for about 70 million years after the Jurassic transgression and deposition of Cretaceous continental deposits, during which time a new significant peneplanation surface evolved. The geological sequence found in Ethiopia with lithologic assemblages is given in Table 10.1.

During the Oligocene Period, a prolific outpouring of trap flood basalts occurred above the eroded (peneplain) land surface of low height, which coincided with the rifting (first phase) of the Afro Arabic plate. The huge shield volcanoes of the Miocene age were overlaid atop flood basalts in the northern Ethiopian plateau. During the end of the Oligocene Period, volcanism wandered toward the Afar depression, which was stretching, and traveled between the southern Ethiopian plateau and the Somali plateau in sequence, coinciding with the creation of the Main Ethiopian Rift (MER).

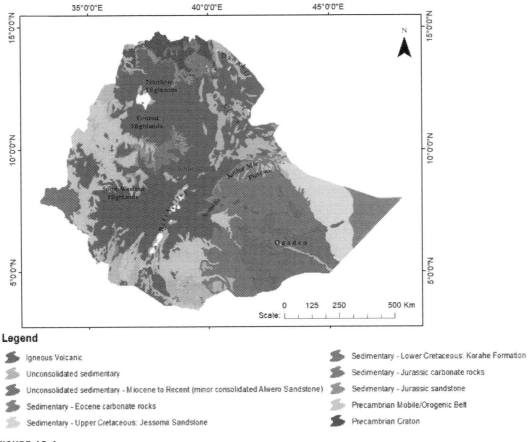

FIGURE 10.4

Geological map of Ethiopia.

Table 10.1 Geological sequence and lithology of Ethiopia (second edition, 1996 geological map of Ethiopia published by Geological Survey of Ethiopia).

Geological Era	Lithological assemblages
QUATERNARY INDIFERENTIATED	Alluvial and lacustrine deposits sand. Silt, clay, diatomite, limestone, and beach sand.
	Basalt flows, spatter cones and hyaloclastites a) Transitional type between alkaline and tholeiitic. b) Alkaline Olivine basalt.
	Plateau basalts: Alkaline basalt and trachyte
	Rhyolitic volcanic centers. Obsidian pitchstone, pumice, ignimbrite, tuffs subordinate trachytic flows (predominantly peralkaline in composition)
	Ghinir formation: Rhyolita with subordinate basalt

Table 10.1 Geological sequence and lithology of Ethiopia (second edition, 1996 geological map of Ethiopia published by Geological Survey of Ethiopia).—cont'd

Geological Era	Lithological assemblages
HOLOCENE	Undifferentiated alluvial lacustrine and beach sediments.
	Hawaiite, mugearite, trachyte, andesine basalt and ferrobasalt.
PLEISTOCENE	Alluvial, lacustrine and marine sediments: Conglomerate, Sand, clay, reef, limestone, marl and gypsum.
	Dino formation: Lgnimbrite, tuff, course pumice waterline pyroclastic rocks with rare intercalations of lacustrine sediments.
PLIOCENE–PLEISTOCENE	Bishoftu formation: Alkaline basalt and trachyte.
	Undivided Lacustrine and fluvial sediments: Sand. Silt gravel Conglomerate (Omo group and Hadar formation).
	Mursi and Bofa basalts: Alkaline basalt
	Chilalo formation (Upper part): Alkaline basalt
	Chilalo formation (lower part): Trachyte, trachy-basalt, peralkaline rhyolite with subordinate alkaline, basalt.
MIOCENE–PLIOCENE	Danakil group (Red sea series): Conglomerate. Sandstone. Siltstone with intercalated basalt flows and lacustrine sediments of the Chorora Formation
	Nazret series: Lgnimbrites unwelded tuffs, ash flows, rhyolitic flows, domes and trachyte.
	Afar series: Alkaline basalt with subordinate alkaline and peralkaline silicic (rhyolitic dome and flows and ignimbrites)
LATE MIOCENE	Dalaha formation: Fissural basalts and hawaiites with some intercalated detrital and lacustrine sediments. With rhyolitic flows and ignimbrites in the upper part.
	Tulu Wolel trachyte with subordinate basalt.
MIDDLE-MIOCENE	Tarmaber Magezez formation: Transitional and alkaline basalt.
	Mabla and Arba Guracha formation: Rhyolitic domes, flows and pyroclastic rocks of dominantly peralkaline composition with subordinate trachyte and basalt flow interstratified at the base.
	Teltele and Surma basalt: Flood basalts.
	Adwa formation: Trachyte and phonolite plug.
OLIGOCENE–MIOCENE	Tarmaber Gussa formation: Alkaline to transitional basalts often form shield volcanoes with minor trachyte and phonolite flows
	Alage formation: Transitional and sub alkaline basalts with minor rhyolite and trachyte eruptives.

Continued

Table 10.1 Geological sequence and lithology of Ethiopia (second edition, 1996 geological map of Ethiopia published by Geological Survey of Ethiopia).—cont'd

Geological Era	Lithological assemblages
	Makonnen basalts: Flood basalts_commonly directly overlaying the crystalline basement.
	Arsi and Bale basalts: Flood basalts are often connected to volcanic edifices sillsic near the upper part.
MIDDLE-LATE OLIGOCENE	Alba basalts: Flood basalts with rare basic tuff.
LATE EDCENE—LATE OLIGOCENE	Jimma volcanics (Upper part): Rhyolite and trachyte flows and tuff with minor basalt
	Jimma volcanics (lower part): Flood basalt with minor salic flows.
EOCENE	Ashangi formation: Deeply weathered alkaline and transitional basalt flows with rare intercalations of tuff often tilted (includes Akobo basalts of SW Ethiopia).
	Karkar formation: Middle-late Eocene limestone with marly intercalations.
	Taleh formation: Early-middle Eocene: Anhydrite, gypsum, dolomite, and clay
	Auradu formation: Late Paleocene—early Eocene limestone.
	Jessoma formation: Late Cretaceous-Paleocene sandstone.
CRETACEOUS	Amba Aradom formation: Sandstone, conglomerate and shale.
	Belet Uen formation: Late Cretaceous (Cenomanian-Turonian) limestone with some sandstone and shale.
	Ferfer formation: Albabian-Cenomanian shale dolomite and anhydrite.
	Mustahil formation: Aptian-Albian limestone, marl and sandstone.
	Korahe formation (main Gypsum formation): (Neocomian/Barremian)
	Upper Korahe (Kg2) gypsum, shale, dolomite and anhydrite intercalation
	Lower Korahe (Kg1) shale and limestone with basal sandstone.
LATE JURASSIC	Gabredarre formation: Kimmeridgia—Tithonian; (Jg2) Upper unit and (Jg1)—lower unit limestone with shaly and gypsiferous units.
	Agula formation: Kimmeridgian shale, marl and limestone. a) Urandab formation (Ju): Oxforclian-

Table 10.1 Geological sequence and lithology of Ethiopia (second edition, 1996 geological map of Ethiopia published by Geological Survey of Ethiopia).—cont'd

Geological Era	Lithological assemblages
EARLY-LATE JURASSIC	Kmmerdgian marl and shaly limestone b) Antalo formation (Jt): Limestone. Abay formation: Middle Jurassic limestone, shale, and gypsum. Hamanlei formation: Oxfordian limestone and shale. Adigrat formation: Triassic-middle Jurassic sandstone.
LATE PALEOZOIC–TRIASSIC	Enticho Sandstone, Edaga Arbi glacials, Gura and Gilo formations: Sandstone, shale, conglomerate, and tillite.
LATE PROTEROZOIC	Upper Proterozoic: Undifferentiated. Shiraro formation: Sandstone and conglomerate. Didikama formation: Slate and dolomite. Tembien group: Chlorite. Sericite and graphite phyllites, limestone, slate and dolomite. Tsaliet group: Metaandesite, metadacite, metarhyolite, chlorite, sericite and graphite phyllite, greenschist limestone, and quartzite Tulu Dimtu group: Metabasalt, metaandesite, greenschist, phyllite, metal conglomerate, quartzite, and marble. Birbir group: Metabasalt, metaandesite, metarhyolit, phyllite, Grephiticschist, marble, quartzite, metal conglomerate, green, Schist, metasandstone, metachert, and amphibolite. Kajimiti Beds: Metaconglomerate and metasandstone.
EARLY PROTEROZOIC	Adola group: Amphibolite, quartzite and graphitic phyllite. Marmora group: Biotite schist.gneiss.marble and graphitic schist. Wadera group: Metasandstone, quartzite, biotite, and muscovite schists
ARCHEAN	Baro group: Biotite, hornblende—biotite, garnet—amphibole, garnet—sillimanite, calc—silicate and muscovite gneisses. Yavello group: Ouartzo—feldspathic gneiss and granulite. Awata group: Biotite, hornblende, sillimanite—garnet, calc-silicate and quartzo—feldspathic gneisses, marble, and granulite. Alghe group: Biotite and hornblende gneisses, granulite and migmatite with minor metasedimentary gneisses.

Continued

Table 10.1 Geological sequence and lithology of Ethiopia (second edition, 1996 geological map of Ethiopia published by Geological Survey of Ethiopia).—cont'd

Geological Era	Lithological assemblages
PRECAMBRIAN AND PHANEROZOIC INTRUSIVE ROCKS	Konso group: Hornbtende. Pyroxene-garnet—pyroxene gneisses and amphibolite with minor metasedimentary gneiss.
	Alkali granite and syenite.
	Post-tectonic granite and syenite.
	Late to post-tectonic granite.
	Late Proterozoic Ultramafic rocks: Serpentinite-peridotite dunite and talc schists.
	Syn - tectonic granite.
	Pretectonic and syntectonic granite.
	Granodiorite and tonalite.
	Granodiorite.
	Tonalite
	Diorite
	Gabbro

The Danakil block and Arabian subcontinent were separated from the Nubian plate, resulting in steep marginal escarpments with flexure and prolonged sedimentary basins. In the Afar depression and MER, new stretching stages resulted in the development of new basins. Many of these Ethiopian basins have been traced to early human evolution and also provide human remains (Lucy) that are capable of reconstructing the early stages of human evolution. The Gulf of Aden and the Red Sea rifts were already advancing toward a link via the volcanic highlands of northern Afar when the MER crossed the Afar region in the Early Pliocene, forming a true triple junction. The creation of the Afar depression along with MER and Ethiopian plateaus is linked to Ethiopia's current physiography and neo-tectonism (1000 ma). The effect of one or more mantle plumes beneath the Afro-Arabian plate is linked to these geological recent events. The high morphology of Ethiopia's plateaus is the result of massive volcanic accumulation and subsequent uplift. The new highland structure caused a reorganization of East Africa's river system as well as substantial changes in atmospheric circulation.

3. Material and methods

A total of 15 surface sediment samples were collected from the north-western part of Chamo Lake on June 26, 2019, using Van Veen Grab Sampler and Etrex Global Positioning System (Fig. 10.5, Table 10.2).

All the sediment samples were properly labeled and transported to the laboratory for detailed analysis. The water depths at each sample location were derived from the contour map of Chamo Lake published by Awulachew (2007). The location of each sample was plotted using Arc GIS.

3. Material and methods

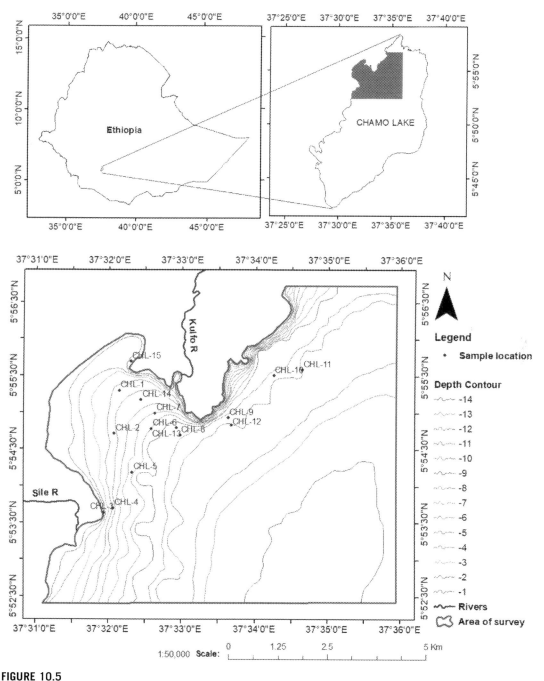

FIGURE 10.5

Sediment sample locations in Chamo Lake, Ethiopia.

Table 10.2 Sediment samples location and physical properties of the sediment.

Sample no.	Latitude (°N)	Longitude (°E)	Water depth (m)	Physical properties of sediment
CHL-1	5 degrees 55′ 17.80″	37 degrees 32′ 09.09″	4.5	Light brown in color, compact after drying
CHL-2	5 degrees 54′ 42.66″	37 degrees 32′ 05.28″	5.0	Light brown in color, compact after drying
CHL-3	5 degrees 53′ 38.07″	37 degrees 31′ 56.48″	3.0	Medium brown in color, fragile after drying
CHL-4	5 degrees 53′ 41.61″	37 degrees 32′ 04.33″	5.2	Light to medium brown, compact after drying
CHL-5	5 degrees 54′ 11.05″	37 degrees 32′ 19.86″	8.5	Dark gray in color, compact after drying
CHL-6	5 degrees 54′ 46.74″	37 degrees 32′ 35.64″	8.0	Light brown in color, compact after drying
CHL-7	5 degrees 54′ 58.18″	37 degrees 32′ 38.50″	7.5	Medium brown, very fragile after drying
CHL-8	5 degrees 54′ 41.43″	37 degrees 32′ 59.13″	10.0	Medium brown, medium compact after drying
CHL-9	5 degrees 54′ 41.43″	37 degrees 32′ 59.13″	10.5	Light brown in color, compact after drying
CHL-10	5 degrees 54′ 55.79″	37 degrees 33′ 38.30″	10.5	Dark gray in color, compact after drying
CHL-11	5 degrees 55′ 35.21″	37 degrees 34′ 38.84″	11.0	Very dark gray in color, compact after drying
CHL-12	5 degrees 54′ 49.74″	37 degrees 33′ 40.88″	11.0	Medium brown in color, compact after drying
CHL-13	5 degrees 54′ 47.81″	37 degrees 32′ 56.38″	9.5	Light brown, medium compact after drying
CHL-14	5 degrees 55′ 10.58″	37 degrees 32′ 27.09″	5.0	Light brown in color, soapy touch, less to medium compact after drying
CHL-15	5 degrees 55′ 41.75″	37 degrees 32′ 18.71″	3.5	Dark gray in color, less compact after drying

3.1 Sediment texture analysis

For textural analysis of the sediment samples, representative samples were dried at 75°C temperature overnight in the hot air oven. About 10–15 g of properly dried and weighed sediment samples were transferred in a 1000 mL beaker. A little amount of water was added to the beakers to dissociate the sediments, and subsequently, the beakers were filled with water, stirred with a glass rod, and kept overnight for settling without any disturbance. The next day, the water was removed using a decanting tube to remove supernatant material. The process was repeated 3 to 4 times till the water in the beaker is clear. To dissociate the clay lumps, 10 mL of 10% of sodium hexametaphosphate ($Na_2(PO_4)_6$) was added to the samples. If the clay lumps in some sediment samples were not dissociated, an additional 10 mL of 10% of sodium hexametaphosphate was added to those samples to dissociate the clay lumps. Thereafter, the sediments were washed out through a 63 μm sieve repeatedly until clear water passes through the sieve. The sediments screened through the sieve were collected in a plastic pan and later transferred to a 1000 mL measuring cylinder using a plastic funnel and wash bottle for silt and clay ratio estimation. Proper care was taken to note that the total volume should not exceed 1000 mL in the measuring cylinder. The residual coarse fraction obtained in the sieve was dried and weighed. This is the amount of sand content in the sediment. The volume of the water in the cylinders was made to 1000 mL. Later, the content in the cylinder was stirred vigorously using a stirrer for 45 s, and the time was noted. As per the room temperature exactly after the given time for eight phi (Φ), a pipette of 20 mL volume capacity connected with a tube was lowered in the cylinder up to 10 cm from the water level. Thereafter, 20 mL of the aliquot containing clay was pipetted out. This aliquot was transferred in a properly washed, dried, and weighed 25 mL beaker and kept in a hot air oven for drying. The dried clay sediments were weighed using an electronic balance, and the amount was recorded. The amount of sodium hexametaphosphate was subtracted from each sample while calculating the amount of clay in sediment samples. The amount of salt present in the sediment was estimated by subtracting the collective weight of sand and clay from the total weight of sediment.

3.2 XRF analysis

10 g of dried sediment samples were grinded using agate mortar and pestle. The powder samples were sieved through ASTM number 120 (125 μm) mesh sieve. The unpassed materials from the sieve were regrinding until the entire fraction was passed through the sieve. All the powdered samples were thoroughly mixed. The powdered samples were transferred into voiles and covered with a transparent plastic sheet. The voil filled with powder samples were placed into a portable Thermo Scientific Niton XRF analyzer connected with the laptop having suitable XRF software. The instruments were run to estimate the concentration of elements in the sample. Thermo Scientific Niton XRF analyzer can determine elements of above 11 atomic numbers. All the figures were plotted using Arc GIS. The percentages of each element were subjected to statistical analysis utilizing SPSS computer software to derive meaningful inferences.

3.3 Organic matter estimation

For organic matter estimation, about 5 g of properly dried and weighed sediment samples were treated with hydrogen peroxide (H_2O_2). Till the effervescence stops, repeatedly H_2O_2 has been added. Then, the H_2O_2 solution was decanted. The sediment samples were washed with distilled water and oven-dried. The weight loss has been considered equivalent to the weight of organic substances.

4. Results and discussion
4.1 Sediment texture analysis

Sediment textures are the results of the interplay of the various hydrodynamic conditions such as the velocity of water, volume of water, wave, currents, etc., hence have been utilized by researchers for classifying the sedimentary environment (Loveson et al., 2007; Rajamanickma et al., 2007; Chaturvedi et al., 2014, 2021; Ojha et al., 2020). The estimation of sand, silt, and clay contents in the surface sediments of Chamo lake reveals that clay is the dominant constituent in the lake followed by silt, whereas sand contributes least to the lake sediment. The sand content in the lake sediments is generally low and varies from 0.04% to 16.34%. Silt content varies between 0% and 64.71%. Similarly, the clay content in the lake was observed to be between 22.14% and 99.89% (Table 10.3).

The overall spatial distribution of sand content in the Chamo Lake sediments is observed to be higher near the lakeshore in the vicinity of the mouth of the river where the slope of the lake floor is comparatively gentle indicating the higher rate of sedimentation brought by the rivers (Fig. 10.6). A higher concentration of clay is observed in the north-east part of the lake at the comparatively deeper water depth indicating a slow rate of sediment deposition, whereas a higher concentration of salt content is observed in the south-western part of the study area in Chamo Lake (Figs. 10.7 and 10.8).

For the textural classification of the surface sediments of Chamo Lake, the revised textural classification proposed by Flemming (2000) for gravel-free muddy sediments is adopted as it provides a detailed textural subdivision of sedimentary environments and identifies different hydrodynamic regimes. Flemming (2000) developed a three-component classification scheme based on sand/silt/clay ratios and identified 25 sediment classes (Table 10.4). The sediment class names are used to identify respective depositional environments or facies.

The percentage distribution of sand and silt clay of the surface sediments of Chamo Lake was plotted in a ternary diagram under a three-component sediment classification scheme (Fig. 10.9). The ternary diagram of the three-component classification scheme of the surface sediments of Chamo Lake reflects that station numbers CHL3 and CHL7 are classified as silty slightly sandy mud, whereas station CHL8 and CHL15 are classified as clayey slightly sandy mud. Similarly, station number CHL14 is classified as clayey silt sediment class, while station numbers CHL4, CHL9, and CHL13 are

Table 10.3 Textural analysis of Chamo Lake sediments.

Sample no.	Sand	Silt	Clay	Organic matter	Textural class (Flemming, 2000)
CHL-1	0.04	7.13	92.83	0.55	Mud
CHL-2	0.11	0.00	99.89	1.92	Mud
CHL-3	16.34	61.52	22.14	0.45	Slightly sandy mud
CHL-4	0.64	38.02	61.34	1.19	Mud
CHL-5	0.76	7.97	91.27	0.56	Mud
CHL-6	0.05	22.16	77.79	1.24	Mud
CHL-7	15.32	46.75	37.93	0.54	Slightly sandy mud
CHL-8	10.29	23.55	66.16	4.50	Slightly sandy mud
CHL-9	2.44	27.03	70.53	4.81	Mud
CHL-10	0.33	0.00	99.67	5.08	Mud
CHL-11	0.59	0.00	99.41	2.86	Mud
CHL-12	2.61	11.86	85.52	5.06	Mud
CHL-13	0.46	49.06	50.48	3.21	Mud
CHL-14	0.19	64.71	35.11	2.39	Mud
CHL-15	14.83	27.93	57.24	1.03	Slightly sandy mud

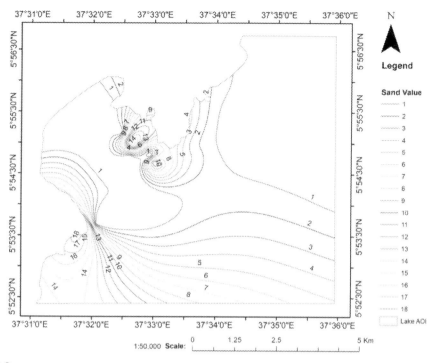

FIGURE 10.6

Distribution pattern of the percentages of sand contents in surface sediments of Chamo Lake.

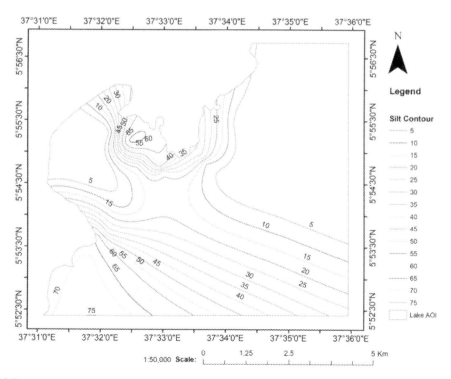

FIGURE 10.7

Distribution pattern of the percentages of silt contents in surface sediments of Chamo Lake.

194 Chapter 10 Potential of the lake sediments study of Ethiopia

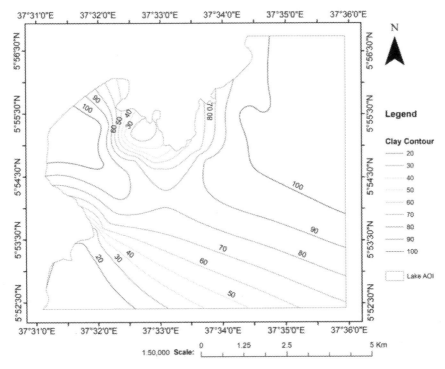

FIGURE 10.8

Distribution pattern of the percentages of clay contents in surface sediments of Chamo Lake.

Table 10.4 Descriptive terminology for the 25 textural classes based on sand/silt/clay ratios (Flemming, 2000).

Code	Textural class	Code	Textural class
S	Sand	D-I	Extremely silty slightly sandy mud
A-I	Slightly silty sand	D-II	Very silty slightly sandy mud
A-II	Slightly clayey sand	D-III	Silty slightly sandy mud
		D-IV	Clayey slightly sandy mud
B–I	Very silty sand	D-V	Very clayey slightly sandy mud
B-II	Silty sand	D-VI	Extremely clayey slightly sandy mud
B-III	Clayey sand		
B-IV	Very clayey sand	E-I	Silt
		E-II	Slightly clayey silt

Table 10.4 Descriptive terminology for the 25 textural classes based on sand/silt/clay ratios (Flemming, 2000).—cont'd

Code	Textural class	Code	Textural class
C–I	Extremely silty sandy mud	E-III	Clayey silt
C-II	Very silty sandy mud	E-IV	Silty clay
C-III	Silty sandy mud	E-V	Slightly silty clay
C-IV	Clayey sandy mud	E-VI	Clay
C–V	Very clayey sandy mud		
C'-VI	Extremely clayey sandy mud		

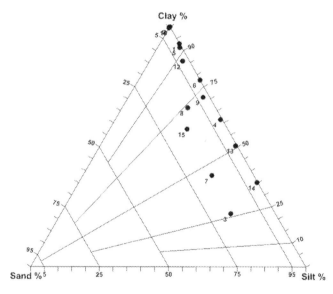

FIGURE 10.9

Textural classification of the surface sediments of Chamo Lake based on the three-component ternary plot of the percentages of the sand silt and clay.

classified as silty clay. Further, station number CHL6 and CHL12 are classified as silty clay, whereas station numbers CHL1, CHL2, CHL5, CHL10, and CHL11 belong to the clay texture class, reflecting different hydrodynamic conditions in the lake.

The organic content in lake sediments generally reflects the biological productivity of the lake. In the surface sediments of Chamo Lake, organic matter varies from 0.45% at sample location CHL3 to 5.08% at sample location CHL10. In general, the organic matter contents in the lake sediments increase with the increasing water depth (Fig. 10.10). Statistically, a good positive correlation between the organic matter and water depth having a coefficient of determination of $r2 = 0.571$ has been observed in Chamo Lake (Chaturvedi et al., 2021). The distribution pattern of organic matter in Chamo Lake shows a higher concentration of organic matter in the north-eastern part of the survey area where the influence of the river in the lake is minimum.

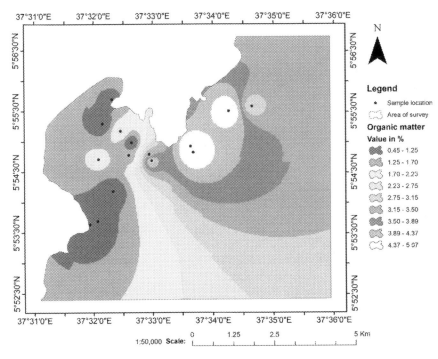

FIGURE 10.10

Distribution map of the percentages of organic matter in surface sediments of Chamo Lake.

4.2 Geochemical analysis of lake sediments
4.2.1 Spatial distribution of elements in lake

Thermo Scientific Niton XRF analyzer can determine elements of above 11 atomic numbers in powder and mining selection mode in ppm units and percentage respectively. The ppm units of elements in tern were calculated into percentages. On average, a total of 48.10% of elemental concentration of the sediments with a maximum of 50.96% and a minimum of 45.97% were determined by the XRF analyzer.

Among all the elements analyzed by the XRF analyzer, seven elements viz., Si, Al, Fe, Ca, Mg, Ti, and K were having their concentration above 1% in the sediments. The concentration of silicon in the sediment ranges from 21.36% to 24.03% having an average concentration of 22.46%, whereas the concentration of iron in the samples was observed to be between 8.86% and 10.66% with an average concentration of 9.61%. Similarly, aluminum concentration ranges between 7.89% and 9.81% with an average concentration of 8.64%. The concentration of 11 elements in the sediments such as P, Mn, S, Cl, Zr, Ba, V, Zn, Cr, Sr, and Co was observed to be between 0.3% and 0.01%. The elements such as Ni, Cu, Sc, Pb, Hg, Th, Sc, Rb, As Mo, and Au were observed to be in traces below the detection limit in a few samples whereas, elements such as Se, Sb, Sn, Cd, Ag, Bi, and W were observed to be below the detection limit in all the samples.

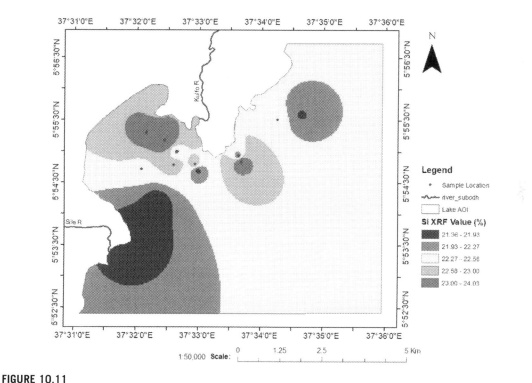

FIGURE 10.11

Spatial distribution of silicon in the Chamo Lake sediments shows its concentration near the river mouth.

The spatial distribution pattern of silicon and aluminum reflects their higher concentration in shallow water depth (Figs. 10.11 and 10.12). The spatial distribution pattern of elements such as iron and titanium is concentrated near the river mouth in the north-western and south-western portions of the lake in the study area (Figs. 10.13 and 10.14), whereas calcium and potassium are concentrated in the northwestern part of the vicinity of the lakeshore of the study area (Figs. 10.15 and 10.16).

4.3 Statistical analysis of elements in sediments

Some statistical techniques like cluster and factor techniques help to obtain meaningful patterns out of a large data set. In view of this, an attempt has been made to apply the cluster and factor techniques to the geochemical data of the study area. A total of 25 elements were utilized for statistical analysis.

4.3.1 Clustering analysis

The Q-mode cluster technique is applied to sort the stations' data into clusters that can be plotted to define regions of comparatively similar elemental composition within the study area. Similarly, R-mode clustering is carried out to produce a dendrogram of variables (elements). The clustering was performed by the computer-based statistical package SPSS. Results of the Q-mode cluster analysis are

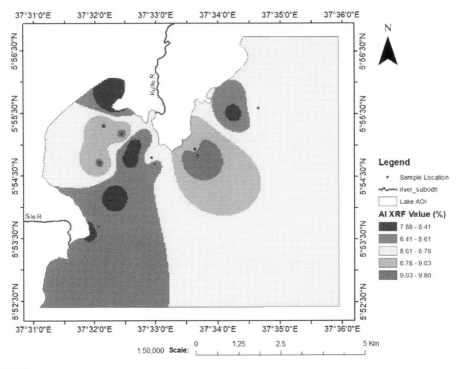

FIGURE 10.12

Spatial distribution of aluminum in the Chamo Lake sediments shows its concentration near the river mouth.

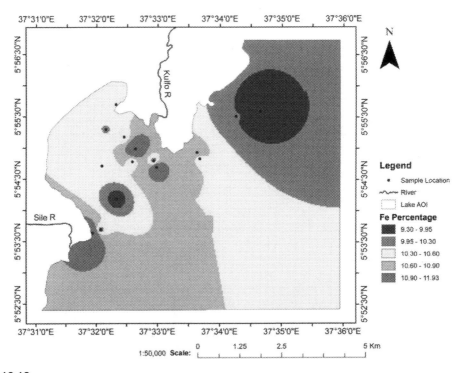

FIGURE 10.13

Spatial distribution of iron in the Chamo Lake sediments shows its concentration near the river mouth.

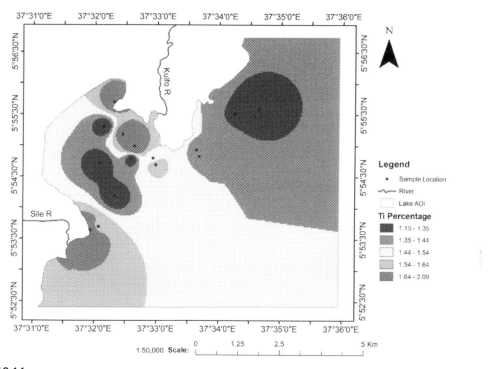

FIGURE 10.14

Spatial distribution of titanium in the Chamo Lake sediments shows its concentration near the river mouth.

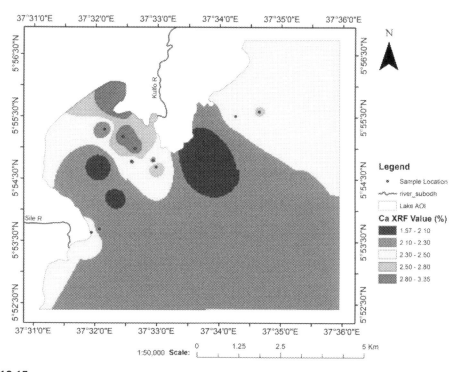

FIGURE 10.15

Spatial distribution of calcium in the Chamo Lake sediments show its concentration in the protected area in the northwestern part near the lake shore.

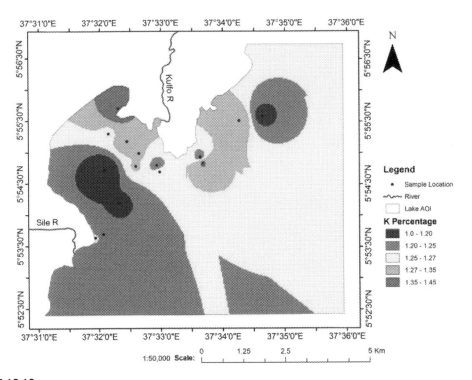

FIGURE 10.16

Spatial distribution of potassium in the Chamo Lake sediments shows its concentration in the protected area in the northwestern part near the lakeshore.

presented in the form of a two-dimensional hierarchy dendrogram in which stations are presented along the horizontal axis and linkage distance values between stations along the vertical axis. Similarly, R-mode clustering produces a dendrogram of variables in which variables are presented along the horizontal axis and linkage distance values between variables along the vertical axis. The geographical distributions of various clusters obtained at different levels of clustering are presented in Figs. 10.17 and 10.18.

In R-mode clustering, several clusters can be identified at the shortest linkage distance; however, at 10 linkage distance, three major clusters can be identified. Most of the elements including major, minor, and trace elements represented by Cluster 1. Only three elements viz., Cl, P, and Sc represent Cluster two whereas, Cluster three is represented by only gold.

Further, in Q-mode clustering, at 11 linkage distances, all the station data can be grouped into three clusters. Cluster one is represented by stations CHL 6, 10,13, 15, three, and seven, while cluster two is represented by clusters 1, 12, and 14. Cluster three represent stations CHL 2, 9, 4, 8, 5, and 11. Overall cluster one represents stations near the lakeshore, while cluster two represents stations at shallower depths, while cluster three represents stations of deeper depths.

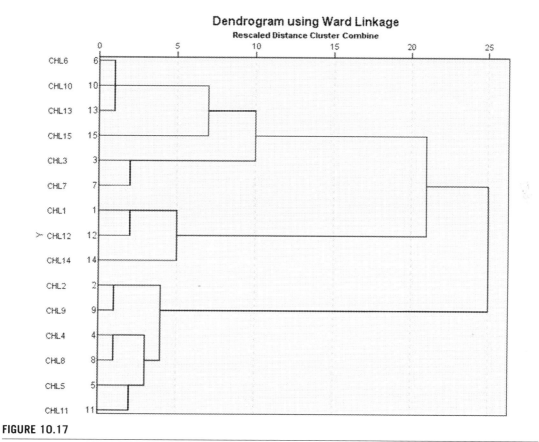

FIGURE 10.17

Dendrogram showing results of Q-mode clustering.

While comparing R-mode cluster data with Q-mode clusters, we can conveniently interpret that overall most of the major, minor, and trace elements are associated with stations near the lakeshore in cluster 1, while elements such as Cl, P, and Sc represent in Cluster two are associated with shallow water depths. Stations of deeper depths in cluster three are represented by gold.

The texture and composition of the lake sediments are strongly influenced by the slope of the terrain, distance, velocity, and volume of river water flow; rainfall, nature of rocks and soil in lake basins; waves, currents, and morphology of the lake. In terms of the hydrodynamics, the location of a data set within the ternary diagram reflects specific hydrodynamic energy conditions. The higher content of sand and silt fraction in the sediments reflect the higher energy level; on the contrary, the higher content of clay reflects lower energy condition. Based on the ternary plot of the percentages of sand silt clay contents of the surface sediments of the Chamo Lake, it can be observed that the data points show the progressive shift toward higher clay contents indicating a more rapidly decreasing energy gradient in this part of the lake. Though the mineralogical and textural composition of sediments reflects the nature and composition of source rocks, however, any source-controlled grain-size

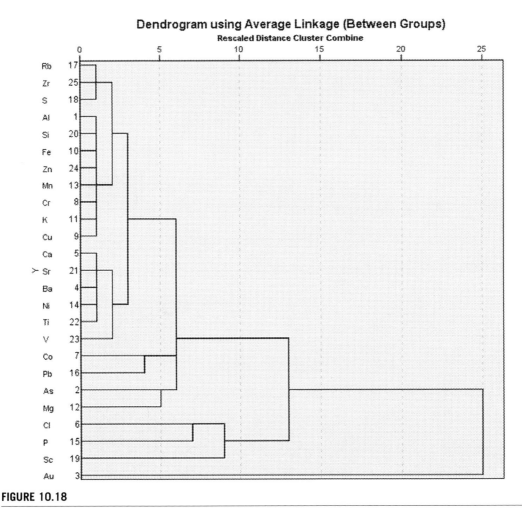

FIGURE 10.18

Dendrogram showing results of R-mode clustering.

sediments would still be subjected to a size-sorting mechanism in the course of hydraulic transport, which result in selective deposition along the local hydrodynamic energy gradient (Bagnold, 1968). In general, the sediments in the lake have been deposited under low energy conditions however, among all the sample locations, sample numbers CHL3, CHL7, CHL8, and CHL15 contain comparatively higher percentages of sand than in the other sediment samples indicating deposition of sediments comparatively in higher energy condition as compared to the rest of the samples. These samples have been classified as slightly silty sandy mud and clayey slightly sandy mud sediment texture, as these samples are in general located closer to lakeshore having greater river influence.

5. Conclusions and recommendation

A total of 15 surface sediment samples collected from the north-western part of the Chamo Lake covering a lake floor area of about 20 km^2 were utilized for the textural and organic content analysis of the sediments. The spatial distribution pattern of sand, silt, clay, and organic matter contents reveal a higher concentration of clay and organic matter in the north-east part of the lake, whereas silt content is observed in the south-western part of the study area in Chamo Lake. A comparatively higher percentage of sand contents in the surface sediments of the lake is generally concentrated near the lakeshore. The sediment and organic content distribution pattern in Chamo Lake are influenced by the river discharge, productivity, and morphology of the lake.

The presence of Fe, Ca, K, Ti, etc., in an appreciable amount in Chamo lake sediments confirms that the sediments are derived from surrounding mafic rocks. The major rivers like the Kulfo and Sile rivers which discharge their water and sediments into the Chamo Lake, together with the hydrodynamic condition of the lake, influence the distribution pattern of sediments and organic matter in the lake.

The study reveals the overall deposition of sediments in Chamo Lake under low energy conditions. The spatial distribution of sediment characteristics and organic matter contents reflects the modern hydrodynamic conditions and organic production in the lake. The present study is the modest beginning to carrying out the scientific study of the Chamo Lake sediments.

An integrated study including the sedimentological, mineralogical, geochemical, and paleontological (paleontological, micropaleontological (Ostracoda) diatom study of the surface sediments of the whole of Chamo Lake is recommended. Such a study will help to monitor changes in environmental conditions and pollution monitoring and can form baseline data over which paleoclimate from the sediment core of the lake can be inferred. Further, it is also recommended to carry out an integrated geophysical and geological study of Chamo Lake to understand the subsurface structure, nature, and thickness of surface and subsurface sediments of the lake. The study will also help to identify the suitable locations to collect the long sediment core for the paleoclimatic study of the region.

Acknowledgments

The authors are grateful to the Vice-Chancellor, University of Lucknow and President of the Arba Minch University for the permission to publish this manuscript. The financial support provided by the College of Natural Sciences, Arba Minch University vide their grant no. RCP/4019/2011 is highly acknowledged.

References

Asrat, A., Demissie, M., Mogessie, A., 2012. Geoheritage conservation in Ethiopia: the Case of the simien mountains. Quaest. Geogr. 31 (1), 7–23.

Awulachew, S.B., 2007. Abaya-Chamo Lakes Physical and Water Resources Characteristics, Including Scenarios and Impacts Catchment and Lake Research LARS 2007, pp. 162–167.

Ayalew, L., Yamagishi, H., 2004. Slope failures in the Blue Nile basin, as seen from a landscape evolution perspective. Geomorphology 57, 95–116.

Bagnold, R.A., 1968. Deposition in the Process of Hydraulic Transport Sedimentology, vol. 10, pp. 45–56.

Capobianco, M., Devriend, J.H., Nichollos, R.J., Stive, M.J.F., 1999. Coastal area impact and vulnerability assessment: the point of view of morphodynamic modeller. J Coastal Res 15, 701–716.

Chaturvedi, S.K., 2001. Distribution and Ecology of Foraminifera in Kharo Creek and Adjoining Shelf Area off Kachchh, Gujarat. Unpublished PhD. Thesis. Goa University, p. 366.

Chaturvedi, S.K., Dharani, K., Umarajeswari, S., Vidhya, B., Bharathi, V., 2014. A comparative study of pre-monsoon and monsoon sediments from the coastal regions of Nagapattinam district, Tamil Nadu, India. In: Proceedings of Advances in Soil, Water and Environmental Engineering. SASTRA University, Thanjavur, pp. 272–283.

Chaturvedi, S.K., Ojha, J.R., Geremu, G., Gosaye, B., Bhowmick, A., Abel, A., 2021. Spatial distributions of sediment texture in Chamo Lake, Southern Nations, Nationalities and People's Region, Ethiopia: it's implications in lake hydrodynamics and paleoclimatic investigations. In: Aslam, M.A.M., Lakhundi, T. (Eds.), Water Resources: Quality and Manageent. Weser Books Publisher, Germany, ISBN 978-3-96492-300-4, pp. 302–313.

deMenocal, P.B., 2001. Cultural responses to climate change during the late Holocene. Science 292 (27), 667–673.

Flemming, B.W., 2000. A revised textural classification of gravel-free muddy sediments on the basis of ternary diagrams. Continent. Shelf Res. 20, 1125–1137.

Juch, D., 1975. Geology of the South-Eastern escarpment of Ethiopia between 39° and 42° long. East. In: Pilger, A., Rosler, A. (Eds.), Afar Depression of Ethiopia. Schweizebart, Stuttgart, pp. 310–316.

Kassa, T.G., 2015. Holocene Environmental History of Lake Chamo, South Ethiopia. Ph.D. Dissertation, der Mathematisch-Naturwissenschaftlichen Fakultät der Universität zu Köln, Germany, 177p.

Kebedew, M.G., Tilahun, S.A., Zimale, F.A., Steenhuis, T.S., 2020. Bottom sediment characteristics of a tropical lake: lake Tana. Ethiopia Hydrol. 7, 1–14.

Kieffer, B., Arndt, N., Lapierre, H., Bastien, F., Bosch, D., Pecher, A., Yirgu, G., Ayelew, D., Weis, D., Jerram, A.D., Keller, F., Meugniot, C., 2004. Flood and shield basalts from Ethiopia: magmas from African superswell. J. Petrol. 54 (4), 793–834.

Last, W.M., 2001. Textural analysis of lake sediments. In: Last, W.M., Smol, J.P. (Eds.), Tracking Environmental Change Using Lake Sediments (41-81), Volume 2: Physical and Geochemical Methods. Kluwer Academic Publishers, Dordrecht, The Netherlands.

Loveson, V.J., Gujar, A.R., Rajamanickam, G.V., Chandrasekar, N., Manickaraj, S.D., Chandrasekaran, R., Chaturvedi, S.K., Mahesh, R., Deepa, P.J.J.V., Sudha, V., Sunderasen, D., 2007. Post tsunami rebuilding of beaches and the texture of sediments. In: Loveson, V.J., Sen, P.K., Sinha, A. (Eds.), Exploration, Exploitation, Enrichment and Environment of Coastal Placer Minerals (PLACER 2007). Macmillan India Ltd., New Delhi, pp. 131–146.

Mazumder, A., Govil, P., Sharma, S., Ravindra, R., Khare, N., Chaturvedi, S.K., 2013. A testimony of detachment of an inland lake from marine influence during the mid-Holocene in the Vestfold Hills region. East Antarctica Limnol. Rev. 13, 209–214.

Merla, G., Abbate, E., Azzaroli, A., Bruni, P., Canuti, P., Fazzuoli, M., Sagri, M., Tacconi, P., 1979. A Geological Map of Ethiopia and Somalia. CNR Italy, Firenze.

Nigam, R., 2003. Problems of global warming and role of micropaleontologists. *Gondwana Geol. Magazine*, Special Issue 6, 1–3.

Nigam, R., Chaturvedi, S.K., 2006. Do inverted depositional sequences and allochthonous foraminifers in sediments along the Coast of Kachchh, NW India, indicate palaeostorm and/or tsunami effects? Geo-Mar Lett. 26, 42–50. https://doi.org/10.1007/s00367-005-0014-y.

Nigam, R., Hashimi, N.H., 1995. Marine sediments and palaeoclimatic variations since the Late Pleistocene: an overview for the Arabian Sea, Quaternary environments and geoarchaeology of India. In: Wadia, S., Korisettar, R., Kale, V.S. (Eds.), Mem. Geol. Soc. India, vol. 32. Geol. Soc. India, pp. 380–390.

Ojha, J.R., Chaturvedi, S.K., Bhowmick, A., 2020. Granulometric studies of the surface sediments of Kulfo River, southern Nations, Nationalities and People's region (SNNPR), Ethiopia. East Africa Int. J. Sci. Technol. Res. 9, 428–434.

Pearman, G.I., 1988. Greenhouse: planning for climate change. Commonwealth Scientific and Industrial Research Organization, Melbourne, Australia.

Rajamanickam, G.V., Chandrasekaran, R., Manickaraj, D.S., Gujar, A.R., Loveson, V.J., Chaturvedi, S.K., Chandrashekhar, N., Mahesh, R., 2007. Source rock indication from the heavy mineral weight percentages, central Tamil Nadu, India. In: Loveson, V.J., Sen, P.K., Sinha, A. (Eds.), Exploration, Exploitation, Enrichment and Environment of Coastal Placer Minerals (PLACER 2007). Macmillan India Ltd., New Delhi, pp. 75–86.

Schutt, B., Thiemann, S., 2006. Kulfo River, south-Ethiopia as the regulator of Lake level changes in the lake Abaya-Lake Chamo system. Zentralblatt fur Geologie Palaontol. 1, 129–143.

Warrick, R.A., Oerlemans, J., Woodworth, P.L., Meier, M.F., Le Provost, C., 1996. Changes in sea level. In: Houghton, J.J., Meira Filho, L.G., Callender, B.A., Harris, N., Kattenburg, A., Maskel, K. (Eds.), Climate change 1995: the science of climate change. Cambridge University Press, Cambridge, pp. 359–405.

Wood, R.B., Talling, J.F., 1988. Chemical and algal relationships in a salinity series of Ethiopian inland waters. Hydrobiologia 158, 29–67.

CHAPTER 11

High-resolution paleoclimatic records from the tropical delta shelf off Ayeyarwady, Myanmar

Rajani Panchang[1,2] and Rajiv Nigam[2]

[1]*Marine Bio-Geo Research-Lab for Climate and Environment (MBReCE), Department of Environmental Science, Savitribai Phule Pune University, Pune, Maharashtra, India;* [2]*CSIR-National Institute of Oceanography, Dona Paula, Goa, India*

> "Can the world put the brakes on global warming and climate change without threatening economic growth that has lifted millions out of poverty?" This was the issue under deliberation at the UN-sponsored climate change conference held in Bali in December 2007. According to the then UN Secretary-General Ban Ki and its climate change experts, "Timing is critical", because UN-produced scenarios project that if an agreement on emissions isn't reached soon, global temperature increase from non-action would produce flooding, droughts, and other devastating consequences around the world with increasing frequency. "Climate is not just an environmental issue, it's a development issue," added Kathy Sierra, Vice President of the Sustainable Development Network at the World Bank. "Any agreement has to take into consideration the need for developing countries to be able to grow, to create jobs, as well as to deal with local and global pollution." Many developing nations are expected to be disproportionately affected by global warming, which, according to UN scenarios, would flood low-lying land where hundreds of millions of people live and disrupt agriculture that feeds additional hundreds of millions.
>
> **Website of the World Bank (2007)**

This meeting followed the 2004 Indian Ocean Tsunami and was soon followed by Cyclone "Nargis" in May 2008, being rated as one of the deadliest cyclones recorded in history, that devastated the isolated and low-lying regions of Myanmar.

1. Preamble

For every economy, climate is an inevitable issue when it comes to resource management. This is all-the-more true for most of the Southeast Asian economies, which every year spend a major chunk of their finances and natural resources in mitigating disasters resulting from natural calamities like floods, cyclones, typhoons, droughts, coastal land erosion, salt water ingression, land inundation, etc. Management of resources is largely dependent on scientific communities for data and predictive climate models, which help pre-empt liabilities in the near future. However, all predictive models require a vast database about past climatic variations. The set of climatic problems varies from region to region; thus, reconstruction of regional climates and sea levels is the need of the hour. But, if such predictions were to be put to the use of mankind, we need very high-resolution data, which would not only help predict climatic events on centennial to decadal time scales but also improve the precision of such predictions.

Datasets for past climate are limited. Instrumental records do not exist beyond a maximum of 100–150 years and are not sufficient to document enough periodicity in events to be used for reliable predictions. Thus, proxies which have existed since time immemorial and documented the past climate such as tree-rings, fossils, glacial varves, etc., are used. Oceanic sediments preserve the best and the longest signatures of climate change. Of the many biotic and abiotic proxies it holds, "Foraminifera" are a very sensitive group of marine microfossils which have emerged as extremely reliable paleoclimatic proxies with wide ranged applicability (Nigam, 2003, 2005; Srinivasan, 2007).

2. Why foraminifera?

Foraminifera are a group of single celled, almost exclusively marine micro-organisms. They are either planktic or benthic in habit. They are excellent paleoecological indicators because they are short-lived protists with rapid generation times and thus are generally more sensitive to changing environmental conditions (McCarthy et al., 1995). They respond to these changes by adapting, perishing, multiplying, diversifying, or changing the chemical composition of their tests, etc. Foraminifera most commonly secrete a calcareous exoskeleton called "test," which is secreted in near-equilibrium with their ambient water. These hard tests thus incorporate environmental signatures prevalent during their lifetime and get preserved as biogenic sediments after their death and serve as paleoclimatic proxies.

Some benthic species make up their tests by gathering available sediments; they are termed agglutinated forms. When abundant, they are generally indicative of calcium carbonate depleted environments rich in terrigenous material or deep marine environments. Likewise, different species with similar ecological requirements tend to form an assemblage. Thus, different environmental settings are characterized by typical assemblages and thus serve as excellent indicators of spatial and temporal change. As stated before, the calcium carbonate tests are secreted in equilibrium with the ambient sea water, whose stable isotopic and elemental concentration are a function of different environmental parameters like temperature and salinity. Thus, geo-chemical analyses of these tests provide qualitative as well as quantitative clues to climatic processes of the past.

Planktic foraminifera, abundant in the deeper waters, float in the water column and incorporate information about the water column, while benthic foraminifera give information about bottom water conditions. The latter are abundant in the coastal regions, which are characterized by high rates of sedimentation and thus serve as better sites for high-resolution paleoclimatic records (Nigam and Nair, 1989).

A thorough knowledge about the distribution and ecology of foraminifera in any study area refines the applicability of foraminifera in paleo-environmental as well as paleoclimatic studies. Such a database compiled as a ready catalog for the usage of future workers provides the basic infrastructure for R&D programs addressing paleoclimatic queries. Recognizing India's dependance on climate, foraminiferal distributions have been extensively documented in the Indian waters. The foraminiferal studies along the west and east coast of India have been reviewed (Bhalla et al., 2007; Khare et al., 2007), from which it is evident that the foraminiferal studies in the Arabian Sea have been extensive and have encompassed many disciplines, from traditional micropaleontology, archeology, and pollution (Nigam et al., 2005, 2006; Panchang et al., 2005) to culture experiments

(Saraswat et al., 2004; Panchang et al., 2006; Linshy et al., 2007). The reconstruction of past monsoon fluctuations at higher resolutions has gained the maximum attention in the recent past (Nigam, 1993; Nigam and Khare, 1992, 1995; Naidu, 1999; Naidu and Malmgren, 2005; Gupta et al., 2003, 2005; Sarkar et al., 1990; Khare et al., 2008). Due to the major role played by the African subcontinent in the origination of the Indian monsoons (which is of global interest), even the African coast has been studied by many international scientific groups globally. This is however not true with the Bay of Bengal.

3. Status of foraminiferal studies from the Bay of Bengal

Khare et al. (2007) summarized the major foraminiferal works (published up to 2006) from the shallow water and nearshore regions of the East Coast of India. Subsequent works were summarized by Rana (2009). Panchang (2008) summarized all the foraminiferal works from the entire Bay of Bengal, not included in Khare's report (2007) into four tables. She divided the Bay of Bengal into four regions (Fig. 11.1) as coastal and near-shore regions along east cost of India (region A), deeper regions of Bay of Bengal (region B), Andaman Sea (region C), and Ayeyarwady Shelf (region D). From the review, she observed that the coastal and nearshore regions all along the East coast of India (Fig. 11.1, region A) had been studied thoroughly in terms of foraminiferal distributions. Many of these studies have been reinforced with live distribution data, seasonal, and ecological data which have been in turn analyzed statistically to produce reliable databases. Studying past sea levels, detection of pollution and mapping tsunami inundated areas along coastal regions have been the main applications of foraminifera from shallow water regions. Bhatia (1956) attempted to study the concept of foramo-geographic provinces, first proposed by Cushman (1948), in the Indian waters. Barik et al. (2022) have studied the impact of metal pollution in the sediments of Chilika lagoon on benthic foraminifera. The delta regions and sedimentary basins such as the Krishna-Godavari and Cauvery basins have been studied to estimate their oil and natural gas bearing potentials.

Panchang (2008) also observed that primarily planktic foraminiferal assemblages and their isotopic analyses have been used to reconstruct paleoclimates on the glacial−interglacial time scales, only from the deeper regions of Bay of Bengal (Fig. 11.1, region B) and Andaman Sea (Fig. 11.1, region C). Interestingly, only Naqvi et al. (1994) and Saraswat et al. (2005) have used a benthic foraminiferal proxy from deeper water regions of Bay of Bengal to reconstruct past environments. Bhadra and Saraswat (2022) studied the foraminiferal dissolution with respect to calcite saturation in the western Bay of Bengal. Suokhrie et al. (2020) has compared the living benthic foraminifera from the Bay of Bengal and Arabian Sea from the oxygen minimum zone and correlated 20 species with the low oxygen. Govil et al. (2022) studied the variation in the southwestern monsoon and surface water hydrography during Holocene in the western Bay of Bengal, correlating Holocene climatic variability to the change in the solar activity. Govil and Naidu (2011) studied the monsoonal variability of last 32,000 years using planktonic foraminifera from the western Bay of Bengal. Burton and Vance (2000) observed the glacial−interglacial variations over the 150,000 years in the Bay of Bengal by studying the neodymium isotope using planktonic foraminifera. Verma et al. (2021) used benthic foraminifera to understand influence of monsoonal variations on productivity and deep-water oxygenation in the western Bay of Bengal in the last 45,000 years. A few sediment trap samples had been used from the deeper regions of Central Bay of Bengal to understand seasonal processes like productivity.

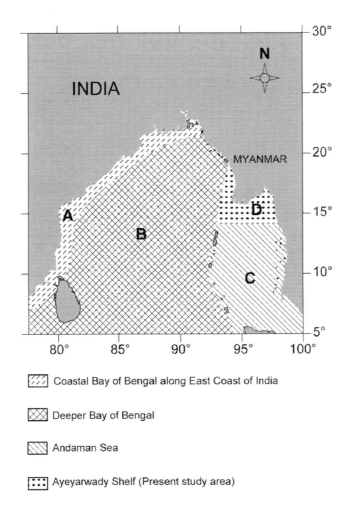

FIGURE 11.1

Map showing the four regions demarcated for reviewing the works on foraminifera in the Bay of Bengal.

After Panchang (2008).

Of the published works from the Andaman Sea, two works (Frerichs, 1971a,b) reported planktic assemblages from the deeper water regions around the Andaman Islands. A series of works from the deeper regions of the Andaman Sea have employed planktonic foraminiferal assemblages in understanding the Indian Summer monsoons and changes in thermocline (Sijin Kumar et al., 2011, 2016a,b).

Very few studies have been done along the Ganges-Bramhaputra coast, Bangladesh, Myanmar, and west coast of Thailand. In all, five reports exist on the foraminiferal distribution close to the coast of Myanmar (Fig. 11.1, region D), of which only one (Frerichs, 1970) is a published work. However, the biggest drawback of Frerichs (1970) report is that it does not contain any illustrations of foraminiferal species. The second drawback was that the species names did not indicate the original authors. In absence of the two, the report has limited applicability and is confusing to new workers in the region, especially because of the tremendous change that the taxonomy of foraminifera has undergone over the past half a century. Additionally, unfortunately, as revealed by published literature, the same foraminiferal species have been identified/named differently in different regions of the world. The work done by Kyaw & Chit Saing (2001) an interim report and the other three are published in Journals of the Union of Myanmar. These literature are inaccessible for reference.

The study of previous literature underscores the dearth of foraminiferal data along the east coast of Bay of Bengal, especially the west coast of Myanmar in the shelf regions off Myanmar. Deeper regions of the Bay of Bengal have been studied for longer time scale paleoclimatic studies but at lower resolution. But as mentioned before, reconstructions to the tune of decadal to centennial scales are useful in the prediction of paleoclimates for the benefit of the current civilization. Though the Ayeyarwady Shelf accounts for one of the highest rates of sedimentation in the world and could serve as the best archives for high-resolution paleoclimatic studies, it has not been exploited.

4. Significance of studying benthic foraminiferal distributions on the Ayeyarwady shelf

- Such work would be the first of its kind which would produce a complete listing of the recent foraminifera and their ecology in the study area. This would help cover up for the hiatus in the data collected till date.
- The comparison of data from the study area with those from the adjoining coasts, especially the east and west coasts of India, should enable us to assign the study area to known faunal provinces of the world.
- The sedimentation rate that is rather high compared to other basins provides a high-resolution record of sedimentary history, important in the study of paleo-monsoonal precipitation and variability.
- This region receives sediments mainly from Ayeyarwady, which in turn is controlled by the Asian monsoons, a phenomenon which is common to India. Especially the East coast of India shares a lot of climatic events with the west coast of Myanmar, and thus studies in this region would help extrapolate the results for the Indian regions too.
- Detailed studies in this region would help in gathering updated data, which in turn would give knowledge about natural climatic variability over which anthropogenic impact can be easily evaluated.
- The off-shore regions of Myanmar are predicted to be extremely promising in terms of petroleum reserves. Detailed foraminiferal studies should prove beneficial for such investigations.

5. The Indo-Myanmar joint oceanographic studies

Recognizing the importance of and the need to explore the Northern Andaman Sea in order to understand its modern bio-geo-chemical processes better, this region was surveyed by the CSIR-National Institute of Oceanography under the "Indo-Myanmar Joint Oceanographic Studies" initiated by the Ministry of External Affairs, Government of India, with active support from the Department of Ocean Development (DoD; now Ministry of Earth Sciences) and Council of Scientific and Industrial

Research (CSIR), India in April 2002. The international program had varied objectives of mutual interest:

- To study the bio-geo-physico-chemical processes in the Northern Andaman Sea, particularly over the continental shelf and slope regions
- To develop baseline data to initiate collaborative investigations with Myanmar in the Andaman Sea
- To impart training to the Myanmar participants in the operation of oceanographic equipment
- To organize lectures and demonstrations on board the sampling vessel ORV Sagar Kanya for the benefit of the Myanmar researchers, academicians, teachers, and students alike

National Institute of Oceanography's flagship project on paleoclimate—"Paleoceanography of the Northern Indian Ocean" collected sediment samples for foraminiferal studies during this Indi-Myanmar Expedition in the Exclusive Economic Zone of Myanmar on its Delta Shelf with a prime objective to understand this region better in terms of modern environmental processes and paleoenvironmental variations. All the surface samples and a gravity core collected off the mouths of the Ayeyarwady have been used for the present work.

6. Objectives of the present study

In view of the foregoing, the present study attempts to identify spatio-temporal signatures of benthic foraminifera to infer paleoclimatic records from one of the least understood delta shelves of the world. Recognizing the limitations of using a single proxy in paleoclimatic reconstruction in a new area of investigation, a multiproxy approach has been used. Attempts have been made to reinforce the foraminiferal findings with data on accessory groups of microfossils like sclerites and geochemical proxies like stable isotopic ratios and elemental ratios.

7. Myanmar Ayeyarwady delta shelf

As evident from the foregoing, the waters surrounding Myanmar have been least studied, though this region is characterized by one of the largest fluvial systems in the world. The large freshwater discharge from the rivers Ayeyarwady and Salween causes significant salinity changes in the Andaman Sea. The Andaman Sea is also characterized by a rather high rate of sedimentation within the Indian Ocean and thus provides a high-resolution record of sediment history. As the major source of sediment influx, here is the River Ayeyarwady, which is in turn controlled by the monsoons; the sediments of the Andaman Basin are ideal proxy for paleomonsoonal precipitation.

7.1 Andaman Sea

The Andaman Sea is a semi-enclosed basin southeast of the Bay of Bengal, south of Myanmar, west of Thailand, and east of the Andaman Islands; it is part of the Indian Ocean. It is roughly 1200 km long (north—south) and 650 km wide (east—west), with an area of 797,700 km^2. Its average depth is 870 m, and the maximum depth is 3777 m. At its southeastern reaches, the Andaman Sea (Fig. 11.2 and 11.3)

7. Myanmar Ayeyarwady delta shelf

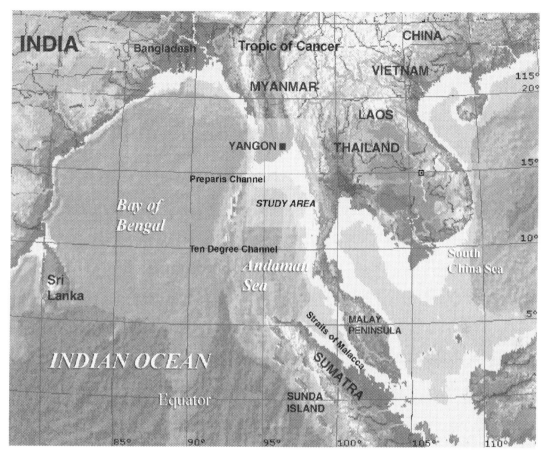

FIGURE 11.2

Regional setting of the study area.

Modified after world Atlas, MSN Encarta.

narrows to form the Straits of Malacca, which separate the Malay Peninsula from the island of Sumatra. It is connected to the Bay of Bengal by several channels among which the Preparis Channel and the 10 Degree Channel are prominent. At its southeastern reaches, the Andaman Sea narrows to form the Straits of Malacca, which separate the Malay Peninsula from the island of Sumatra (Wikipedia).

For the present study, a total area of 125,000 km^2 of the continental shelf and slope of Myanmar was sampled. Emphasis was given to cover the regions being influenced by the three main Burmese rivers. A few samples were collected off the Rakhine coast in the Bay of Bengal. To be able to interpret the data generated during the present study, it was important to have a thorough knowledge of this little-known region. The same is discussed in the next section.

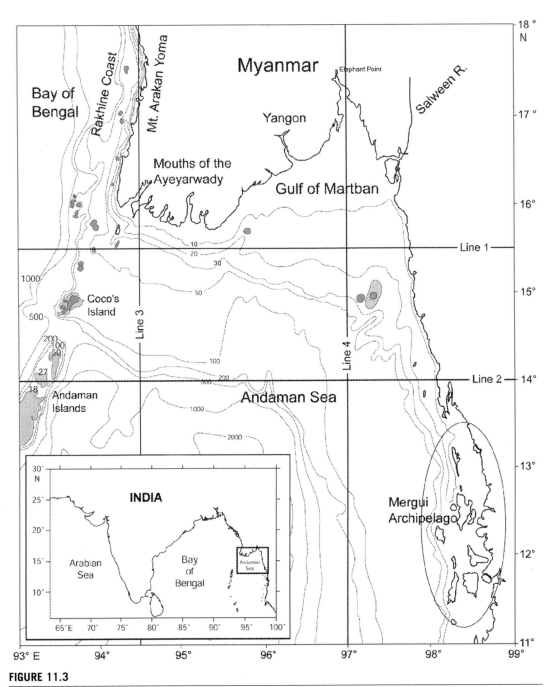

FIGURE 11.3

The bathymetric map of the study area. Line 1-4 depict the transects along which hydrographic data has been obtained from Levitus Climatology (Levitus and Boyer, 1994).

7.2 Location and extent of the Hinterland

Myanmar (formerly known as Burma) is a Southeast Asian country and very much a part of the Asian Continent (Fig. 11.2). Located along the eastern flanks of Bay of Bengal, it shares 1463 km of its land border with India along the Naga Hills. Its other neighbors with whom its shares its borders are Bangladesh in the northwest (193 km), China to the north (2185 km), Laos to the east (235 km), and Thailand to the southeast (1800 km).

7.3 Coastline

It has a 1930 km long uninterrupted coastline, which accounts for one third of its total perimeter. The Bay of Bengal occupies its southwest coastline while the Andaman Sea lies to the south.

7.4 Drainage

Three major rivers drain Myanmar and join the Andaman Sea along the coastline in the south (Fig. 11.4).

7.4.1 River Ayeyarwady (former Irrawaddy)

With a length of 2170 km is the principal river of Myanmar. It rises from the glaciers of the high and remote mountains of northern Myanmar and flows through western Myanmar, draining the eastern slope of the country's western mountain chain. The Ayeyarwady River bisects the country from north to south and empties into the Andaman Sea after forming a remarkable nine-armed delta. With its base width of nearly 200 km, the Ayeyarwady Delta is one of the world's major rice-growing areas. The flow in the Ayeyarwady is at its lowest in February and March, while there is a sharp rise in level in April—May as a result of melting snow in the upper catchment, and a further steep rise in May—June with the onset of the monsoon. The maximum flow occurs in July or August. Near the head of the Delta, a mean low and flood discharge of 2300 m^3/s and 32,600 m^3/s have been determined (Encyclopedia Britannica, 2000). The Ayeyarwady River discharges >430 km^3 of fresh water and >260 million ton of sediment annually, of which more than 80% is during the southwest monsoon (Rudolfo, 1969).

7.4.2 River Salween

With a total length of 2820 km, the Salween is the world's 26th longest river, and its basin covers a total area of 32,000 km^2 (Longcharoen, 2003). The Salween River originates in the eastern Tibetan highlands, part of the Himalayan mountain range, flows through China's Yunnan province into Burma before entering Thailand. After entering Myanmar, it forms the border with Thailand for about 110 km and continues through eastern Myanmar to empty into the Andaman Sea. Although the catchment area of Salween is limited and sheltered from seasonal rains, its water volume fluctuates considerably from season to season (Encyclopedia Britannica, 2000; FAO, 2000). It is the longest undammed river in the mainland Southeast Asia.

7.4.3 River Sittang

It is located in the south of the country between the Ayeyarwady and Salween Rivers. The Pegu Yoma range separates its basin from that of the Ayeyarwady. The river originates at the edge of the Shan

FIGURE 11.4

Major rivers in Myanmar.

Plateau southeast of Mandalay and flows southward to the Gulf of Martaban. Its length is 420 km, and its mean annual discharge is around 50 km^3 per year. The annual sediment discharge from the Salween is about 100 million ton (Meade, 1996).

The Ayeyarwady is the fifth largest river in terms of suspended sediment discharge and together with the Salween and the Sittang annually deposits 350 million tons of sediment into the Northern Andaman Sea (Ramaswamy et al., 2004). The Ayeyarwady and Salween catchments adjoin each other, debouching into the Indian Ocean over a length scale similar to the deltas of the Ganges-Brahmaputra or the Amazon. Therefore, the Ayeyarwady and Salween rivers could be considered a single point source contributing to the global ocean. The implied organic carbon yield from the catchments is 8.4–12.9 t/km^2/yr, which is clearly among the highest in the world among rivers of similarly large size (Bird et al., 2008).

7.5 The Ayeyarwady Delta

The most dominant feature of the country is the Ayeyarwady River system, the surrounding valleys and the river's massive delta in the south. The most densely populated part of the country is the valley of the Ayeyarwady River which, with its vast delta, is one of the main rice-growing regions of the world. The delta system of the Ayeyarwady River extends in a great alluvial fan from the limit of tidal influence near Myanaung (18°15′N) and extends to the Bay of Bengal and Andaman Sea, 290 km to the south. It lies between latitudes 15°40′ and 18°30′N approximately and between longitudes 94°15′ and 96°15′E and has an area of 35,135 km^2. Fig. 11.5 gives an idea about the proportion of land area of the country occupied by the delta. Its altitude above MSL varies between 0 and 5 m. This alluvial plain is bounded to the west by the southern Arakan Yoma range and to the east by the Pegu Yoma.

Drainage of the Ayeyarwady River is directly into the Andaman Sea through nine major river mouths, the Bassein, Thetkethaung, Ywe, Pyamalaw, Ayeyarwady, Bogale, Pyapon, China Bakir, and Rangoon. These rivers carry a heavy silt load, and their waters are very turbid. The delta is actively

FIGURE 11.5

The extent of the Delta on land.

accreting seawards, and as a result, the sea is very shallow for some distance out to sea. Water depths are less than 5.5 m across the whole coastline fronting the delta and up to 28 km offshore in the east. The present rate of advance of the delta is estimated at 5–6 km per 100 years, equivalent to about 1000 ha per year. Several small islands, some of which are visible only at low tide, have developed offshore.

7.6 Climate

Myanmar has a monsoonal climate, with an average annual rainfall of about 1500–2000 mm in the north increasing to 2500 mm in the southeast and 3500 mm in the southwest. Over 90% of the rain falls between mid-May and mid-November. During the monsoon season, the maximum and minimum temperatures in the coastal zone are about 37° and 22 °C, respectively. The seas may be very rough, and there are often strong winds from the south and southwest. The period from mid-October to mid-February is generally dry and cool. Temperatures rise after February, and April and early May are characterized by hot, variable weather with pre-monsoon squalls (Website of the ASEAN Regional Center for Biodiversity Conservation).

7.7 Coastal geomorphology

The coastline of Myanmar can be divided into four major physiographic divisions-

7.7.1 The Rakhine coast

Fringing the Bay of Bengal to its east, the Rakhine coast is characterized by the Rakhine Yoma (Arakan Mountains) range in the west. Between the Bay of Bengal and the hills of the Arakan Yoma is Rakhine State, a narrow coastal plain. The coast drops steeply toward the Bay of Bengal and in interrupted by occasional shoals, rocks, cliffs, submerged islands, and topographic highs, all of which follow the trend of the Andaman-Nicobar chain of Islands, which can be traced up to the Naga Hills in Northeastern India.

7.7.2 The Ayeyarwady continental shelf

The shelf region off the Ayeyarwady delta has complex geological setting in the Andaman Basin (Curray et al., 1979). The shelf width is about 170 km off the Ayeyarwady River mouths and increases to more than 250 km in the center of the Gulf of Martaban (Fig. 11.3). Bathymetric data acquired during the present study shows that the shelf break is at 110 m isobath (Fig. 11.3). Beyond the shelf break, the depth increases rapidly to approximately 2000 m, except in the bathymetric low. The seafloor within the bathymetric low is riddled with erosion channels and "V-shaped" notches (Rao et al., 2005).

7.7.3 The Gulf of Martaban

An arm of the Andaman Sea, it lies on the east of the Ayeyarwady delta, indenting South Myanmar and receiving the waters of the Sittang and Salween rivers (Fig. 11.3). A complex system of N–S trending dextral strike slip faults runs through the Gulf of Maratban and the Ayeyarwady shelf; the most prominent of these is the Sagaing Fault System that extends southwards and joins the Central Andaman Rift (Curray et al., 1979; Raju et al., 2004). An N–S trending 120 km wide bathymetric low is present between the above fault systems and the Malay continental margin. The Martaban Canyon lies within this bathymetric low and appears to be controlled by the N—S trending fault systems (discussed in Section 7.10). Ayeyarwady adjacent inner shelf is generally smooth, whereas the outer shelf has a rough surface with relief of 2—20 m and has topographic features such as pinnacles, highs and valleys, buried channels, and scarps (Rao et al., 2005).

7.7.4 Islands

The Mergui Archipelago (also **Myeik Archipelago**) is an archipelago in far southern Myanmar (Burma) (Fig. 11.3). It consists of more than 800 islands, varying in size from very small to hundreds of square kilometers, all lying in the Andaman Sea off the western shore of the Malay Peninsula near its landward (northern) end where it joins the rest of Indo-China. Geologically, the islands are characterized mainly by limestone and granite. They are as a general rule covered with thick tropical growth, including rainforests and their shorelines are punctuated by beaches, rocky headlands, and in some places, mangrove swamps. Offshore are extensive reefs. Some islands have huge boulders, soft corals, and sea fans. Most of them are completely uninhabited. The archipelago's isolation is such that much of it has not even yet been thoroughly explored (Wikipedia).

Coco's Island is a group of three small islands in the Bay of Bengal located about 50 km northeast of the Indian Andaman Islands and 300 km south of mainland Myanmar. Geographically, they are a part of the Andaman Islands archipelago and separated from the North Andaman Island (India) by the 20 km wide Coco channel (Wikipedia).

A few other islands are located in the Bay of Bengal along the Rakhine coast. The Cheduba Island with an area of approximately 500 km^2 and the Ramree Island is among the better known among these (Wikipedia).

7.8 Sediment distribution on the Ayeyarwady Shelf

Based on grain size variations, three distinct areas of sediment texture have been delineated on the Ayeyarwady shelf (i) near shore muds, (ii) outer shelf relict sands, and (iii) mixed sediments in the Martaban (Fig. 11.6) (Rao et al., 2005). The outer shelf has been known to be a zone of nondeposition and starved of modern fine-grained sediments. The relict sands cover an area of about 50,000 km^2 suggesting probable deposition during the Holocene transgression (Rudolfo, 1969). Most of the sediments discharged by the Ayeyarwady are displaced eastwards by the prevailing westerly currents into the Gulf of Martaban. The Gulf of Martaban acts as a sediment trap. The Martaban Canyon is a conduit for terrigenous sediments reaching deep Andaman Sea. The Gulf is characterized by modern muds and mixed sediments. At the center of the Gulf, the mud belt is as wide as 250 km and ranks among the largest modern mud belts of the world oceans (Rao et al., 2005).

7.9 Coastal hydrography

Depth profiles of the annual mean temperature, salinity and oxygen concentrations in the study area were obtained along two latitudes (14° N and 15.5° N, that is, Line 1 and Line 2 in Fig. 11.3) and two longitudes (94.5° E and 97° E, that is, Line 3 and Line 4 in Fig. 11.3), from Levitus Climatology (Levitus and Boyer, 1994). These profiles have been illustrated as Figs. 11.7—11.10 and discussed below.

220 Chapter 11 High-resolution paleoclimatic records

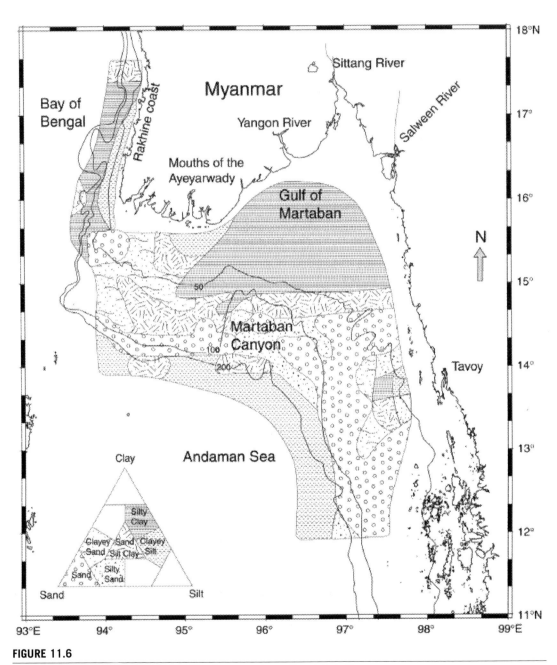

FIGURE 11.6

Textural variation in the sediments of the Ayeyarwady continental shelf.

After Rao et al. (2005).

7. Myanmar Ayeyarwady delta shelf   221

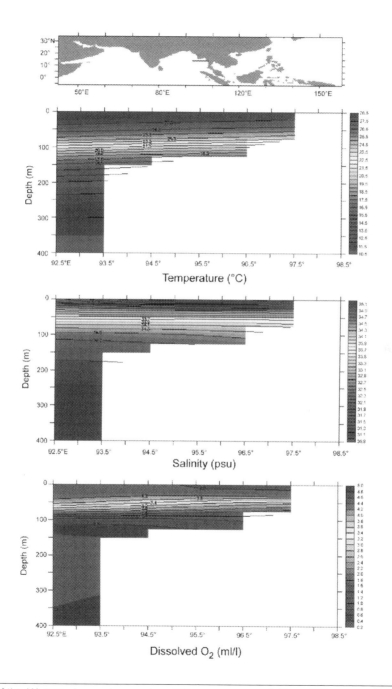

FIGURE 11.7

Depth profiles of the (A) annual mean temperature (B) annual mean salinity and (C) annual mean dissolved oxygen concentration, obtained from NOAA/PMEL TMAP FERRET ver. 5.22; data set: levannual.nc, along latitude 14° N (interpolated). Profile corresponds to line 2 in Fig. 11.3.

222 Chapter 11 High-resolution paleoclimatic records

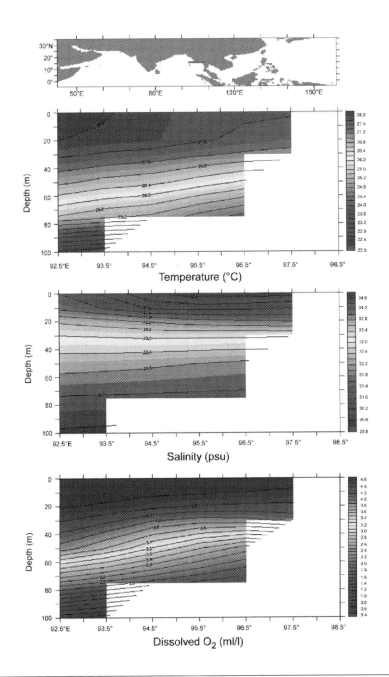

FIGURE 11.8
Depth profiles of the (A) annual mean temperature (B) annual mean salinity and (C) annual mean dissolved oxygen concentration, obtained from NOAA/PMEL TMAP FERRET ver. 5.22; data set: levannual.nc, along latitude 15.5° N (interpolated). Profile corresponds to line 1 in Fig. 11.3.

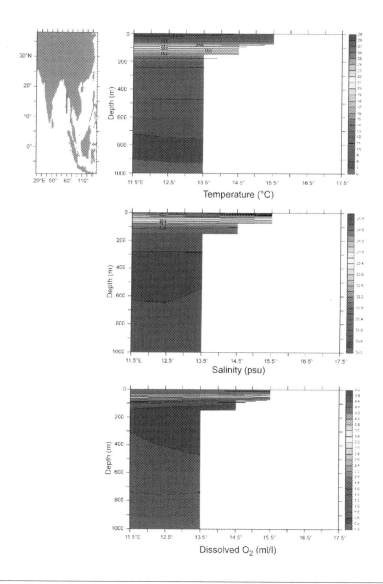

FIGURE 11.9

Depth profiles of the (A) annual mean temperature (B) annual mean salinity and (C) annual mean dissolved oxygen concentration, obtained from NOAA/PMEL TMAP FERRET ver. 5.22; dataset: levannual.nc, along longitude 94.5° E (interpolated). Profile corresponds to line 3 in Fig. 11.3.

7.9.1 Temperature

The annual mean surface temperature in the study area is ∼28 °C and drops gradually to ∼27 °C up to a depth of 40 m, after which the temperature gradient is very high. The temperature drops to about 22 °C at a depth of 100 m. Again, beyond the depth of 200 m, the gradient becomes gentle. The temperature is about 10 °C at 400 m depth and drops to about 5 °C at 1000 m (Levitus & Boyer 1994).

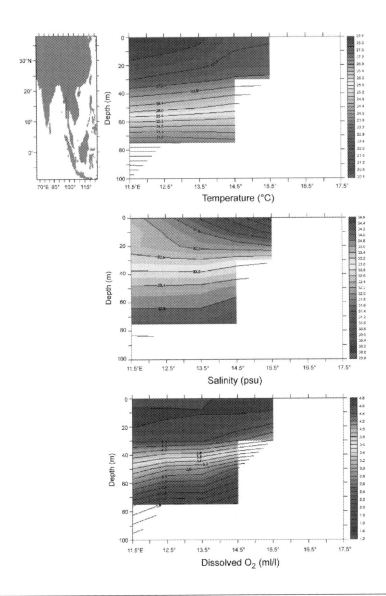

FIGURE 11.10

Depth profiles of the (A) annual mean temperature (B) annual mean salinity and (C) annual mean dissolved oxygen concentration, obtained from NOAA/PMEL TMAP FERRET ver. 5.22; dataset: levannual.nc, along longitude 97° E (interpolated). Profile corresponds to line 4 in Fig. 11.3.

7.9.2 Salinity

Closer to the coastline along the 15.5° N latitude the surface salinity varies between 31.4 psu west of the shelf to 30.2 psu in the Gulf of Martaban. The water column is well stratified in terms of salinity, and a horizontal is attained at a shallow depth of ∼15 m. However, the salinity increases rapidly up to a depth of 30 m, after which the waters are thickly stratified. At about 100 m, the salinity drops to 34.5 psu and to a maximum of 35 psu at about 200 m and remains same up to a depth of 600 m (Levitus and Boyer, 1994). At shllower depths, the salinities are lower towards the Gulf of Martaban.

7.9.3 Dissolved oxygen
The surface waters across the delta shelf seem to be well oxygenated (~ 5 mL/L dissolved O_2), especially toward the Gulf of Martaban. The drop in oxygen content is very gradual and falls below 1 mL/L beyond 90 m. The dissolved oxygen concentration drops to 0.2 mL/L at depths between 140 m and 350 m in the water column (Levitus Climatology, Levitus & Boyer, 1994).

7.9.4 Tides and current circulation pattern
Tides at the mouth of the Ayeyarwady are semi-diurnal and have a range of 2.0–2.5 m along the outer coast. At Yangon, 72 km from the open sea, the tidal range is 3.5–5.1 m. Sea dykes have been constructed in some areas to prevent tidal inundation, and the government has recently carried out several polderization schemes in the outer delta. The Andaman Sea experiences the seasonally reversing Asian monsoon (Wyrtki, 1973). Circulation in the Andaman Sea is cyclonic during southwest monsoon (May–September) and anticyclonic during northeast monsoon (December–February). The Gulf of Martaban is a macro-tidal area with its highest tidal range of nearly 7 m recorded at the Elephant Point (Indian Tide Tables, 2002). Near the mouths of the Ayeyarwady, tidal range is between 2 and 4 m and can be classified as meso-tidal. The tidal currents are strongest during spring tide, reaching as high as 3 m/s in the Gulf of Martaban (Bay of Bengal Pilot, 1978).

During the SW monsoon, the surface currents flow eastward and prevent sediments escaping into the Bay of Bengal (Rodolfo, 1969). The general circulation reverses during the NE monsoon period (November–January) and the surface currents flow toward west. These currents may push some of the suspended sediment-tongues westwards into the Bay of Bengal. Satellite images obtained during November–December reveal tongues of suspended sediments heading westward into the Bay of Bengal. The sediments carried into the Bay of Bengal may move northwards by the anticyclonic circulation of NE monsoon (Shetye and Gouveia, 1998) and probably reach the shelf region off Rakhine coast westward into the eastern Bay of Bengal (Fig. 11.10).

7.10 Tectonic setting of the study area
Situated on one of the most active continental margins of the world, the tectonic framework of the study area cannot be neglected. This is true especially because it could have implications while making estimates about paleoclimatic events such as sea-level fluctuations. The presence of relict sands in on the shelf are signatures enough of such events in the past.

The tectonic and geological history of the Andaman Sea cannot be separated from the tectonics and geological histories of Myanmar (Burma) on the north, the Andaman and Nicobar Islands part of the accretionary prism on the western side of the Andaman Sea, and Sumatra on the south (Curray, 2005). Running in a rough north–south line on the seabed of Andaman Sea is the boundary between two tectonic plates, the Burma plate and the Sunda Plate (Fig. 11.11). These plates (or microplates) are believed to have formerly been part of the larger Eurasian Plate but were formed when transform fault activity intensified as the Indian Plate began its substantive collision with the Eurasian continent. As a result, a seafloor spreading center was created, which began to form the marginal basin which would

226 Chapter 11 High-resolution paleoclimatic records

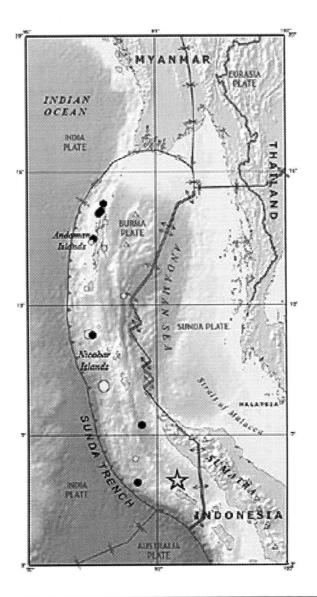

FIGURE 11.11

Tectonic framework of the study area (Wikipedia).

become the Andaman Sea, the current stages of which commenced approximately 3—4 million years ago (Ma). Within the sea to the east of the main Great Andaman Island group is Barren Island, an active volcano (the only presently active volcano associated with the Indian subcontinent). Its volcanic activity is due to the ongoing subduction of the India Plate beneath the Andaman Island arc, which forces magma to rise in this location of the Burma Plate.

As on date, the Andaman Sea is an active backarc basin lying above and behind the Sunda subduction zone where convergence between the overriding Southeast Asian plate and the subducting Australian plate is highly oblique. As stated before, a complex system of N–S trending dextral strike slip faults runs through the Gulf of Maratban and the Ayeyarwady shelf; the most prominent of these is the Sagaing Fault System that extends southward and joins the Central Andaman Rift (Curray et al., 1979 and Raju et al., 2004). Vigny et al. (2003) estimate total strike-slip plate motion in Myanmar as 35 mm/yr, with <20 mm/yr along the Sagaing Fault itself. Curray (2005) estimates that N–S motion in the Central Andaman Basin is 27 mm/yr. Sieh and Natawidjaja (2000) and Genrich et al. (2000) estimate 25 mm/yr at northwest Sumatra, decreasing to 10–20 mm/yr in southeast Sumatra. The E–W component of opening of the Central Andaman Basin at the present time is 12 mm/yr. This compares with the rate of convergence between the Andaman Islands and mainland eastern India of 15 mm/yr reported by Paul et al. (2001) from GPS surveys.

8. Data collection and analysis: the systematic approach to paleoclimatic studies

8.1 On board sampling and data collection

8.1.1 Sediment sample collection

During the 175th cruise of the ORV *Sagar Kanya* in the Exclusive Economic Zone (EEZ) of Myanmar in April 2002, a total of 126 grab samples (surface sediments) using a modified Peterson Grab were collected on the continental shelf and slope off Myanmar from depths ranging between 10 and 1130 m. The sampling locations have been plotted on the study area map (Fig. 11.12). Details such as the geographic co-ordinates and water depth at which samples have been collected are included in Panchang and Nigam (2012 and 2014). One core, GC-5 (37 m water depth, length 1.76 m) was selected for the down-core analysis of benthic foraminiferal content. It was sub-sampled at 2 cm intervals throughout its length. Thus, the core yielded 88 samples.

8.1.2 Echo sounding profiles

Geo-physical data were acquired all along the cruise track with the help of Elac deep sea Echo Sounder operating at 12 KHz. Some interesting records helped explain distribution of relict foraminifera off Myanmar.

8.1.3 CTD profiles

Temperature and salinity are very important ecological parameters for the study of foraminifera as well as climate. Though CTD profiles were obtained at a few stations onboard, temperatures and salinities could not be collected at every station onboard during the sampling. Thus, station wise data were obtained from Levitus Climatology (NOAA/PMEL TMAP FERRET Ver. 5.22, Dataset: levannual.ac). In order to assess the compatibility of the Levitus data with observed measurements at sea, the CTD profiles are presented in Fig. 11.13 At deep station 18, the temperature was 30 °C at the surface and dropped to 10 at 150 m depth and gradually to 2.5 °C at 2000 m depth (Fig. 11.13A). The salinity increased rapidly from 31.2 psu at the surface to 34.8 psu at 100 m depth after which it remained constant along the profile (Fig. 11.13B). However, the profiles obtained at shallow deltaic stations were

228 Chapter 11 High-resolution paleoclimatic records

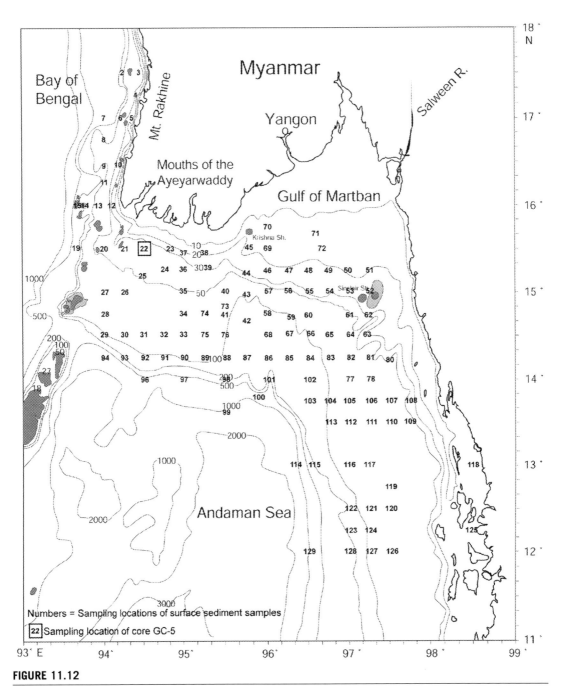

FIGURE 11.12

Map of the study area showing the sampling locations and bathymetry.

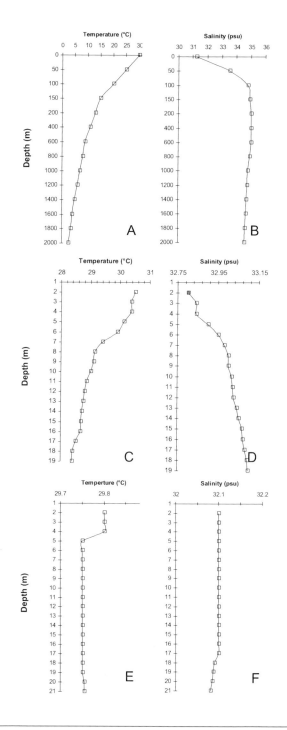

FIGURE 11.13

Temperature and salinity profiles recorded through CTD operations during the cruise: Profiles A and B were obtained at deep station 18; profiles C and D were obtained at station 46; and profiles E and F at station 48; both the shallow stations have comparable depths (20 and 22 m, respectively) but show significant differences from the top to the bottom.

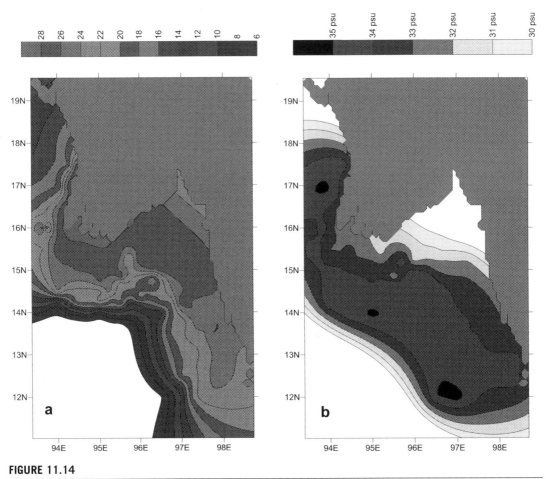

FIGURE 11.14

Spatial distribution of (A) Bottom water temperature in °C and (B) bottom water salinity in psu.

different. Station 46 and 48, though the two stations have comparable depths (20 m and 22 m, respectively), they showed different temperature and salinity profiles.

At station 46, the temperature (Fig. 11.13C) and salinity (Fig. 11.13D) range between 30.5 °C and 31.2 psu at the top to 28.4 °C and 34.8 psu at the bottom (19 m), respectively. However, at station 48, it remains constant throughout (Fig. 11.13E and F). Comparing these values with annual average values obtained from Levitus Climatology, it is inferred that the two are comparable.

Thus, the station-wise values obtained for temperature and salinity were plotted on contour diagrams to reveal their spatial variation in the benthic environment (Fig. 11.14).

8.2 Laboratory analysis
8.2.1 Sediment processing
Totally, 15 g of sediments from each of the 124 surface samples was dried overnight at 60 °C. Each dried sample was weighed and soaked in distilled water. They were subsequently treated with 10 mL

of 10% sodium hexa-metaphosphate to dissociate the clay particles followed by 5 mL of 10% hydrogen peroxide to oxidize the organic matter, if present. The treated samples were wet sieved through 63 μm (250 mesh) size sieve. The sand residue retained over the sieve was dried at 60 °C to get the weight of the sand fraction. The filtrate collected in a measuring cylinder was used for pipette analysis to determine the silt and clay fraction in each sample. The sand fraction was coned and quartered to obtain a representative aliquot from which a minimum of 300 benthic foraminiferal specimens were picked to obtain statistically adequate numbers to portray the faunal distribution (Ujjie, 1962; Dennison and Hay, 1967; Chang, 1967). They were mounted on micropaleontological slides and percentages of the individual benthic species were calculated.

8.2.2 Statistical analysis

Linear correlation coefficients (r) among various water and sedimentological parameters and the abundance of *A. trispinosa* recorded at different sampling stations in the study area, showing at least some abundance, were computed individually. Looking at the trends in the abundance pattern, four the complete dataset (also including zero abundances), was subjected to stepwise regression by removing outliers in the model by using SigmaStat for Windows (ver. 3.5). As a measure of goodness of fit, the values of the coefficient of determination (R2) were calculated. To ensure the statistical significance of the computed regression coefficients, they were subjected to a detailed t-test statistical analysis.

8.2.3 Sub-surface samples

For studying the sub-surface signatures, core GC-5 with a length of 178 cm was selected. It was collected at the delta front at a water depth of 37 m (Fig. 11.12, location same as that of surface sample no. SK-175/22). Four sub-samples were obtained from the first 10 cm of the core and the rest was sub-sampled at every 2 cm interval. Thus, 88 sub-samples were used for the sub-surface studies. All the procedures carried out on surface sediments for foraminiferal studies were also repeated for sub-surface samples. However, for reconstructing past environmental events in this region, a multiple proxy approach was adopted. The downcore variation in foraminiferal distribution was supplemented with the morphometric analysis of the species *A. trispinosa*, stable isotopic (oxygen) and elemental (Mg/Ca and Sr/Ca) studies. Results of all downcore parameters were plotted along with their three-point running averages in order to smoothen large variations.

8.2.4 Morphometric analysis

Of the picked fauna, the downcore variation in the abundance of the species *A. trispinosa* was recorded. The size of the proloculus for each specimen of *A. trispinosa* picked was measured using an Olympus stereozoom microscope "SZX12," with a maximum magnification of 180x and a precision of ±3 μm, by keeping the specimens in dorsal view. A total of 1398 specimens from 88 samples were thus measured. A value for abundance and Mean Proloculus Size (MPS) was derived nearly for every 2 cm interval (no specimen of *A. trispinosa* was found at level 72 cm within the core). These values were plotted to understand their downcore variation. The ecological information of this species derived from its surface distribution has been used to interpret the downcore data in GC-5.

8.2.5 Stable oxygen isotopic ratios

As the abundance and morphometric plots of *A. trispinosa* were showing interesting downcore variations, the same species was used to generate stable isotopic ratios throughout the length of GC-5.

Approximately 5-8 specimens of *A. trispinosa* were picked from every sub-sample of the core (Fig. 11.15) and subjected to standard sample preparation procedures. The stable isotopic analysis was performed at the newly set up facility at the Freie Universitat, Berlin, Germany on "Finnigan MAT 253 Isotope Ratio Mass Spectrometer" coupled to an automatic carbonate preparation device (Kiel I) (Fig. 11.16). All measurements were normalized to Vienna Pee Dee Belemnite (VPDB) using Laaser Marmor (internal standard), NBS-18 and NBS-19, and IAEA−CO−1 (international standards). The values are given in δ-notation versus VPDB. Precision of oxygen isotope measurements based on repeat analysis of a laboratory standard was better than 0.05‰.

FIGURE 11.15

Sample vials.

FIGURE 11.16

Samples being loaded on the gas bench.

8.2.6 Elemental studies

Elemental (Ca–Mg–Sr) concentrations were measured on 20–25 specimens of *Hanzawaia concentrica* (per sub-section) to determine the variations in temperature (Mg/Ca) and salinity (Sr/Ca) in core GC-5, using the VISTA-MPX ICP-OES (inductively coupled plasma optical emission spectrometer) (Figs. 11.17 and 11.18) at GeoForschungsZentrum (GFZ) Potsdam, Germany. This species was chosen because it was available in enough numbers throughout the length of the core. Mg/Ca analysis requires a systematic cleaning procedure which includes five important steps: (rinse,

FIGURE 11.17

Simultaneous data log of the analysis.

FIGURE 11.18

Samples being fed through the inductively coupled plasma optical emission spectrometer (ICP-OES) sample inlet.

reduction, oxidation, chelation, and leaching) (Martin and Lea, 2002; Barker et al., 2003). It has been established and reported (Martin et al., 1999; Martin and Lea, 2002) that there is no difference in the mean Sr/Ca values measured from samples subjected to the simplest procedure (rinse and leach) versus those subjected to more aggressive cleaning. However, the reproducibility of the measurement improves with rigorous cleaning. Because rigorous cleaning does not introduce any artifact into Mg/Ca and Sr/Ca datasets, it is acceptable to use the full trace metal cleaning protocol (Martin and Lea, 2002). Thus, same samples were used for simultaneous measurements of Mg/Ca and Sr/Ca because they gave the advantage of measuring more parameters on the same sample. Results were obtained as an average of three replicate measurements of the samples which gave a mean reproducibility of ±0.03 Mg/Ca mmol/mol and ±0.05 Sr/Ca mmol/mol.

8.3 Geochronology

Depending upon the findings, both surface and subsurface samples have been dated to assign the geochronology of events observed. As no reservoir age exists for the Ayeyarwady shelf, those previously used for the Andaman region, that is, Delta R value 12 (±34) years, reported for Stewart Sound (13° N 93° E), North Andaman (Dutta et al., 2001) has been used to calibrate all samples. The conventional radiocarbon ages were calibrated using the Marine-04 dataset by Hughen et al. (2004) as used by the online version of the standard radiocarbon calibration program CALIB 5.0.2 (Stuiver and Reimer, 1993).

8.3.1 Surface sediment dating

As stated before, relict foraminiferal assemblages were encountered in the study area. Thus to establish their significance, relict specimens from seven samples were picked out and dated at the Ion Beam Lab of the Institute of Physics, Bhubaneswar, using accelerator mass spectrometry (AMS). This is the first set of AMS dates ever to be generated in India. The details of the samples dated and the AMS dates obtained have been listed in Table 11.1.

Table 11.1 Details of surface samples dated and AMS dates obtained.

Geographic location		Sample number	Depth of sample (m)	^{14}C age (yr. BP)	Calibrated age
Lat. °N	Long. °E				
17°29′	94°15′	SK-175/2	137	9968 (±130)	8969 (±183) BC
16°00′	93°73′	SK-175/14	160	13,840 (±150)	13,840 (±273) BC
16°00′	93°68′	SK-175/15	29	996 (±95)	1376 (±71) AD
15°49′	94°00′	SK-175/20	33	2997 (±85)	825 (±101) BC
14°74′	93°99′	SK-175/28	76	7354 (±100)	5855 (±105) BC
14°50′	97°00′	SK-175/64	52	3394 (±110)	1276 (±150) BC
14°25′	94°26′	SK-175/93	122	11,567 (±100)	11,092 (±95) BC

Table 11.2 Details of sub-surface samples dated and AMS dates obtained.

Depth in core (cm)	^{14}C age yr. BP	Calibrated age
80–82	635 ± 30 BP	1676 (±48) AD
176–178	800 ± 25 BP	1513 (±46) AD

8.3.2 Subsurface sediment dating

Pb-210 dating of the top of core GC-5 done at the Geochronology Lab at the National Institute of Oceanography, Goa, revealed that the top was intact. Subsequently three radiocarbon AMS dates were obtained on recent foraminiferal specimens picked at two different levels within the core GC-5, at the Leibniz Labor für Altersbestimmung und Isotopenforschung, Christian-Albrechts-Universität Kiel, the details of which are included in Table 11.2.

Considering the fact that the top was intact on the basis of Pb-210 activity in the top of the core, it was assigned to the year 2002 (year of collection). On the basis of the two AMS dates obtained, the ages for the remaining levels within the core were intrapolated (Fig. 11.19) Thus, in the top 82 cm length of

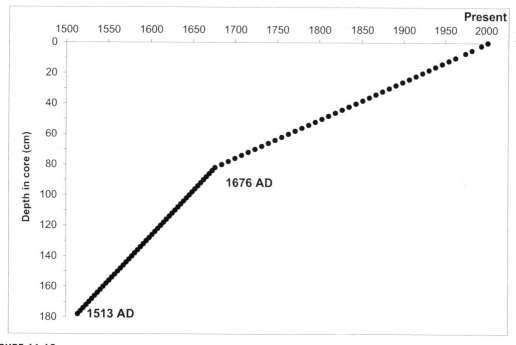

FIGURE 11.19

Intrapolation of dates within core GC-5 on the basis of Pb-210 dating and ^{14}C accelerator mass spectrometry (AMS) dates. X-axis represents age in cal. Yrs.

the core, the resolution of each 2-cm sub-section is 7.95 years and 3.38 years for the rest of the core. This chronology has been used to describe the different climatic events reflected in core GC-5.

8.4 Coral sclerites: associated microfauna with environmental significance

Associated fauna lead to key ecological data that could reinforce the foraminiferal interpretations. The samples with relict foraminiferal fauna evidently contained soft coral sclerites which accounted for 10%—15% of sand fraction. Most soft corals form "spicules" in their soft tissues (endoskeleton), which get preserved as fossils. The soft corals are an extremely diverse group and show variation in distribution based on their depth preferences. The sclerites are typical to different species and thus can be used in identifying the soft coral species that produced them (Gosliner et al., 1996). As the coral sclerites seem to indicate significant paleoecology in the study area, the different kinds of sclerites were picked and were identified to decode sea level history along the West Coast of Myanmar.

9. Comparison of the foraminiferal assemblages from Ayeyarwady delta shelf with adjoining coasts: insights into global ocean-atmospheric linkages of the study area

A total of 737 foraminiferal species belonging to 235 genera have been reported from the study area (Panchang, 2008; Panchang and Nigam, 2014). The taxa were compared with adjoining coasts, with special reference to the Indian Coasts (they being of immediate relevance). Table 11.3 enlists the reports which have been used to compare the present data with.

Table 11.3 List of literature with which present assemblages are compared.

Region	Author	Study area
Arabian Sea	Mazumder (2005)	Off central west coast of India
Bay of Bengal	Subbarao et al. (1979)	Visakhapatnam shelf, central east coast of India
	Rao (1998)	Inner shelf off Karikkattukuppam, near Chennai
	Saraswat (2006)	Distal Bay of Bengal fan
Ayeyarwady shelf	Frerichs (1970)	Complete expanse of Andaman sea
	Theingi Kyaw and Chit Saing, 2001	Off the mouth of the Ayeyarwady
South China sea	Lambert (2003)	Mahakam delta, east Kalimantan, Indonesia
	Renema (2006)	Shelf off east Kalimantan
	Saidova (2007)	South China sea

It can be summarized that of the 737 species reported from the Myanmar, 132 have already been reported before from the study area indicating that the remaining 605 species are being reported from the study area for the first time from the study area (of which 57 have been identified up to the generic level). However, it was noticed during a thorough literature review, during the process of identification that majority of these species have been reported before from the Pacific region, while many others are cosmopolitan species. Of those reported here, 150 species have been reported from the Arabian Sea before and 230 species are common between the Bay of Bengal and the study area. 35 of these species reported here have been reported from the South China Sea before.

Foraminifera lend themselves very well to zoogeographic investigations since they tend to be representative of particular geographic areas. They are rather sensitive to changes in the environment in terms of species diversity and abundance (Boltovskoy and Wright, 1976). Cushman (1948) was the first one to propose the concept of foram-geographic provinces and he assigned the recent warm water fauna all over the world to four main geographical provinces: East African, Indo-Pacific, West Indian, and Mediterranean. He assigned the west coast of India and part of the east coast to the east African province. Ever since, the assignment of the Indian regions to the known provinces became debatable. Bhatia (1956) assigned both the coasts to the Indo-Pacific province. Bhalla (1968, 1970) postulated that almost the entire east coast belonged to a "mixed zone." Later, Bhalla & Nigam (1988) selected the assemblages at three beaches each, along the east (Puri beach in Orissa, Visakhapatnam beach in Andhra Pradesh, and Marina beach in Tamil Nadu) and west coasts of India (Bhogat beach in Gujrat, Juhu beach in Maharashtra, and Calangute Beach in Goa). These assemblages were subjected to cluster analysis. They obtained two individual clusters representing the two different coasts. On the basis of these results, they postulated that the two coasts belonged to different foram-geographic provinces. They attributed this difference to the possible difference in ecological factors of Bay of Bengal and the Arabian Sea.

Comparing the foraminiferal assemblages from the study area with those from the West Coast of India, East Coast of India, Eastern Kalimantan, and South China Sea, it is evident that the study area has greater number of species (230) that are common to the Bay of Bengal and only 150 species common to the Arabian Sea. At the same time, the study area has 154 species common to the Bay of Bengal, South China Sea, and eastern Kalimantan, which have not been reported from the Arabian Sea before. Thus, it seems like the study area shares more similarity with the faunal realms east of the Indian sub-continent.

In view of the foregoing, it seems like this region has a strong affinity toward the "mixed zone" proposed by Bhalla (1970) and/or Indo-Pacific foramogeographic province of Cushman (1948). At the same time, the faunal assemblages from the Bay of Bengal and the Arabian Sea are quite different. The species *Asterorotalia trispinosa* has never been reported from the Arabian Sea (Rao, 1998; Mazumder, 2005). It however occurs all along the East Coast of India and also further east of the study area, up to Eastern Kalimantan. This indicates a foramogeographic boundary between the east and west coast of India, which could be attributed to ecological differences in the two water bodies (Panchang and Nigam, 2012).

The station-wise foraminiferal taxonomic data subjected to cluster analysis revealed the existence of different microenvironments within the study area in the form of clusters viz.: Cluster A- representing coral reef environment; cluster B- representing extremely low salinity regimes in regions close to the mouths of the Ayeyarwady and the Gulf of Martaban; Cluster C—representing the shelf environments with higher salinity regimes. This environment is further broadly clustered into three major sub-clusters, to reveal microenvironments within the shelf. However, Panchang (2008) and Panchang and Nigam (2014) concluded that the benthic foraminiferal assemblages are governed by the strong salinity gradient prevalent in the study area.

10. Relict benthic foraminifera and sea level history along west coast of Myanmar

Of the 126 surface sediment samples, the sand fractions of 22 samples contained relict Foraminifera along with associated relict fauna. These samples have a depth range of 29–160 m and occur almost parallel to the coastline, but their spatial distribution is dispersed and discontinuous (Fig. 11.20). The depth of collection of these samples, and the percentage of relict foraminiferal assemblage at such locations is incorporated in Table 11.4.

10.1 Relict benthic foraminifera

Those foraminiferal specimens that looked quite different from the normal recent benthic foraminifera were termed "relict." They could be identified (Murray, 1973) by their dull luster, yellowish to gray color and slightly broken to abraded surfaces, with identifying features nearly intact (Fig. 11.21). Their appearance indicated their relict nature suggesting that they have been lying on the surface of the ocean bottom, without being buried under sediments. The relict faunal species are enlisted in Table 11.5.

10.2 Larger benthic foraminifera

The relict foraminiferal assemblage commonly comprises of "larger benthic foraminifera," which is characteristic of coral reef environments. They are algal symbiont-bearing foraminifera. They protect their own protoplasm by secreting calcium carbonate tests, microscopically working as greenhouses for symbiotic micro-algae, the zooxanthellae. The algal partner provides nutrition and oxygen to the host and also removes waste matter, and in return receives shelter, protection, and organic nutrients from the host. In housing symbiotic microalgae, foraminifers become independent of food in nutrition-depleted environments like coral reefs (Hallock, 2002). Foraminifers being single-celled are generally micro- or meio-fauna in size. However, the foraminiferal assemblage of a coral reef ecosystem is unique. High salinity and pH, in shallow tropical and subtropical marine environments, enriches the amount of carbonate-ions $(CO_3)^{2-}$ in seawater, favoring the foraminifera to grow to large sizes, sometimes reaching a few centimeters (Hohenegger, 1996). Thus, they are commonly termed *"Larger Foraminifera."* The genera which flourish in these environments include *Amphistegina, Operculina, Alveolinella, Heterostegina, Calcarina, Marginopora, Amphisorus, Borelis, Cycloclypeus, Peneroplis,* and *Sorites* (Saraswati, 2002). Similar to stony corals, all Larger Foraminifera depend on light that is essential for photosynthesis of their algal symbionts (Hohenegger, 1996). Thus, coral reef environments are known for prolific development of larger foraminifera. Thus, the occurrence of the larger foraminifera encountered in the study area and enlisted in Table 11.5 is interpreted with reference to the ecology discussed above.

10. Relict benthic foraminifera and sea level history 239

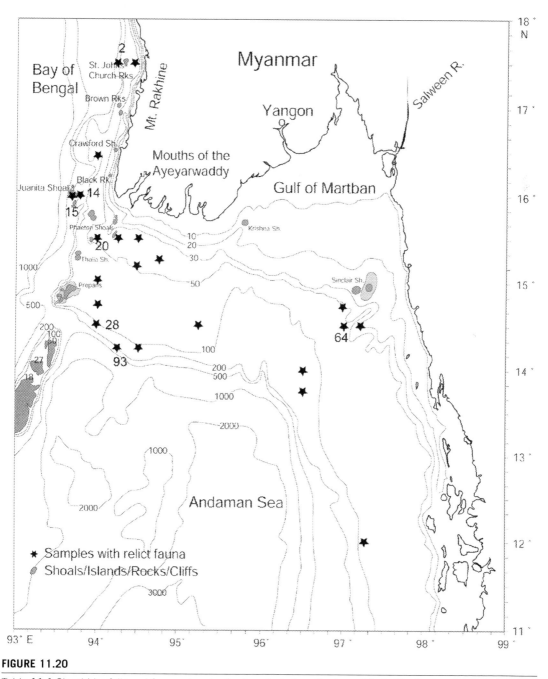

FIGURE 11.20

Table 11.1 Should be followed for precise depth of each station. Contours filled with shades of gray indicate submerged islands and shoals. The stations accompanied by station numbers indicate the locations at which accelerator mass spectrometry (AMS) dates have been procured.

Table 11.4 Details of relict foraminiferal abundance in samples containing relict fauna. Depths mentioned have been recorded using echo-sounder, onboard during sample collection.

Sample no.	Depth (m)	Percentage abundance of relict foraminifera
SK-175/2	137	84%
SK-175/3	49	39%
SK-175/9	132	7%
SK-175/14	160	11%
SK-175/15	29	36%
SK-175/20	33	82%
SK-175/21	32	45%
SK-175/22	37	13%
SK-175/24	35	30%
SK-175/25	47	15%
SK-175/27	70	22%
SK-175/28	76	15%
SK-175/29	105	22%
SK-175/61	40	2%
SK-175/63	35	40%
SK-175/64	52	70%
SK-175/75	73	15%
SK-175/92	97	63%
SK-175/93	122	40%
SK-175/102	110	35%
SK-175/103	138	5%
SK-175/127	92	8%

10.3 Associated relict fauna

The station with relict foraminiferal assemblage also showed the presence of some other relict fauna. This comprised soft coral sclerites (Fig. 11.22A), solitary caryophyllids (Fig. 11.22B and C), coralline algae (Fig. 11.22D), coral debris, and coarse coral sand. Each one is discussed in brief to enumerate their ecological preferences.

10.3.1 Soft coral sclerites and caryophyllids

Soft corals and solitary corals are frequently a conspicuous component of shallow, Indo-Pacific tropical reef communities. They have been reported to provide ~95% of the total living animal cover on some reefs (Wylie and Paul, 1989). Like all corals, even they have skeletons made of calcium carbonate which occur in the form of small needles, flowers, dumbbells, or plates, all with the general name **"sclerites"** (Website of NOAA Ocean Explorer). Calcified sclerites are common in soft corals and gorgonians and are frequently used as taxonomic indicators (Van Alstyne et al., 1992). 10%−15% of sand fraction with relict fauna comprised soft coral sclerites, which were identified after a close

10. Relict benthic foraminifera and sea level history

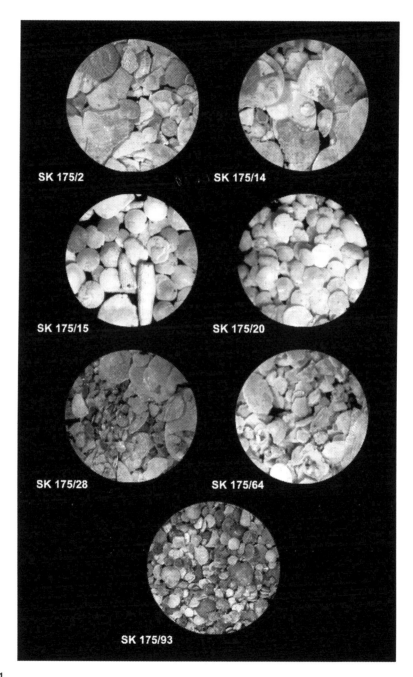

FIGURE 11.21

The relict foraminiferal assemblages encountered in the study area. Accelerator mass spectrometry (AMS) dates were obtained on the illustrated assemblages picked from seven different samples.

Table 11.5 List of species constituting the relict foraminiferal assemblage in the study area; species have been listed alphabetically and classified on the basis of their abundance. The larger foraminiferal species is indicated by an asterix.

Occurrence	Species
Abundant:	*Alveolinella quoii**
	*Alveolinella Sp.**
	*Amphistegina bicirculata**
	*Amphistegina papillosa**
	*Amphistegina radiata**
	*Calcarina hispida**
	*Calcarina spengleri**
	*Gypsina vesicularis**
	*Heterostegina operculinoides**
	*Heterostegina sub-orbicularis**
	*Nummulites cummingi**
	*Operculina ammonoides**
	*Operculina granulosa**
	*Operculina heterosteginoides**
	*Parasorites orbitolitoides**
	*Planorbulinella acervalis**
	*Planorbulinella larvata**
Common:	*Asterorotalia dentata*
	Cycloforina semiplicata
	Elphidium crispum
	Elphidium macelliforme
	*Peneroplis pertusus**
	Quinqueloculina elongata
	Quinqueloculina kerimatica
	Quinqueloculina laevigata
	Quinqueloculina parkeri
	Quinqueloculina partschii
	Quinqueloculina rariformis
	Rotalidium annectens
	Septotextularia rugosa
	Siphogenerina raphanus
	Siphogenerina transversus
	*Sphaerogypsina globulus**
	Textularia pseudocarinata

Table 11.5 List of species constituting the relict foraminiferal assemblage in the study area; species have been listed alphabetically and classified on the basis of their abundance. The larger foraminiferal species is indicated by an asterix.—cont'd

Occurrence	Species
Rare:	*Cornunuspiroides compressa Amma**
	*Coscinospira hemprichii Ehrenberg 1840**
	Hauerina elongata
	Pseudohauerina ornatissima
	*Siphoniferoides siphoniferus**
	*Spincterules anaglyptus**
	Spiroloculina subimpressa

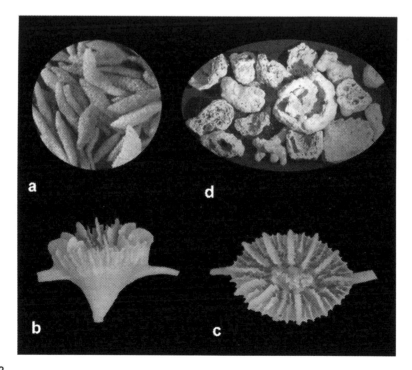

FIGURE 11.22

Associated coral fauna; (A) Microphotograph showing bulk sclerites (25x); (B) side view of a caryophyllid specimen (7x); (C) top view of the same caryophyllid specimen; (D) Coral rubble and debris; transverse sections of a few specimens show the concentric growth of coralline algae.

examination of their morphology with the help of SEM pictures (Fig. 11.23). The soft coral sclerites found in these samples belong to soft coral assemblage *"Siphonogorgia* (5a), *Chironephthya* (5b), *Subergorgia* (5c), *Acalycigorgia* (5d), and *Ctenocella* (5e)" assemblage.

10.3.2 Coralline algae

Coralline algae are widespread in all of the world's oceans, where they often cover close to 100% of rocky substrata. They are primary producers and provide the basic food supply for the entire reef ecosystem. Not only are these algae food for some coral reef animals, but they also contribute to the making of the limestone framework of the reef. They are characterized by a thallus that is hard as a result of calcareous deposits contained within the cell walls. Coralline algae are made up of masses of very fine thread-like filaments that spread out in thin layers over the reef rock surface. These filaments produce calcium carbonate thus giving the algae an appearance more like a rock than a plant. Many are typically encrusting and rock-like, found in tropical marine waters all over the world. The encrusting filaments trap sediments of sand, as well as cement the particles of sand together. Thus, coralline algae help to stabilize the coral reef structure (William and Edwards, 1993). Calcification by crustose coralline algae is crucial to the formation and maintenance of Coral reefs (Wray, 1971; Littler, 1972). Coralline algae bind adjacent substrata and provide a calcified tissue barrier against erosion (Bak, 1976).

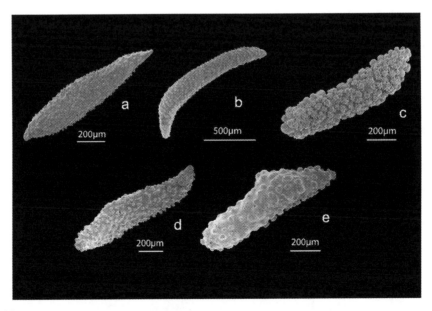

FIGURE 11.23

Sclerites of the soft coral assemblage encountered during the present study; (A) *Siphonogorgia*, (B) *Chironephthya*, (C) *Subergorgia*, (D) *Acalycigorgia*, and (E) *Ctenocella*.

10.4 Integration of relict faunal proxy data

Three groups of organisms are responsible for the construction of coral reefs. Besides corals and coralline algae, the Foraminifera are extremely abundant in coral reef environments. The coral reef environments are known for prolific development of algal symbiont-bearing Foraminifera, commonly termed *"Larger Foraminifera"* reported to inhabit modern reefs (Cockey et al., 1996; Hallock, 2000; Hart and Kaesler, 1986; Hohenegger et al., 1999; Langer and Lipps, 2003; Saraswati, 2002; Venee-Peyre, 1985). But those Foraminifera found in present study are "relict" as indicated by their dull, earthy luster, gray to brown color, and aging, biofouling, and deterioration of the tests over the time (Fig. 11.21). They are interpreted to indicate the presence of former reefs in the study area.

Another important observation is that these relict Foraminiferal tests show no indications of transport or reworking such as fracturing, breakage, polishing, rounding abrasion, and secondary replacement of test material or obliteration of identifying features. This eliminates the possibility of their being transported from another location.

Secondly, a similar relict reef-foraminiferal assemblage (*Alveolinella, Amphistegina,* and *Operculina*) indicative of coral reefs has been encountered at a comparable depth range of 90–135 m along the west coast of India (Mazumder, 2005). Mazumder has suggested the occurrence of a paleo-reef off the coast of Goa, dated to 11,000 years ago. He reinforces the postulation by Vora et al. (1996) who using east-west echo-sounding profiles along the length of the west coast confirmed the existence of a north-south trending paleo-reef parallel to the west-coast of India. Furthermore, both studies argue that the postulated reef was destroyed during the Holocene sea-level rise.

A comparison of echo-sounding records in the two areas only enhanced the degree of similarity along the two coastlines. The echograms obtained along E–W profiles off Rakhine coast of Myanmar exhibited "peaks" and "pinnacles" (Fig. 11.24A) at several locations (Eg. Station 2, 9, 13, 14, 15, 20, 28) comparable to those obtained by Vora et al. (1996) along the West coast of India (Fig. 11.24B) and interpreted as paleo-reefs.

The existence of paleo-reefs is further indicated by the abundant occurrence of soft coral sclerites. Most soft corals form "spicules" in their soft tissues (endoskeleton) which get preserved as fossils. The soft corals are an extremely diverse group and show variation in distribution based on their depth preferences. The sclerites are typical to different species and thus can be used in identifying the soft coral species that produced them (Van Alstyne et al., 1992; Gosliner et al., 1996). In the present study, the sclerites were identified to belong to the *"Siphonogorgia, Chironephthya, Subergorgia, Acalycigorgia, and Ctenocella"* assemblage (Fig. 11.23A–E). These genera of soft corals prefer hard rocky substrates and are restricted to shallow depths (maximum 32 m). They are common inhabitants of the Indo-Pacific coral reefs and occur on reef slopes, barrier walls, and other vertical surfaces (Gosliner et al., 1996).

Based on the relict larger Foraminifera, the echo-sounding records, soft coral sclerites, calcareous algae, and coral debris in the samples along the west coast of Myanmar, it is logical to postulate the existence of a paleo-reef parallel to much of the coast of Myanmar. I further discuss why the relict fauna occurs in patches and at variable depths.

The Indian Naval Hydrographic Chart No.41 (published in 1979 by the National Hydrographic Office, Dehra Dun, India) was referred for an overview of the bathymetry in the region. The Rakhine

246 Chapter 11 High-resolution paleoclimatic records

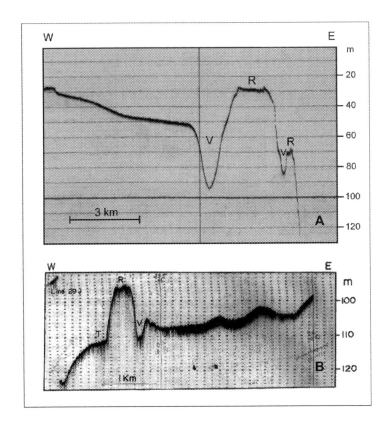

FIGURE 11.24

Comparison between E−W echograms obtained along the west coast of (A) Myanmar (station 14 and 15 during the present study) and (B) India (Vora et al., 1996).

coast slopes abruptly and the shelf is extremely narrow. But toward the south, the Ayeyarwady deltaic sediments extend far into the sea, giving rise to a gentle slope. A series of topographic highs, with rather small aerial extents, can be noticed at certain locations. From North to South these can be identified as St. John's or Church Rocks (35 m water depth), Brown Rocks (40 m), Hnget-taung Kyun (20 m), Crawford Shoal (25 m), Black Rocks (23 m), Juanita Shoal with a surrounding group (29−58 m), Phaeton shoals (6−23 m), and Thalia Shoal (18−44 m) and groups of shoals within the Preparis Channel. All these shoals are shown in gray in Fig. 11.20. Similar small shoals also occur in the outer shelf parallel to the deltaic coastline to the south. Most of these are also the sites where the relict fauna occurs. It is has been reported earlier that the outer shelf has a rough topography and is characterized by features such as pinnacles, highs, and valleys, buried channels and scarps and is a zone of nondeposition, starved of modern fine-grained sediments (Rao et al., 2005). Based on the presence of relict foraminifera in this region, Rodolfo (1969) suggested that the outer relict shelf sands were probably deposited during the Holocene transgression. The exposure of the relict sands near the Ayeyarwady River mouths points to the lack of cross shelf transport of the river sediments. The suspended sediment concentration and the sediment distribution pattern show that most of the river discharge is transported eastward along the coast into the Gulf of Martaban (Ramaswamy et al., 2004).

The echograms (Fig. 11.24A) show "pinnacles" only across the shoaling sites; in between two such locations the sea-floor depicted by the echogram is flat. The echo-sounding data and bathymetry thus portray that the ocean bottom at such locations steeply slopes giving rise to small submerged islands or groups of islands which are restricted in aerial extent. The chain of islands between Myanmar and the Andaman-Nicobar Archipelago are well known though have been hardly studied. But the small cliffs/ islands/shoals close to the coastline of Myanmar are inconspicuous and seem to be neglected as they do not fall into the purview of extensive navigation. In view of the foregoing, the occurrence and distribution of the relict fauna in the study area can be explained with the help of a schematic diagram Fig. 11.25 as follows:

The flat reliefs have been sites of sediment accumulation, whereas the pinnacles, formed due to tectonic accretion (Bender, 1983; Brunnschweiler, 1974; Subramanian, 2005), obviously have escaped sedimentation. It has been reported that corals do not develop without a stable substratum to attach and optimum depth-light conditions (Fox, 2001; The Coral Reef Alliance, 2004). Given the conditions, the corals found these so-called "submerged islands" ideal sites to nucleate and grow. This explains the intermittent absence of relict coral reef fauna along the tract. It is a well-known fact that the sea level has been lower than the present quite a few times in the geological past (Hashimi et al., 1995). After each standstill, the sea level has risen in episodes, sometimes gradually and sometimes rapidly. Thus, Panchang et al. (2008) postulated that the relict fauna at different depths suggest the colonization of corals during different sea stands and thus would show different ages.

10.5 Proposition of a regional sea-level curve

So as to establish the depth to age relationship of these samples, specimens of the relict foraminiferal assemblage from seven stations representing different depths were identified for ^{14}C AMS dating and dated as discussed in the previous section. The calibrated ages were plotted against the depth at which the samples were collected. Based on the average of the depth preference of the different soft coral species, 17.5 is taken as the common depth and added to the depth at which the relict fauna occur (Fig. 11.26). The error bars used in Fig. 11.27; however, account for a variation from 5 to 30 m, that is, ± 12.5 m from the assigned sea level. Thus, considering the coral ecology in the region, the sea level curve (Fig. 11.27) for the west coast of Myanmar has been constructed.

10.6 Comparison with other sea-level curves

The sea-level curve proposed for the study area shows significant differences from other global and regional sea-level curves published before (Fig. 11.28). Many global sea-level curves (Waelbroeck et al., 2002; Stanley, 1995; Toushingham and Peltier, 1991) and the one those proposed for Barbados (Fairbanks, 1989), all indicate the sea level to be at ~ 110 m below present MSL $\sim 16,000$ yr. BP. However, present study shows that the sea level was at ~ 140 m below present MSL in the study area at the same time. The curve has similar shifts throughout the 16,000-year record.

However, it is worth noticing that at 13,000 yr BP, the sea level in the study area is comparable with the regional sea level at Barbados and the global sea level proposed by Stanley (1995). Along the east coast of India, $\sim 14,500$ and $\sim 12,500$ yr. BP, the sea level was reported to be 110 m (Vaz, 1996) and

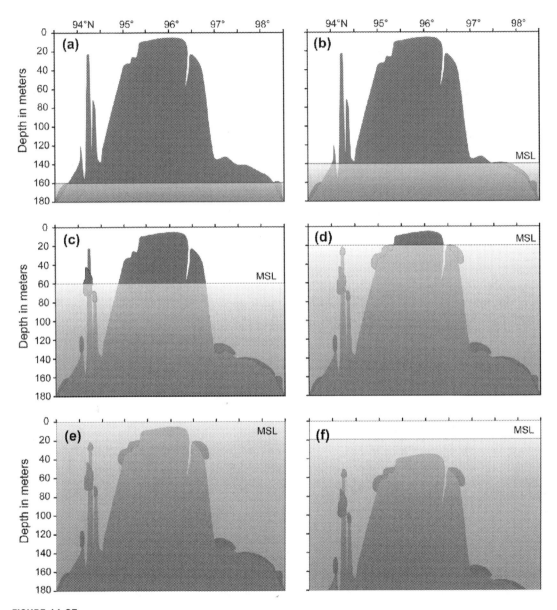

FIGURE 11.25

Schematic explaining the proliferation of coral patches at different depths and different times in the geological past. The green patches depict live coral and the red patches depict dead coral. (A) Coral colonies develop at depths with optimum light on hard substratum and hard ground. (B) When sea level rises, those which can adapt continue to prolifer while those which cannot die out. (C and D) corals continue to opportunistically nucleate; however cannot cope up with steep slopes and rapid sea level rise. (E) In the present study, coral fauna also occurring at shallow depths are relict, representing dead coral. This could be attributed either to eustatic sea level rise (F) or this could be a result of tectonic subsidence in the region.

10. Relict benthic foraminifera and sea level history

	Depth distribution in m.					
Name of Soft Coral Species	0-5m	6-10	11-15	16-20	20-25	25-30
Siphonogorgia godeffroyi Kolliker, 1874			▨	■		
Chironephthya cf. macrospiculata Thomson & Henderson, 1906			▨	■		
Subergorgia suberosa (Pallas, 1766)				■	▨	
Acalycigorgia sp.			▨			
Ctenocella (Ellisella) *sp.*			▨			

FIGURE 11.26

The ecological depth preferences of soft coral assemblage that existed in the study area is shaded in gray. The diagonal-line shading represents the depth zone where all the genera seem to have co-existed. Thus considering their occurrence from 6 to 30 m, 17.5 m is used as the average height to which the past sea-level has been assigned.

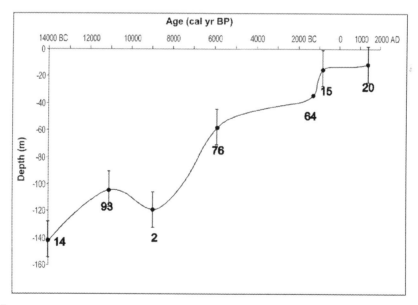

FIGURE 11.27

Sea-level curve proposed for the west coast of Myanmar based on the ecology of the soft coral assemblage (sample numbers indicated with corresponding numbers) and ^{14}C dates of relict foraminiferal assemblage.

100 m (Mohana Rao and Rao, 1994) below present MSL, respectively. In the study area, the sea level was at −115 m at 14,000 yr. BP and at 100 m and −100 m at 13,000 yr. BP. These levels thus seem very close to being comparable. When the west coast of India was experiencing a prolonged sea stand between 11,500 and 10,000 yr. BP (Hashimi et al., 1995), the west coast of Myanmar seems to have undergone significant subsidence after which the sea level rose as reported globally, except

250 Chapter 11 High-resolution paleoclimatic records

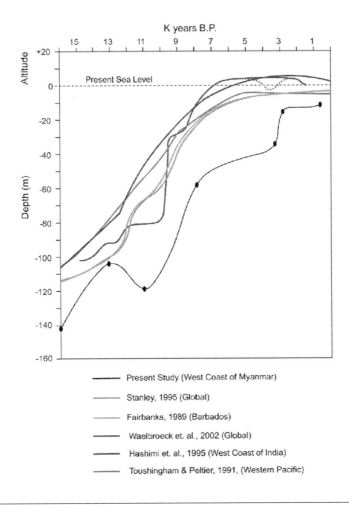

FIGURE 11.28

Figure showing comparison of the sea-level curve proposed for the west coast of Myanmar with previously published sea-level curves.

maintaining a level ∼30 m lower than elsewhere. Another major subsidence event is recorded at ∼3000 to 2800 yr. BP after which the sea level gradually attained the present level.

Though the general rising trend of the sea level in the study area can be correlated with the Holocene sea level rise, the destruction of the coral patches along the west coast of Myanmar cannot be entirely due to the same. Sea level rise apart from eustasy is also largely dependent on rates of sedimentation and the tectonic setting of the area. The present study indicates that in addition to the global Holocene Sea-level rise, the subsidence of the region west coast off Myanmar seems to have led to destruction of patch coral reefs in this region. A comparison of this curve with other published global and regional sea-level curves suggests that the vertical tectonic component of displacement, especially during the Late Pleistocene—Holocene, has a direct bearing on the local sea levels of Myanmar.

Most of the studies on the tectonic evolution of the Andaman basin consider time scales of millions of years and propose the opening of the basin along numerous but small fault systems. However, the vertical tectonic component of displacement, especially during the late Pleistocene—Holocene, has a direct bearing on the local sea levels of Myanmar. The present study demonstrates the utility of ^{14}C AMS dating of relict coral reef Foraminifera in deciphering changing sea levels along the west coast of Myanmar and proposing a local sea level curve. Neither has work been done along the west coast of Myanmar to understand the fluctuating sea levels nor have any such events been dated. This work offers a baseline data for future workers to pursue this line of research. Coral colonies have only been known far from the deltaic environments and south of Myanmar, close to the Thai border. This study provides much evidence for the existence of late Pleistocene—Holocene soft corals, along the west coast of Myanmar which have been destructed due to sea level rise.

10.7 Significance of sea level fluctuations off Myanmar

Global climate change has become synonymous with global warming of which sea-level rise is most dreaded consequence. At the direct risk of assured inundation are the coastal low-lying regions all over the globe, which are the most populated. Thus estimation of past, present, and thereby future sea level fluctuations remains a major challenge to paleoclimatologists. The results brought out by the IGCP Project 61 demonstrated that determination of a single sea-level curve of global applicability was an illusionary task. Thus, the need was felt for individual regions to be studied for local sea-level histories considering all local influencing factors (Pirazzoli, 1991). Several attempts have been made to reconstruct past sea levels in regions adjoining the study area, namely, West Coast of India (Bruckner, 1989; Merh, 1992; Hashimi et al., 1995; Rao et al., 1996; Vora et al., 1996; Mazumder, 2005), East Coast of India (Vaz, 1996, 2000; Vaz and Bannerjee, 1997; Banerjee, 2000; Rana et al., 2007), Thailand—(Sinsakul, 1992) and South China Sea (Yim et al., 2006). However, the coast of Myanmar had remained unexplored in terms of signatures of high-resolution sea-level history before the present study. Being a promising basin in terms of petroleum and natural gas reserves as well as a tectonically active margin, the study area faces far reaching consequences of sea level fluctuations. Thus, the present results do not only form baseline data for future workers but also is very promising for the academia and industry. In order to refine the present preliminary curve proposed, more surface and subsurface samples covering the entire Holocene period needs to be collected in order to generate more data as well as dates.

11. Subsurface foraminiferal signatures on Ayeyarwady Delta shelf: intradecadal paleoclimatic records for the past five centuries

GC-5, the 1.78 m long sediment core collected at 37 m water depth right on the delta front was used for paleoclimatic reconstructions as discussed in previous sections. The core represents 489 years before present and was calibrated to represent time since 1513 AD till 2002. Thus, all downcore distributions have been described with respect to time. In order to eliminate noise in the data, 3-point and 5-point averages of all abundances/data points have also been plotted.

11.1 Sub-surface distribution of foraminifera

The total foraminiferal number (TFN) varies between a maximum of 65,185 at 1779 AD and a minimum of 470 at 1708 AD (Fig. 11.29A). The abundances show a distinct accent in numbers after ~1700 AD. Before 1700 AD, the abundances have remained lower than ~30,000, and fluctuations have been very minor (amplitudes range between 10% and 30%). Since 1700 AD to present, the fluctuations are higher in amplitude and significant.

The benthic numbers peak to 56,883 at 1795 AD and drop to 263 at 1708 AD (Fig. 11.29B). The planktonic numbers vary between 11,481 at 1779 AD and 245 at 1708 AD (Fig. 11.29C). Both the parameters follow the trend same as that shown by TFN, that is, significant low abundances prior to 1700 AD after which they show large fluctuations in abundance. Thus, ~1700 AD seems to mark an environmental boundary.

Comparing the relative percentage abundance of benthic and planktonic foraminifera (Fig. 11.30A and B), it is seen that both show the same boundary at ~1650 AD. Prior to 1650 AD, both planktonic

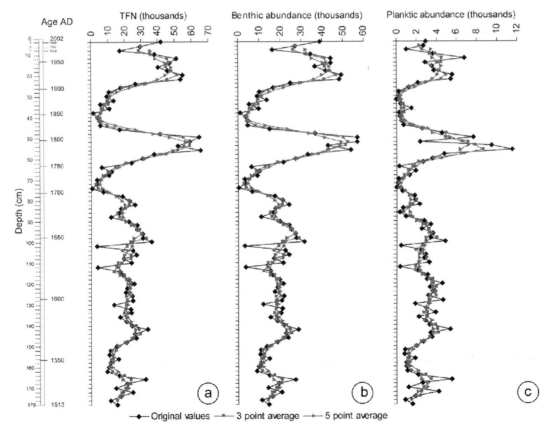

FIGURE 11.29

Downcore abundances of (A) total foraminiferal number (B) benthic and (C) planktonic foraminifera in GC-5.

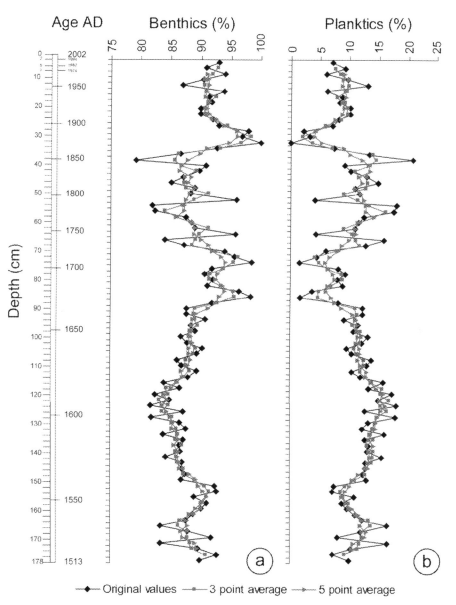

FIGURE 11.30

Comparison of the downcore percentage abundances of (A) benthic and (B) planktonic foraminifera.

and benthic foraminifera show a gradual trend with very little change in amplitude. This period is characterized by relatively higher planktonic percentages corresponding to lower benthic percentages (as compared to their abundances in the younger portion of the core). Generally higher salinities support higher planktonic numbers and thus the evidenced climatic boundary could be indicative of the termination of a prolonged and pronounced high salinity period between 1550 and 1650 AD. Post 1650 AD, there are frequent and large variations in the amplitudes in both cases.

The TFN is largely composed of benthic foraminifer, which, in turn, is largely contributed by low salinity assemblages. Four genera, namely, *Asterorotalia, Hanzawaia, Ammonia,* and *Elphidium* show an inverse relationship with salinity and are characteristic of low-salinity regimes (Panchang, 2008). Thus, in order to assess the downcore variation in the relative abundances of the low-salinity assemblages, their collective percentage abundances were also plotted (Fig. 11.31). The low salinity assemblage shows three peaked abundances at three occasions post 1675 AD at ∼1675, 1765, and 1850 AD, apart from showing frequent high-amplitude abundance variations. These events are evident in almost all down core plots and are marked as I, II, and III.

11.2 Asterorotalia trispinosa as an indicator of paleomonsoons

During the species distribution studies, it was noticed that *Asterorotalia trispinosa* (Fig. 11.32) was not reported from the Arabian Sea, whereas it is reported all along the east coast. Of the total benthic foraminifera picked from all the surface sediments, proportion of *Asterorotalia trispinosa* in each sample was calculated. A correlation analysis was performed using STATISTICA 5.0 to determine the relationship between the distribution of *Asterorotalia trispinosa* and the sedimentological and water property parameters measured at the respective sample localities. The studies (Panchang and Nigam, 2012) revealed that:

- *A. trispinosa* prefers low salinities (<32 psu), silty-clay substrates and a depth zone of 18–40 m.
- It is characteristic of the Indo-pacific realm, occurring in deltaic and high sedimentation regions.
- Its sensitivity to salinity makes it a promising proxy for indicating past monsoons in subsurface records.

As *A. trispinosa* prefers low-salinity regimes, Panchang (2008) proposed that higher abundances of *A. trispinosa* are an indicator of low salinities in the past, that is, higher freshwater influx or stronger monsoons. In order to test the proposed hypothesis, downcore variation in the abundances of the species was computed and plotted in Fig. 11.33A. Different specimens of the same species are known to have different size, diameter, coiling direction, and proloculus sizes. Thus, morphometric analyses of ecologically sensitive species have proven to be very useful in the past. These are many times the function of the reproductive behavior of the species in response to climatic variations. It has been reported in earlier studies that Mean Proloculus Sizes (MPS) show inverse relationship to salinity (Nigam, 1986; Nigam and Rao, 1987; Nigam and Khare, 1995). Thus, downcore values for MPS have been plotted in Fig. 11.33B.

As observed in all the downcore abundances discussed above, even the *A. trispinosa* abundances (Fig. 11.33A) shows a marked difference in its abundance patterns in the older and younger part of the core. Its abundances plotted in can be enumerated as follows:

- Abundances of *A. trispinosa* vary between a maximum of ∼31% at 1835 AD and a minimum of 0.8% at 1820 AD. Its abundances are maintained at an average of ∼5% abundances prior to 1675 AD after which the abundances fluctuate with large amplitudes. As larger abundances of *A. trispinosa* are taken as an indicator of lower salinities, its lower abundances seem to indicate dry climates, with increased salinities in the study area.

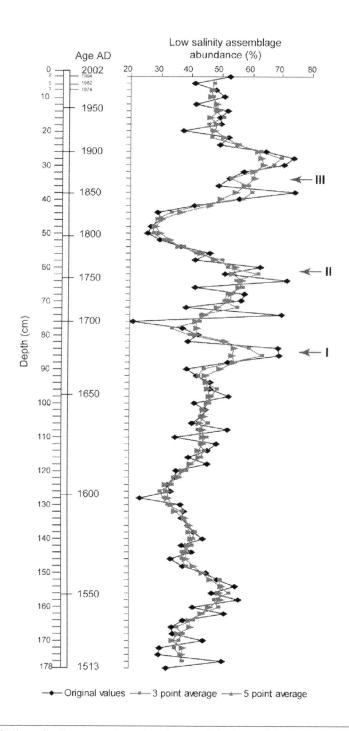

FIGURE 11.31

Downcore variation in the collective percentage abundances of the low salinity species viz. *Ammonia, Asterorotalia, Elphidium* and *Hanzawaia*. I, II and III indicate the three major events of fresh water pulses in the study area.

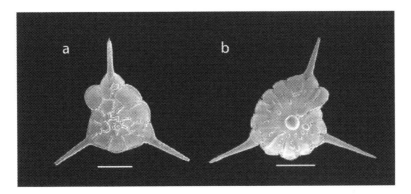

FIGURE 11.32

Benthic foraminiferal species *Asterorotalia trispinosa* (A) ventral view; (B) dorsal view.

- Maximum fluctuations are seen between a ∼1675 to 1850 AD, within which 2 events of opposite extremities are recorded.
- Highest salinity and thus lowest fresh water influx indicated by lowest *A. trispinosa* abundances at ∼1710 AD (indicated as "A").
- Lowest salinity and highest freshwater influx, indicated by highest abundances of *A. trispinosa* between 1827 and 1850 AD (indicated as "B").

On the basis of the *A. trispinosa* abundances, we can delineate freshwater pulses in the study area since 1513 AD, of which 3 significant events occur after 1650 AD. The 4 freshwater pulses prior to 1650 AD are very small and less significant (indicated as a, b, c and d in Fig. 11.33B) and are reflected better by the reproductive behavior of *A. trispinosa* (discussed in the next section). Approximately, the periods around 1675, 1765, and 1850 AD are characterized by peaked fresh water pulses in the study area indicated by the annotations I, II, and III in Fig. 11.33A. Two other peaks of lesser significance are recorded at ∼1900 and 1938 AD indicated as IV and V.

Fig. 11.33B illustrates the downcore variation in the MPS which varies between 37.55 μm at ∼1810 AD and ∼19 μm at ∼1650. MPS shows direct relationship with *A. trispinosa* abundances and thus an inverse relationship with salinity. It is interesting to note that the climatic events reflected by the *A. trispinosa* abundances (especially the major freshwater pulses) are also reflected in the downcore variations in MPS (though not with the same amplitude) and are indicated in Fig. 11.33B.

The climatic events indicated by *A. trispinosa* abundances and MPS are also well reflected by the additional proxies. A climatic transition from dry to wet is reflected in all the parameters discussed above. Even the three major freshwater pulses are well reflected by all. Sometimes, there seems to be a lead and lag in events as reflected by different proxies. For example, *A. trispinosa* abundances peak at ∼1845 AD indicating highest low salinity. The same is reflected in MPS at ∼1820 AD. Such a signature is obvious given the fact that environmental changes first tend to affect the reproductive behavior of any species which is a very sensitive parameter (Bradshaw, 1955, 1957). Similarly, planktonic foraminiferal abundances show the same event at ∼1835 AD.

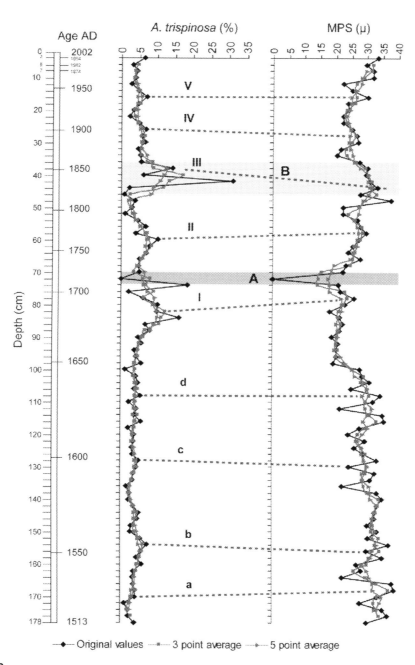

FIGURE 11.33

Downcore variations in (A) abundance of *A. trispinosa* and (B) MPS.

12. Periodicity in paleoclimatic records

The maximum abundances of *A. trispinosa* indicate periods of lowest salinity and thereby high freshwater influx in the study area. The periods of larger freshwater pulses in the study area are ~1675, 1765, and 1850 AD. Considering these three major wet events, they seem to occur at intervals of 90 and 85 years, respectively. These events are very evident. However, the dataset does show several smaller fluctuations, especially in the older part of the core.

Thus, in order to verify if any cyclicity exists in climatic events, the MPS data of *A. trispinosa* was analyzed for cyclicity using the software Redfit 3.8. The same parameter has been used by previous workers to document cyclicity in monsoons in the past few hundred years (Nigam and Khare, 1995). The MPS data revealed a ~93-year cyclicity at a 90% level of reliability (Fig. 11.34).

Cycles of almost similar length have been noticed in different climatic records as summed up in Table 11.6. They include phenomenon such as changes in the flooding of the River Nile and monsoon variations along the central west coast of India. Variations in the radius of the sun are believed to modulate these cycles at a periodicity of 80 ($\pm$10) years, known as the "Gleissberg Cycle."

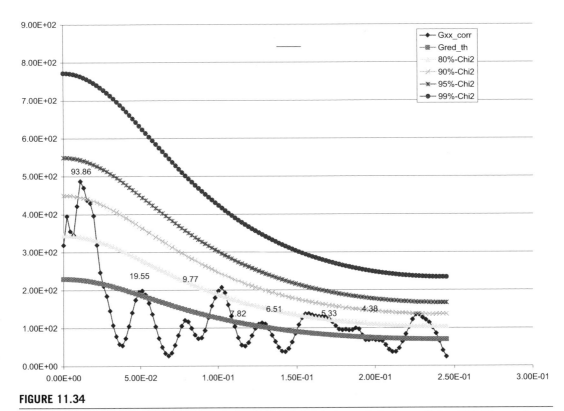

FIGURE 11.34

Plot of varied cyclicities obtained from analysis of MPS of *A. trispinosa*. Only the cyclicity of 93.86 years is indicated at a very high level of significance.

Table 11.6 List of various climatic phenomenon observed to vary on a ~80-year cyclicity in solar activity.

Authors	Event observed	Duration of cycle	Study area
Parkinson et al. (1980)	Solar radius	80	
Gilliland (1981)		76	
Landschit (1981)		79	
Gleissberg (1965)	Aurora	88	
Siscoe (1980)		80	
Feynman and Fougere (1984)		88	
Stuiver (1980)	Atmospheric ^{14}C	70	
Stuiver and Quay (1980)		79	
Agee (1980)	Temperature	90	N. Hemisphere
Schlesinger and Ramankutty (1994)		65–70	
Joseph and Amatya (1986)		90	Lower latitudes with special reference to south India
Reid (1987)	Average SST	90	Global
Hameed et al. (1983)	Rainfall	78–80	Beijing, China
Currie and Fairbridge (1985)		78–80	
Fairbridge (1984)	River flood levels	78	River Nile, Africa
Johnsen et al. (1970)	Oxygen isotopic ratios	78	Greenland ice sheets
Nigam et al. (1995)	Monsoon	77	Central west coast of India
Present study	Freshwater pulses	85–90	Off Myanmar

12.1 Asian signatures of the Little Ice Age (LIA)

The records seem to account for the period of cooling called the Little Ice Age, that is, ~16th to the mid-19th century. Though there is no agreed consensus on an agreed date of its beginning, it is believed that it ended at ~1850 AD after which warming began. It is also agreed that there were three sunspot minima beginning at ~1650, 1770, and 1850 AD, each separated by slight warming intervals (Website of the NASA Earth Observatory; UNEP/GRID-Arendal). Coincidentally, the present datset which offers very-high-resolution climatic records seems to clearly document Asian signatures of the Little Ice Age, which was most pronounced in the climatic records of Europe.

The three fresh water pulses recorded at ~1675, 1765, and 1950 AD (Fig. 11.31 & Fig. 11.33) in the current dataset collected on the shelf off Myanmar seem to be in response to the three sunspot minima reported under the period under consideration. However, the little lead/lag as seen in the data is expected as noticed elsewhere (Gilliland, 1981; Nigam et al., 1995). There are reports indicating changes in the incoming energy from the sun to the earth either due to changes in the radius of the sun and/or number of sunspots. These have been reported to bring about climatic changes, including monsoonal rains over India (thus the fresh water discharge through rivers) and adjoining areas (Nigam et al., 1995; Gupta et al., 2005; Nigam et al., 2006; Saraswat et al., 2007).

The Little Ice Age is a commonly discussed phenomenon in Europe, where sunspots varied considerably and brought about significant climatic changes in the last few hundred years. It is worth exploring if this phenomenon affected the present study area in any way.

12.2 Geochemical proxies as modern reinforcement to traditional micropaleontology

Modern trends suggest a multiproxy approach toward paleoclimatic analyses, especially supplementation with geochemical data which provide quantitative, reliable, and precise measurements of climatic events. In order to confirm the wet and dry events suggested by the foraminiferal studies, elemental and stable isotopic ratios were obtained on the same core as described previously.

12.2.1 Elemental ratios

The long residence times of Mg, Ca ($\sim$1Ma each) and Sr ($\sim$5 Ma) (Broecker and Peng, 1982) suggests that there should be very limited variability of seawater Sr/Ca on the relatively short timescales of glacial—interglacial change (Martin et al., 1999) or even less on very high-resolution records of the Holocene. Therefore variations in Sr/Ca recorded in marine carbonates on the short timescales have been attributed to environmental parameters (Elderfield et al., 2000). Coupled numerical models of the Sr and Ca budgets of the ocean reveal that large changes in river fluxes and carbonate accumulation rates can produce seawater Sr/Ca variations that approximate both the shape and amplitude of foraminiferal Sr/Ca variations. Mg/Ca in marine carbonates varies with latitude suggesting temperature dependence.

Taking advantage of the fact that all three elements have longer residence times and could thus give reliable signatures for fresh water discharge in the region, simultaneous measurements of Ca, Mg, and Sr were done. The Mg/Ca and Sr/Ca plots are illustrated in Fig. 11.35A and B respectively. The Mg/Ca ratios normally vary between 3.5 and 5.5 mmol/mol. The Sr/Ca values have a very narrow range from 1.27 to 1.37 mmol/mol varying significantly only occasionally. As observed in the foraminiferal plots, prior to 1650 AD the Mg/Ca ratios have remained constant at $\sim$4 mmol/mol for a prolonged duration in contrast to values > 5 mmol/mol during the warmer phase later. As seen in all the other parameters, even Mg/Ca and Sr/Ca ratios show significant fluctuations in the warmer phase. Highest Mg/Ca ratios and the lowest Sr/Ca values are observed between $\sim$1800 and 1850 AD. This is the period when *A. trispinosa* shows highest abundances (Fig. 11.33) and thus lowest salinity. Thus, the elemental concentrations support the hypothesis that *A. trispinosa* can be effectively used as an indicator of freshwater influx.

The higher Mg/Ca ratios and lower Sr/Ca ratios at these intervals seem to indicate a significant warming resulting in heavy fresh water discharge causing significant drop in salinities. **As stated above, because these elements have an extremely long residence times, significant environmental changes/salinity gradients are required to cause a variation in their amplitude over smaller time scales. Thus, large amplitude changes are seen only during the warming events.**

12. Periodicity in paleoclimatic records   261

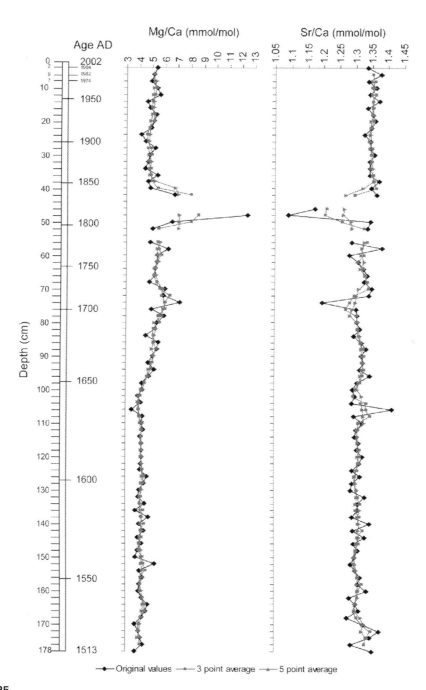

FIGURE 11.35

Downcore variation of (A) Mg/Ca ratios and (B) Sr/Ca ratios obtained on foraminiferal species *Hanzawaia concentrica*.

12.2.2 Stable isotopic ratios

Oxygen isotopic concentrations of foraminiferal shells serve as a measure of ambient water temperatures. However, these can also vary because of fresh water influx (glacial ice melt or evaporation–precipitation), especially in the study area, it being largely governed by high gradients in the salinity regimes.

Five sub-samples were merged to get a single measurement between 40 and 50 cm of the core as enough specimens were not available. The oxygen isotopic ratios vary between -2.4 and $-3.25°/_{oo}$. As seen in the foraminiferal plots, the core shows an initial prolonged dry phase, followed by significant warm events (characterized by lower $\delta^{18}O$ values) characterized by three major fluctuations (Fig. 11.36).

The $\delta^{18}O$ values remain constantly at $\sim -2.6°/_{oo}$ and above for a major portion of the drier phase. Most prominent between 1550 and 1600 AD, the driest event is seen at ~ 1575 AD. The transition from dry to wet can be set at ~ 1650 AD after which the highly fluctuating $\delta^{18}O$ values have remained lower than $-2.8°/_{oo}$. Significant drop in the $\delta^{18}O$ values is seen at approximately 1675, 1765 and 1875 AD, reinforcing the major fresh water pulses implied in the *A. trispinosa* abundances.

These climatic events documented by *A. trispinosa* are not only evidenced by foraminiferal assemblages but are also reinforced by geochemical proxies namely, stable isotopic and elemental ratios. This validates the climatic interpretations drawn and discussed in Section 8.3 above. *Asterorotalia trispinosa* is thus validated as a reliable proxy for paleomonsoons, as proposed.

13. Conclusions

The Ayeyarwady (former Irrawaddy) Delta Shelf off Myanmar is characterized by one of the largest freshwater influxes and rates of sedimentation in the world. It is influenced by both the SW summer monsoon as well as the NE winter monsoon. Despite being a site of thick, long-term, high-resolution sediment archives, it is one of the most unexplored sites in terms of paleoclimatic studies. The present work was a maiden successful attempt in reconstructing 14,000 years long sea-level history along the west coast of Myanmar as well as reconstructing intradecadal paleoclimates at the mouths of the Ayeyarwady.

Apart from the normal recent benthic foraminifera, relict fauna comprising larger foraminiferal assemblage, soft coral sclerites and coral caryophyllids, and coralline algae are found along the west coast of Myanmar. This fauna in combination with echo-sounding records and ^{14}C AMS dates provide numerous evidences for the existence of Late Pleistocene–Holocene soft coral patches, along the west coast of Myanmar, which have been destructed due to sea level rise. A regional sea-level curve for the west coast of Myanmar is proposed which illustrates the sea level history of the past 16,000 years.

Reconstructions of past climates on centennial to decadal scales are the need of the time. The highlight of this study is that these datasets offer high-resolution records for the past 489 years; a resolution of ~ 4 years since 1513 to 1676 AD and since then 8 years resolution unto present.

1. Downcore distribution of foraminifera, benthic-planktonic percentage abundances, abundances and MPS of *Asterorotalia trispinosa*, additional microfossil abundances and geochemical data generated on foraminiferal species, all indicate a major climatic boundary at $\sim 1650–75$ AD.

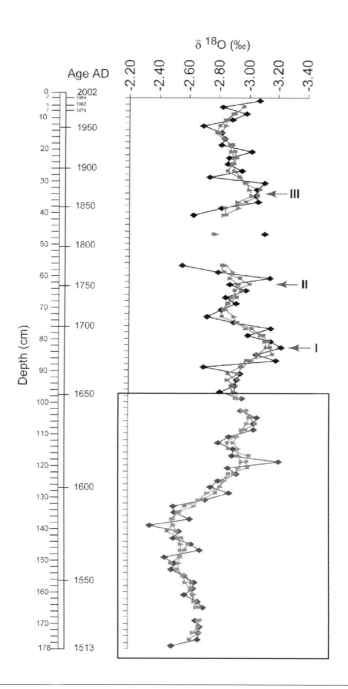

FIGURE 11.36

Downcore variation in stable oxygen isotopic ratios obtained on foraminiferal species *A. trispinosa*. A climatic boundary is seen at 1650 AD. I, II, and III indicate the three major events of fresh water pulses in the study area.

2. Downcore variations in the abundance and MPS of *A. trispinosa* indicate:
   * Two significant climatic conditions in the study area since 1513 AD; a dry climate prior to 1650 AD and warm and wet climate since 1650 to present.
   * The period between 1800 and 1850 AD experienced maximum rainfall/fresh water influx
   * Since 1650 AD, three major freshwater pulses are recorded in the core at ∼1675, 1750 and 1850 AD.
   * MPS data shows a 93.86 year cyclicity in climatic events in the study area which falls into the scope of the Gleissberg Cycle (viz. 80 ± 10years).
3. Significant drop in the $\delta^{18}O$ values is seen at ∼1675, 1765, and 1875 AD, reinforce the major fresh water pulses implied in the *A. trispinosa* abundances.
4. Mg/Ca and Sr/CA ratios show signs of significant fluctuations post ∼1675 AD. Highest Mg/Ca ratios and corresponding lowest Sr/Ca ratios seem to indicate a significant warming resulting in heavy fresh water discharge causing significant drop in salinities.
5. The support data reinforce the results for three major fresh water pulses in the study area, there by proving the efficiency and reliability of *A. trispinosa* as a proxy for paleomonsoons.
6. The records seem to represent the Asian signatures of the climatic events of the Little Ice Age in Europe, that is, the duration from the 16th to the mid-19th century.
7. Sunspot minimas recorded at ∼1650, 1770, and 1850 AD responsible for the Little Ice Age in Europe seem to have influenced the monsoons over the Indian sub-continent, during the same period, manifesting in the form of three major fresh water pulses recorded in the study area.

Acknowledgments

Dr. P.S. Rao and Dr. V. Ramaswamy of National Institute of Oceanography, Goa, India are thanked for providing samples for the present study. The first author (RP) is thankful to the Deutcher Academischer Austausch Dienst, Bonn, Germany for the DAAD Fellowship, under which part of this work was carried out at the Interdisciplinary Center for Ecosystem Dynamics in Central Asia, Freie Universität, Berlin and to Council of Scientific and Industrial Research (CSIR) India for financial support in the form of Senior Research Fellowship (SRF). We are also thankful to Ministry of Earth Sciences for financial support (DOD/18/PMN/EFG/BENFAN/96). We are greatly thankful to Prof. Frank Riedel and Asst. Prof. Uwe Wiechert from Freie Universitat, Berlin and Dr. Sabine Tonne from Deutsche GeoForschungsZentrum (GFZ), Potsdam for providing us the infrastructure and facilities to carry out the isotope analysis and elemental analysis.

References

Agee, R.N., 1980. Present climatic cooling and proposed causative mechanism. Bull. Am. Meteorol. Soc. 61, 1356–1367.

Bak, R.P.M., 1976. The growth of coral colonies and the importance of crustose coralline algae and burrowing sponges in relation with carbonate accumulation. Neth. J. Sea Res. 10, 285–337.

Banerjee, P.K., 2000. Holocene and Late Pleistocene relative sealevel fluctuations along the east coast of India. Mar. Geol. 167, 243–260.

Barik, S.S., Singh, R.K., Tripathy, S., Farooq, S.H., Prusty, P., 2022. Geochemical Pooling and Accumulation of Metals in Coastal Lagoon Sediments and Their Influence on Benthic Foraminifera. Science of the Total Environment. https://doi.org/10.1016/j.scitotenv.2022.153986.

References

Barker, S., Greaves, M., Elderfield, H., 2003. A study of cleaning procedures used for foraminiferal Mg/Ca paleothermometry. G-cubed 4. https://doi.org/10.1029/2003GC000559.

Bay of Bengal Pilot, 1978. Hydrographer to the Navy. Somerset, England.

Bender, F., 1983. In: Bender, F., Jacobshagen, V., de Jong, J.D., Lutig, G. (Eds.), Geology of Burma. Gebruder Borntraeger, Berlin, Stuttgart.

Bhadra, S.R., Saraswat, R., 2022. Exceptionally high foraminiferal dissolution in the western Bay of Bengal. Anthropocene 40, 100351.

Bhalla, S.N., 1968. Recent Foraminifera from Vishakhapatnam Beach Sands and its Relation to the Known Foramgeographical Provinces in the Indian Ocean. Bulletin National Institute of Sciences, India, pp. 376–392.

Bhalla, S.N., 1970. Foraminifera from Marina beach sands of Madras and faunal provinces of the Indian ocean. Contrib. Cushman Found. Foraminifer. Res. 21, 156–163.

Bhalla, S.N., Nigam, R., 1988. Cluster analysis of the foraminiferal fauna from the beaches of the east and west coasts of India with reference to foramgeographical provinces of the Indian Ocean. J. Geol. Soc. India 32, 516–521.

Bhalla, S.N., Khare, N., Shanmukha, D.H., Henriques, P.J., 2007. Foraminiferal studies in nearshore regions of western coast of India and Laccadives Islands: a review. Indian J. Mar. Sci. 36, 272–287.

Bhatia, S.B., 1956. Recent foraminifera from the shore sands of western India. Contrib. Cushman Found. Foraminifer. Res. 7, 15–24.

Bird, M.I., Robinson, R.A.J., Oo, N.W., Aye, M.M., Lu, X.X., Higgitt, D.L., Swe, A., Tun, T., Win, S.L., Aye, K.S., Hoey, T.B., 2008. Organic carbon transport by the Ayeyarwady (Irrawaddy) and Thanlwin (Salween) rivers, Myanmar. Geophys. Res. Abstr. 10. EGU2008-A-03406.

Boltovskoy, E., Wright, R., 1976. Recent Foraminifera. Dr. W. Junk Publishers, The Hague, p. 515.

Bradshaw, J.S., 1955. Preliminary laboratory experiments on ecology of foraminiferal populations. Micropaleontology 1, 351–358.

Bradshaw, J.S., 1957. Laboratory studies on the rate of growth of foraminifera. J. Palaeont. 31, 1138–1147.

Broecker, W.S., Peng, T.H., 1982. Tracers in the Sea. Lamont-Doherty Geological Observatory, Palisades, NewYork, p. 690.

Bruckner, H., 1989. Late quaternary shorelines in India. In: Schott, D.B., Pirazzoli, P.A., Honig, C.A. (Eds.), Late Quaternary Sea Level Correlation and Applications. Kluver, Dordrecht, pp. 169–194.

Brunnschweiler, R.O., 1974. Indoburman ranges-Data for Orogenic studies. In: Spencer, A.M. (Ed.), Mesozoic-Cenozoic Orogenic Belts, vol. 4. Journal of the Geological Society of London Special Publication, pp. 279–299.

Burton, K.W., Vance, D., 2000. Glacial–interglacial variations in the neodymium isotope composition of seawater in the Bay of Bengal recorded by planktonic foraminifera. Earth Planet Sci. Lett. 176 (3–4), 425–441.

Chang, Y.M., 1967. Accuracy of fossil percentage estimation. J. Palaeontol. 41, 500–502.

Cockey, E.M., Hallock, P.M., Lidz, B.H., 1996. Decadal-scale changes in benthic foraminiferal assemblages off Key Largo, Florida. Coral Reef. 15, 237–248.

Curray, J.R., 2005. Tectonics and history of the Andaman Sea region. J. Asian Earth Sci. 25, 187–232.

Curray, J.R., Moore, D.G., Lawver, L.A., Emmel, F.J., Raitt, R.W., Henry, M., Kieckhefer, R., 1979. Tectonics of the Andaman Sea and Burma. Mem. Am. Assoc. Petrol. Geol. 29, 189–198.

Currie, R.G., Fairbridge, R.W., 1985. Periodic 18.6 year and cyclic 11 year induced drought and flood in north eastern China and some global implications. Quat. Sci. Rev. 23, 109–134.

Cushman, J.A., 1948. Foraminifera, Their Classification and Economic Use, fourth ed., p. 605

Dennison, J.M., Hay, W.W., 1967. Estimating the needed sampling area for subaquatic ecologic studies. J. Palaeontol. 41, 706–708.

Dutta, K., Bhushan, R., Somayajulu, B.L.K., 2001. ^{14}C reservoir ages and ΔR correction values for the northern Indian Ocean. Radiocarbon 43, 483–488.

Elderfield, H., Cooper, M., Ganssen, G., 2000. Sr/Ca in multiple species of planktonic foraminifera: implications of reconstructions of seawater Sr/Ca. G-cubed 1, 1999GC000031, ISSN:1525-2027.

Fairbridge, R.W., 1984. The Nile floods as a global climatic/solar proxy. In: Climatic Changes on a Yearly to Millennial Basis, pp. 181–190.

Fairbanks, R.G., 1989. Glacio-eustatic sea level record 0-17,000 years before present: influence of glacial melting rates on Younger Dryas event and deep ocean circulation. Nature 342, 637–642.

Feynman, J., Fougere, P.F., 1984. Eighty eight year periodicity in solar-terrestrial phenomena confirmed. J. Geophys. Res. 89, 3023–3027.

Frerichs, W.E., 1970. Distribution and ecology of Benthonic foraminifera in the sediments of the Andaman Sea. Contrib. Cushman Found. Foraminifer. Res. 21 (4), 123–147.

Frerichs, W.E., 1971a. Paleobathymetric trends of Neogene foraminiferal assemblages and sea floor tectonism in the Andaman Sea area. Mar. Geol. 11, 159–173.

Frerichs, W.E., 1971b. Planktic foraminifera in the sediments of the Andaman Sea. J. Foraminifer. Res. 20 (2), 117–127.

Genrich, J.F., Bock, Y., McCaffrey, R., Prawirodirdjo, L., Stevens, C.W., Puntodewo, S.S.O., Subarya, C., Wdowinski, S., 2000. Distribution of slip at the northern Sumatran fault system. J. Geophys. Res. Solid Earth 105 (B12), 28327–28341.

Gilliland, R.L., 1981. Solar radius variations over the past 265 years. Astrophys. J. 248, 1144.

Gleissberg, W., 1965. The eighty years solar cycle in auroral frequency numbers. J. Br. Astron. Assoc. 75, 227–231.

Gosliner, T.M., Behrens, D.W., Williams, G.C., 1996. Coral Reef Animals of the Indo-Pacific. Sea Challengers, Monterey, California, p. 314.

Govil, P., Naidu, P.D., 2011. Variations of Indian monsoon precipitation during the last 32 kyr reflected in the surface hydrography of the Western Bay of Bengal. Quat. Sci. Rev. 30 (27–28), 3871–3879.

Govil, P., Mazumder, A., Agrawal, S., Azharuddin, S., Mishra, R., Khan, H., Verma, D., 2022. Abrupt changes in the southwest monsoon during mid-late Holocene in the western Bay of Bengal. J. Asian Earth Sci. 227, 105100.

Gupta, A.K., Anderson, D.M., Overpeck, J.T., 2003. Abrupt changes in the Asian southwest monsoon during the Holocene and their links to the North Atlantic ocean. Nature 421, 354–357.

Gupta, A.K., Das, M., Anderson, D.M., 2005. Solar influence on the Indian summer monsoon during the Holocene. Geophys. Res. Lett. 32, L17703. https://doi.org/10.1029/2005GL022685.

Hallock, P., 2000. Larger Foraminifera as indicators of coral-reef vitality. Environ. Micropalaentol. 15, 121–150.

Hallock, P., 2002. Foraminifers as Bioindicators in coral reef Management. In: Micropaleontological Applications to Problems of Urbanization, Denver Annual Meeting at Colarado Convention Center. October 27-30, 2002).

Hameed, S., Yeh, W.M., Cess, R.D., Wang, W.C., 1983. An analysis of periodicities in the 1470–1974, Beijing precipitation record. Geophys. Res. Lett. 10, 436–439.

Hart, A.M., Kaesler, R.L., 1986. Temporal changes in Holocene lagoonal assemblages of foraminifera from northeastern Yucatan Peninsula, Mexico. J. Foraminifer. Res. 16 (2), 98–109.

Hashimi, N.H., Nigam, R., Nair, R.R., Rajgopalan, G., 1995. Holocene sea level fluctuations on the Western Indian continental margin: an update. J. Geol. Soc. India 46, 157–162.

Hohenegger, J., Yordanova, E., Nakano, Y., Tatzreiter, F., 1999. Habitats of larger foraminifera at the upper reef slope of Sesoko island, Okinawa, Japan. Marine Micropalaeontol. 36 (2), 109–168.

Hughen, K.A., Baillie, M.G.L., Bard, E., Bayliss, A., Beck, J.W., Bertrand, C., Blackwell, P.G., Buck, C.E., Burr, G., Cutler, K.B., Damon, P.E., Edwards, R.L., Fairbanks, R.G., Friedrich, M., Guilderson, T.P.,

Kromer, B., McCormac, F.G., Manning, S., Ramsey, C.B., Reimer, P.J., Reimer, R.W., Remmele, S., Southon, J.R., Stuiver, M., Talamo, S., Taylor, F.W., van der Plicht, J., Weyhenmeyer, C.E., 2004. Radiocarbon 46, 1059–1086.

Khare, N., Chaturvedi, S.K., Mazumder, A., 2007. An overview of foraminiferal studies in near shore regions off eastern coast of India and Andaman and Nicobar Islands. Indian J. Mar. Sci. 36, 288–300.

Johnsen, S.J., Dansgaard, W., Clausen, H.B., Lanfway, C.C., 1970. Climatic oscillations 1200–2000 AD. Nature 227, 482.

Joseph, C.P., Amatya, B.V.S., 1986. Influence of long term changes in sunspot activity on annual mean temperature over south India. Mausam 37, 276–278.

Khare, N., Nigam, R., Hashimi, N.H., 2008. Revealing monsoonal variability of the last 2,500 years over India using sedimentological and foraminiferal proxies. Facies 54, 167–173.

Kyaw, T., Chit Saing, U., 2001. Distribution of Recent Foraminifera from the Ayeyarwady Delta Shelf, vol. 20. Yangon University 81st Anniversary Paper Reading Session on Science and Technology Physics Theatre.

Lambert, B., 2003. Micropaleontological investigations in the modern Mahakam delta, east Kalimantan (Indonesia). Carn. de Geologie/Notebooks on Geol., Article 2003/02 1–21.

Landscheidt, T., 1981. Swinging sun,79 year cycle and climatic change. J. Interdiscip. Cycle Res. 12, 3–19.

Langer, M.R., Lipps, J.H., 2003. Foraminiferal distribution and diversity, Madang reef and lagoon, Papua New Guinea. Coral Reefs 22, 102–132.

Levitus, S., Boyer, T., 1994. World Ocean Atlas 1994 Volume 4: Temperature. NOAA Atlas NESDIS 4. U.S. Department of Commerce, Washington, DC. http://iridl.ldeo.columbia.edu/SOURCES/.LEVITUS94/.

Linshy, V.N., Rana, S.S., Kurtarkar, S., Nigam, R., 2007. Appraisal of laboratory culture experiments on benthic foraminifera to assess/develop paleoceanographic proxies. Indian J. Mar. Sci. 36, 301–321.

Longcharoen, L., 2003. Egos and Scams, EGAT and the Salween Dams. Water 9 (2), 31–38.

Martin, P.A., Lea, D.W., 2002. A simple evaluation of cleaning procedures on fossil benthic foraminiferal Mg/Ca. G-cubed 3, 8401.

Martin, P.A., Lea, D.W., Mashiotta, T.A., Papenfus, T., Sarnthein, M., 1999. Variation of foraminiferal Sr/Ca over quaternary glacial-interglacial cycles: evidence for changes in mean Ocean Sr/Ca? G-cubed 1, 1999GC0006. ISSN:1525–2027.

Mazumder, A., 2005. Paleoclimatic Reconstruction through the Study of Foraminifera in Marine Sediments off Central West Coast of India. Unpublished Ph.D. thesis, Goa University.

McCarthy, F.M.G., Collins, E.S., McAndrews, J.H., Kerr, H.A., Scott, D.B., Medioli, F.S., 1995. A comparison of postglacial arcellacean ("Thecamoebians") and pollen succession in Atlantic Canada, illustrating the potential of arcellaceans for paleoclimatic reconstruction. J. Paleontol. 69 (5), 980–993.

Meade, R.H., 1996. River-sediment inputs to major deltas. In: Milliman, J.D., Haq, B.U. (Eds.), Sea-level Raise and Coastal Subsidence. Kluwer Academic Publishers, The Netherlands, pp. 63–85.

Merh, S.S., 1992. Quaternary sea level changes along Indian Coasts. Proc. Ind. Nat.l Sci. Acad. 58, 461–472.

Mohana Rao, K., Rao, T.C.S., 1994. Holocene sea levels of Vishakhapatnam shelf, East coast of India. J. Geol. Soc. India 44, 685–689.

Murray, J.W., 1973. Distribution and Ecology of Living Benthic Foraminiferids. Heinemann Educational Books, London, p. 274.

Naidu, P.D., 1999. A review of the Holocene climatic changes in the Indian subcontinent. Mem. Geol. Soc. India 42, 303–314.

Naidu, P.D., Malmgren, B.A., 2005. Seasonal sea surface temperature contrast between the Holocene and last glacial period in the western Arabian Sea (Ocean Drilling Project Site 723A): Modulated by monsoon upwelling. Paleoceanography 20, PA1004. https://doi.org/10.1029/2004PA001078.

Naqvi, W.A., Charles, C.D., Fairbanks, R.G., 1994. Carbon and oxygen isotopic records of benthic foraminifera from the Northeast Indian Ocean: implications on glacial-interglacial atmospheric CO_2 changes. Earth Planet Sci. Lett. 121 (1–2), 99–110.

Nigam, R., 1986. Foraminiferal assemblages and their use as indicators of sediments movement: a study in the shelf region off Navapur, India. Continent. Shelf Res. 5, 421–430.

Nigam, R., Rao, A.S., 1987. Proloculus size variation in recent benthic foraminifera: implications for paleoclimatic studies. Estuar. Coast. Shelf Sci. 24, 649–655.

Nigam, R., 1993. Foraminifera and changing pattern of monsoon rainfall. Curr. Sci. 64 (11–12), 935–937.

Nigam, R., 2003. Problems of global warming and role of micropaleontologists - Presidential Address. Gondwana Geol. Mag. 6, 1–3.

Nigam, R., 2005. Addressing environmental issues through foraminifera: case studies from the Arabian Sea. J. Palaeontol. Soc. India 50 (2), 25–36.

Nigam, R., Khare, N., 1992. The reciprocity between coiling direction and dimorphic reproduction in benthic foraminifera. J. Micropalaeontol. 11 (2), 221–228.

Nigam, R., Khare, N., 1995. Significance of correspondence between river discharge and proloculus size of benthic foraminifera in paleomonsoonal studies. Geo Mar. Lett. 15 (1), 45–50.

Nigam, R., Nair, R.R., 1989. Cyclicity in monsoon. Bullet. Sci. 5, 42–44.

Nigam, R., Khare, N., Nair, R.R., 1995. Foraminiferal evidences for 77-year cycles of droughts in India and its possible modulation by the Gleissberg solar cycle. J. Coast Res. 11 (4), 1099–1107.

Nigam, R., Saraswat, R., Panchang, R., 2006. Application of foraminifers in ecotoxicology: retrospect, perspect and prospect. Environ. Int. 32 (2), 273–283.

Panchang, R., 2008. Ecology and Distribution of Benthic Foraminifera off Myanmar. Doctoral dissertation, Goa University.

Panchang, R., Nigam, R., 2012. High resolution climatic records of the past ~ 489 years from Central Asia as derived from benthic foraminiferal species, *Asterorotalia trispinosa*. Mar. Geol. 307, 88–104.

Panchang, R., Nigam, R., 2014. Benthic ecological mapping of the Ayeyarwady delta shelf off Myanmar, using foraminiferal assemblages. J. Palaeontol. Soc. India 59 (2), 121–168.

Panchang, R., Nigam, R., Baig, N., Nayak, G.N., 2005. A foraminiferal testimony for the reduced adverse effects of mining in Zuari Estuary, Goa. Int. J. Environ. Stud. 62 (5), 579–591.

Panchang, R., Nigam, R., Linshy, V., Rana, S.S., Ingole, B.S., 2006. Effect of oxygen manipulations on benthic foraminifera: a preliminary experiment. Indian J. Mar. Sci. 35 (3), 235–239.

Panchang, R., Nigam, R., Raviprasad, G.V., Rajagopalan, G., Ray, D.K., Hla, U., 2008. Relict faunal testimony for sea-level fluctuations off Myanmar (Burma). J. Palaeonl. Soc. India 53 (2), 185–195. December 2008.

Parkinson, J.H., Morrison, L.V., Stephenson, F.R., 1980. The constancy of the solar diameter over the past 250 years. Nature 288, 548–551.

Paul, J., Burgemann, R., Gaur, V.K., Dulham, R., Larsen, K.M., Ananda, M.B., Jade, S., Mukal, M., Anupama, T.S., Satyal, G., Kumar, D., 2001. The motion and active deformation of India. Geophys. Res. Lett. 28, 647–650.

Pirazzoli, P.A., 1991. World Atlas of Holocene Sea-Level Changes. Elsevier, Amsterdam.

Raju, K., Ramprasad, K.A., Rao, T.P.S., Ramalingeswara Rao, B., Varghese, J., 2004. New insights into the tectonic evolution of the Andaman basin, northeast Indian Ocean. Earth Planet Sci. Lett. 221, 145–162.

Ramaswamy, V., Rao, P.S., Rao, K.H., Swe Thwin, N., Srinivasa, R., Raiker, V., 2004. Tidal influence on suspended sediment distribution and dispersal in the northern Andaman Sea and Gulf of Martaban. Mar. Geol. 208, 33–42.

Rana, S.S., 2009. Study of Foraminiferal Distribution in Surface and Subsurface Sediments off Central East Coast of India and Their Paleoecological Significance (Ph.D. thesis).

Rana, S.S., Nigam, R., Panchang, R., 2007. Relict benthic foraminifera in surface sediments off Central East Coast of India as indicator of sea level changes. Indian J. Mar. Sci. 36 (4), 355–360.

Rao, N.R., 1998. Recent Foraminifera from the Inner Shelf Sediments of the Bay of Bengal, off Karikkattukuppam. Ph. D. Thesis, Madras University, near Madras, South India.

Rao, V.P., Veerayya, M., Thamban, M., Wagle, B.G., 1996. Evidences of Late Quaternary neo-tectonic activity and sea level changes along the western continental margin of India. Curr. Sci. 71, 213−219.

Rao, P.S., Ramaswamy, V., Thwin, S., 2005. Sediment texture, distribution and transport on the Irrawaddy continental shelf, Andaman Sea. Mar. Geol. 216, 239−247.

Reid, G.C., 1987. Influence of solar variability on global sea surface temperatures. Nature 329, 142−143.

Renema, W., 2006. Habitat variables determining the occurrence of larger benthic foraminifera in the Berau area (East Kalimantan, Indonesia). Coral Reef. 25 (3), 351−359.

Rodolfo, K.S., 1969. Sediments of the Andaman Basin, northeastern Indian ocean. Mar. Geol. 7, 371−402.

Saidova, K.M., 2007. Benthic foraminiferal assemblages of the South China sea. Oceanology 47 (5), 653−659.

Saraswat, R., 2006. Development and Evaluation of Proxies for High Resolution Paleoclimatic Reconstruction with Special Reference to North Eastern Indian Ocean. Ph.d. Thesis, Goa University, India.

Saraswat, R., Kurtarkar, S.R., Mazumder, A., Nigam, R., 2004. Foraminifers as indicators of marine pollution: a culture experiment with. Rosalina leei. Marine Pollut. Bullet. 48 (1−2), 91−96.

Saraswat, R., Nigam, R., Barreto, L., 2005. Palaeoceanographic implications of abundance and mean proloculus diameter of benthic foraminiferal species *Epistominella exigua* in sub-surface sediments from distal Bay of Bengal fan. J. Earth Syst. Sci. 114 (5), 453−458.

Saraswat, R., Nigam, R., Weldeab, S., Mackensen, A., 2007. A tropical warm pool in the Indian Ocean and its influence on ENSO over the past 137, 000 yrs BP. Curr. Sci. 92 (8), 1153−1156.

Saraswati, P.K., 2002. Growth and habitat of some recent Miliolid foraminifera. Curr. Sci. 82 (1), 81−84.

Sarkar, A., Ramesh, R., Bhattacharya, S.K., Rajagopalan, G., 1990. Oxygen isotope evidence for stronger winter monsoon current during the last glaciation. Nature 343 (6258), 549−551.

Schlesinger, M.E., Ramakutty, N., 1994. An oscillation in the global climate system of period 65−70 years. Nature 367, 723−726.

Shetye, S.R., Gouveia, A., 1998. Coastal circulation in the north Indian Ocean. In: Robinson, A.R., Brink, K.H. (Eds.), The Sea, vol. 11. John Wiley and Sons Inc., New York, pp. 523−556.

Sieh, K., Natawidjaja, D., 2000. Neotectonics of the Sumatran fault, Indonesia. J. Geophys. Res. 105 (B12), 28295−28326.

Sijinkumar, A.V., Nath, B.N., Possnert, G., Aldahan, A., 2011. Pulleniatina minimum events in the Andaman Sea (NE Indian Ocean): implications for winter monsoon and thermocline changes. Mar. Micropaleontol. 81 (3−4), 88−94.

Sijinkumar, A.V., Nath, B.N., Clemens, S., 2016a. Rapid climate changes in the North Atlantic is reflected in the foraminiferal abundance in the Andaman Sea. Palaeogeogr. Palaeoclimatol. Palaeoecol. 446, 11−18.

Sijinkumar, A.V., Nath, B.N., Clemens, S., 2016b. North Atlantic climatic changes reflected in the Late Quaternary foraminiferal abundance record of the Andaman Sea, north-eastern Indian Ocean. Palaeogeogr. Palaeoclimatol. Palaeoecol. 446, 11−18.

Sinsakul, S., 1992. Evidence of Quaternary Sea level changes in the coastal areas of Thailand: a review. J. Southeast Asian Earth Sci. 7 (1), 23−37.

Siscoe, G.L., 1980. Evidence in the auroral for secular solar variability. Rev. Geophys. Space Phys. 18, 647−658.

Srinivasan, M.S., 2007. A journey through morphological micropaleontology to molecular micropaleontology. Indian J. Mar. Sci. 36 (4), 251−271.

Stanley, D.J., 1995. A global sea-level curve for the late Quaternary: the impossible dream? Mar. Geol. 125, 1−6.

Stuiver, M., 1980. Solar variability and climatic change during the current millennium. Nature 286, 868−871.

Stuiver, M., Quay, P.D., 1980. Changes in atmospheric ^{14}C attributed to a variable sun. Science 207, 11−19.

Stuiver, M., Reimer, P.J., 1993. Extended ^{14}C data base and revised CALIB 3.0 ^{14}C Age calibration program. Radiocarbon 35 (1), 215−230.

Subbarao, M., Vedantam, D., Nageshwar Rao, J., 1979. Distribution and ecology of benthic foraminifera in the sediments of the Vishakhpatnam shelf, East Coast of India. Paleogeog., Paleoclimat., Paleoecol. 27, 349–369.

Subramanian, K.S., 2005. The Sunda trench and the Andaman-Nicobar islands. J. Geol. Soc. India 66, 255.

Suokhrie, T., Saraswat, R., Nigam, R., 2020. Lack of denitrification causes a difference in benthic foraminifera living in the oxygen deficient zones of the Bay of Bengal and the Arabian Sea. Mar. Pollut. Bull. 153, 110992.

The Coral Reef Alliance, 2004. What Corals Need Most. http://www.coralreefalliance.org/aboutcoralreefs/coralneeds.html.

Thomson, J.A., Henderson, W.D., 1906. An Account of the Alcyonarians Collected by the Royal Indian Marine Survey Ship 'Investigator' in the Indian Ocean. I. The Alcyonarians of the Deep Sea. Trustees of the Indian Museum, University of Aberdeen, 1–128, plates 1–7, pp. 7–8. https://www.biodiversitylibrary.org/page/12037643.

Toushingham, A.M., Peltier, W.R., 1991. ICE-3G: a new model of late Pleistocene deglaciation based upon geophysical predictions of post-glacial sea level change. J. Geophys. Res. 96, 4497–4523.

Ujjié, H., 1962. Introduction to statistical foraminiferal zonation. J. Geol. Soc. Jpn. 803, 431–450.

Van Alstyne, K., Wylie, C.R., Paul, V.J., Meyer, K., 1992. Anti-predator defenses in tropical pacific soft corals (Ceolenterata: Alcyonacea). I. Sclerites as Defenses against generalist carnivorous fishes. Bullet. Biol. 182, 231–240.

Vaz, G.G., 1996. Relict coral reef and evidence of Pre-Holocene sea level stand off Mahabalipuram, Bay of Bengal. Curr. Sci. 71 (3), 240–242.

Vaz, G.G., 2000. Age of relict coral reef from the continental shelf off Karaikal, Bay of Bengal: evidence of last glacial maximum. Curr. Sci. 79 (2), 228–230.

Vaz, G.G., Banerjee, P.K., 1997. Middle and late Holocene sea level changes in and around Pulicat lagoon, Bay of Bengal, India. Mar. Geol. 138, 261–271.

Venee-Peyre, M.T., 1985. The study of the living foraminiferan distribution in the lagoon of the high volcanic island of Moorea (French Polynesia). Proc. 5th Intern. Coral Reef Congress, Tahita 5, 227–232.

Verma, K., Singh, H., Singh, A.D., Singh, P., Satpathy, R.K., Naidu, P.D., 2021. Benthic foraminiferal response to the Millennial-scale variations in monsoon-Driven productivity and deep-water oxygenation in the western Bay of Bengal during the last 45 ka. Front. Mar. Sci. 8, 733365.

Vigny, C., Socquet, A., Rangin, C., Chamot-Rooke, N., Pubellier, M., Bouin, M.-N., Bertrand, G., Becker, M., 2003. Present-day crustal deformation around Sagaing Fault, Myanmar. J. Geophys. Res. 108. ETG 6-1-10.

Vora, K.H., Wagle, B.G., Veerayya, M., Almeida, F., Karisiddaiah, S.M., 1996. 1300 Km. Long late Pleistocene-Holocene shelf edge barrier system along the western continental shelf of India: occurrence and significance. Mar. Geol. 134, 145–162.

Waelbroeck, C., Labeyrie, L., Michel, E., Duplessy, J.C., McManus, J.F., Lambeck, K., Balbon, E., Labracherie, M., 2002. Sea-level and deep water temperature changes derived from benthic foraminifera isotopic records. Quat. Sci. Rev. 21, 295–305.

Williams, E., Edwards, A., 1993. Coral and Coral Reefs in the Caribbean: A Manual for Students, St. Michael (Barbados), second ed. Caribbean Conservation Association.

Wray, J.L., 1971. Algae in reefs through time. In: Proceedings of the North American Paleontological Convention, Part J: Reef Organisms through Time. Allen Press, Lawrence, KS, pp. 1358–1373.

Wylie, C.R., Paul, V.J., 1989. Chemical defenses in three species of *Sinularia* (Coelenterata, Alcyonacea): effects against generalist predators and the butterflyfish *Chaetodon unimaculatus* Bloch. J. Exp. Mar. Biol. Ecol. 129, 141–160.

Wyrtki, K., 1973. Physical oceanography of the Indian ocean. In: Zeitschel, B., Gerlach, S.A. (Eds.), The Biology of the Indian Ocean. Springer Verlag, Berlin, pp. 18–36.

Yim, W.W.S., Huang, G., Fontugne, M.R., Hale, R.E., Paterne, M., Pirazzoli, P.A., Ridley Thomas, W.N., 2006. Postglacial sea-level changes in the northern South China Sea continental shelf: evidence for a post-8200 calendar yr BP meltwater pulse. Quat. Int. 145–146, 55–67.

Electronic references

Encyclopaedia Britannica, 2000. http://www.britannica.com/.

FAO, 2000. Aquastat, an On-Line Database of Statistics on Freshwater Availability. http://www.fao.org/waicent/faoiFnfo/agricult/agl/aglw/aquastat/aquastat.html.

Fox, H.E., 2001. Enhancing reef recovery at Komodo National Park, Indonesia: a proposal for coral reef rehabilitation at ecologically significant scales. http://www.komodonationalpark.org/downloads/CorRehabProp.PDF.

Hohenegger, J., 1996. Larger foraminifera as important calcium-carbonate producers in coral reef environments and constituting the main components of the carbonate beach sands: examples from Ryukyu archipelago. In: Proceedings of FORAMS'98- International Symposium on Foraminifera. Monterrey, Mexico. http://www.okinawa-conference.unibonn.de/Hohenegger.html.

Littler, M.M., 1972. The Crustose Corallinaceae, Oceanogr. Mar. Biol. Ann. Rev. Ed. Harold Barnes, vol. 10. Publ: George Allen and Unwin Ltd, London, pp. 311–347. http://hdl.handle.net/10088/2475.

Further reading

Elderfield, H., Ganssen, G., 2000. Past temperature and delta ^{18}O of surface ocean waters inferred from foraminiferal Mg/Ca ratios. Nature 405, 442–445.

Ellis and Messina, 2007. Catalogues for Foraminifera, Ostracoda and Diatom. Micropaleontology Press, New York, USA. http://micropress.org/em/login.php.

http://en.wikipedia.org/wiki/Coco_Islands.

http://en.wikipedia.org/wiki/Mergui_Archipelago.

Irrawaddy Delta Website of the ASEAN Regional Centre for Biodiversity Conservation. http://www.arcbc.org.ph/wetlands/myanmar/mmr_irrdel.htm.

Lea, D.W., Mashiotta, T.A., Spero, H.J., 1999. Controls on magnesium and strontium uptake in planktonic foraminifera determined by live culturing. Geochem. Cosmochim. Acta 63, 2369–2379.

Lear, C.H., Elderfield, H., Wilson, P.A., 2000. Cenozoic deep-sea temperatures and global ice volumes from Mg/Ca in benthic foraminiferal calcite. Science 287, 269–272.

Loeblich, A.R., Tappan, H., 1988. Foraminiferal Genera and Their Classification. Von Nostrand Reinhold Company, New York.

Mashiotta, T.A., Lea, D.W., Spero, H.J., 1999. Glacial-interglacial changes in Subantarctic sea surface temperature and ^{18}O water using foraminiferal Mg. Earth Planet Sci. Lett. 170, 417–432.

Pielou, E.C., 1966. The measurement of diversity in different types of biological collections. J. Theor. Biol. 13, 131–144.

Website of NOAA Ocean Explorer. http://oceanexplorer.noaa.gov/explorations/03mountains/background/octocorals/media/sclerite.html.

Website of the NASA Earth Observatory; http://eobglossary.gsfc.nasa.gov/Library/glossary.php3?xref=Little%20Ice%20Age.

Website of the World Atlas, MSN-Encarta http://encarta.msn.com/map_701514912/myanmar.html.

Website of the World Bank: http://web.worldbank.org/WBSITE/EXTERNAL/NEWS/0,,menuPK:51062077~pagePK:64001221~piPK:64001770~theSitePK:4607~topicMDK:473883~PageNo:9~PageSize:20~startIndex:161,00.html.

Wikipedia for Ayeyarwady River. http://en.wikipedia.org/wiki/Ayeyarwady_River.

Wikipedia for Islands of Myanmar. http://en.wikipedia.org/wiki/Cheduba_Island.

CHAPTER 12

Sedimentation model for the Pinjor Formation of the Upper Siwalik, Dehradun Sub-basin, Garhwal Himalaya, India: a response of climate and tectonics

Bijendra Pathak and U.K. Shukla

Centre of Advanced Study in Geology, Banaras Hindu University, Varanasi, Uttar Pradesh, India

1. Introduction

Siwalik molasse is one of the most extensive clastic sequences on the Earth, deposited in the Himalayan Foreland Basin between 18 Ma and 0.5 Ma (Johnson et al., 1985; Kotla et al., 2018), that dominantly represents the Outer Himalaya forming the southern front of the lofty Himalayas. About 6000 m thick Neogene succession represents a fairly continuous record of fluvial sedimentation, which varied greatly in lithology with space and time (Prakash et al., 1980), and attributed to different auto cyclic and allocyclic processes acting together in the basin as well as in the source terrain (Kumar et al., 2003a,b).

Pilgrim (1913) based on vertebrate fossil assemblages proposed a threefold classification of the Siwalik succession, viz: the Lower Siwalik Subgroup, the Middle Siwalik Subgroup, and the Upper Siwalik Subgroup, which are further divided into seven formations, for these extensive fluvial successions (Table 12.1). There are several sub-basins in the Himalayan Foreland Basin system where the Upper Siwalik was deposited in a time-transgressive manner (Raiverman et al., 1983). The Upper Siwalik succession was deposited between the periods of nearly 5 Ma and 0.22 Ma (Table 12.1). Kumar et al. (2003a,b, 2004) have magneto stratigraphically dated the boundary between the Middle and the Upper Siwalik at 5.23 Ma in the Mohand Rao section of the Dehradun Sub-basin. The Upper Siwalik succession shows an overall coarsening upward trend in the stratigraphy with the conglomerate percentage increasing the up section in the down-dip direction. Dehradun sub-basin bounded between Ganga and Yamuna tear fault in East and West, respectively, exhibits a continuous stretch of the Middle and the Upper Siwalik Subgroup formations. Extensive studies have been done by many workers on the underlying Middle Siwalik succession of the area for gross-lithology and magnetostratigraphy (Kumar et al., 2003a,b, 2004; Kumaravel et al., 2005a,b; Rathi et al., 2007). The Upper Siwalik succession of the Dehradun Sub-

Table 12.1 Stratigraphic classification of Siwalik Group.

	Sub group	Formation	Calibrated age (Ma)	Geological range
Siwalik Group	Upper Siwalik	Boulder Conglomerate	1.6–0.22	Pliocene to middle pleistocene
		Pinjor	2.5–1.6	
		Tatrot	5–2.5	
	Middle Siwalik	Dhok pathan	7.9–5.1	Upper miocene
		Nagri	10.5–7.9	
	Lower Siwalik	Chingi	14.3–11.5	Middle to upper miocene
		Kamalial	18.3–14.3	

Modified after Pilgrim (1913); Johnson et al. (1982); Opdyke et al. (1979).

basin remained unattended though some work has been done in the adjacent Mohand Rao section on sedimentology and petrography (Kumar and Ghosh, 1991; Kumar et al., 1999; Ghosh and Kumar, 2000) (Fig. 12.1), yet research gap exists in terms of detailed sedimentological evolution and modeling of the Upper Siwalik Formation of the study area. The Upper Siwalik represents the terminal phases of the Himalayan orogeny and is crucial to understanding the synsedimentary climatic and tectonic phases in the Himalayan Foreland Basin. The present work attempts in detail lithofacies, architectural elements, petrography, and paleocurrent analyses of the Pinjor Formation of the Upper Siwalik Subgroup, along Badshahibagh Rao (stream) located in the western section of Dehradun Sub-basin. This is to be emphasized that this section is new and hitherto untouched by any type of geological investigation.

2. Geology of the area

The Yamuna Transverse Fault in the west and the Ganga Transverse Fault in the east define the Dehradun re-entrant of the Himalayan Foreland Basin. The Middle and the Upper Siwalik Subgroup rocks are exposed along various river/stream sections flowing southwest on the southern slopes of the mountain front of the Siwalik Mountain chain in the Dehradun Sub-basin. In these stream, sections of the Middle Siwalik grade to the Upper Siwalik sequence with a conformable contact from south-west to northeast along with the down dip (Fig. 12.1). The studied stream section is located at the border of the Dehradun district of the Uttarakhand and Saharanpur district of Uttar Pradesh, India (Fig. 12.1). In Dehradun re-entrant, the Siwalik Group stratigraphically shows two formations of the Middle Siwalik Subgroup (Nagri and the Dhokpathan formations) and the Pinjore and the Boulder Conglomerate formations of the Upper Siwalik Subgroup (Singh, 2020). Characteristically, the Tatrot Formation of the Upper Siwalik is not developed in the area. Pinjor Formation in the studied section is represented by, alternate sandstone, mudstone, and pebble–cobble conglomerate in the lower part and pebble–cobble conglomerate along with subordinate sandstone in the upper part. The base of this formation can be marked with a pebble–cobble conglomeratic bed of about 2.50 m, which gradationally overly Dhokpathan Formation of the Middle Siwalik. Along the Himalayan Frontal Thrust (HFT), the southern boundary is in faulted contact and thrust over the Quaternary alluvium of the Gangetic Plains.

Quaternary gravels cover the northern slopes of the Siwalik ranges, which marks the northernmost boundaries for the Upper Siwalik rocks outcropping in the study area. Another significant tectonic structure in the study area is the Mohand anticline (Fig. 12.1).

3. Methodology

The current study is based on a combination of field observations made along the Badshahibagh Rao (stream) section located in the western part of the Siwalik ranges of the Dehradun Sub-basin. In this section, being gravelly sandstone near the upper levels in the stratigraphy, the Middle Siwalik shows a gradational transition with the Upper Siwalik, which are conglomerate dominated. Therefore, the first conglomerate bed revealed along the examined stream segment (with embedded pebble size of up to 12 cm axial length) was taken as the base of the Upper Siwalik formations while traveling from, south to north, that is, from the Middle to the Upper Siwalik formations along with the upstream. Vertical sections, lateral profiles, and field photographs were used to study sedimentary facies and associated architectural elements. Thin sections of gravelly sandstones and mudstones were prepared for petrography, and gravels being larger (2 mm – 12 cm in size) were measured in the field for their size, lithology, and imbrications. Different geomorphic zones of sedimentation were identified using sedimentary structures, textures, litho-unit geometry and paleocurrents (Shukla, 2009; Shukla et al., 1999), and lateral facies transitions emphasizing genetically significant surfaces (Miall, 1985). Lithological successions were subjected to a thorough lithofacies examination, which included a comparison to present equivalents (Shukla and Bora, 2003, 2005; Shukla et al., 2009). To determine the depositional environments, lithofacies were identified and interpreted using a process–response technique (de Raaf et al., 1965; Walker, 1984; Shukla et al., 2010). Paleocurrent based on imbricated gravels in conglomerate has been carried out following the processes described by Curray (1956) and Shukla et al. (1999), to diagnose sediment dispersal patterns and the paleoslope.

4. Lithofacies analysis

Sedimentological data were collected from the continuous outcrops exposed along the studied stream section and that enabled the identification of constituent lithofacies of the Pinjore Formation. As the stream runs mainly across the strike of the said formation, only vertical profiles having steep tectonic dips were available for the study. Following Miall (1985), channels (CH), sandy bedforms (SB), laminated sand sheets (LS), lateral accretion macro form (LA), gravel bars and bedforms (GB), and overbank fines (OF) are the characteristic architectural elements recognized in the Pinjor Formation of the studied area. Various lithofacies characteristics and interpretations are summarized in Table 12.2.

4.1 Horizontal and low-angle planar gravelly sandstone

The Gravelly Sandstone facies is characterized by thick and thin tabular Sandstone unit, which shows the typical fining upward sequence of a channel deposit. These sandstone units, 2.5–10 m thick, with erosional bases (CH element), are multistoried and laterally exhibit sheet-like or tabular geometry. The sandstone unit is internally composed of horizontal or low-angle planar structures (LS elements). Fining upward sequences are recorded which terminate into fine sediments (mudstone, OF element) at the top (Figs. 12.2B and 12.9). Sandstone is friable due to poor calcite cementation and contains

Table 12.2 Lithofacies of the Pinjor Formation, their characters, and depositional settings.

Lithofacies	Description	Interpretation
Matrix to clast supported conglomerate	Pebble to cobble-grade clasts, poorly sorted, individual units 40 cm to 5 m thick, sharp and erosional bases, sandstone lenses embedded, and conglomerates show imbrication locally (Fig. 12.4A and B)	Poor sorting and unstratified nature suggest rapid deposition during hyper-concentrated flood flow or catastrophic flood events. (Kumar et al., 2007)
Clast supported conglomerate	Unsorted, pebble to cobble grade clast in sand or silty sand matrix with individual units ranging from 2.5 m to 10 s of meter with lower erosional contacts, planar cross-stratification (Fig. 12.4C and D)	Product of 2D high relief linguoid and transverse bar migration (Teisseyre, 1975; Koster and Steel, 1984; Shukla, 2009; Miall, 1977).
Planar cross-bedded gravelly sandstone	45 cm-10 m thick fine to coarse gravelly sand units, 20—100 cm thick planar cross-bedding, and units exhibit lenticular or wedge-shaped geometry (Fig. 12.2E and Fig. 12.3A)	Formed by migration of two-dimensional bedforms in a lower flow regime. (Shukla et al., 2001)
Horizontally bedded gravelly sandstone (facies S2)	Fine to medium sand with individual units ranging in thickness from 50 cm to 5 m. beds show tabular, planar or sheet geometry, gravel size 2 mm — 5 cm (Fig. 12.2C)	Deposited during planar bed flow in upper and lower flow regime during floods (Miall, 1977).
Low angle planar gravelly sandstone	2.5—10 m thick fine to coarse gravelly sand units, forests dipping at 5 degrees to 10 degrees (Fig. 12.2A)	Deposited during planar bed flow in upper and lower flow regime during floods (Miall, 1977), or scour fill crevasse splays or antidunes (Miall, 1977).
Ripple bedded sandstone	Fine sand, 50 cm — 60 cm thick units of ripple cross-bedding or climbing ripple lamination (Fig. 12.2D)	Deposited in a lower flow regime by migrating small ripples.
Laminated mudstone	Gray-colored laminated sand silt clay, of 15—20 cm thick units (Fig. 12.2B)	Lacustrine or back swamp deposit.

nodules and concretions. Basal lag is also recorded at the base of the unit. Sandstone is ill-sorted with some rock fragments and is coarse- to fine-grained. Occasionally, there is the presence of matrix in clast-supported conglomerate unit, containing pebble to cobble size clasts that show imbrication in patches (Figs. 12.4B and 12.5).

The multistoried sandstone units are channel deposits of small to large rivers (Visher, 1965; Reineck and Singh, 1980; Thomas et al., 2002; Kumar et al., 2021). The horizontal bedded sandstone units can be interpreted as the planar bed of the lower and the upper flow regimes (Miall, 2016). The vertical association of sedimentary structures and increased sandstone to mudstone ratio within these lithofacies suggests deposition in a braided channel system (Kumar et al., 1991; Miall, 2014, 2016).

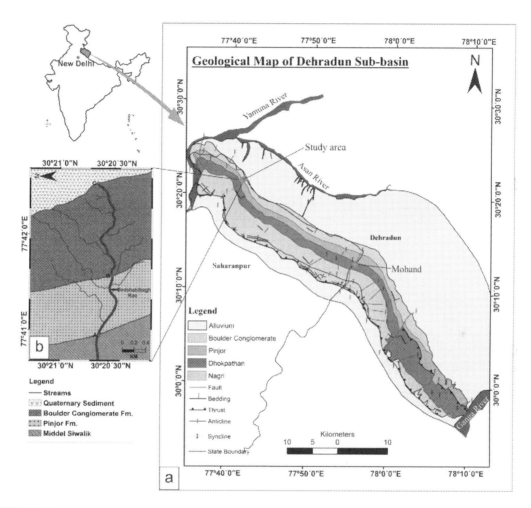

FIGURE 12.1

(A). Geological map of study area showing various lithotectonic units and major fault/thrust planes modified after map published by geological survey of India (Singh, 2020) geological map showing close up of Badshahibagh Rao.

However, the sandstone unit also shows lateral accretion surfaces (LA elements). Lensoidal sand bodies showing erosional base and spaced lateral accretion surfaces made up of clast to matrix-supported conglomerate represent a deposit of shifting channel in response to which point bar is migrated in the down-dip direction (Fig. 12.2F). The gravelly LA elements represent the channel lag deposits formed under energetic turbulent flows at the thalweg of the channel (Miall, 1985; Bridge, 1985). Within this unit, the point bar thickness increases along with gravel in the down current direction, which indicates that progressively the meandering river became larger and more energetic.

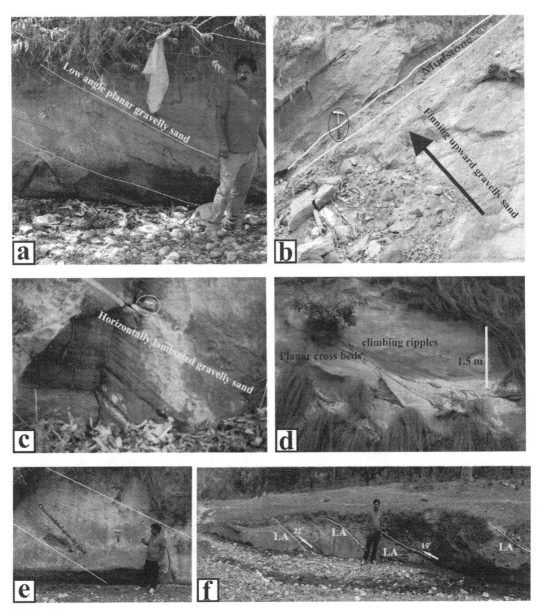

FIGURE 12.2

(A) Low-angle planar sand unit in of pinjor formation; a person for scale, 160 cm. (B) Gravelly sandstone finning upward into laminated mudstone, followed by low-angle planar gravelly sandstone units; hammer length 32 cm. (C) Horizontal bedded gravelly sandstone unit; diameter of measuring tape 11 cm. (D) Planar cross-bedded gravelly sandstone followed by ripple-cross laminated sandstone unit. (E) Planar cross-bedding with well-defined reactivation surface at the top; hammer length 32 cm. (F) Gravelly sandstone units showing lateral accretion surface with gravel at top of each accretion unit; a person for scale, 160 cm.

4. Lithofacies analysis 279

FIGURE 12.3

(A) Multistorey complex having two channels 1 and 2 separated by an erosional base. Channel 1 is large-scale cross-beds showing finning upward sequence and channel 2 is buff in color showing large cross-beds. (B) Sand–gravel complex showing superposition of constituent lithofacies in vertical superposition.

FIGURE 12.4

(A) Clast-to-matrix-supported conglomerate showing convex upward sand lenses; length of hammer 32 cm. (B) Clast-to-matrix-supported conglomerate showing imbrications in patches; a person for scale 158 cm. (C) Clast supported conglomerate finning up to sand units. (D) Planar cross-bedded supported conglomerate and sand unit.

4. Lithofacies analysis 281

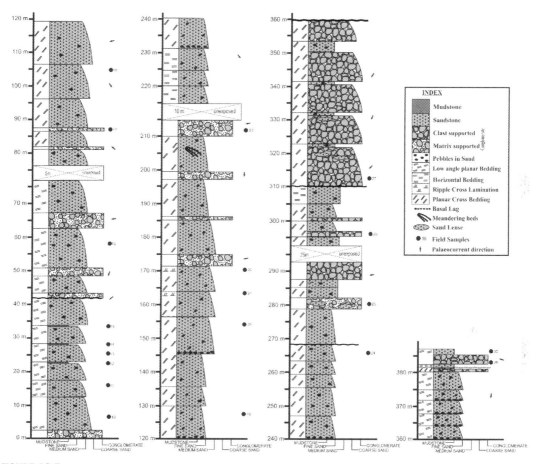

FIGURE 12.5

Litholog prepared from the measured outcrop section of the pinjor formation in Badshahibagh Rao (stream) of Dehradun sub-basin. The paleocurrent azimuths plotted alongside lithology show paleoflow changing from SW in the lower levels to NW and SE in the upper levels of the succession.

There is the presence of matrix to clast-supported conglomerate units within the body of sandstones representing recurring flood deposits. The lateral accretion suggests the presence of sinuous (meandering) streams, and low-angle planar beds abundantly developed are deposited at Channel banks as longitudinal bars (Visher, 1965; Reineck and Singh, 1980; Miall, 1985; Singh, 2020).

4.2 Planar cross-bedded gravelly sandstone

The planar cross-bedded gravelly sandstone facies is medium- to fine-grained sandstone units of 45 cm to 10 s of meter thickness, showing tabular and wedge-shaped geometries with erosional bases (Fig. 12.5). Large-scale planar cross-bedding is recorded which shows thickness up to 2.5 m (SB element) (Figs. 12.2E and 12.3A). Some sandstone units show cross-beds and grade upward into climbing ripple laminations (Fig. 12.2D). These sandstone units are poorly sorted and pebbly too, where pebbles are arranged along with the cross-bed forests and studded at the bottom and top sets of the cross-beds. These cross-bedded sandstone units are occasionally truncated by the clast-to-matrix-supported conglomeratic units containing pebble-cobble clasts.

The planar cross-bedded sandstone units (Sp element) can be interpreted as the migration of 2-D bedforms (Shukla et al., 2001, 2009) in a lower flow regime while ripples formed during waning flood conditions. Gravels were brought during the recurring floods and were deposited with sand (Shukla and Singh, 2004, 2005).

4.3 Matrix-to-clast-supported conglomerate

These facies constitute of unorganized fabric, 0.5−5 m thick units of clast-to-matrix-supported conglomerates that grade upward in sandstone units (CH and GB elements) (Fig. 12.5). Clasts are subangular to rounded and very poorly sorted (Fig. 12.4A and B). These conglomeratic units show sheet geometry in the stratigraphically lower part of the Pinjor, while tabular and wedge shape geometries in stratigraphic upper portions. Sandstone lenses are embedded within the conglomerate beds (Fig. 12.4A). These are patches of imbricated gravel within the conglomerate sequences (Fig. 12.4B). Toward the top, the thickness of sandstone units decreases. Several sand lenses show convex-up geometry (Fig. 12.5A).

The disorganized gravels are the result of diffused and dispersed bedforms by turbulent flow (Sinha et al., 2007; Shukla, 2009) and sand lenses formed during waning flood stages. The convex upward sandstone lenses at the top of the gravely unit suggest bar top morphology (Miall, 1985; Shukla et al., 2001). Patches of imbricated gravel or within bed pavement suggest low-stage discharge (Teisseyre, 1975).

4.4 Clast-supported conglomerate

This lithofacies is made up of about 1−3 m thick clast-supported conglomerate units, capped by sandstone units (Figs. 12.4C and 12.5). These conglomerate units have sharp to erosional bases and show well-developed imbrications and planar cross-beds (GB and CH elements). The constituent gravels are pebble to cobble size clast and are poorly sorted and subangular to round in nature. Planar cross-beds are evident at several places in outcrop at the intersection of sandy and gravelly beds (Figs. 12.4D and 12.5), which might be deposited at the margin of a gravelly bar slip face (Sinha et al., 2007).

These facies are the resultant of 2-dimensional high-relief bedforms within channels (Teisseyre, 1975; Shukla, 2009). These cross-bedded gravelly units with sand capping can be attributed to the transverse bars and sand waves migrating in aggradational channels (Teisseyre, 1975).

4.5 Ripple cross-bedded sandstone

Generally occurring at the top of cross-bedded sandstone units, ripple cross-laminated sandstone units are 50−60 cm thick fine-grained sandstone units (Fig. 12.2D). These thin sandstone unit shows sheet to lensoidal geometry and shows gradational contact with the lower-lying medium to fine planar cross-bedded sandstone unit and erosional relationship with the overlying gravelly sandstone unit. These units were formed during migrating ripple movements in lower flow regimes, at very shallow depths (Allen, 1973; Shukla et al., 2001).

4.6 Laminated mudstone

Present in a very minor amount in outcrop, these 15−20 cm gray mudstone units generally form the top of finning upward gravelly sandstone sequences (Fig. 12.2B). These very thin mudstones show sheet-like geometry and occur at the top of finning upward medium to fine sandstone units, and the overlying sandstone unit shows a sharp contact relationship with it (Fig. 12.2B). Some calcite precipitation can be noticed within interlamination spaces. Less presence of these units might be a result of high erosional processes that might be active during their time of deposition in the basin. These facies were also developed in local standing bodies of water or back swamps (Kumar and Tandon, 1985; Schieber, 1999).

5. Lithofacies association

The studied succession of Pinjor Formation leads to the identification of lithofacies vertical organization in different parts of the outcrops (Fig. 12.6). Following seven lithofacies associations were observed.

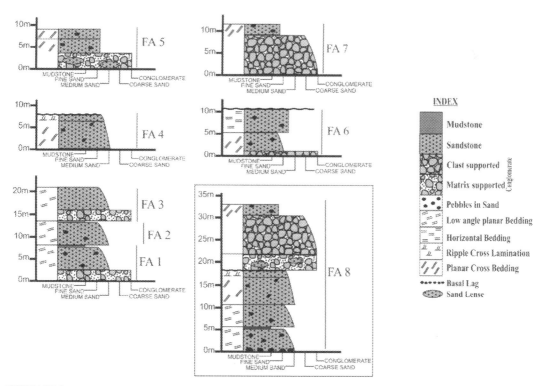

FIGURE 12.6

Representative vertical section of lithofacies association (FA 1 to FA 7), with an idealized lithofacies association (FA 8) observed in different parts of outcrop from the Pinjor Formation of Badshahibagh Rao.

1. Matrix-to-clast-supported conglomerate » Low-angle planar gravelly sandstone » Laminated mudstone (F1 in Fig. 12.6).
2. Low-angle planar gravelly sandstone » Laminated mudstone (FA 2 in Fig. 12.6).
3. Matrix-to-clast-supported conglomerate » Low-angle planar gravelly sandstone (FA 3 in Fig. 12.6).
4. Planar cross-bedded gravelly sandstone—Ripple cross-laminated sandstone (FA 4 in Fig. 12.6).
5. Matrix-to-clast-supported conglomerate » Planar cross-bedded gravelly sandstone (FA 5 in Fig. 12.6).
6. Clast-supported conglomerate » Planar cross-bedded gravelly sandstone » Horizontally laminated gravelly sandstone (FA 6 in Fig. 12.6).
7. Clast-supported conglomerate » Planar cross-bedded gravelly sandstone (FA 7 in Fig. 12.6).

Based on the aforementioned seven lithofacies associations, an idealized lithofacies association is identified, which represents coarsening upward (CU) character representing the progradation of proximal fan conglomerates over the mid-fan gravelly sandstones (FA 8 in Fig. 12.6).

6. Palaeocurrent pattern

Paleoflow during deposition of the Pinjor Formation around Badshahibagh Rao Section was inferred mainly by observing the orientation and imbrications of long axes of the gravels of conglomerate units and also from the cross-bed foreset azimuths. Temporal variation of the paleocurrent is plotted against sandstone/conglomerate units showing changes in paleoflow (Fig. 12.5). Representative azimuthal readings were collected stratigraphically from bottom to top of the succession. Readings were analyzed by plotting rose diagrams (Fig. 12.7) and calculating statistical parameters following Curray (1956) and Shukla et al. (1999) (Table 12.3).

FIGURE 12.7

Rose diagram showing SW paleoflow, drawn from the data obtained by measuring the orientation of imbricated pebbles and cross-beds azimuths.

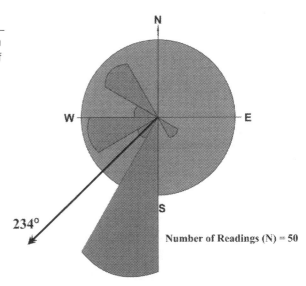

Table 12.3 Statistical parameters of the paleocurrent study from the Pinjor Formation.

Total number of readings (N)	50
Vector mean (in degree) (Θ_v)	234 degrees
Vector strength (L)	64%
Standard deviation (S^2)	2823
Variance (σ)	53

The paleocurrent study shows a radiating or fanning out pattern ranging from SE to NW directions with a subordinate paleoflow directed to NNW. The mean paleo flow is directed to SW (234 degrees) direction, indicating sediment supply from the NE and a paleoslope toward the SW (Fig. 12.7). However, NW-directed modes may imply sedimentation at the meander bends of the channels. The higher variance of flow implies sedimentation on the fan surface where different channel patterns were present and which were laterally migrating in response to sedimentation, erosion, and channel abandonment (Tandon, 1976; Kumar et al., 1991; Shukla et al., 2001; Shukla, 2009).

7. Petrography

In general, the Pinjor Formation is petrographically characterized by fine- to medium-grained sandstones with angular to subangular, moderately sorted bimodal grain size distribution, set in calcareous cement (Fig. 12.8). These sandstones are gravelly, and gravel is scattered throughout the sandstone units. Being larger (2 mm — 12 cm in size), they are identified lithologically in the field, and they are made up of primary quartzite, with rare occurrences of sandstone and mudstone clasts. In sandstones, the predominant framework grains are quartz, feldspar, and mica, where K-feldspar dominates over plagioclase. Microcline, muscovite, and biotite are other constituent mineral components set in the matrix (Fig. 12.8A and B). Lithic fragments include fine-grained quartzite, gneiss, schist, sandstone, and siltstone (Fig. 12.8A—D). Matrix constitutes altered mica and recrystallized calcite. Two types of quartz grain are identified, one having straight or parallel extinction under crossed-nicol, the other exhibiting undulose extinction. This reveals a bimodal nature of the source material being the igneous and metamorphic origin. Grains typically reveal point contact. However, they were infrequently seen floating in matrix and cement (Fig. 12.8A and B). Cryptocrystalline to crystalline calcite cement is most common. Thin sections of muddy units are not much revealing under a petrological microscope; however, silt-sized grains of quartz can be identified (Fig. 12.8C and D). Mud units are silty and turn out to be silty mudstones when analyzed in a laser particle grain size analyzer.

The angularity of grains with significant lithic fragments and an appreciable amount of matrix suggests textural immaturity of the sandstones (Pettijohn et al., 1973; Lindholm, 1987). Euhedral grains of quartz and felspar suggest less transportation of sediment and point out the nearness of the source area, which most probably can be the Himalayan hinterland in the north. Sediments derived from orogenic belts are more feldspathic (Pettijohn et al., 1973; Dickinson, 1985). The presence of a significant amount of feldspar such as orthoclase and microcline suggests a granitic source area (Pettijohn et al., 1973). Moreover, the presence of strained quartz grain showing undulose extinction in thin slides indicates a metamorphic rock origin. The presence of fresh and partly weathered feldspars

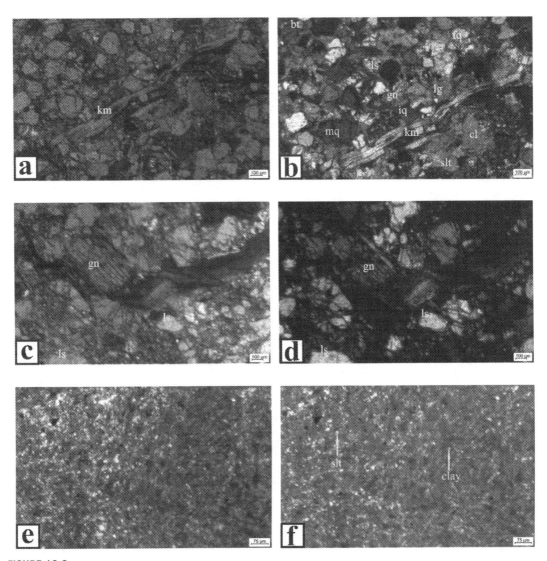

FIGURE 12.8

Photomicrograph of a thin section of rocks from pinjor Formation of Badshahibagh Rao area. (A) Fine to medium-grained sandstone in plane polarized light. (B) Fine-to-medium-grained sandstone in crossed polars. Biotite (bt), kinked mica (km), gneiss (gn), igneous quartz (iq), metamorphic quartz (mq), fine-grained quartzite (fq), lithic sandstone (ls), silt (slt), and calcite (cl). (C) Fine- to medium-grained sandstone in plane polarized light. (D) Fine- to medium-grained sandstone in crossed polars. Gneiss (gn) and lithic sandstone (ls). (E) Thin section of mudstone in plane polarized light. (F) Thin section of mudstone in crossed polars; silt (slt).

implies a high relief and fast-eroding source terrain for the Pinjor basin (Krynine, 1948). It may also imply rapid sedimentation and basin subsidence. No quartz and feldspar overgrowth and high intergranular porosity indicate that the rock units have been subjected to very less burial and compaction. Thus, both texturally and mineralogically, these rocks are immature. The presence of fractured grains and bends in mica flakes indicates that it has been subjected to postdiagenetic deformation. The presence of strained quartz and other aforementioned metamorphic lithic fragments along with igneous quartz and feldspar suggests a source area comprising both metamorphic and igneous rock suites, which might be sourced from northerly located Central crystalline and the Lesser Himalaya. Lithic sandstone and siltstone suggest uplifted part of Lower Siwalik also contributed detritus to the Pinjor basin (Fig. 12.8). Paleocurrent analysis reflects an SW paleoflow, which also confirms the Himalayan hinterland as the source (Fig. 12.7).

8. Depositional model

The Pinjor Formation consists of (a) medium- to fine-grained pebbly sandstone, which is white to gray, with occasional hard cemented layers and calcareous nodules (Fig. 12.3A), (b) conglomeratic beds, containing pebble-cobble sized clasts which are matrix-to-clast-supported embedded in gritty,

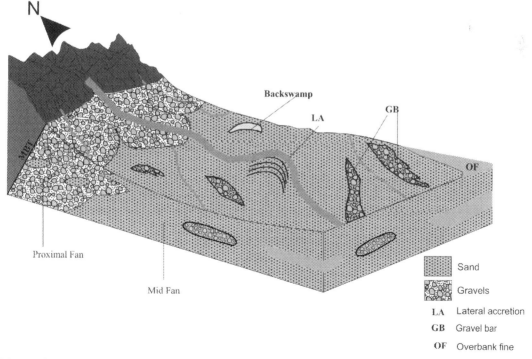

FIGURE 12.9

Schematic depositional model of the Pinjore Formation based on the Badshahibagh Rao (stream). The paleocurrent azimuths plotted alongside litholog show paleoflow changing from SW in the lower levels to NW and SE in the upper levels of the succession.

medium-to-coarse-grained sandy matrix (Figs. 12.4 and 12.5), and (c) subordinate mudstones (Fig. 12.2B). Clast reported from this formation is extramarital, constituting mainly of quartzite which is derived from the nearby hinterland and randomly oriented, that is, unorganized, showing some minor patches of imbrications within it, and have a subrounded to subangular nature. Sandstone units show fining upward trend, ending up in fine sediments, and forming a thin to a thick bed of 2.5—10 m. There is an increase in the amount and size of pebbles and a decrease in the amount of sandstone mudstone in the upper stratigraphic levels in the section. The dominant gravelly sandstone deposits in the lower part of the succession change to conglomeratic deposits in the upper part. Therefore, the succession exhibits a graphic (grain-size) coarsening upward character, which is typical of propagating fans (Fig. 12.5) (Blair et al., 1994; Nemec, 1995; Shukla et al., 2001; Shukla, 2009; Ventra and Clarke, 2018). It is envisaged that during the deposition of the Pinjor Formation, many rivers were emanating from the Lesser Himalaya and uplifted parts of the Lower Siwalik, and they were carrying a huge amount of sediment from the source and forming small- to moderate-sized piedmont fans at the Himalayan foot. Under the humid climate, rivers have enough water and energy to carry sand and gravel, and depending upon the sediment—water ratio and relatively high gradient, the channels acquired different channel patterns ranging from braided and straight in the proximal fan to meandering in the mid- and-distal fan areas.

The Pinjor Succession shows low-angle planar beds and horizontal planar beds as the most dominant features developed in gravelly sandstone, which are most abundant in the lower part of the succession between 0 and 280 m levels (Fig. 12.5). Depositional conditions of planar beds took place in the lower and the upper flow regimes in relatively shallow channels. Cross-bedded sandstone deposits generally appearing at the erosional base of the sandstone bodies lying below the horizontally bedded parts represent longitudinal bar deposits under surging flood conditions (Visher, 1965; Kirk, 1983; Miall, 1985; Shukla et al., 2001).

Above 280—390 m levels, depositional cycles are mostly gravelly, which turn out to be deposits formed by persistent gravelly bed-load rivers (Fig. 12.5). The cross-stratified nature of conglomeratic and sandstone beds at this level reflects the picture of transverse bars along with deltaic growth of older bar remnants (Miall, 1985). Hyper-concentrated flood flows and catastrophic flood events lead to deposition of matrix-to-clast-supported conglomerate and with slowing intensity and velocity of water as time passes the deposition changes to sandy forming fining upward channel fill deposits (Figs. 12.2B and 12.5) (Visher, 1965; Reineck and Singh, 1980; Miall, 1985). Climbing ripple cross-lamination and parallel lamination were formed on the levees under the lower flow regime conditions. Locally prevailing lacustrine condition or back swamps leads to the formation of mudstone deposits, which is rare in succession but evident. The rarity of fine-grained muddy lithofacies in the Pinjor Formation implies that the sand and gravel deposited in river channels were freely migrating, and mudstones deposited on the floodplains were continuously eroded (Shukla et al., 2001; Shukla, 2009).

Lithofacies character and the paleocurrent pattern with higher variance recorded collectively suggest that deposition of the major part of the Pinjor Formation took place under a prograding fan environment (Figs. 12.5 and 12.9). The lower part made up of gravelly sandstone was deposited in the midfan setting, whereas the conglomeratic upper part was sedimented in the proximal part of the fans, by fluid gravity processes (Blair, 1987; Shukla, 2009). It is to be noted that debris flow deposits are not encountered anywhere in the succession, implying that most of the sediments were water-laid and carried in the channelized form. It is also important to understand that the distal fan deposits are also not encountered in the succession, and they are most probably concealed below the Quaternary gravels on the northern slopes of the valley.

Textural and mineralogical immaturity of the Pinjore sandstone suggests that the sediments had undergone less transportation (Fig. 12.8). The presence of soft lithic fragments and angular to subspherical grains, as well as a high percentage of the matrix, indicate the proximity of the source (Tandon, 1972; Pettijohn et al., 1973; Dickinson, 1985). The petrological evidence such as the presence of strained quartz, different metamorphic lithic fragments, and igneous quartz and felspar suggest the dominance of metamorphic rocks and subordinate igneous rock in provenance. The presence of calcite cement on matrix materials suggests that it is of intrastriatal origin (Pettijohn et al., 1973; Tandon, 1976; Dickinson, 1985). The lack of quartz and feldspar overgrowth, as well as the considerable intergranular porosity, indicate that the rock units were only lightly buried and compacted.

9. Discussion

Although many workers have attempted the Upper Siwalik in different parts of the Himalayan Foreland Basin, the Upper Siwalik of Dehradun Sub-basin has not yet been studied much in detail, except for some work done on sedimentology and petrography in the Mohand Rao section (Kumar and Ghosh, 1991; Kumar et al., 1999; Ghosh and Kumar, 2000) (Fig. 12.1). A detailed introspection of the whole Dehradun Sub-basin still needs to be done to come out with a holistic evolutionary model. This paper incorporates detailed sedimentology of the Badshahibagh Rao section, which is about 30 km west of the well-documented Mohand Rao section in the Dehradun Sub-basin (Fig. 12.1). The study is a part of a bigger project intending to investigate the Upper Siwalik Subgroup in said Sub-basin between the Ganga and the Yamuna tear faults, which shall able us to propose a depositional model for the Upper Siwalik as a whole in response to allogenic forcing including tectonics, climate, and eustatic changes.

The Upper Siwalik is a time transgressive depositional succession, and it shows a gradation contact with the underlying Middle Siwalik in the study area. For field studies, the occurrence of the first conglomeratic bed with pebbles to cobble size clast has been taken as the boundary between the Middle and the Upper Siwalik (Fig. 12.5). The clasts occurring in the Upper Siwalik are dolomite, dolomitic limestones, arenitic sandstones, slate, ferruginous siltstone, shale, quartzite, meta basics, granite, and gneiss clasts in composition. The Middle Siwalik in the measured transect exhibits granule-sized clasts of quartzite, sandstone, siltstone, and mudstone composition in its upper stratigraphic levels making the contact with overlying the Upper Siwalik gradational. The deposition of the Upper Siwalik around the Dehradun Sub-basin is 5.23 Ma old (Kumar et al., 2004). The distinction between the Pinjor Formation and the succeeding Boulder Conglomerate Formation has also been done based on assimilated clasts. Clasts constituents present in the Boulder Conglomerate Formation are polymictic in nature being dolomites, limestones, sandstones, slate, siltstone, shale, quartzite, meta basics, granite, and gneiss, whereas mainly quartzite clasts are yielded from the Pinjor Formation.

9.1 Sedimentation

The Pinjor Formation was deposited as sandy braided to gravelly braided deposits. However, the presence of the lateral accretion (LA) elements in channelized conglomeratic and sandstone complexes suggests the existence of sinuous meandering channels as well. Also, domination of horizontal bedding and low-angle planar cross-bedding implies sedimentation in low-sinuosity straight channels. The paucity of the trough cross-bedding indicates that conditions for the development of the transverse bars with cuspate lee face were not suitable (Allen, 1963; Ashley, 1990; Mckee and Weir, 1953). The channels were not confined and freely migrated over the fan surface. However, domination of the planar cross-beds in the succession implies migration of 2-D bedforms under the high energy conditions both in sandy and gravelly channels. Sandstone lenses in the gravel bodies were formed during channel abandonment. Backswamp and small lacustrine conditions are evident in the mudstone deposits. The conglomeratic lithofacies association architectural elements, that is, gravel bars and bedforms (GB) in the section conclude the presence of high-energy flow deposits. Poorly sorted grains and clasts, major changes in lateral and vertical facies association, absence of fossils or any organic matter, the limited suite of sedimentary structures, and lenticular and wedge-shaped geometry of beds imply, that the succession was deposited in mid alluvial fan to proximal alluvial fan setting (Figs. 12.5 and 12.9) (Nilson, 1982; Shukla et al., 2001; Shukla and Bora, 2003, 2005).

At the same time, the multistoreyed clast-supported conglomerates with localized sandstone lenses and showing mainly horizontal bedding with clast imbrications, large-scale planar cross-bedding and lateral accretion elements imply deposition in the proximal fan setting in freely migrating channels on the fan surface (Figs. 12.5 and 12.9). The channels were mostly braided and shallow having a relatively higher gradient. They were transforming into the meandering pattern at the proximal-to-mid -fan boundary due to loss of gradient and decreased sediment budget (Nilsen, 1982; Shukla and Bora, 2003, 2005).

9.2 Climate versus tectonics control on sedimentation

Upward coarsening character of the Pinjor Formation, its conglomerate-sandstone dominated lithofacies character, fanning out nature of palaeocurrent pattern with higher variance and poorly sorted gravelly nature of sandstones with angular clasts indicate that the 390 m thick succession was deposited under proximal-to- mid-fan setting (Figs. 12.5 and 12.9). Mainly sandstones in the lower part and gravelly sandstone with some conglomeratic deposits in the upper part of succession indicate advancement of the proximal fan facies over the mid-fan deposits. The total 390 m succession seems to have been evolved by the propagation of smaller fan-building events generated in response to changing tectonic and climatic events. At least four to five such coarsening upward events starting from sandstone and gravelly sandstone and ending with conglomerates are identified from bottom to top of the succession, viz. 0–65 m, 70–200 m, 220–290 m, 300–360 m, and 300–360 m levels (Fig. 12.5). Most of these fan building cycles are gravelly sandstone dominated and covered by 4–6 m thick clast-matrix-supported conglomerate layers. However, in the topmost cycle occurring between 300 and 390 m levels, the conglomerates are 50–60 m thick. Based on the observations made in the present-day Ganga Plain (Shukla and Bora, 2003; Shukla et al., 2001; Shukla, 2009), it is envisaged that during the deposition of the Pinjor Formation, several small to moderately sized fans having the radial length of 20–70 km were forming at the Himalayan Mountain front, and with time they laterally coalesced to form 60 to 90 m thick coarsening upward cycles. In each fan building cycle, the gravelly sandstone was deposited in the mid-fan setting, and the covering conglomerates were deposited in the proximal fan setting. Because of repeated humid climatic cycles associated with enhanced tectonics and resulting sediment release, the gravelly proximal fan prograded over the gravelly sandstone of the mid-fan setting. Nevertheless, the top cycles in which conglomerates are dominating and clasts are larger also imply the renewed tectonic activity in the hinterland. Magnetostratigraphic studies also proved a fair rapid increase in sedimentation rate, in response to increased MBT activity in the Dehradun Sub-Basin at around 5.23 Ma, that is, at the boundary of the Middle and the Upper Siwalik (Sangode et al., 1999; Kumar et al., 2003a,b). MBT in Himalayan foothills and MCT in the Nepal Himalayas also show significant tectonic reactivity at around 3.3 and 2.4 Ma (Sorkhabi et al., 1996).

Kumaravel et al. (2005a,b) magnetostratifically estimated the age of the Pinjor Formation, in the Ghaggar section, the type locality, between 2.7 Ma and 1.79 Ma. The lower boundary of the Pinjor Formation coincides with Gauss and Matuyama polarity reversal event (2.58 Ma), while its upper boundary marks the polio-Pleistocene limit. Gaur (1987) suggested a well-drained landscape during the deposition of the Pinjor Formation based on large mammalian fossils. Rao and Patnaik (2001) based on palynology suggested tropical-subtropical humid climate conditions during the sedimentation of the Pinjor Formation. Patnaik and Nanda (2010) indicated warm and humid conditions with unpredictable monsoonal climates based on fungal spores. Further, Kotla et al. (2018), using isotopic studies of carbon and oxygen isotopes, suggested the prevalence of warm and humid climates during the deposition of the Pinjor Formation. Characteristically, the debris flows and mud-supported breccias/conglomerate deposits are missing in the Pinjor Formation and that testifies during sedimentation the climate was always humid, rivers have enough water to carry gravel and sand, and the dry climatic phases were singularly absent. Similar fan building cycles generated in response to changing climate tectonics are described from the Gaula Fan of the piedmont zone of Kumaun Himalaya (Shukla and Bora, 2003; Shukla, 2009). It is suggested that the Pinjor Formation was deposited in Piedmont Fan setting in the actively subsiding Himalayan Foreland Basin. Clast from the Pinjor Formation is mainly Quartzite with minor sandstone and siltstone. The Quartzite clast can be attributed to the Nagthat Formation and sandstones from the uplifted Siwalik in the hinterland area of the fat deposition. However, the sandstone petrography of the Pinjor Formation yielded metamorphic clasts comprising gneiss and schist as well as implying sediment contribution from the Higher Himalayan crystallines.

10. Conclusions

Based on the sedimentological investigation in the Badshahibagh Rao section of Garhwal Himalaya, following conclusions are made:

Contact between the Middle Siwalik and the Upper Siwalik is gradational. Upper Siwalik of the study area is comprised of only the Pinjor and the Boulder conglomerate formations, while the Tatrot Formation is not developed. The contact between the Pinjor and the Boulder conglomerate formations is also gradational but can be demarcated based on clast composition. The Pinjor Formation is composed of clasts of quartzite, sandstone, siltstone, and mudstone, whereas the Boulder conglomerates are polymictic and composed of limestones, sandstones, slate, siltstone, shale, quartzite, meta basics, granite, and gneiss. The Pinjor Formation is the product of several fan-building events deposited under the dominantly humid climate in the Himalayan Foreland Basin as piedmont deposits. Dry phases are not encountered in the succession. Upper levels of the succession imply intensified tectonics at the provenance.

Acknowledgments

We thank the Head, Department of Geology, BHU, Varanasi for providing working facilities and support of the Department. We also acknowledge Ms Prinsi Singh, Department of Geology, BHU, for generously processing Map work in the Arc GIS platform for the manuscript. This study has been financed by CSIR, New Delhi, a JRF grant (File no. 019/013/(0723)/2017-EMR-1) to Mr. Bijendra Pathak.

References

Allen, J.R., 1963. The classification of cross-stratified units. With notes on their origin. Sedimentology 2 (2), 93−114.

Allen, J.R., 1973. A classification of climbing-ripple cross-lamination. J. Geol. Soc. 129 (5), 537−541.

Ashley, G.M., 1990. Classification of large-scale subaqueous bedforms; a new look at an old problem. J. Sediment. Res. 60 (1), 160−172.

Blair, T.C., McPherson, J.G., 1994. Alluvial fans and their natural distinction from rivers are based on morphology, hydraulic processes, sedimentary processes, and facies assemblages. J. Sediment. Res. 64 (3a), 450−489.

Blair, T.C., 1987. Tectonic and hydrologic controls on cyclic alluvial fan, fluvial, and lacustrine rift-basin sedimentation, Jurassic-Lowermost Cretaceous Todos Santos Formation, Chiapas, Mexico. J. Sediment. Res. 57 (5), 845−862.

Bridge, J.S., 1985. Paleochannel patterns inferred from alluvial deposits; a critical evaluation. J. Sediment. Res. 55 (4), 579−589.

Curray, J.R., 1956. The analysis of two-dimensional orientation data. J. Geol. 64 (2), 117−131.

De Raaf, J.F.M., Reading, H.G., Walker, R.G., 1965. Cyclic sedimentation in the lower westphalian of no Devon, England. Sedimentology 4 (1-2), 1−52.

Dickinson, W.R., 1985. Interpreting provenance relations from detrital modes of sandstones. In: Provenance of Arenites. Springer, Dordrecht, pp. 333−361.

Gaur, R., 1987. Environment and Ecology of Early Man in Northwest India. BR Pub Corp, p. 107.

Ghosh, S.K., Kumar, R., 2000. Petrography of Neogene Siwalik sandstone of the Himalayan foreland basin, Garhwal Himalaya: implications for source-area tectonics and climate. J. Geol. Soc. India 55 (1), 1−15.

Johnson, N.M., Opdyke, N.D., Johnson, G.D., Lindsay, E.H., Tahirkheli, R.A.K., 1982. Magnetic polarity stratigraphy and ages of Siwalik Group rocks of the Potwar Plateau, Pakistan. Palaeogeography, Palaeoclimatology, Palaeoecology 37 (1), 17−42.

Johnson, N.M., Stix, J., Tauxe, L., Cerveny, P.F., Tahirkheli, R.A., 1985. Paleomagnetic chronology, fluvial processes, and tectonic implications of the Siwalik deposits near Chinji village, Pakistan. J. Geol. 93 (1), 27−40.

Kirk, M., 1983. Bar development in a fluvial sandstone (Westphalian 'A'), Scotland. Sedimentology 30 (5), 727−742.

Koster, E.H., Steel, R.J. (Eds.), 1984. Sedimentology of gravels and conglomerates, 10. Canadian Society of Petroleum Geologists, Calgary, p. 441.

Kotla, S.S., Patnaik, R., Sehgal, R.K., Kharya, A., 2018. Isotopic evidence for ecological and climate change in the richly fossiliferous Plio-Pleistocene Upper Siwalik deposits exposed around Chandigarh, India. J. Asian Earth Sci. 163, 32−42.

Krynine, P.D., 1948. The megascopic study and field classification of sedimentary rocks. J. Geol. 56 (2), 130−165.

Kumar, P., Shekhar, S., Shukla, A., Chakraborty, P.P., 2021. Facies architecture and spatio-temporal depositional variability in the pliocene sandhan fluvial system, Kutch basin, India. J. Earth Syst. Sci. 130 (4), 1−23.

Kumar, R., Suresh, N., Sangode, S.J., Kumaravel, V., 2007. Evolution of the Quaternary alluvial fan system in the Himalayan foreland basin: Implications for tectonic and climatic decoupling. Quaternary international 159 (1), 6–20.

Kumar, R., Tandon, S.K., 1985. Sedimentology of plio-pleistocene late orogenic deposits associated with intraplate subduction—the Upper Siwalik Subgroup of a part of Panjab Sub-Himalaya, India. Sediment. Geol. 42 (1–2), 105–158.

Kumar, R., Ghosh, S.K., 1991. Sedimentological studies of the upper Siwalik Boulder Conglomerate Formation, Mohand area, district Saharanpur, UP. J. Himal. Geol. 2 (2), 159–167.

Kumar, R., Ghosh, S.K., Sangode, S.J., 1999. Evolution of a Neogene Fluvial System in a Himalayan Foreland Basin, India. Special papers-Geological Society of America, pp. 239–256.

Kumar, R., Ghosh, S.K., Sangode, S.J., 2003a. Mio-Pliocene sedimentation history in the northwestern part of the Himalayan foreland basin, India. Curr. Sci. 1006–1013.

Kumar, R., Ghosh, S.K., Mazari, R.K., Sangode, S.J., 2003b. Tectonic impact on the fluvial deposits of Plio-Pleistocene Himalayan foreland basin, India. Sediment. Geol. 158 (3–4), 209–234.

Kumar, R., Sangode, S.J., Ghosh, S.K., 2004. A multistorey sandstone complex in the Himalayan Foreland Basin, NW Himalaya, India. J. Asian Earth Sci. 23 (3), 407–426.

Kumaravel, V., Sangode, S.J., Kumar, R., Siddaiah, N.S., 2005a. Magnetic polarity stratigraphy of plio–pleistocene Pinjor Formation (type locality), Siwalik Group, NW Himalaya, India. Curr. Sci. 1453–1461.

Kumaravel, V., Sangode, S.J., Siddaiah, N.S., Kumar, R., 2005b. Rock magnetic characterization of pedogenesis under high energy depositional conditions: a case study from the Mio–Pliocene Siwalik fluvial sequence near Dehra Dun, NW Himalaya, India. Sediment. Geol. 177 (3–4), 229–252.

Lindholm, R.C.l, 1987. A Practical Approach to Sedimentology. Springer, Netherlands, p. 276.

McKee, E.D., Weir, G.W., 1953. Terminology for stratification and cross-stratification in sedimentary rocks. Geol. Soc. Am. Bull. 64 (4), 381–390.

Miall, A.D., 1977. A review of the braided-river depositional environment. Earth Sci. Rev. 13 (1), 1–62.

Miall, A.D., 1985. Architectural-element analysis: a new method of facies analysis applied to fluvial deposits. Earth Sci. Rev. 22 (4), 261–308.

Miall, A., 2014. The facies and architecture of fluvial systems. In: Fluvial Depositional Systems. Springer, Cham, pp. 9–68.

Miall, A.D., 2016. Facies models. In: Stratigraphy: A Modern Synthesis. Springer, Cham, pp. 161–214.

Nemec, W., Postma, G., 1995. Reply-Quaternary alluvial fans in southwestern Crete: sedimentation processes and geomorphic evolution. Sedimentology 42 (3), 535–549.

Nilsen, T.H., 1982. Alluvial fan deposits. In: Sandstone Depositional Environments. AAPG Memoir 31, pp. 49–86.

Opdyke, N.D., Lindsay, E., Johnson, G.D., Johnson, N., Tahirkheli, R.A.K., Mirza, M.A., 1979. Magnetic polarity stratigraphy and vertebrate paleontology of the Upper Siwalik Subgroup of northern Pakistan. Palaeogeography, Palaeoclimatology, Palaeoecology 27, 1–34.

Parkash, B., Sharma, R.P., Roy, A.K., 1980. The Siwalik Group (molasse)—sediments shed by the collision of continental plates. Sediment. Geol. 25 (1–2), 127–159.

Patnaik, R., Nanda, A.C., 2010. Early Pleistocene mammalian faunas of India and evidence of connections with other parts of the world. In: Out of Africa I. Springer, Dordrecht, pp. 129–143.

Pettijohn, F.J., Potter, P.E., Siever, R., 1973. Sand and Sandstone. Springer, Verlag, p. 618.

Pilgrim, G.E., 1913. The correlation of the Siwalik with mammal horizons of Europe. Record Geol. Surv. India 43, 264–325.

Raiverman, V., 1983. Foreland Sedimentations: Himalayan Tectonic Regime: A Relook at the Orogenic Process. Bisen Singh Mahendra Pal Singh Publication, p. 371.

Rao, M.R., Patnaik, R., 2001. Palynology of the late pliocene sediments of Pinjor Formation, Haryana, India. Palaeobotanist 50, 267—286.

Rathi, G., Sangode, S.J., Kumar, R., Ghosh, S.K., 2007. Magnetic fabrics under the high-energy fluvial regime of the Himalayan Foreland Basin, NW Himalaya. Curr. Sci. 933—944.

Reineck, H.E., Singh, I.B., 1980. Depositional Sedimentary Environments. Springer, Berlin, Heidelberg, p. 551.

Sangode, S.J., Kumar, R., Ghosh, S.K., 1999. Palaeomagnetic and Rock Magnetic Perspectives on the Post-collision Continental Sediments of the Himalaya. Geol. Soc. India Memoir, India, pp. 221—248.

Schieber, J., 1999. Distribution and deposition of mudstone facies in the upper Devonian sonyea Group of New York. J. Sediment. Res. 69 (4), 909—925.

Shukla, U.K., Bora, D.S., 2003. Geomorphology and sedimentology of Piedmont zone, Ganga plain, India. Curr. Sci. 1034—1040.

Shukla, U.K., Bora, D.S., 2005. Sedimentation model for the quaternary intermontane Bhimtal—Naukuchiatal lake deposits, Nainital, India. J. Asian Earth Sci. 25 (6), 837—848.

Shukla, U.K., 2009. Sedimentation model of gravel-dominated alluvial piedmont fan, Ganga Plain, India. Int. J. Earth Sci. 98 (2), 443—459.

Shukla, U.K., Bachmann, G.H., Singh, I.B., 2010. Facies architecture of the stuttgart formation (schilfsandstein, upper Triassic), central Germany, and its comparison with the modern Ganga system, India. Palaeogeogr. Palaeoclimatol. Palaeoecol. 297 (1), 110—128.

Shukla, U.K., Singh, I.B., Sharma, M., Sharma, S., 2001. A model of alluvial megafan sedimentation: Ganga Megafan. Sediment. Geol. 144 (3—4), 243—262.

Shukla, U.K., Singh, I.B., 2004. Signatures of palaeofloods in sandbar-levee deposits, Ganga Plain, India. J. Geol. Soc. India 64 (Spl Iss 4), 455—460.

Shukla, U.K., Singh, I.B., Srivastava, P., Singh, D.S., 1999. Paleocurrent patterns in braid-bar and point-bar deposits; examples from the Ganga River, India. J. Sediment. Res. 69 (5), 992—1002.

Singh, R. (Ed.), 2020. Lithostratigraphy and Biostratigraphy of Siwalik Group of Rocks of Northwest Himalaya. Special publication, pp. 1—232.

Sinha, S., Kumar, R., Ghosh, S.K., Sangode, S.J., 2007. Controls on expansion-contraction of late Cenozoic alluvial architecture: a case study from the Himalayan Foreland Basin, NW Himalaya, India. Himal. Geol. 28 (1), 1—22.

Sorkhabi, R.B., Stump, E., Foland, K.A., Jain, A.K., 1996. Fission-track and 40Ar39Ar evidence for episodic denudation of the Gangotri granites in the Garhwal higher Himalaya, India. Tectonophysics 260 (1—3), 187—199.

Tandon, S.K., 1972. Shape analysis of Middle and upper Siwalik sediments around Ramnagar, district, Nainital Kumaun Himalaya. J. Indian Acad. Geosci. 14, 61—73.

Tandon, S.K., 1976. Siwalik sedimentation in a part of the Kumaun Himalaya, India. Sediment. Geol. 16 (2), 131—154.

Teisseyre, A.K., 1975. Pebble fabric in braided stream deposits with examples from Recent and "frozen" Carboniferous channels (Intrasudetic Basin, Central Sudetes). Geol. Sudet. 10 (1), 7—56.

Thomas, J.V., Parkash, B., Mohindra, R., 2002. Lithofacies and palaeosol analysis of the Middle and upper Siwalik Groups (Plio—Pleistocene), Haripur-Kolar section, Himachal Pradesh, India. Sediment. Geol. 150 (3—4), 343—366.

Ventra, D., Clarke, L.E., 2018. Geology and geomorphology of alluvial and fluvial fans: current progress and research perspectives. Geol. Soci., London, Spec. Pub. 440 (1), 1—21.

Visher, G.S., 1965. Use of vertical profile in environmental reconstruction. AAPG (Am. Assoc. Pet. Geol.) Bull. 49 (1), 41—61.

Walker, R.G., Cant, D.J., 1984. Sandy fluvial systems. Facies Mod. 1, 71—89.

CHAPTER 13

A catalogue of Quaternary diatoms from the Asian tropics with their environmental indication potential for paleolimnological applications

Mital Thacker[1,2] and Balasubramanian Karthick[1,2]

[1]Biodiversity & Paleobiology Group, Agharkar Research Institute, Pune, Maharashtra, India; [2]Department of Botany, Savitribai Phule Pune University, Pune, Maharashtra, India

1. Introduction

Diatoms (Bacillariophyceae) are microscopic, unicellular, eukaryotic algae that are ubiquitously distributed in all aqueous habitats (Seckbach and Kociolek, 2011). Diatoms are a dominant part of the phytoplankton community in freshwater and marine environments and play an essential role in these aquatic ecosystems' food web structure by holding a significant share of primary production in these ecosystems (Kock et al., 2019). The frustule of diatoms consists of two valves and girdle elements that link the two valves together (Round et al., 1990). The taxonomy of diatoms is traditionally based upon the structure of their frustule and the symmetry of the cell wall. They exhibit diverse life forms like epilithic (on rock), epipelic (living on fine bottom sediments), episammic (attached to sediment), epiphytic (on plant surface), and epibryophytic (on moss) (Tiffany and Lange, 2002; Tiffany, 2011), which enable diatoms to exist in a wide variety of environments, even found on damp walls in caves (Poulickova and Hasler, 2007), on the bark of a tree, on leaves, on wet walls and rocks (Rybak et al., 2018). It is estimated that there are between 30,000 and probably c.a. 100,000 species of diatoms worldwide (Mann and Vanormelingen, 2013). Because of this wide distribution, high abundance, taxonomic diversity, and sensitivity to environmental changes, diatoms have been proven robust and reliable environmental indicators (Battarbee et al., 2002; Smol and Stoermer, 2010). A good understanding of the relationship between diatoms and environmental changes (physico-chemical parameters) and the narrow ecological optima and tolerances of different diatom taxa makes them ideal biological indicators to infer the water quality in lakes and rivers (Battarbee et al., 2001; Julius and Theriot, 2010). Some examples of water quality parameters that have been associated with changes in diatom assemblages are pH and historical lake acidity (Birks et al., 1990; Battarbee et al., 2010), conductivity (Tibby and Reid, 2004), lake warming (Rühland et al., 2015), nutrient concentrations (Winter and Duthie, 2000; Hall and Smol, 2010), water levels (Wolin and Stone, 2010), lake eutrophication (Kelly and Whitton, 1995; Potapova and Charles, 2007), and climatic factors (Pajunen et al., 2016; Soininen, 2019).

Diatom frustules can be preserved for long periods after cell death. Even when the system has dried out, their cell walls made of silica assure their preservation inside the lake sediments, making them an excellent and successful proxy indicator compared to other biological proxies (Dixit, 1992). Their preservation within the sediments provides a continuous record to reconstruct past climate (Smol and Cumming, 2000; Verschuren, 2003). Each diatom can be distinguished from others based on its siliceous cell walls' size, shape, and sculpturing and harbors a unique taxonomic position. So, changes in diatom species composition can be beneficial for monitoring aquatic environments' status and understanding past ecological conditions in both freshwaters (Flower et al., 1997; Witon and Witkowski, 2003) and marine (Shukla et al., 2009) environments. Apart from monitoring present and past ecological conditions, diatoms also played a significant role in interpreting the strengthening and weakening of monsoon intensity (Chen et al., 2014a; Li et al., 2018).

Lake and wetland sediments are natural archives for paleoclimatic and paleoenvironmental history. They can be used to reconstruct past climatic and environmental records. Sediment that accumulates at the bottom of the lake over time provides us with evidence for past climate changes, such as local precipitation, humidity, evaporation, and temperature (Battarbee, 2000). According to the Global Water Bodies database (GLOWABO), there are approximately 117 million lakes worldwide; about 40% of lakes on Earth lie within tropical latitudes (Verpoorter et al., 2014).

Nevertheless, most paleolimnological studies have been carried out in high-latitude regions. Studies in the tropics still lag far behind in comparison to higher latitudes. These limited studies primarily focused on lake ontogeny, climate, pollution, biogeography of aquatic and terrestrial plants, human migration and colonization, fire history, and societal changes in response to climate change (Escobar et al., 2020).

As we know that global monsoon dominates mainly in the tropics and subtropics (Trenberth et al., 2000), lake sediments in the tropics should be focused on understanding the past monsoon variability. Keeping this importance in mind, the Asian Summer Monsoon (ASM), including the Indian Summer Monsoon (ISM) and the East Asian Summer Monsoon (EASM), influences more than 60% of the world's population over much of South and East Asia (Webster et al., 1998; Wang, 2006), prompted us to review the "monsoon" phenomenon focusing in tropical regions.

Many recent studies originate from the tropics that deal with reconstructing the past monsoonal changes in the tropical areas. Paleomonsoon variability over here has been studied using various proxies like magnetic susceptibility (χ) and speleothem $\delta^{18}O$ (An et al., 1991; Liu et al., 2014), marine sediments (Govil and Naidu, 2011; Cao et al., 2015), lake sediments (Juyal et al., 2009; Achyuthan et al., 2016), peats (Rühland et al., 2006), speleothems (Fleitmann et al., 2007; Kotlia et al., 2015), corals (Ahmad et al., 2011), Foraminifera (Suokhrie et al., 2018), Pollen (Li et al., 2017; Quamar and Bera, 2020; Zhang et al., 2020), stalagmite oxygen isotopes (Zhao et al., 2015; Band et al., 2018), tree ring widths, and stable $\delta^{18}O$ oxygen isotopes (Cook et al., 2010; Sun et al., 2019).

Keeping the above in mind that the diatoms constitute excellent climatic conditions and hold huge applications as palaeoecological proxies, several studies carried out by Wang et al. (2013); Chen et al. (2014a); Li et al. (2018); Ma et al., 2018) used diatom as a proxy to reconstruct paleoclimate and monsoonal variability (Fig. 13.1). The studies that use diatoms as a proxy to reconstruct the past climate and monsoon variability often use autecological data as a critical factor in arriving at any diatom-based inference (Van Dam et al., 1994). Various studies such as Lowe (1974), Beaver (1981), de Wolf (1982), and Denys (1991) have compiled this autecological information, which is taken from the scattered literature. However, most of the observation has arisen from the temperate region only. Moreover, various diatom databases such as Diatoms of North America (Spaulding et al., 2020), a

Principle of Diatom-Based Reconstruction Techniques

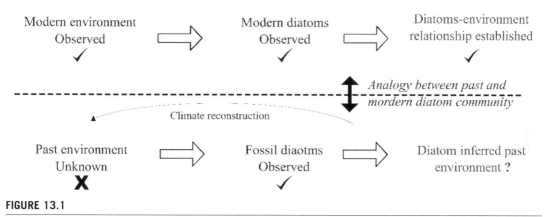

FIGURE 13.1

Schematic representation of the principle underlying diatom-based climate reconstruction approaches. The modern relationships between climate and diatom are used to infer past climate under the assumptions that similar diatom compositions are the result of the same climatic conditions.

Figure modified from Chevalier et al. (2020).

global diatom database (Leblanc et al., 2012), and R-Syst database (Rimet et al., 2016) also contain a lot of useful literature data on the ecology of freshwater and marine diatom taxa. However, all these studies deal with current diatom taxa. Only a few studies like Van Dam et al. (1994) and Circumpolar Diatom Database (CDD) (Pienitz and Cournoyer, 2017) deal with the list of fossil diatom taxa and ecological inference inferred using it. However, all these compilations have specific limitations such as, (1) most of the observations come from the temperate region, as stated before; now, if we want to use these diatoms as a proxy to reconstruct past monsoon variability, a significant part of the monsoon dominated regions lies in the tropical area. In this case, observations from the temperate region will miss the input of the run-off, water level, and monsoon-related diatoms. (2) Most of the compilations are reasonably older, lacking the inputs from many modern and recently discovered diatom taxa. (3) Most of the database contains information from the current diatom taxa; knowledge about fossil diatom taxa inferred ecology is scarce and almost nil from the tropics.

In the present time, when a remarkable number of paleo-monsoonal studies are originating from the monsoon-dominated regions of the tropics using diatom as a proxy to study past monsoon and its variability, it is a dire need to have a compilation of tropics-based fossil diatom taxa and ecological inference derived from them. Therefore, after reviewing all possible literature from diatom inferred monsoon from the tropics, we present a checklist of fossil diatom taxa and their ecological inference reported from the paleolimnological studies originating from the freshwater and marine environment of the tropics.

2. Materials and methods

For this review, we have searched in Google Scholar, Web of Science, and Scopus with the following keywords, "Indian summer monsoon," "diatom," "monsoon variability," "precipitation," "ecology," "environmental inference," and "diatom inferred monsoon." We also did a back referencing of the paper that we collected from the search engines. Fifty-one research papers on Asian monsoon precipitation and its variability across time were collated and used for this review.

We specified our study by collecting the paleo-monsoon records, mainly inferred by biological proxies, particularly "diatoms." Observations for the survey were recorded primarily from the monsoon-dominated regions of the tropics (recorded specifically from East Asia, Central Asia, the Indian subcontinent, the Northern Indian Ocean, and the South China Sea) to ascertain the monsoonal changes during various geological periods. Primarily the papers that published the studies on fossil diatoms in which they incorporated their autecological knowledge were consulted and given preference for this review over many articles that dealt with only taxonomy and presence-absence of any diatom taxa. The central part of the work consists of a literature survey of the ecological requirements attributed to the taxa.

We synthesized the list of diatom taxa from each paper, environmental conditions inferred using those taxa, and understanding of this diatom-based inference for the monsoon reconstruction. Finally, we prepared the catalogue of fossil diatom species, their distribution across space and time, and the geological period when they existed and environmental optima inferred by them. To ensure sufficient data quality, we provided that all records included in our analyses were required to meet the following criteria.

1. An analyzed paleo-monsoon-related study should have an exclusive table or list dealing with the number of diatom taxa provided in the paper.
2. The study preferably has dealt with diatom taxa with species level, a list of indicator species (abundant diatom taxa), and its ecological indication.
3. All ecological inferences based on diatoms should be supported with respective references.

To further examine the habitat-specific preference of the diatom taxa, we have divided our records into freshwater and marine and further analyzed them based on the geological period during which they dominated. Additionally, we plotted the study sites' spatial distribution derived from the papers using QGIS (Fig. 13.2).

3. Result

For this review, 51 research papers were studied on Asian monsoon precipitation and its variability across time using diatoms as a proxy to identify the monsoon signals across space and time. Of these 51 studies, only 25 papers dealt with the autecological information derived from the diatom taxa described. After going into every paper's detail, we selected the papers that dealt with the list of diatom taxa at a species level. The article should have provided the autecological information about that used fossil diatom taxa. The reviewed literature covered the Late Pleistocene and Holocene periods. The reviewed studies span tropical areas like East Asia, Central Asia, and South Asia. After studying the diatom taxa and their ecological inference from each paper, we present a checklist of all fossil freshwater and marine diatoms from tropical areas. The list below includes 266 species of fossil diatom taxa belonging to 64 genera.

The ecological information is presented in separate paragraphs, covering conductivity, pH, trophic conditions, saprobity, oxygen requirement, salinity, current, and temperature. The environmental

3. Result

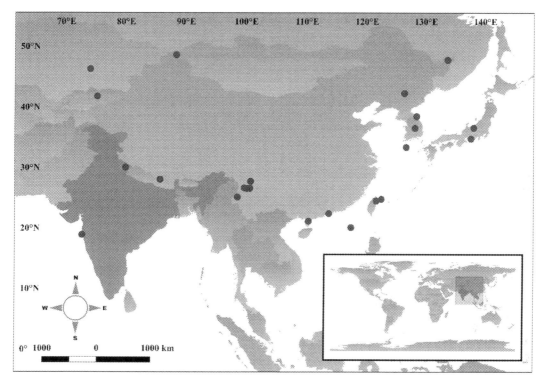

FIGURE 13.2

Spatial distribution of diatom-based paleoclimatic records used in this study.

facies inferred from diatom taxa over here is broadly classified into two categories freshwater/terrestrial taxa, and marine taxa. Freshwater/terrestrial taxa further can be divided into 10 sub-classes based on the various environmental conditions inferred by them: (1) Wet/high rainfall condition indicating taxa, (2) drought/aridity condition indicating taxa, (3) eutrophic (high nutrient level) condition indicating taxa, (4) oligotrophic (nutrient-poor) condition indicating taxa, (5) acidic condition indicating taxa, (6) alkaline condition indicating taxa, (7) circumneutral condition indicating taxa, (8) high energy/high water turbulence condition indicating taxa, (9) low temperature indicating taxa, and (10) taxa indicating shallow water depth condition. Marine taxa can be divided into three different three groups (1) based on their salinity level preference; they can be divided into three different categories as Polyhalobous, Mesohalobous, and Oligohalobous (2) elevated temperature indicating taxa, and (3) taxa indicating alkaline condition.

3.1 Freshwater taxa

3.1.1 Rainfall

Understanding the dynamics and variability of the global monsoon has been a critical objective for climate scientists because monsoon anomalies may cause droughts, floods, and other extreme weather or climate events. Monsoon variability has profound societal and economic influences on more than 70% of the world's population (An et al., 2015).

Diatoms are sensitive to climatic changes, including temperature increases and rainfall variability (Kienel et al., 1999; Brugam and Swain, 2000). Diatoms also show a relatively high sensitivity to water level changes (Birks et al., 1990). As the diatom species show significant responses to changes in dry and wet conditions, diatom fossils represent powerful indicators that can be applied in reconstructing paleo-hydrological conditions in the region (Ma et al., 2018).

In the present review, studies focusing on diatom-inferred changes in moisture regimes during the past have identified significant relationships between water levels and diatom compositions. From the current study, based on diatom autecology and reconstruction of past water levels, diatom taxa can be categorized into three different groups.

1. Wet/high rainfall conditions indicating taxa
2. Aridity/drought conditions indicating taxa
3. Taxa insensitive to water level

Ma et al. (2018), and Li et al. (2015a) from East Asia (NE and SW China), reported the wet condition indicating diatom taxa belonging to the genera *Aulacoseira, Cyclotella, Cyclostephanos,* and *Achnanthes (Achnanthidium). Aulacoseira crenulata, Cyclostephanos dubius,* and *Amphora libyca* indicated wet conditions, whereas *Achnanthes nodosa,* along with high wetness, suggested a preference for oligotrophic cold waters. In particular, *A. crenulata* is an indicator of wet conditions and occurs mainly in wet places (Van Dam et al., 1994; Stanek-Tarkowska et al., 2015).

Cyclotella meneghiniana indicated a high-water level along with nutrient-rich/eutrophic conditions (Van Campo and Gasse, 1994; Van Dam et al., 1994; Krammer and Lange Bertalot, 1999). Taxa such as *Eunotia formica, Eunotia incisa* var. *incisa, Stauroneis anceps,* and *Stauroneis phoenicenteron* mainly occur in moist and wet places; the occurrence of these diatoms indicates the relatively high-water level and high rainfall condition along with oligotrophic (nutrient-poor), acidic to circum-neutral condition (Van Campo and Gasse, 1994; Van Dam et al., 1994; Van de Vijver and Beyens, 1997, 1999). Whereas *Gomphonema truncatum, Navicula cuspidata,* and *Pinnularia brevicostata* indicated high rainfall conditions along with eutrophic (nutrient-rich) and alkaline environmental preference (Van Dam et al., 1994; Gaiser and Johansen, 2000; Cantonati et al., 2011; Chen et al., 2012, 2014b; Hargan et al., 2015). All these taxa were mainly reported during the late Pleistocene to Holocene period. Table 13.1 represents the autecology of different diatom species indicating the wet/high rainfall condition during the Quaternary period.

Ma et al. (2018) reported the drought condition indicating taxa across East Asia during Mid Holocene (6500 yr) period. Drought condition in general was indicated by *Cymbella amphioxys, Navicula tridentula,* and *Navicula ignota* var. *ignota* (Van Campo and Gasse, 1994; Van Dam et al., 1994; Liu, 2007). In addition, taxa such as *Eunotia bilunaris* var. *mucophila* and *Cymbella paucistriata* indicated dry environmental conditions and a preference to thrive in acidic, oligotrophic lakes, and sphagnum bogs (Van Campo and Gasse, 1994; Cantonati et al., 2011; Hargan et al., 2015). Apart from this, diatoms like *Hantzschia amphioxys, Navicula gallica* var. *perpusilla, Navicula mutica* var. *mutica,* and *Nitzschia terrestris* preferred dry environment and strongly indicated terrestrial and aerial to subaerial conditions. Mostly these taxa are found in widespread distribution among lakes, bogs, and swamps, with acidic conditions (Van Campo and Gasse, 1994; Van Dam et al., 1994; Cantonati et al., 2011).

Table 13.1 Summary of diatom taxa indicating the wet/high rainfall condition during the Quaternary period.

Sr. no	Diatom taxa	Site name	Geological time	Environmental facies inferred from diatom taxa	Study references	Additional references
1	*Cyclostephanos dubius*	Heqing Basin, SW China Yunnan (East Asia) 26°33′43″ N 100°10′14″ E	Late Pleistocene period (140–35 kyr)	Correlated positively with nitrogen concentration and negatively with Secchi depth and silica concentration, associated with deep water	Li et al. (2015a)	Bradshaw and Anderson (2003), Wang et al. (2012c)
2	*Achnanthes nodosa*	Honghe Peatland, NE China (East Asia) 47°47.335′ N 133°37.65′ E	Holocene period (6500 yr. BP)	Wet condition indicating taxa, reported from oligotrophic cold waters	Ma et al. (2018)	Van Campo and Gasse (1994); Van Dam et al. (1994); Van de Vijver and Beyens (1997, 1999); Krammer and Lange-Bertalot (1999); Gaiser and Johansen (2000); Weilhoefer and Pan (2007); Cantonati et al. (2011); Chen et al. (2012, 2014b); Hargan et al. (2015)
3	*Amphora libyca*	Honghe Peatland, NE China (East Asia) 47°47.335′ N 133°37.65′ E	Holocene period (6500 yr. BP)	Wet condition indicating taxa, found in small numbers in epilithon in rivers, but also in the mud at the edges of a pond	Ma et al. (2018)	Van Campo and Gasse (1994); Van Dam et al. (1994); Van de Vijver and Beyens (1997, 1999); Krammer and Lange-Bertalot (1999); Gaiser and Johansen (2000); Weilhoefer and Pan (2007); Cantonati et al. (2011); Chen et al. (2012, 2014b); Hargan et al. (2015)

Continued

Table 13.1 Summary of diatom taxa indicating the wet/high rainfall condition during the Quaternary period.—cont'd

Sr. no	Diatom taxa	Site name	Geological time	Environmental facies inferred from diatom taxa	Study references	Additional references
4	Aulacoseira crenulata	Honghe Peatland, NE China (East Asia) 47°47.335′ N 133°37.65′ E	Holocene period (6500 yr. BP)	Wet condition preferring taxa, common in streams, ponds and swamps mostly epiphytic, an indicator of wet conditions and occurs mostly in wet places	Ma et al. (2018)	Van Campo and Gasse (1994); Van Dam et al. (1994); Van de Vijver and Beyens (1997, 1999); Krammer and Lange-Bertalot (1999); Gaiser and Johansen (2000); Weilhoefer and Pan (2007); Cantonati et al. (2011); Chen et al. (2012, 2014b); Hargan et al. (2015)
5	Cyclotella meneghiniana	Honghe Peatland, NE China (East Asia) 47°47.335′ N 133°37.65′ E	Holocene period (6500 yr. BP)	Wet condition preferring taxa, widespread in shallow, eutrophic waters	Ma et al. (2018)	Van Campo and Gasse (1994); Van Dam et al. (1994); Van de Vijver and Beyens (1997, 1999); Krammer and Lange-Bertalot (1999); Gaiser and Johansen (2000); Weilhoefer and Pan (2007); Cantonati et al. (2011); Chen et al. (2012, 2014b); Hargan et al. (2015)
6	Eunotia formica	Honghe Peatland, NE China (East Asia) 47°47.335′ N 133°37.65′ E	Holocene period (6500 yr. BP)	Wet condition preferring taxa, widespread and frequently reported from acidic lakes and bogs	Ma et al. (2018)	Van Campo and Gasse (1994); Van Dam et al. (1994); Van de Vijver and Beyens (1997, 1999); Krammer and Lange-Bertalot (1999); Gaiser and Johansen (2000); Weilhoefer and Pan (2007); Cantonati et al. (2011); Chen et al. (2012, 2014b); Hargan et al. (2015)

7	*Eunotia incisa* var. *incisa*	Honghe Peatland, NE China (East Asia) 47°47.335′ N 133°37.65′ E	Holocene period (6500 yr. BP)	Wet condition preferring taxa, common in northern lakes and bogs, common in acidic, oligotrophic freshwater environments, often found associated with mosses	Ma et al. (2018)	Van Campo and Gasse (1994); Van Dam et al. (1994); Van de Vijver and Beyens (1997, 1999); Krammer and Lange-Bertalot (1999); Gaiser and Johansen (2000); Weilhoefer and Pan (2007); Cantonati et al. (2011); Chen et al. (2012, 2014b); Hargan et al. (2015)
8	*Stauroneis anceps*	Honghe Peatland, NE China (East Asia) 47°47.335′ N 133°37.65′ E	Holocene period (6500 yr. BP)	Wet condition preferring taxa, found in lakes and ponds with circumneutral pH and very low specific conductance	Ma et al. (2018)	Van Campo and Gasse (1994); Van Dam et al. (1994); Van de Vijver and Beyens (1997, 1999); Krammer and Lange-Bertalot (1999); Gaiser and Johansen (2000); Weilhoefer and Pan (2007); Cantonati et al. (2011); Chen et al. (2012, 2014b); Hargan et al. (2015)
9	*Stauroneis phoenicenteron*	Honghe Peatland, NE China (East Asia) 47°47.335′ N 133°37.65′ E	Holocene period (6500 yr. BP)	Wet condition preferring taxa, most common in samples from ditches, small pools and bogs with slightly acid to circumneutral pH, oligohalobous (indifferent)	Ma et al. (2018)	Van Campo and Gasse (1994); Van Dam et al. (1994); Van de Vijver and Beyens (1997, 1999); Krammer and Lange-Bertalot (1999); Gaiser and Johansen (2000); Weilhoefer and Pan (2007); Cantonati et al. (2011); Chen et al. (2012, 2014b); Hargan et al. (2015)

Continued

Table 13.1 Summary of diatom taxa indicating the wet/high rainfall condition during the Quaternary period.—cont'd

Sr. no	Diatom taxa	Site name	Geological time	Environmental facies inferred from diatom taxa	Study references	Additional references
10	*Gomphonema truncatum*	Honghe Peatland, NE China (East Asia) 47°47.335′ N 133°37.65′ E	Holocene period (6500 yr. BP)	Wet condition preferring taxa, alkalibiontic, live in mesotrophic to eutrophic waters	Ma et al. (2018)	Van Campo and Gasse (1994); Van Dam et al. (1994); Van de Vijver and Beyens (1997, 1999); Krammer and Lange-Bertalot (1999); Gaiser and Johansen (2000); Weilhoefer and Pan (2007); Cantonati et al. (2011); Chen et al. (2012, 2014b); Hargan et al. (2015)
11	*Navicula cuspidata*	Honghe Peatland, NE China (East Asia) 47°47.335′ N 133°37.65′ E	Holocene period (6500 yr. BP)	Wet condition preferring taxa, prefer to live in eutrophic and medium to the high conductivity of salt waters, never lives outside water bodies, these species are all reported to prefer relatively humid habitats	Ma et al. (2018)	Van Campo and Gasse (1994); Van Dam et al. (1994); Van de Vijver and Beyens (1997, 1999); Krammer and Lange-Bertalot (1999); Gaiser and Johansen (2000); Weilhoefer and Pan (2007); Cantonati et al. (2011); Chen et al. (2012, 2014b); Hargan et al. (2015)
12	*Pinnularia brevicostata*	Honghe Peatland, NE China (East Asia) 47°47.335′ N 133°37.65′ E	Holocene period (6500 yr. BP)	Wet condition preferring taxa, cosmopolitan distribution often found in large numbers in streams and highly eutrophic ponds	Ma et al. (2018)	Van Campo and Gasse (1994); Van Dam et al. (1994); Van de Vijver and Beyens (1997, 1999); Krammer and Lange-Bertalot (1999); Gaiser and Johansen (2000); Weilhoefer and Pan (2007); Cantonati et al. (2011); Chen et al. (2012, 2014b); Hargan et al. (2015)

Table 13.2 represents the summary of diatom taxa indicating the drought/aridity condition during the Quaternary period.

Ma et al. (2018) from East Asia reported the diatom taxa that are insensitive to water levels. Table 13.3 represents the list of diatom taxa insensitive to water level. Taxa such as *Eunotia implicata, Gomphonema gracile,* and *Nitzschia perminuta* are insensitive to water levels, widely distributed in lakes and swamps, and found in rivers. In addition to this, *Gomphonema parvulum* and *Pinnularia streptoraphe* var. *minor* are insensitive, most common in rivers with running water, also recorded in rivers and creeks with stagnant or near stagnant water or in ponds (Van Campo and Gasse, 1994; Van Dam et al., 1994; Krammer and Lange-Bertalot, 1999).

In conclusion, as the diatom species show significant responses to changes in dry and wet conditions, diatom fossils act as powerful indicators that can be applied to reconstruct paleohydrological conditions. The key diatom species that prefer deep water are *A. crenulata, N. cuspidata,* and *P. brevicostata*. In contrast, those that indicate shallow water are *Hantzschia amphioxys* and *E. bilunaris* var. *mucophila*.

3.1.2 Nutrient level

Diatoms are highly sensitive to nutrient-level changes and can be effectively used to reconstruct lake environment changes. An elevated level of nutrients (N and P) leads to eutrophication. Diatoms are well suited to study the nutrient level of any aquatic body because each diatom has its optimum nutrient tolerance level and is sensitive to changes in concentrations, supply rates, and ratios of nutrients (Hall and Smol, 2010). With this ability to quantify responses of individual taxa to nutrient concentrations and preference for the nutrients level, diatom taxa can be categorized into two groups.

1. Eutrophic (high nutrient level) condition indicating taxa
2. Oligotrophic (nutrient-poor) condition indicating taxa

From the late Pleistocene to the Anthropocene period, studies across the various part of Asia like Nagayasu et al. (2017), Cho et al. (2019), Cho et al. (2018) (Japan), Li et al. (2015a), Wang et al. (2014), Li et al. (2018) (SW China), Krstić et al. (2012) (Nepal), Schwarz et al. (2017) (Kyrgyzstan, Central Asia), Lee et al. (2018) (Korea), and Wang et al. (2012b) (NE China) reconstructed the eutrophic (elevated nutrient level) condition using various diatom taxa.

Eutrophic condition across various part of Asia like East Asia (NE China, SW China, Japan, Korea) and Central Asia (Kyrgyzstan) was inferred using 24 diatom taxa. Species like *Aulacoseira ambigua, Aulacoseira granulata, Cyclotella rhomboideo-elliptica, Stephanodiscus minutulus, Stephanodiscus* spp., *Asterionella formosa, Eunotia* cf. *pseudopapilio, Fragilaria construens* var. *subsaline, Fragilaria elliptica, Fragilaria tenera, Gomphonema subclavatum, Sellaphora pupula, Cyclotella atomus, Cyclotella atomus* var. *gracilis, Gyrosigma acuminatum,* and *Staurosirella pinnata* indicated high conductivity and meso to eutrophic condition in general with high nutrient availability (Hustedt, 1930, 1942; Bradbury, 1975; Anderson, 1990; Padisák et al., 2009; Joh, 2010; Li et al., 2012).

On the other side, taxa such as *Cyclotella ocellata, Cyclotella dubius,* and *Discotella* sp. indicated a wide range of trophic conditions, from cold, oligotrophic to warm eutrophic lakes with a preference for higher pH conditions (Haworth and Hurley, 1984; Bracht et al., 2008). *Lindavia radiosa* was observed under turbulent conditions along with higher nutrient supply and eutrophic conditions of the lakes (Rioual et al., 2007). Table 13.4 represents the summary of diatom taxa indicating eutrophic (high nutrient level) conditions across various parts of Asia.

Table 13.2 Summary of diatom taxa indicating the drought/aridity condition during the Quaternary period.

Sr. no	Diatom taxa	Site name	Geological time	Environmental facies inferred from diatom taxa	Study references	Additional references
1	*Cymbella amphioxys*	Honghe Peatland, NE China (East Asia) 47°47.335′ N 133°37.65′ E	Holocene period (6500 yr. BP)	Drought preferring taxa commonly lives in relatively shallow, swampy water	Ma et al. (2018)	Liu (2007), Van Campo and Gasse (1994); Van Dam et al. (1994); Van de Vijver and Beyens (1997, 1999); Krammer and Lange-Bertalot (1999); Gaiser and Johansen (2000); Weilhoefer and Pan (2007); Cantonati et al. (2011); Chen et al. (2012, 2014b); Hargan et al. (2015)
2	*Cymbella paucistriata*	Honghe Peatland, NE China (East Asia) 47°47.335′ N 133°37.65′ E	Holocene period (6500 yr. BP)	Drought-preferring taxa, widespread in the northern hemisphere, prefer to live in low pH swamps also reported from oligotrophic waters and sphagnum bogs	Ma et al. (2018)	Van Campo and Gasse (1994); Van Dam et al. (1994); Van de Vijver and Beyens (1997, 1999); Krammer and Lange-Bertalot (1999); Gaiser and Johansen (2000); Weilhoefer and Pan (2007); Cantonati et al. (2011); Chen et al. (2012, 2014b); Hargan et al. (2015)
3	*Eunotia bilunaris* var. *mucophila*	Honghe Peatland, NE China (East Asia) 47°47.335′ N 133°37.65′ E	Holocene period (6500 yr. BP)	Drought-preferring taxa, widespread in acidic lakes and bogs, usually associated with other algae or aquatic plants as an epiphyte	Ma et al. (2018)	Van Campo and Gasse (1994); Van Dam et al. (1994); Van de Vijver and Beyens (1997, 1999); Krammer and Lange-Bertalot (1999); Gaiser and Johansen (2000); Weilhoefer and Pan (2007); Cantonati et al. (2011); Chen et al. (2012, 2014b); Hargan et al. (2015)

4	*Hantzschia amphioxys*	Honghe Peatland, NE China (East Asia) 47°47.335′ N 133°37.65′ E	Holocene period (6500 yr. BP)	Found in drought conditions, prefer to live in low pH swamps, most common in rivers with running water, also recorded in rivers and creeks with stagnant, near stagnant or slowly running water and soils, aerophilous, oligohalobous (indifferent), terrestrial or xerophytic diatoms	Ma et al. (2018)	Van Campo and Gasse (1994); Van Dam et al. (1994); Van de Vijver and Beyens (1997, 1999); Krammer and Lange-Bertalot (1999); Gaiser and Johansen (2000); Weilhoefer and Pan (2007); Cantonati et al. (2011); Chen et al. (2012, 2014b); Hargan et al. (2015)
5	*Navicula gallica var. perpusilla*	Honghe Peatland, NE China (East Asia) 47°47.335′ N 133°37.65′ E	Holocene period (6500 yr. BP)	Observed in drought conditions, cosmopolitan species, oligohalobous, typical in aerial to subaerial habitat, prefer biotopes characterized by reduced light intensity and slightly brackish waters	Ma et al. (2018)	Van Campo and Gasse (1994); Van Dam et al. (1994); Van de Vijver and Beyens (1997, 1999); Krammer and Lange-Bertalot (1999); Gaiser and Johansen (2000); Weilhoefer and Pan (2007); Cantonati et al. (2011); Chen et al. (2012, 2014b); Hargan et al. (2015)
6	*Navicula ignota var. ignota*	Honghe Peatland, NE China (East Asia) 47°47.335′ N 133°37.65′ E	Holocene period (6500 yr. BP)	Observed in drought conditions, continuously high dissolved oxygen level	Ma et al. (2018)	Van Campo and Gasse (1994); Van Dam et al. (1994); Van de Vijver and Beyens (1997, 1999); Krammer and Lange-Bertalot (1999); Gaiser and Johansen (2000); Weilhoefer and Pan (2007); Cantonati et al. (2011); Chen et al. (2012, 2014b); Hargan et al. (2015)

Continued

Table 13.2 Summary of diatom taxa indicating the drought/aridity condition during the Quaternary period.—cont'd

Sr. no	Diatom taxa	Site name	Geological time	Environmental facies inferred from diatom taxa	Study references	Additional references
7	*Navicula mutica* var. *mutica*	Honghe Peatland, NE China (East Asia) 47°47.335′ N 133°37.65′ E	Holocene period (6500 yr. BP)	Observed in drought conditions, widespread distribution among lakes, bogs, rivers and estuaries, often dominates damp soil and moss collections, terrestrial or xerophytic diatoms	Ma et al. (2018)	Van Campo and Gasse (1994); Van Dam et al. (1994); Van de Vijver and Beyens (1997, 1999); Krammer and Lange-Bertalot (1999); Gaiser and Johansen (2000); Weilhoefer and Pan (2007); Cantonati et al. (2011); Chen et al. (2012, 2014b); Hargan et al. (2015)
8	*Navicula tridentula*	Honghe Peatland, NE China (East Asia) 47°47.335′ N 133°37.65′ E	Holocene period (6500 yr. BP)	Found in drought conditions, widespread in freshwater	Ma et al. (2018)	Van Campo and Gasse, 1994; Van Dam et al. (1994); Van de Vijver and Beyens (1997, 1999); Krammer and Lange-Bertalot (1999); Gaiser and Johansen (2000); Weilhoefer and Pan (2007); Cantonati et al. (2011); Chen et al. (2012, 2014b); Hargan et al. (2015)
9	*Nitzschia terrestris* (Petersen) Hustedt	Honghe Peatland, NE China (East Asia) 47°47.335′ N 133°37.65′ E	Holocene period (6500 yr. BP)	Observed in drought conditions, Subaerial, prefer to live in high acidic Sphagnum bogs near lakes, rarely in running waters	Ma et al. (2018)	Van Campo and Gasse (1994); Van Dam et al. (1994); Van de Vijver and Beyens (1997, 1999); Krammer and Lange-Bertalot (1999); Gaiser and Johansen (2000); Weilhoefer and Pan (2007); Cantonati et al. (2011); Chen et al. (2012, 2014b); Hargan et al. (2015)

Table 13.3 Summary of diatom taxa insensitive to water level conditions during the Quaternary period.

Sr. no	Diatom taxa	Site name	Geological time	Environmental facies inferred from diatom taxa	Study references	Additional references
1	*Eunotia implicata*	Honghe Peatland, NE China (East Asia) 47°47.335′ N 133°37.65′ E	Holocene period (6500 yr. BP)	Insensitive, widely distributed in coastal areas of lakes and swamps, also found in the stagnant water of rivers	Ma et al. (2018)	Van Campo and Gasse (1994); Van Dam et al. (1994); Van de Vijver and Beyens (1997, 1999); Krammer and Lange-Bertalot (1999); Gaiser and Johansen (2000); Weilhoefer and Pan (2007); Cantonati et al. (2011); Chen et al. (2012, 2014b); Hargan et al. (2015)
2	*Gomphonema gracile*	Honghe Peatland, NE China (East Asia) 47°47.335′ N 133°37.65′ E	Holocene period (6500 yr. BP)	Insensitive taxa, widespread in oligotrophic waters, mostly found attached to epiphytes but can be found in epilithic and epipelic assemblages	Ma et al. (2018)	Van Campo and Gasse (1994); Van Dam et al. (1994); Van de Vijver and Beyens (1997, 1999); Krammer and Lange-Bertalot (1999); Gaiser and Johansen (2000); Weilhoefer and Pan (2007); Cantonati et al. (2011); Chen et al. (2012, 2014b); Hargan et al. (2015)
3	*Gomphonema parvulum*	Honghe Peatland, NE China (East Asia) 47°47.335′ N 133°37.65′ E	Holocene period (6500 yr. BP)	Insensitive taxa, widely distributed in a variety of habitats but rarely in the acidic swamp common in ponds and underground seepage waters	Ma et al. (2018)	Van Campo and Gasse (1994); Van Dam et al. (1994); Van de Vijver and Beyens (1997, 1999); Krammer and Lange-Bertalot (1999); Gaiser and Johansen (2000); Weilhoefer and Pan (2007); Cantonati et al. (2011); Chen et al. (2012, 2014b); Hargan et al. (2015)

Continued

Table 13.3 Summary of diatom taxa insensitive to water level conditions during the Quaternary period.—cont'd

Sr. no	Diatom taxa	Site name	Geological time	Environmental facies inferred from diatom taxa	Study references	Additional references
4	*Nitzschia perminuta*	Honghe Peatland, NE China (East Asia) 47°47.335′ N 133°37.656′ E	Holocene period (6500 yr. BP)	Insensitive, prefer to live in Sphagnum bogs	Ma et al. (2018)	Van Campo and Gasse (1994); Van Dam et al. (1994); Van de Vijver and Beyens (1997, 1999); Krammer and Lange-Bertalot (1999); Gaiser and Johansen (2000); Weilhoefer and Pan (2007); Cantonati et al. (2011); Chen et al. (2012, 2014b); Hargan et al. (2015)
5	*Pinnularia streptoraphe* var. *minor*	Honghe Peatland, NE China (East Asia) 47°47.335′ N 133°37.65′ E	Holocene period (6500 yr. BP)	Insensitive, most common in rivers with running water, also recorded in rivers and creeks with stagnant or near stagnant water or in ponds and on wet mosses halophobous	Ma et al. (2018)	Van Campo and Gasse (1994); Van Dam et al. (1994); Van de Vijver and Beyens (1997, 1999); Krammer and Lange-Bertalot (1999); Gaiser and Johansen (2000); Weilhoefer and Pan (2007); Cantonati et al. (2011); Chen et al. (2012, 2014b); Hargan et al. (2015)

Table 13.4 Summary of diatom taxa indicating eutrophic (high nutrient level) condition during the Quaternary period.

Sr. no	Diatom taxa	Site name	Geological time	Environmental facies inferred from diatom taxa	Study references	Additional references
1	*Aulacoseira ambigua*	Takano river basin, Central Japan (East Asia) 36°32′55″N 138°02′07″ E	Late Pleistocene period (156.8–38 kyr)	Euplanktonic species in eutrophic lakes	Nagayasu et al. (2017)	Hustedt (1930, 1942)
2	*Aulacoseira granulata*	Heqing Basin, SW China, Yunnan (East Asia) 26°33′43″ N 100°10′14″ E	Late Pleistocene period (140–35 kyr)	Associated with high trophic status and with shallow lakes or in the layer of continuous or semi-continuous mixing in stratified lakes. Turbulence is essential for this species, as it keeps the heavy cells in the photic zone, so it is a strong indicator for increased mixing. It is also known as a thermophilic diatom, associated with water in excess of 15°C, and tolerance of low light conditions linked to its preference for turbid waters	Li et al. (2015a)	Padisák et al. (2009), Wolin and Duthie (2010), Rioual et al. (2007)
3	*Cyclotella rhomboideo-elliptica*	Heqing Basin, SW China, Yunnan (East Asia) 26°33′43″ N 100°10′14″ E	Late Pleistocene period (140–35 kyr)	Planktonic diatom indicate eutrophication and climate warming, an indicator of stable summer stratification, regarded as typical of clear and deep lakes	Li et al. (2015a)	Li et al. (2012), Chen et al. (2014b), Wang et al. (2014), Padisák et al. (2009)

Continued

Table 13.4 Summary of diatom taxa indicating eutrophic (high nutrient level) condition during the Quaternary period.—cont'd

Sr. no	Diatom taxa	Site name	Geological time	Environmental facies inferred from diatom taxa	Study references	Additional references
4	*Stephanodiscus minutulus*	Heqing Basin, SW China, Yunnan (East Asia) 26°33′43″ N 100°10′14″ E	Late Pleistocene period (140–35 kyr)	Typical of mesotrophic and medium-sized lakes and is sensitive to the onset of stratification, a strong indicator of long and deep spring circulation and low Si:P conditions	Li et al. (2015a)	Padisák et al. (2009), Bradbury and Diederich-Rurup (1993)
5	*Asterionella formosa*	Lugu Lake, SW China, Yunnan, (East Asia) 27°43′08.4″ N 100°46′33.9″ E	Late Pleistocene period (30–10 kyr)	Indicative of elevated nutrient conditions	Wang et al. (2014)	Bradbury (1975)
6	*Cyclotella dubius*	Lugu Lake, SW China, Yunnan (East Asia) 27°43′08.4″ N 100°46′33.9″ E	Late Pleistocene period (30–10 kyr)	High nutrient, alkaline waters	Wang et al. (2014)	Anderson (1990), 1997; Anderson et al. (1993); Håkansson and Regnéll (1993); Bradbury et al. (1994); Bradshaw and Anderson (2003); Yang et al. (2008)
7	*Stephanodiscus* spp.	Lugu Lake, SW China, Yunnan, (East Asia) 27°43′08.4″ N 100°46′33.9″ E	Late Pleistocene period (30–10 kyr)	Indicative of elevated nutrient conditions	Wang et al. (2014)	Bradbury (1975); Anderson (1990); Bradbury and Diederich-Rurup (1993)

8	*Aulacoseira ambigua*	Lake Qinghai, SW China. (East Asia) 25°07′48″–25°08′6″N 98°34′11″–98°34′16″E	Late Pleistocene period (18.5 kyr–till present)	Generally, prefer colored, humic waters with high DOC concentrations and are known to be adapted to low light conditions; often appears in high nutrient water bodies and is favored by an increase in nutrient availability; prefers higher temperature relative to other members of the genus	Li et al. (2018)	Yoshikawa et al. (2000); Velez et al. (2003); Merilainen et al. (2003); Ginn et al. (2007); Korsman and Birks (1996); Houk (2003); Köster and Pienitz (2006); Buczko et al. (2010); Shear et al. (1976)
9	*Eunotia* cf. *pseudopapilio*	Lake Panch Pokhari, Nepal Himalaya. (South Asia) 28°02.533′N 85°42.822′E	Late Pleistocene period (16 kyr–till present)	Increased conductivity and/or higher nutrient inputs likely due to elevated nutrient deposition	Krstić et al. (2012)	
10	*Fragilaria construens* var. *subsalina*	Lake Panch Pokhari, Nepal Himalaya. (South Asia) 28°02.533′N 85°42.822′E	Late Pleistocene period (16 kyr–till present)	Increased conductivity and/or higher nutrient inputs likely due to elevated nutrient deposition	Krstić et al. (2012)	
11	*Fragilaria elliptica*	Lake Panch Pokhari, Nepal Himalaya. (South Asia) 28°02.533′N 85°42.822′E	Late Pleistocene period (16 kyr–till present)	Mesotrophic to eutrophic conditions	Krstić et al. (2012)	
12	*Fragilaria tenera*	Lake Panch Pokhari, Nepal Himalaya. (South Asia) 28°02.533′N 85°42.822′E	Late Pleistocene period (16 kyr–till present)	Occurs in increased conductivity and/or higher nutrient inputs likely due to elevated nutrient deposition	Krstić et al. (2012)	

Continued

Table 13.4 Summary of diatom taxa indicating eutrophic (high nutrient level) condition during the Quaternary period.—cont'd

Sr. no	Diatom taxa	Site name	Geological time	Environmental facies inferred from diatom taxa	Study references	Additional references
13	*Gomphonema subclavatum*	Lake Panch Pokhari, Nepal Himalaya. (South Asia) 28°02.533′N 85°42.822′E	Late Pleistocene period (16 kyr–till present)	Increased conductivity and/or higher nutrient inputs likely due to elevated nutrient deposition	Krstić et al. (2012)	
14	*Sellaphora pupula*	Lake Panch Pokhari, Nepal Himalaya. (South Asia) 28°02.533′N 85°42.822′E	Late Pleistocene period (16 kyr–till present)	Mesotrophic to eutrophic conditions	Krstić et al. (2012)	
15	*Cyclotella ocellata*	Lake Chenghai, SW China Tibetian plateau (East Asia) 26°27′–26°38′ N 100°38′–100°41′ E	Holocene period (7800 yr. BP)	Prefers nitrogen limited conditions and is abundant in oligo- to eutrophic conditions, greater light penetration and declining dissolved organic carbon (DOC) concentration and water temperatures	Li et al. (2015b)	Bracht et al. (2008), Kirilova et al. (2010); Wang (2012c), Saros et al. (2012), Fishman et al. (2010), Wang et al. (2015b)
16	*Lindavia radiosa*	Son Kol Lake, Kyrgyzstan (Central Asia) 41°50′ N 75°10′ E	Holocene period (6000 yr. BP)	Frequently observed at higher nutrient supply and under turbulent conditions at low water transparency in late summer and autumn	Schwarz et al. (2017)	Rioual et al. (2007)

17	*Gyrosigma acuminatum*	Artificial Reservoir Gonggeomji in Sangju City, Korea (East Asia) 36°30′56.65″ N 128°09′26.4″ E	Holocene period (2000 yr. BP)	Found in circumneutral environments and lives mainly in eutrophic lakes, found in inland waters of low or, at most, moderate electrolyte content, best developed in shallow streams, lakes, and rivers but is also found in swamps or shallow ponds	Lee et al. (2018) Hwang et al. (2014), Sterrenburg (1995)
18	*Aulacoseira ambigua*	Lake Hamana, Central Japan (East Asia) 34°45.912′ N 137°34.995′ E	Anthropocene/ Holocene period (last 1800 yr)	Abundant in generally mesotrophic to eutrophic condition	Cho et al. (2019) Joh (2010)
19	*Aulacoseira granulata*	Lake Hamana, Central Japan (East Asia) 34°45.912′ N 137°34.995′ E	Holocene period (last 1800 yr)	Abundant in generally mesotrophic to eutrophic condition	Cho et al. (2019) Joh (2010)
20	*Cyclotella atomus*	Hwajinpo Lagoon, Korea (East Asia)	Holocene period (last 1800 yr)	Freshwater to a brackish taxon that thrives under eutrophic conditions	Cho et al. (2018)
21	*Cyclotella atomus*	Lake Hamana, Central Japan (East Asia) 34°45.912′ N 137°34.995′ E	Holocene period (last 1800 yr)	Freshwater to brackish, eutrophic and halotolerant diatom	Cho et al. (2019) Joh (2010)
22	*Cyclotella atomus* var. *gracilis*	Lake Hamana, Central Japan (East Asia) 34°45.912′ N 137°34.995′ E	Holocene period (last 1800 yr)	Freshwater to brackish, prefer euryhaline and eutrophic condition	Cho et al. (2019) Genkal and Kiss (1993)

Continued

Table 13.4 Summary of diatom taxa indicating eutrophic (high nutrient level) condition during the Quaternary period.—cont'd

Sr. no	Diatom taxa	Site name	Geological time	Environmental facies inferred from diatom taxa	Study references	Additional references
23	*Discotella* sp.	Maar lake, Erlongwan NE China (East Asia) 42°18′ N 126°21′ E	Anthropocene/Holocene period (last 1000 yr)	Reported in a wide range of trophic conditions, from oligotrophic to eutrophic lakes, and can also develop in response to eutrophication	Wang et al. (2012b)	Haworth and Hurley (1984); Wunsam et al. (1995); Houk et al. (2010); Gregory and Oerlemans (1998)
24	*Staurosirella pinnata*	Mulyoungari swamp, Jeju Island, South Korea (East Asia) 33°22′09″ N 126°41′36″ E	Anthropocene/Holocene period (last 1000 yr)	Commonly found in shallow eutrophic lakes	Park et al. (2017)	

Apart from nutrient-rich conditions preferring taxa, various studies from East Asia by Nagayasu et al. (2017) (Japan), Wang et al. (2012a), and Wang et al. (2008) (SE China) recorded oligotrophic (nutrient-poor) conditions indicating taxa covering the late Pleistocene period. The list of oligotrophic conditions indicating taxa has been described in Table 13.5. Oligotrophic state across the various part of East Asia was mainly inferred using two diatom genera, *Aulacoseira* and *Cyclotella*. Species like *Aulacoseira alpigena*, *Cyclotella stelligera*, and *Cyclotella* sp. indicated oligotrophic lakes with low nutrient availability and thermally stratified lake during warm periods with weak wind (Battarbee et al., 2002; Rühland et al., 2003; Watanabe, 2005).

3.1.3 pH

pH is probably the single most crucial controlling variable on species composition in freshwater systems, and its significance has been long recognized in the diatom literature, for example, (Hustedt, 1942; Battarbee et al., 2001). Considering that diatom assemblages are very sensitive to changes in pH, they are being used as excellent indicator groups to reconstruct past lake pH (Battarbee et al., 2010) and surface water acidification (Battarbee et al., 1990; Cumming et al., 1994) based on estimates of the pH optima of individual taxa. The strong relationship between diatoms and pH gave rise to a pH classification of diatom taxa into three different groups.

1. Acidic condition (low pH) indicating taxa
2. Alkaline condition (high pH) showing taxa
3. Circumneutral condition indicating taxa

Studies across the various parts of Asia by Li et al. (2018), Ma et al. (2018), Wang et al. (2015a), Lee et al. (2018), and Wang et al. (2013) from Central and East Asia (NE and SW China, South Korea, Taiwan), Krstić et al. (2012) (Nepal) from South Asia reported the acidic condition indicating diatom taxa belonging to the genera *Eunotia*, *Frustulia*, *Pinnularia*, *Stauroneis*, *Brachysira*, *Tabellaria*, and *Neidium*.

Table 13.6 represents the summary of diatom taxa indicating the Acidic condition during the Quaternary period. Species like *Eunotia paludosa*, *Eunotia incisa*, *Eunotia* spp., *Frustulia rhomboides*, *Gomphonema gracile*, *Pinnularia braunii*, *Pinnularia gibba*, *Stauroneis phoenicenteron*, *Neidium alpinum*, *Neidium ampliatum*, *Anomoeoneis vitrea*, *Anomoeoneis brachysira*, *Caloneis silicula*, and *Surirella linearis* indicated a strong preference for acidic conditions (Krasske, 1949; Johansson, 1982; Whitmore, 1989; Stevenson et al., 1991; Racca et al., 2001; Bigler and Hall, 2003; Wang et al., 2013). At the same time, taxa such as *Achnanthes subatomoides*, *Cyclotella stelligera*, *Cymbella gracilis*, *Fragilaria exigua*, *Fragilaria nanana*, and *Navicula pupula* var. *pupula* indicated the slightly less acidic condition of the water (Philibert and Prairie, 2002; Jones and Birks, 2004; Leira et al., 2007).

Along with the acidic environment, diatom taxa also indicated a preference for the nutrients, such as the taxa preferring an acidic condition with a pristine environment and low nutrient availability. Diatom taxa such as *Eunotia flexuosa*, *Cymbopleura naviculiformis*, *Encyonema silesiacum*, *Pinnularia* aff. *viridiformis* var. *minor*, *Pinnularia rhombarea*, *Aulacoseira distans*, *Frustulia rhomboides* var. *crassinervia*, *Frustulia* spp., *Pinnularia microstauron*, *Pinnularia gibba*, and *Pinnularia* sp. indicated the pristine, acidic environment of the water along with less nutrient/oligotrophic condition (Patrick and Reimer, 1966; Van Campo and Gasse, 1994; Van Dam et al., 1994; Van de Vijver et al., 2004; Cantonati et al., 2011; Krstić et al. (2012); Wang et al., 2013).

Table 13.5 Summary of diatom taxa indicating oligotrophic (nutrient-poor) conditions during the Quaternary period.

Sr. no	Diatom taxa	Site name	Geological time	Environmental facies inferred from diatom taxa	Study references	Additional references
1	*Aulacoseira alpigena*	Takano river basin, Central Japan (East Asia) 36°32′55″ N 138°02′07″ E	Late Pleistocene period (156.8–38 kyr)	Occurs often in oligotrophic lakes in Japan	Nagayasu et al. (2017)	Watanabe (2005)
2	*Discostella stelligera*	Takano river basin, Central Japan (East Asia) 36°32′55″ N 138°02′07″ E	Late Pleistocene period (156.8–38 kyr)	Trophic indifferent, favorable for oligosaprobous condition	Nagayasu et al. (2017)	Stockner (1971)
3	*Cyclotella* taxa	Huguang Maar Lake, Guangdong Province, SE China (East Asia) 21°09′ N 110°17′ E	Late Pleistocene period (15 kyr–till present)	Euplanktonic species, which most commonly occur when lakes are thermally stratified during warm periods with weak wind, are commonly observed in higher abundances in lakes with low nutrients (i.e., oligotrophic)	Wang et al. (2012a)	Battarbee et al. (2002); Sorvari et al. (2002); Rühland et al. (2003)
4	*Cyclotella stelligera*	Huguang Maar Lake, Guangdong Province, SE China (East Asia) 21°09′ N 110°17′ E	Late Pleistocene period (15 kyr–till present)	Found in low sinking rates and that prefers low nutrient environments	Wang et al. (2012a)	
5	*Cyclotella* sp.	Huguang Maar Lake, Guangdong Province, SE China (East Asia) 21°09′ N 110°17′ E	Late Pleistocene period (17.5–6 kyr)	Euplanktonic species commonly occur when lakes are thermally stratified during warm periods with weak wind, commonly observed in higher abundances in lakes with low nutrients (i.e., oligotrophic), during warm periods with weak wind, lake stratification causing nutrients (P, N and Si) to become depleted in the epilimnion, have lower sinking rates which allow them to remain in suspension in the lake column much longer	Wang et al. (2008)	Battarbee et al. (2002); Sorvari et al. (2002); Rühland et al. (2003); Fahmenstiel and Glime (1983); Tilman et al. (1982); Petrova (1986); Rühland et al. (2008)

Table 13.6 Summary of diatom taxa indicating acidic condition during the Quaternary period.

Sr. no	Diatom taxa	Site name	Geological time	Environmental facies inferred from diatom taxa	Study references	Additional references
1	*Tabellaria flocculosa*	Takano river basin, Central Japan (East Asia) 36°32′55″ N 138°02′07″ E	Late Pleistocene period 156.8–38 kyr	Often occurred in sapropheric/acidic swamps, ponds and lakes of oligo-to mesotrophic water	Nagayasu et al. (2017)	Patrick and Reimer (1966)
2	*Anomoeoneis vitrea*	Lake Qinghai, SW China, (East Asia) 25°07′48″–25°08′6″ N 98°34′11″–98°34′16″ E	Late Pleistocene period 18.5 kyr–till present	Acidophilous	Li et al. (2018)	Whitmore (1989)
3	*Anomoeoneis brachysira*	Lake Qinghai, SW China, (East Asia) 25°07′48″–25°08′6″ N 98°34′11″–98°34′16″ E	Late Pleistocene period 18.5 kyr–till present	Acidophilous	Li et al. (2018)	Johansson (1982)
4	*Achnanthes subatomoides*	Lake Qinghai, SW China, (East Asia) 25°07′48″–25°08′6″ N 98°34′11″–98°34′16″ E	Late Pleistocene period 18.5 kyr–till present	Slight acidophilous	Li et al. (2018)	Jones and Birks (2004)
5	*Caloneis silicula*	Lake Qinghai, SW China, Asia) 25°07′48″–25°08′6″ N 98°34′11″–98°34′16″ E	Late Pleistocene period (18.5 kyr–till present)	Acidophilous	Li et al. (2018)	Bigler and Hall (2003); Jones and Birks (2004)

Continued

Table 13.6 Summary of diatom taxa indicating acidic condition during the Quaternary period.—cont'd

Sr. no	Diatom taxa	Site name	Geological time	Environmental facies inferred from diatom taxa	Study references	Additional references
6	*Cyclotella stelligera*	Lake Qinghai, SW China, (East Asia) 25°07′48″–25°08′6″N 98°34′11″–98°34′16″E	Late Pleistocene period (18.5 kyr–till present)	Slight Acidophilous	Li et al. (2018)	Chen and Wu (1999); Philibert and Prairie (2002)
7	*Cymbella gracilis*	Lake Qinghai, SW China, (East Asia) 25°07′48″–25°08′6″N 98°34′11″–98°34′16″E	Late Pleistocene period (18.5 kyr–till present)	Slight acidophilous	Li et al. (2018)	Philibert and Prairie (2002)
8	*Eunotia incisa*	Lake Qinghai, SW China, (East Asia) 25°07′48″–25°08′6″N 98°34′11″–98°34′16″E	Late Pleistocene period (18.5 kyr–till present)	Acidophilous	Li et al. (2018)	Stevenson et al. (1991); Thoms et al. (1999)
9	*Eunotia* spp.	Lake Qinghai, SW China, (East Asia) 25°07′48″–25°08′6″N 98°34′11″–98°34′16″E	Late Pleistocene period (18.5 kyr–till present)	Acidophilous	Li et al. (2018)	Dixit et al. (1988); Krammer and Lange-Bertalot (1991a); Dixit et al. (1993); Pan et al. (1996); Wunsam et al. (2002)
10	*Fragilaria exigua*	Lake Qinghai, SW China, (East Asia) 25°07′48″–25°08′6″N 98°34′11″–98°34′16″E	Late Pleistocene period (18.5 kyr–till present)	Slight Acidophilous	Li et al. (2018)	Dixit et al. (1988); Leira et al. (2007)

11	*Fragilaria nanana*	Lake Qinghai, SW China, (East Asia) 25°07′48″–25°08′6″ N 98°34′11″–98°34′16″ E	Late Pleistocene period (18.5 kyr–till present)	Slight acidophilous	Li et al. (2018)	Philibert and Prairie (2002)
12	*Frustulia rhomboides*	Lake Qinghai, SW China, (East Asia) 25°07′48″–25°08′6″ N 98°34′11″–98°34′16″ E	Late Pleistocene period (18.5 kyr–till present)	Acidophilous	Li et al. (2018)	Krasske (1949); Foged (1968); Cameron et al. (1998); Racca et al. (2001); Cattaneo et al. (2004)
13	*Gomphonema gracile*	Lake Qinghai, SW China, (East Asia) 25°07′48″–25°08′6″ N 98°34′11″–98°34′16″ E	Late Pleistocene period (18.5 kyr–till present)	Acidophilous	Li et al. (2018)	Bigler and Hall (2003); Jones and Birks (2004)
14	*Navicula pupula var. pupula*	Lake Qinghai, SW China, (East Asia) 25°07′48″–25°08′6″ N 98°34′11″–98°34′16″ E	Late Pleistocene period (18.5 kyr–till present)	Very slight acidophilous	Li et al. (2018)	Haworth and Offer (1986); Bigler and Hall (2002); Philibert and Prairie (2002)
15	*Pinnularia braunii*	Lake Qinghai, SW China, (East Asia) 25°07′48″–25°08′6″ N 98°34′11″–98°34′16″ E	Late Pleistocene period (18.5 kyr–till present)	Acidophilous	Li et al. (2018)	Jones and Birks (2004)
16	*Pinnularia gibba*	Lake Qinghai, SW China, (East Asia) 25°07′48″–25°08′6″ N 98°34′11″–98°34′16″ E	Late Pleistocene period (18.5 kyr–till present)	Acidophilous	Li et al. (2018)	Racca et al. (2001)

Continued

Table 13.6 Summary of diatom taxa indicating acidic condition during the Quaternary period.—cont'd

Sr. no	Diatom taxa	Site name	Geological time	Environmental facies inferred from diatom taxa	Study references	Additional references
17	*Stauroneis phoenicenteron*	Lake Qinghai, SW China, (East Asia) 25°07′48″–25°08′6″ N 98°34′11″–98°34′16″ E	Late Pleistocene period (18.5 kyr—till present)	Acidophilous	Li et al. (2018)	Racca et al. (2001)
18	*Tabellaria flocculosa*	Lake Qinghai, SW China, (East Asia) 25°07′48″–25°08′6″ N 98°34′11″–98°34′16″ E	Late Pleistocene period (18.5 kyr—till present)	Acidophilous	Li et al. (2018)	Denys (1991); Selby and Brown (2007)
19	*Cymbopleura naviculiformis*	Lake Panch Pokhari, Nepal Himalaya. (South Asia) 28°02.533′N 85°42.822′E	Late Pleistocene period (16 kyr—till present)	Indicates a pristine acidic environment deprived of significant nutrient loads	Krstić et al. (2012)	
20	*Encyonema silesiacum*	Lake Panch Pokhari, Nepal Himalaya. (South Asia) 28°02.533′N 85°42.822′E	Late Pleistocene period (16 kyr—till present)	Pristine acidic environment deprived of significant nutrient loads	Krstić et al. (2012)	
21	*Pinnularia* aff. *viridiformis* var. *minor*	Lake Panch Pokhari, Nepal Himalaya. (South Asia) 28°02.533′N 85°42.822′E	Late Pleistocene period (16 kyr—till present)	Indicates a pristine acidic environment deprived of significant nutrient loads	Krstić et al. (2012)	
22	*Pinnularia rhombarea*	Lake Panch Pokhari, Nepal Himalaya. (South Asia) 28°02.533′N 85°42.822′E	Late Pleistocene period (16 kyr—till present)	Indicator taxa of pristine acidic environment deprived of significant nutrient loads	Krstić et al. (2012)	

23	*Eunotia bilunaris* var. *bilunaris*	Honghe Peatland, NE China (East Asia) 47°47.335′ N 133°37.656′ E	Holocene period (6500 yr. BP)	Insensitive, widespread in low conductivity and oligotrophic to mesotrophic waters, mostly epiphytic	Ma et al. (2018)	Van Campo and Gasse (1994); Van Dam et al. (1994); Van de Vijver and Beyens (1997, 1999); Krammer and Lange-Bertalot (1999); Gaiser and Johansen (2000); Weilhoefer and Pan (2007); Cantonati et al. (2011); Chen et al. (2012, 2014b); Hargan et al. (2015)
24	*Eunotia flexuosa*	Honghe Peatland, NE China (East Asia) 47°47.335′ N 133°37.656′ E	Holocene period (6500 yr. BP)	Mainly distributed in swamps and stagnant or almost stagnant water oligohalobous and acidophilous	Ma et al. (2018)	Van Campo and Gasse (1994); Van Dam et al. (1994); Van de Vijver and Beyens (1997, 1999); Krammer and Lange-Bertalot (1999); Gaiser and Johansen (2000); Weilhoefer and Pan (2007); Cantonati et al. (2011); Chen et al. (2012, 2014b); Hargan et al. (2015)
25	*Tabellaria flocculosa*	Honghe Peatland, NE China (East Asia) 47°47.335′ N 133°37.656′ E	Holocene period (6500 yr. BP)	Insensitive, cosmopolitan and more common in northern latitudes (temperate to arctic regions) commonly found in peat bogs, running water or lakes grow in freshwater, slightly acidic to acidic waters, though it can have a wide pH tolerance	Ma et al. (2018)	Van Campo and Gasse (1994); Van Dam et al. (1994); Van de Vijver and Beyens (1997, 1999); Krammer and Lange-Bertalot (1999); Gaiser and Johansen (2000); Weilhoefer and Pan (2007); Cantonati et al. (2011); Chen et al. (2012, 2014b); Hargan et al. (2015)

Continued

Table 13.6 Summary of diatom taxa indicating acidic condition during the Quaternary period.—cont'd

Sr. no	Diatom taxa	Site name	Geological time	Environmental facies inferred from diatom taxa	Study references	Additional references
26	*Pinnularia gibba*	Tsuifong Lake, North-eastern Taiwan (East Asia) 24°30′ N 121°36′ E	Holocene period (3500 yr. BP)	Favors freshwater with low mineral content	Wang et al. (2015a)	
27	*Eunotia bilunaris*	artificial Reservoir Gonggeomji Korea (East Asia) 36°30′56.65″ N 128°09′26.41″ E	Holocene period (2000 yr. BP)	Extends from circumneutral to slightly acidophilous sites, commonly observed in the freshwater of low mineral content but may also occur in slightly alkaliphilous sites, frequently found in swamps or shallow ponds and lakes but also streams and rivers, predominantly in epiphytic (or epipelic) algal assemblages in swamps and bogs	Lee et al. (2018)	Joh (2011), Chung and Noh (1987), Kim et al. (2007)
28	*Pinnularia* sp.	Artificial Reservoir Gonggeomji, Korea (East Asia) 36°30′56.65″ N 128°09′26.41″ E	Holocene period (2000 yr. BP)	Essentially freshwater diatoms, mainly restricted to low acidic waters, and prefer oligotrophic waters and saproxenous taxa waters and prefer oligotrophic waters and saproxenous taxa	Lee et al. (2018)	Hwang et al. (2014)

29	*Stauroneis* sp.	Artificial Reservoir Gonggeomji, Korea (East Asia) 36°30′56.65″ N 128°09′26.41″ E	Holocene period (2000 yr. BP)	Habitats of Stauroneis taxa tend to be small, standing or slowly moving water (e.g., lakes, ponds, pools, fens, springs, seeps), prefer oligotrophic or dystrophic, circumneutral, or mildly acidic environments with low dissolved solids	Lee et al. (2018)	Van de Vijver et al. (2004); Bahls (2010)
30	*Aulacoseira distans*	Tsuifong Lake, NE Asia (Taiwan, China) 24°30′ N 121°36′ E	Anthropocene/ Holocene period (last 1500 yr)	Acidophilus, oligotrophic, planktonic in shallow waters <4 m deep, but can be classified as a benthic species	Wang et al. (2013)	Brugam et al. (1998); Moos et al. (2005)
31	*Eunotia paludosa*	Tsuifong Lake, Taiwan, China (NE Asia) 24°30′ N 121°36′ E	Anthropocene/ Holocene period (last 1500 yr)	Acidophilous	Wang et al. (2013)	
32	*Eunotia bilunaris* var. *mucophila*	Tsuifong Lake, Taiwan, China (NE Asia) 24°30′ N 121°36′ E	Anthropocene/ Holocene period (last 1500 yr)	Acidophilus, eutrophic	Wang et al. (2013)	
33	*Frustulia rhomboides* var. *crassinervia*	Tsuifong Lake, Taiwan, China (NE Asia) 24°30′ N, 121°36′ E	Anthropocene/ Holocene period (last 1500 yr)	Acidophilus, oligotrophic	Wang et al. (2013)	
34	*Neidium alpinum*	Tsuifong Lake, Taiwan, China (NE Asia) 24°30′ N 121°36′ E	Anthropocene/ Holocene period (last 1500 yr)	Acidophilus	Wang et al. (2013)	

Continued

Table 13.6 Summary of diatom taxa indicating acidic condition during the Quaternary period.—cont'd

Sr. no	Diatom taxa	Site name	Geological time	Environmental facies inferred from diatom taxa	Study references	Additional references
35	*Neidium ampliatum*	Tsuifong Lake, Taiwan, China (NE Asia) 24°30′ N 121°36′ E	Anthropocene/Holocene period (last 1500 yr)	Acidophilus	Wang et al. (2013)	
36	*Pinnularia gibba*	Tsuifong Lake, Taiwan, China (NE Asia) 24°30′ N 121°36′ E	Anthropocene/Holocene period (last 1500 yr)	Acidophilus, broad	Wang et al. (2013)	
37	*Pinnularia microstauron*	Tsuifong Lake, Taiwan, China (NE Asia) 24°30′ N 121°36′ E	Anthropocene/Holocene period (last 1500 yr)	Acidophilus, oligotrophic	Wang et al. (2013)	
38	*Surirella linearis*	Tsuifong Lake, Taiwan, China (NE Asia) 24°30′ N 121°36′ E	Anthropocene/Holocene period (last 1500 yr)	Acidophilus, broad	Wang et al. (2013)	
39	*Frustulia* spp	Mulyoungari swamp, Jeju Island, South Korea (East Asia) 33°22′09″ N 126°41′36″ E	Anthropocene/Holocene period (last 1000 yr)	Prefers oligotrophic acidic waters	Park et al. (2017)	

Apart from the above-recorded taxa, various studies also reported cosmopolitan diatom taxa such as *Eunotia bilunaris* var. *bilunaris*, which was commonly observed in circumneutral to slightly acidic conditions, with low mineral content, and frequent in swamps, bogs, or shallow ponds and lakes (Kim et al., 2007; Joh, 2011). Whereas *Stauroneis* sp. and *Tabellaria flocculosa* are common in slow-moving water (e.g., lakes, ponds, pools, fens, springs, seeps) and prefer oligotrophic or dystrophic, circumneutral, or mildly acidic environments with low dissolved solids (Van Dam et al., 1994; Van de Vijver et al., 2004; Bahls, 2010).

Alkaline condition across various parts of Asia was inferred by Nagayasu et al. (2017), Li et al. (2015a), Wang et al. (2014), Li et al. (2018), Chen et al. (2014a), Schwarz et al. (2017), Chiba et al. (2016), Lee et al. (2018), Cho et al. (2019), and Wang et al. (2013) mainly covering (NE and SW China, Japan, South Korea, Taiwan, Kyrgyzstan, Kazakhstan).

The alkaline condition was inferred from the late Pleistocene to Anthropocene period using 53 diatom taxa belonging to the genera *Achnanthes, Cocconeis, Cyclotella, Fragilaria, Amphora, Cymbella, Epithemia, Navicula,* and *Gomphonema*. Species like *Cocconeis placentula, Achnanthidium minutissimum, Amphora libyca, Fragilaria brevistriata, Fragilaria leptostauron, Fragilaria pinnata* var. *pinnata, Gomphonema truncatum,* and *Achnanthidium minutissimum* were the main taxa indicating alkaline condition (Gasse and Tekaia, 1983; Vos and de Wolf, 1993; Rosén et al., 2000; Bigler and Hall, 2003; Watanabe et al., 2005; Wang et al., 2013; Wang et al., 2014). *Amphora copulata, Amphora inariensis, Amphora ovalis, Amphora pediculus, Aneumastus minor, Aneumastus tusculus, Anomoeoneis sphaerophora* f. *costata, Aulacoseira canadensis, Aulacoseira granulata, Caloneis bacillum, Caloneis schumanniana, Cymatopleura elliptica, Cymbella tumida, Cymbella turgidula, Epithemia adnata, Epithemia sorex, Epithemia turgida, Luticola goeppertiana, Navicula cryptotenella, Navicula elginensis, Nitzschia amphibia,* and *Planothidium lanceolatum,* found widespread distribution in freshwater, alkaliphilous water condition (Vos and de Wolf, 1993; Van Dam et al., 1994; Watanabe et al., 2005; Kiss et al., 2007). In addition, *Rhopalodia gibba, Sellaphora bacillum, Staurosirella pinnata,* and *Thalassiosira lacustris* indicated the brackish, alkaline condition of the streams (Vos and de Wolf, 1993; Van Dam et al., 1994; Watanabe et al., 2005). Species like *Encyonopsis microcephala* and *Pseudostaurosira brevistriata* indicated a preference for alkaline and oligotrophic conditions. They thrived in stable thermal stratification with high water transparency and low nutrient load (Joh, 2012; Wang et al., 2013).

Along with alkalinity, diatom taxa also indicated a preference for nutrients. Species such as *Achnanthes joursacense, Achnanthes lanceolata, Cyclotella ocellata, Cyclotella radiosa,* and *Pantocsekiella costei* indicated alkaline water along with oligo-to mesotrophic condition (Krammer and Lange-Bertalot, 1991b; Van Dam et al., 1994). In contrast, taxa such as *Cyclotella dubius, Navicula oblonga, Staurosira elliptica,* and *Staurosira construens* were abundant in alkaline eutrophic (nutrient-rich) conditions with high electrolyte content (Anderson et al., 1993; Van Dam et al., 1994; Anderson, 1997). *Cyclotella distinguenda* and *Fragilarioid* species prefer alkaline conditions with low temperature and prolonged ice cover and can survive in harsh climatic conditions (Kiss et al., 2007; Wang et al., 2014). Table 13.7 lists all the diatom taxa indicating alkaline conditions across the various parts of Asia.

Circum-neutral condition across the various part of Asia was inferred using 22 diatom taxa. *Achnanthes lacus-vulcani, Navicula indifferens, Gomphonema parvulum, Hantzschia amphioxys, Navicula lanceolata, Pinnularia borealis, Pinnularia subgibba, Pinnularia viridis, Reimeria sinuata, Ulnaria ulna,* and *Surirella robusta* indicated benthic, freshwater, circumneutral condition (Vos and de

Table 13.7 Summary of diatom taxa indicating alkaline condition during the Quaternary period.

Sr. no	Diatom taxa	Site name	Geological time	Environmental facies inferred from diatom taxa	Study references	Additional references
1	*Achnanthes joursacense*	Takano river basin, Central Japan (East Asia) 36°32′55″N 138°02′07″ E	Late Pleistocene period (156.8–38 kyr)	Indicates alkaliphilous and oligo- to mesotrophic condition	Nagayasu et al. (2017)	Van Dam et al. (1994)
2	*Achnanthes lanceolata*	Takano river basin, Central Japan (East Asia) 36°32′55″N 138°02′07″ E	Late Pleistocene period (156.8–38 kyr)	Alkaliphilous and oligo- to mesotrophic species	Nagayasu et al. (2017)	Van Dam et al. (1994)
3	*Cyclotella ocellata*	Takano river basin, Central Japan (East Asia) 36°32′55″N 138°02′07″ E	Late Pleistocene period (156.8–38 kyr)	Alkaliphilous and oligo- to mesotrophentic species	Nagayasu et al. (2017)	Krammer and Lange-Bertalot (1991a)
4	*Cyclotella radiosa*	Takano river basin, Central Japan (East Asia) 36°32′55″N 138°02′07″ E	Late Pleistocene period (156.8–38 kyr)	Alkaliphilous and oligo- to meso-trophentic species	Nagayasu et al. (2017)	Van Dam et al. (1994)
5	*Navicula cryptotenella*	Takano river basin, Central Japan 36°32′55″N, 138°02′07″E	Late Pleistocene period (156.8–38 kyr)	Adapted to alkaliphilous and oligo- to eu-trophentic one	Nagayasu et al. (2017)	Van Dam et al. (1994)
6	*Staurosira construens*	Takano river basin, Central Japan (East Asia) 36°32′55″ N 138°02′07″ E	Late Pleistocene period (156.8–38 kyr)	Adapted to alkaliphilous and meso-eutrophic conditions	Nagayasu et al. (2017)	Van Dam et al. (1994)

7	*Cocconeis placentula*	Heqing Basin, South-west China, Yunnan East Asia 26°33′43″ N 100°10′14″ E	Late Pleistocene period (140–35 kyr)		Li et al. (2015a)	Wang et al. (2014)
8	*Cyclotella dubius*	Lugu Lake, south-west China, Yunnan, East Asia 27°43′08.4″ N 100°46′33.9″ E	Late Pleistocene period (30–10 kyr)	High nutrient, alkaline waters	Wang et al. (2014)	Anderson (1990, 1997); Anderson et al. (1993); Håkansson and Regnéll (1993); Bradbury et al. (1994); Bradshaw and Anderson (2003); Yang et al. (2008)
9	*Amphora libyca*	Lake Qinghai, SW China, (East Asia) 25°07′48″–25°08′6″ N 98°34′11″–98°34′16″ E	Late Pleistocene period (18.5 kyr–till present)	Alkaliphilous	Li et al. (2018)	Bigler and Hall (2003); Jones and Birks (2004)
10	*Cocconeis placentula*	Lake Qinghai, SW China, (East Asia) 25°07′48″–25°08′6″ N 98°34′11″–98°34′16″ E	Late Pleistocene period (18.5 kyr–till present)	Alkaliphilous	Li et al. (2018)	Gasse and Tekaia (1983); Shennan et al. (1993); Kovacs et al. (2006)
11	*Fragilaria brevistriata*	Lake Qinghai, SW China, (East Asia) 25°07′48″–25°08′6″ N 98°34′11″–98°34′16″ E	Late Pleistocene period (18.5 kyr–till present)	Alkaliphilous	Li et al. (2018)	Rosen et al. (2000); Jones and Birks (2004)
12	*Fragilaria leptostauron*	Lake Qinghai, SW China, (East Asia) 25°07′48″–25°08′6″ N 98°34′11″–98°34′16″ E	Late Pleistocene period (18.5 kyr–till present)	Alkaliphilous	Li et al. (2018)	Bigler and Hall (2002)

Continued

Table 13.7 Summary of diatom taxa indicating alkaline condition during the Quaternary period.—cont'd

Sr. no	Diatom taxa	Site name	Geological time	Environmental facies inferred from diatom taxa	Study references	Additional references
13	*Fragilaria pinnata* var. *pinnata*	Lake Qinghai, SW China, (East Asia) 25°07′48″–25°08′6″ N 98°34′11″–98°34′16″ E	Late Pleistocene period (18.5 kyr–till present)	Alkaliphilous	Li et al. (2018)	Gasse and Tekaia (1983); Shennan et al. (1993); Kovacs et al. (2006)
14	*Cyclotella distinguenda*	Tiancai Lake, south-west China, Tibetian plateau (East Asia) 26°38′3.8″ N 99°43′00″ E	Late Pleistocene period (12 kyr–till present)	Can tolerate cold, alkaline, and high solar irradiance conditions	Chen et al. (2014a)	Kiss et al. (2007); Wunsam et al. (1995)
15	*Fragilarioid* species	Tiancai Lake, south-west China, (East Asia) 26°38′3.8″ N, 99°43′00″ E	Late Pleistocene period (12 kyr–till present)	Prefer water columns with high light penetration, high alkalinity, low temperature, prolonged ice cover, able to survive in harsh climatic conditions	Chen et al. (2014a)	Wang et al. (2014), Laing and Smol (2000); Weckström et al. (1997); Schmidt et al. (2004), Lotter and Bigler (2000)
16	*Pseudostaurosira brevistriata*	Tiancai Lake, south-west China, Tibetian plateau (East Asia) 26°38′3.8″,99°43′00″ E	Late Pleistocene period (12 kyr–till present)	Higher pH optimum	Chen et al. (2014a)	Rosén et al. (2000)
17	*Cocconeis placentula*	Hwajinpo Lagoon, Korea (East Asia)	Holocene period (8 kyr –till present)	Freshwater to brackish species and a benthic diatom in streams	Cho et al. (2018)	Joh (2012)

18	*Navicula oblonga*	Son Kol Lake, Kyrgyzstan (Central Asia) 41°50′N 75°10′E	Holocene period (6000 yr. BP)	Occurs in stagnant, alkaline waters with higher electrolyte content or slightly brackish waters	Schwarz et al. (2017)	Krammer and Lange Bertalot (2010)
19	*Pantocsekiella costei*	Son Kol Lake, Kyrgyzstan (Central Asia) 41°50′N 75°10′E	Holocene period (6000 yr. BP)	Common in the littoral or pelagial of alkaline, oligo- to mesotrophic lakes, develops in late spring to summer during stable thermal stratification with high water transparency and low nutrient load	Schwarz et al. (2017)	Houk et al. (2010); Rioual et al. (2007)
20	*Amphora copulata*	Balkhash Lake, Kazakhstan, (Central Asia) 46°17′–21.2″ N 74°00′–33.6″ E	Anthropocene/ Holocene period (last 2000 yr)	Benthic, freshwater, alkaliphilous	Chiba et al. (2016)	Vos and de Wolf (1993); Van Dam et al. (1994); Watanabe et al. (2005)
21	*Amphora inariensis*	Balkhash Lake, Kazakhstan, (Central Asia) 46°17′–21.2″ N 74°00′–33.6″ E	Anthropocene/ Holocene period (last 2000 yr)	Benthic, freshwater, alkaliphilous	Chiba et al. (2016)	Vos and de Wolf (1993); Van Dam et al. (1994); Watanabe et al. (2005)
22	*Amphora ovalis*	Balkhash Lake, Kazakhstan, (Central Asia) 46°17′–21.2″ N 74°00′–33.6″ E	Anthropocene/ Holocene period (last 2000 yr)	Benthic, freshwater, alkaliphilous	Chiba et al. (2016)	Vos and de Wolf (1993); Van Dam et al. (1994); Watanabe et al. (2005)
23	*Amphora pediculus*	Balkhash Lake, Kazakhstan, (Central Asia) 46°17′–21.2″ N 74°00′–33.6″ E	Anthropocene/ Holocene period (last 2000 yr)	Benthic, freshwater, alkaliphilous	Chiba et al. (2016)	Vos and de Wolf (1993); Van Dam et al. (1994); Watanabe et al. (2005)

Continued

Table 13.7 Summary of diatom taxa indicating alkaline condition during the Quaternary period.—cont'd

Sr. no	Diatom taxa	Site name	Geological time	Environmental facies inferred from diatom taxa	Study references	Additional references
24	*Aneumastus minor* Lange-Bertalot	Balkhash Lake, Kazakhstan, (Central Asia) 46°17′–21.2″ N 74°00′–33.6″ E	Anthropocene/ Holocene period (last 2000 yr)	Benthic, freshwater, alkalibiontic	Chiba et al. (2016)	Vos and de Wolf (1993); Van Dam et al. (1994); Watanabe et al. (2005)
25	*Aneumastus tusculus*	Balkhash Lake, Kazakhstan, (Central Asia) 46°17′–21.2″ N 74°00′–33.6″ E	Anthropocene/ Holocene period (last 2000 yr)	Benthic, freshwater, alkalibiontic	Chiba et al. (2016)	Vos and de Wolf (1993); Van Dam et al. (1994); Watanabe et al. (2005)
26	*Anomoeoneis sphaerophora* f. *costata*	Balkhash Lake, Kazakhstan, (Central Asia) 46°17′–21.2″ N 74°00′–33.6″ E	Anthropocene/ Holocene period (last 2000 yr)	Benthic, brackish, alkaliphilous	Chiba et al. (2016)	Vos and de Wolf (1993); Van Dam et al. (1994); Watanabe et al. (2005)
27	*Aulacoseira canadensis*	Balkhash Lake, Kazakhstan, (Central Asia) 46°17′–21.2″ N 74°00′–33.6″ E	Anthropocene/ Holocene period (last 2000 yr)	Euplanktonic, freshwater, alkaliphilous	Chiba et al. (2016)	Vos and de Wolf (1993); Van Dam et al. (1994); Watanabe et al. (2005)
28	*Aulacoseira granulata*	Balkhash Lake, Kazakhstan, (Central Asia) 46°17′–21.2″ N 74°00′–33.6″ E	Anthropocene/ Holocene period (last 2000 yr)	Euplanktonic, freshwater, alkaliphilous	Chiba et al. (2016)	Vos and de Wolf (1993); Van Dam et al. (1994); Watanabe et al. (2005)
29	*Caloneis bacillum*	Balkhash Lake, Kazakhstan, (Central Asia) 46°17′–21.2″ N 74°00′–33.6″ E	Anthropocene/ Holocene period (last 2000 yr)	Benthic, freshwater, alkaliphilous	Chiba et al. (2016)	Vos and de Wolf (1993); Van Dam et al. (1994); Watanabe et al. (2005)

30	*Caloneis schumanniana*	Balkhash Lake, Kazakhstan, (Central Asia) 46°17′–21.2″ N 74°00′–33.6″ E	Anthropocene/ Holocene period (last 2000 yr)	Benthic, freshwater, alkaliphilous	Chiba et al. (2016)	Vos and de Wolf (1993); Van Dam et al. (1994); Watanabe et al. (2005)
31	*Cocconeis placentula*	Balkhash Lake, Kazakhstan, (Central Asia) 46°17′–21.2″ N 74°00′–33.6″ E	Anthropocene/ Holocene period (last 2000 yr)	Benthic, freshwater, alkaliphilous	Chiba et al. (2016)	Vos and de Wolf (1993); Van Dam et al. (1994); Watanabe et al. (2005)
32	*Cymatopleura elliptica*	Balkhash Lake, Kazakhstan, (Central Asia) 46°17′–21.2″ N 74°00′–33.6″ E	Anthropocene/ Holocene period (last 2000 yr)	Benthic, freshwater, alkaliphilous	Chiba et al. (2016)	Vos and de Wolf (1993); Van Dam et al. (1994); Watanabe et al. (2005)
33	*Cymbella tumida*	Balkhash Lake, Kazakhstan, (Central Asia) 46°17′–21.2″ N 74°00′–33.6″ E	Anthropocene/ Holocene period (last 2000 yr)	Benthic, freshwater, alkaliphilous	Chiba et al. (2016)	Vos and de Wolf (1993); Van Dam et al. (1994); Watanabe et al. (2005)
34	*Cymbella turgidula*	Balkhash Lake, Kazakhstan, (Central Asia) 46°17′–21.2″ N 74°00′–33.6″ E	Anthropocene/ Holocene period (last 2000 yr)	Benthic, freshwater, alkaliphilous	Chiba et al. (2016)	Vos and de Wolf (1993); Van Dam et al. (1994); Watanabe et al. (2005)
35	*Epithemia adnata*	Balkhash Lake, Kazakhstan, (Central Asia) 46°17′–21.2″ N 74°00′–33.6″ E	Anthropocene/ Holocene period (last 2000 yr)	Benthic, freshwater, alkalibiontic	Chiba et al. (2016)	Vos and de Wolf (1993); Van Dam et al. (1994); Watanabe et al. (2005)
36	*Epithemia sorex*	Balkhash Lake, Kazakhstan, (Central Asia) 46°17′–21.2″ N 74°00′–33.6″ E	Anthropocene/ Holocene period (last 2000 yr)	Benthic, freshwater, alkalibiontic	Chiba et al. (2016)	Vos and de Wolf (1993); Van Dam et al. (1994); Watanabe et al. (2005)

Continued

Table 13.7 Summary of diatom taxa indicating alkaline condition during the Quaternary period.—cont'd

Sr. no	Diatom taxa	Site name	Geological time	Environmental facies inferred from diatom taxa	Study references	Additional references
37	*Epithemia sorex*	Artificial Reservoir Gonggeomji Korea (East Asia) 36°30′56.65″N 128°09′26.41″E	Holocene period (2000 yr. BP)	Widespread in low alkalinity and saproxenous taxa	Lee et al. (2018)	Watanabe et al. (2005)
38	*Epithemia turgida*	Balkhash Lake, Kazakhstan, (Central Asia) 46°17′–21.2″ N 74°00′–33.6″ E	Anthropocene/ Holocene period (last 2000 yr)	Benthic, freshwater, alkalibiontic	Chiba et al. (2016)	Vos and de Wolf (1993); Van Dam et al. (1994); Watanabe et al. (2005)
39	*Gomphonema truncatum*	Artificial Reservoir Gonggeomji Korea (East Asia) 36°30′56.65″N 128°09′26.41″E	Holocene period (2000 yr. BP)	Widespread in low alkalinity and saproxenous taxa	Lee et al. (2018)	Watanabe et al. (2005)
40	*Luticola goeppertiana*	Balkhash Lake, Kazakhstan, (Central Asia) 46°17′–21.2″ N 74°00′–33.6″ E	Anthropocene/ Holocene period (last 2000 yr)	Benthic, freshwater, alkaliphilous	Chiba et al. (2016)	Vos and de Wolf (1993); Van Dam et al. (1994); Watanabe et al. (2005)
41	*Navicula cryptotenella*	Balkhash Lake, Kazakhstan, (Central Asia) 46°17′–21.2″ N 74°00′–33.6″ E	Anthropocene/ Holocene period (last 2000 yr)	Benthic, brackish, alkaliphilous	Chiba et al. (2016)	Vos and de Wolf (1993); Van Dam et al. (1994); Watanabe et al. (2005)
42	*Navicula elginensis*	Balkhash Lake, Kazakhstan, (Central Asia) 46°17′–21.2″ N 74°00′–33.6″ E	Anthropocene/ Holocene period (last 2000 yr)	Benthic, freshwater, alkaliphilous	Chiba et al. (2016)	Vos and de Wolf (1993); Van Dam et al. (1994); Watanabe et al. (2005)

43	*Nitzschia amphibia*	Balkhash Lake, Kazakhstan, (Central Asia) 46°17′–21.2″ N 74°00′–33.6″ E	Anthropocene/ Holocene period (last 2000 yr)	Benthic, freshwater, alkaliphilous	Chiba et al. (2016)	Vos and de Wolf (1993); Van Dam et al. (1994); Watanabe et al. (2005)
44	*Planothidium lanceolatum*	Balkhash Lake, Kazakhstan, (Central Asia) 46°17′–21.2″ N 74°00′–33.6″ E	Anthropocene/ Holocene period (last 2000 yr)	Benthic, freshwater, alkaliphilous	Chiba et al. (2016)	Vos and de Wolf (1993); Van Dam et al. (1994); Watanabe et al. (2005)
45	*Pseudostaurosira brevistriata*	Balkhash Lake, Kazakhstan, (Central Asia) 46°17′–21.2″ N 74°00′–33.6″ E	Anthropocene/ Holocene period (last 2000 yr)	Tycho-planktonic, freshwater-brackish, alkaliphilous	Chiba et al. (2016)	Vos and de Wolf (1993); Van Dam et al. (1994); Watanabe et al. (2005)
46	*Rhopalodia gibba*	Balkhash Lake, Kazakhstan, (Central Asia) 46°17′–21.2″ N 74°00′–33.6″ E	Anthropocene/ Holocene period (last 2000 yr)	Benthic, brackish, alkalibiontic	Chiba et al. (2016)	Vos and de Wolf (1993); Van Dam et al. (1994); Watanabe et al. (2005)
47	*Sellaphora bacillum*	Balkhash Lake, Kazakhstan, (Central Asia) 46°17′–21.2″ N 74°00′–33.6″ E	Anthropocene/ Holocene period (last 2000 yr)	Benthic, brackish, alkaliphilous	Chiba et al. (2016)	Vos and de Wolf (1993); Van Dam et al. (1994); Watanabe et al. (2005)
48	*Pseudostaurosira brevistriata*	Lake Hamana, Central Japan (East Asia) 34°45.912′ N 137°34.995 E	Holocene period (last 1800 yr. BP)	Oligotrophic condition	Cho et al. (2019)	Joh (2012)
49	*Staurosira elliptica*	Tsuifong Lake, Taiwan, China (Northeast Asia) 24°30′ N 121°36′ E	Anthropocene/ Holocene period (last 1500 yr)	Alkaliphilous, Eutrophic	Wang et al. (2013)	

Continued

Table 13.7 Summary of diatom taxa indicating alkaline condition during the Quaternary period.—cont'd

Sr. no	Diatom taxa	Site name	Geological time	Environmental facies inferred from diatom taxa	Study references	Additional references
50	*Staurosirella pinnata*	Balkhash Lake, Kazakhstan, (Central Asia) 46°17′–21.2″ N 74°00′–33.6″ E	Anthropocene/ Holocene period (last 2000 yr)	Tycho-planktonic, freshwater-brackish, alkaliphilous	Chiba et al. (2016)	Vos and de Wolf (1993); Van Dam et al. (1994); Watanabe et al. (2005)
51	*Thalassiosira lacustris*	Balkhash Lake, Kazakhstan, (Central Asia) 46°17′–21.2″ N 74°00′–33.6″ E	Anthropocene/ Holocene period (last 2000 yr)	Eu-planktonic, brackish, alkaliphilous	Chiba et al. (2016)	Vos and de Wolf (1993); Van Dam et al. (1994); Watanabe et al. (2005)
52	*Encyonopsis microcephala*	Tsuifong Lake, Taiwan, China (NE Asia) 24°30′ N 121°36′ E	Anthropocene/ Holocene period (last 1500 yr)	Alkaliphilous, oligotrophic	Wang et al. (2013)	
53	*Achnanthidium minutissimum*	Tsuifong Lake, Taiwan, China (NE Asia) 24°30′ N 121°36′ E	Anthropocene/ Holocene period (last 1500 yr)	Alkaliphilous, broad	Wang et al. (2013)	

Wolf, 1993; Van Dam et al., 1994; Watanabe, 2005). At the same time, taxa such as *Sellaphora pupula*, *Staurosira construens* var. *binodis*, *Staurosira construens* var. *construens*, *Staurosira venter*, *Staurosirella lapponica*, and *Staurosirella martyi* indicated benthic, brackish, circumneutral condition (Vos and de Wolf, 1993; Van Dam et al., 1994; Watanabe, 2005). In addition, *Eunotia praerupta* and *Gomphonema lagerheimii* indicated circum-neutral, oligotrophic conditions (Van Dam et al., 1994; Van de Vijver and Beyens, 1997; Krammer and Lange-Bertalot, 1999), while *Gomphonema gracile* indicated circum-neutral, eutrophic condition (Wang et al., 2013). Table 13.8 summarizes the list of diatom taxa indicating circumneutral conditions during the Quaternary period.

3.1.4 Water turbulence
Well, mixed water, along with high water turbulence in the water, indicates a high nutrient supply into the water body, which is indirectly related to run-off and high rainfall/increased water supply into the stream or lake. Wang et al. (2008), Wang et al. (2012a), Chen et al. (2014a), Rudaya et al. (2009), Li et al. (2015b), and Schwarz et al. (2017), across various parts of East and Central Asia covering mainly (NE and SW China, Mongolia and Kyrgyzstan) reported the high-water turbulence conditions indicating taxa. *Aulacoseira* sp. such as *Aulacoseira ambigua* and *Aulacoseira granulata*, *Aulacoseira distans*, *Asterionella formosa*, *Cyclostephanos dubius*, and *Puncticulata praetermissa* indicated the well-mixed condition of the water with high water turbulence, low light availability, higher nutrient concentration, and indirect paleoenvironmental indicator of the persistence of robust and seasonal wind stress and resultant turbulent water column mixing and nutrient upwelling conditions (Kilham and Kilham, 1975; Kilham, 1990; Anderson et al., 1993; Rioual et al., 2007). Table 13.9 shows the diatom taxa summary indicating high energy/water turbulence conditions during the Quaternary period.

3.1.5 Temperature
Various studies have used diatoms as a potential tool for reconstructing past temperatures (Vyvermann and Sabbe, 1995; Lotter et al., 1997). Temperature change significantly influences the behavior of other physical and chemical variables in lake systems, such as ice cover, stratification, pH, nutrient cycling, etc (Battarbee, 2000).

Table 13.10 represents the list of diatom taxa indicating low temperature and shallow water depth conditions during the Quaternary period. Species like *Staurosira construens* var. *venter* and *Stephanodiscus hantzschii* f. *tenuis* indicated a colder environment with low temperatures across East and South Asia (Douglas and Smol, 1999). *Staurosirella pinnata* and *Aulacoseira alpigena* indicates slightly acidic, cold, oligotrophic condition with low to moderate specific conductance values (Laing et al., 1999; Lotter et al., 2010; Wolin and Stone, 2010). *Fragilarioid* species indicate low temperature, prolonged ice cover, and high alkalinity. They can survive in harsh climatic conditions (Lotter and Bigler, 2000; Wang et al., 2014).

Cho et al. (2018) and Li et al. (2015a,b), across East Asia, reported the shallow water depth indicating taxa covering the Holocene period. *Cocconeis scutellum* var. *parva*, *Fragilaria brevistriata*, and *Fragilaria pinnata* indicated shallow water depth conditions in small lakes and ponds (Bradbury, 1989). *Luticola dismutica* showed an exclusive preference for aerophytic habitats (Poulickova, 2008), and *Stauroneis pinnata* was found primarily in the disturbed glacial environment.

Apart from the above-listed taxa, various studies also recorded the indifferent cosmopolitan taxa, and their ecological preference is unknown. Table 13.11 lists all such insensitive taxa or shows

Table 13.8 Summary of diatom taxa indicating circumneutral conditions during the Quaternary period.

Sr. no	Diatom taxa	Site name	Geological time	Environmental facies inferred from diatom taxa	Study references	Additional references
1	*Navicula papula*	Takano river basin, Central Japan (East Asia) 36°32′55″ N 138°02′ 07″ E	Late Pleistocene period (156.8–38 kyr)	Adapted to circumneutral pH and mesotrophic conditions	Nagayasu et al. (2017)	Van Dam et al. (1994)
2	*Achnanthes lacus-vulcani*	Lake Qinghai, SW China, (East Asia) 25°07′48″–25°08′6″ N 98°34′11″–98°34′16″ E	Late Pleistocene period (18.5 kyr–till present)	Circumneutral taxa	Li et al. (2018)	Rühland et al. (2003)
3	*Navicula indifferens*	Lake Qinghai, SW China, (East Asia) 25°07′48″–25°08′6″ N 98°34′11″–98°34′16″ E	Late Pleistocene period (18.5 kyr–till present)	Circumneutral, generally indicators of cold-water	Li et al. (2018)	Bigler and Hall (2002); Jones and Birks (2004); Weckstrom et al. (1997)
4	*Gomphonema lagerheimii*	Honghe Peatland, NE China (East Asia) 47°47.335′ N 133°37.65′ E	Holocene period (6500 yr. BP)	Insensitive taxa, distributed in the northern and arctic swamps, live in oligotrophic, circumneutral pH and low conductivity waters with continuously high dissolved oxygen level	Ma et al. (2018)	Van Campo and Gasse (1994); Van Dam et al. (1994); Van de Vijver and Beyens (1997, 1999); Krammer and Lange-Bertalot (1999); Gaiser and Johansen (2000); Weilhoefer and Pan (2007); Cantonati et al. (2011); Chen et al. (2012, 2014b); Hargan et al. (2015)

5	Gomphonema parvulum	Balkhash Lake, Kazakhstan, 46°17′–21.2″ N 74°00′–33.6″ E	Anthropocene/Holocene period (last 2000 yr)	Benthic, freshwater, circumneutral	Chiba et al. (2016)	Vos and de Wolf (1993); Van Dam et al. (1994); Watanabe et al. (2005)
6	Hantzschia amphioxys	Balkhash Lake, Kazakhstan, (Central Asia) 46°17′–21.2″ N 74°00′–33.6″ E	Anthropocene/Holocene period (last 2000 yr)	Benthic, freshwater, circumneutral	Chiba et al. (2016)	Vos and de Wolf (1993); Van Dam et al. (1994); Watanabe et al. (2005)
7	Pinnularia borealis	Balkhash Lake, Kazakhstan, (Central Asia) 46°17′–21.2″ N 74°00′–33.6″ E	Anthropocene/Holocene period (last 2000 yr)	Benthic, freshwater, circumneutral	Chiba et al. (2016)	Vos and de Wolf (1993); Van Dam et al. (1994); Watanabe et al. (2005)
8	Pinnularia subgibba	Balkhash Lake, Kazakhstan, (Central Asia) 46°17′–21.2″ N 74°00′–33.6″ E	Anthropocene/Holocene period (last 2000 yr)	Benthic, freshwater, circumneutral	Chiba et al. (2016)	Vos and de Wolf (1993); Van Dam et al. (1994); Watanabe et al. (2005)
9	Pinnularia viridis	Balkhash Lake, Kazakhstan, (Central Asia) 46°17′–21.2″ N 74°00′–33.6″ E	Anthropocene/Holocene period (last 2000 yr)	Benthic, freshwater, circumneutral	Chiba et al. (2016)	Vos and de Wolf (1993); Van Dam et al. (1994); Watanabe et al. (2005)
10	Reimeria sinuata	Balkhash Lake, Kazakhstan, (Central Asia) 46°17′–21.2″ N 74°00′–33.6″ E	Anthropocene/Holocene period (last 2000 yr)	Benthic, freshwater, circumneutral	Chiba et al. (2016)	Vos and de Wolf (1993); Van Dam et al. (1994); Watanabe et al. (2005)
11	Sellaphora pupula	Balkhash Lake, Kazakhstan, (Central Asia) 46°17′–21.2″ N 74°00′–33.6″ E	Anthropocene/Holocene period (last 2000 yr)	Benthic, brackish, circumneutral	Chiba et al. (2016)	Vos and de Wolf (1993); Van Dam et al. (1994); Watanabe et al. (2005)

Continued

Table 13.8 Summary of diatom taxa indicating circumneutral conditions during the Quaternary period.—cont'd

Sr. no	Diatom taxa	Site name	Geological time	Environmental facies inferred from diatom taxa	Study references	Additional references
12	*Stauroneis* spp.	Balkhash Lake, Kazakhstan, (Central Asia) 46°17′–21.2″ N 74°00′–33.6″ E	Anthropocene/ Holocene period (last 2000 yr)	Benthic, brackish, circumneutral	Chiba et al. (2016)	Vos and de Wolf (1993); Van Dam et al. (1994); Watanabe et al. (2005)
13	*Staurosira construens* var. *binodis*	Balkhash Lake, Kazakhstan, (Central Asia) 46°17′–21.2″ N 74°00′–33.6″ E	Anthropocene/ Holocene period (last 2000 yr)	Benthic, brackish, circumneutral	Chiba et al. (2016)	Vos and de Wolf (1993); Van Dam et al. (1994); Watanabe et al. (2005)
14	*Staurosira construens* var. *construens*	Balkhash Lake, Kazakhstan, (Central Asia) 46°17′–21.2″ N 74°00′–33.6″ E	Anthropocene/ Holocene period (last 2000 yr)	Benthic, brackish, circumneutral	Chiba et al. (2016)	Vos and de Wolf (1993); Van Dam et al. (1994); Watanabe et al. (2005)
15	*Staurosira venter*	Balkhash Lake, Kazakhstan, (Central Asia) 46°17′–21.2″ N 74°00′–33.6″ E	Anthropocene/ Holocene period (last 2000 yr)	Benthic, brackish, circumneutral	Chiba et al. (2016)	Vos and de Wolf (1993); Van Dam et al. (1994); Watanabe et al. (2005)
16	*Staurosirella lapponica*	Balkhash Lake, Kazakhstan, (Central Asia) 46°17′–21.2″ N 74°00′–33.6″ E	Anthropocene/ Holocene period (last 2000 yr)	Benthic, brackish, circumneutral	Chiba et al. (2016)	Vos and de Wolf (1993); Van Dam et al. (1994); Watanabe et al. (2005)
17	*Staurosirella martyi*	Balkhash Lake, Kazakhstan, (Central Asia) 46°17′–21.2″ N 74°00′–33.6″ E	Anthropocene/ Holocene period (last 2000 yr)	Benthic, brackish, circumneutral	Chiba et al. (2016)	Vos and de Wolf (1993); Van Dam et al. (1994); Watanabe et al. (2005)

18	Surirella robusta	Balkhash Lake, Kazakhstan, (Central Asia) 46°17′–21.2″ N 74°00′–33.6″ E	Anthropocene/ Holocene period (last 2000 yr)	Benthic, freshwater, circumneutral	Chiba et al. (2016)	Vos and de Wolf (1993); Van Dam et al. (1994); Watanabe et al. (2005)
19	Ulnaria ulna	Artificial Reservoir Gonggeomji Korea (East Asia) 36°30′56.65″ N 128°09′26.41″ E	Holocene period (last 2000 yr)	Occurs as free-living or epiphytic forms in freshwater around the world and is rarely found in brackish water, this species is widely distributed and seems to prefer circumneutral water and α-meso/poly saprobous, it prefers β-mesosaprobic regions, this species lives in water with a pH of 6.2–9.3 in Korea	Lee et al. (2018)	Van Dam et al. (1994), Watanabe et al. (2005), Joh et al. (2010)
20	Eunotia praerupta	Tsuifong Lake, Taiwan, China (NE Asia) 24°30′ N, 121°36′ E	Anthropocene/ Holocene period (last 1500 yr)	Circumneutral, oligotrophic	Wang et al. (2013)	
21	Gomphonema gracile	Tsuifong Lake, Taiwan, China (NE Asia) 24°30′ N, 121°36′ E	Anthropocene/ Holocene period (last 1500 yr)	Circumneutral, eutrophic	Wang et al. (2013)	
22	Navicula lanceolata	Tsuifong Lake, Taiwan, China (NE Asia) 24°30′ N, 121°36′	Holocene period (last 1500 yr)	Circumneutral, broad	Wang et al. (2013)	

Table 13.9 Summary of diatom taxa indicating high energy/high water turbulence conditions during the Quaternary period.

Sr. no	Diatom taxa	Site name	Geological time	Environmental facies inferred from diatom taxa	Study references	Additional references
1	*Aulacoseira* sp. such as *A. ambigua* and *A. granulata*	Huguang Maar Lake, Southeast China (East Asia) 21°9′ N 110°17′ E	Late Pleistocene period 17.5–6 kyr	Meroplanktonic species enter the water when mixing conditions are such that they can be suspended and maintained in the water column, can adapt to low-light levels, and have resting stages that allow them to survive in the dark when they sink to the sediment surface at the lake bottom, have high nutrient requirements, an indirect paleoenvironmental indicator of the persistence of strong, seasonal wind stress and resultant turbulent water column mixing and nutrient upwelling conditions. taxa are thickly silicified and form filamentous colonies, which make them relatively heavy and more likely to sink out of the photic zone as stratification develops unless turbulent conditions help them to remain in suspension	Wang et al. (2008)	Kilham (1990); Talling (1957); Schelske et al. (1995); Kilham and Kilham (1975); Pilskaln and Johnson (1991); Kilham et al. (1996); Pannard et al. (2008)
2	*Aulacoseira granulata*	Huguang Maar Lake, Guangdong Province, SE China (East Asia) 21°09′ N 110°17′ E	Late Pleistocene period (15 kyr—till present)	Abundant under conditions with high water turbulence	Wang et al. (2012a)	Kilham and Kilham (1975); Pilskaln and Johnson (1991)

3	*Aulacoseira* genus	Tiancai lake, SW China (East Asia) 26°38′3.8″ N 99°43′00″ E	Late Pleistocene period (12 kyr—till present)	Requires turbulence to maintain its presence in the water column, low light	Chen et al. (2014a)	Bradbury (1975), Gibson et al. (2003)
4	*Aulacoseira distans*	Hoton-Nur Lake, Mongolia (Central Asia) 48°40′ N 88°18′ E	Holocene period (11,500 yr. BP)	Coincides with the development of a steppe landscape that is more open and drier than before; thrive in lakes with wind-induced turbulence	Rudaya et al. (2009)	Rühland et al. (2003)
5	*Cyclostephanos dubius*	Lake Chenghai, SW China, Tibetian plateau (East Asia) 26°27′–26°38′ N 100°38′–100°41′E	Holocene period (7800 yr. BP)	Abundant in the well-mixed water column, higher nutrient concentration and low light availability	Li et al. (2015b)	Anderson (1990, 1997; Anderson et al. (1993); Fritz et al. (1993); Bradbury et al. (1994); Edlund and Stoermer (2000); Bradshaw and Anderson (2003); Yang et al. (2008)
6	*Aulacoseira granulata*	Son Kol Lake, Kyrgyzstan (Central Asia) 41°50′N 75°10′ E	Holocene period (6000 yr. BP)	Requires high turbulence to keep its strongly silicified valves in the photic zone and is an indicator of increased nutrients and low light availability	Schwarz et a (2017)	Kilham et al. (1986); Round et al. (1990); Rioual et al. (2007)
7	*Asterionella formosa*	Maar Lake, Erlongwan, NE China (East Asia) 42°18′ N 126°21′ E	Anthropocene/ Holocene period (last 1000 yr)	Abundant under conditions with high water turbulence	Wang et al. (2012b)	Morabito et al. (2002); Pracnik et al. (2003)
8	*Puncticulata praetermissa*	Maar lake, Erlongwan NE China (East Asia) 42°18′ N,126°21′ E	Anthropocene/ Holocene period (last 1000 yr)	Blooms seasonally in spring and autumn, when windy conditions cause a turnover and the water column is well mixed, found in mesotrophic lakes	Wang et al. (2012b)	Rioual et al. (2009); Wunsam et al. (1995)

Table 13.10 Summary of diatom taxa indicating low temperature and shallow water depth conditions during the Quaternary period.

Sr. no	Diatom taxa	Site name	Geological time	Environmental facies inferred from diatom taxa	Study references	Additional references
Low temperature indicating taxa						
1	*Staurosirella pinnata*	Heqing Basin, SW China (East Asia) 26°33′43″ N 100°10′14″ E	Late Pleistocene period (140–35 kyr)	Can develop large populations in high latitude/altitude lakes under cold and oligotrophic conditions, but can also be abundant in warm, shallow, eutrophic lakes which have a high pH	Li et al. (2015a)	Laing et al. (1999); Lotter et al. (2010), Bennion (1994); Schmidt et al. (2004)
2	*Aulacoseira alpigena*	Lake Qinghai, SW China, (East Asia) 25°07′48″–25°08′6″ N 98°34′11″–98°34′16″ E	Late Pleistocene period (18.5 kyr–till present)	Slight acidophilous, generally prefer colored, humic waters with high DOC concentrations and are known to be adapted to low light conditions; an indicator of a slightly acidic, cold, oligotrophic lake with low to moderate specific conductance values	Li et al. (2018)	Rosen et al. (2000); Van de Vijver et al. (2002); Jones and Birks (2004); Ginn et al. (2007); Korsman and Birks (1996); Krammer and Lange-Bertalot (1986, 1988, 1991a, 1991b); Kagan (2001); (2002)
3	*Fragilarioid* species	Tiancai Lake, SW China Tibetian plateau (East Asia) 26°38′3.8″ N 99°43′00″ E	late Pleistocene period (12 kyr–till present)	Prefer water columns with high light penetration, high alkalinity, low temperature, prolonged ice cover, able to survive in harsh climatic conditions	Chen et al. (2014a)	Wang et al. (2014), Laing and Smol (2000), Weckström et al. (1997), Schmidt et al. (2004), Lotter and Bigler (2000)

4	*Stephanodiscus hantzschii* f. *tenuis*	Hwajinpo Lagoon, Korea (East Asia)	Holocene period (8000 yr. BP)	Low-temperature diatom that lives in water temperatures below 12°C	Cho et al. (2018)	
5	*Staurosirella (Fragilaria) pinnata*	Lake Chenghai, SW China Tibetan plateau (East Asia) 26°27′ to 26°38′ N 100°38′ to 100° 41′ E	Holocene period (7800 yr. BP)	Cold, oligotrophic conditions with high light penetration and high alkalinity can also be abundant in warm, shallow and eutrophic lakes, which have a high pH	Li et al. (2015b)	Wolin and Stone (2010), Laing et al. (1999); Lotter et al. (1999), Wang et al. (2014), Laing and Smol (2000), Bennion (1994); Schmidt et al. (2004), Lotter and Bigler (2000)
6	*Staurosira construens* var. *venter*	Pinder Valley, northwest India (South Asia) 30.05°N, 79.93°E	Holocene period (3500 yr. BP)	Abundant in colder environments	Rühland et al. (2006)	Douglas and Smol (1999)
7	*Aulacoseira alpigena*	PinderValley, northwest India (South Asia) 30.05°N 79.93°E	Holocene period (3500 yr. BP)	Free-floating diatom typically found in deeper waters	Rühland et al. (2006)	

Diatoms indicating shallow water depth condition

8	*Cocconeis scutellum* var. *parva*	Hwajinpo Lagoon, Korea (East Asia)	Holocene period (8000 yr. BP)	Benthic species that live in weeds and are indicative of shallow water depths	Cho et al. (2018)	Joh (2012)
9	*Fragilaria brevistriata*	Hwajinpo Lagoon, Korea (East Asia)	Holocene period (8000 yr. BP)	Epipelic-euryhaline species appear mainly in shallow, freshwater lakes	Cho et al. (2018)	Bradbury (1989); Colombaroli et al. (2007)
10	*Fragilaria pinnata*	Southern Okinawa Trough, East China Sea 24°48.04′ N 122°29.35′ E	Holocene period (6800 yr. BP)	Benthic taxa and salt-intolerant commonly occur as epiphytes, growing in freshwater environments such as shallow marshes, lakes and small ponds	Li et al. (2015c)	Shennan et al. (2009), Yang et al. (2005)

Continued

Table 13.10 Summary of diatom taxa indicating low temperature and shallow water depth conditions during the Quaternary period.—cont'd

Miscellaneous taxa

Sr. no	Diatom taxa	Site name	Geological time	Environmental facies inferred from diatom taxa	Study references	Additional references
11	*Fragilaria nanana*	Maar lake, Erlongwan NE China (East Asia) 42°18′ N 126°21′ E	Anthropocene/ Holocene period (last 1000 yr)	This can be indicative of periods of generally weak water column mixing	Wang et al. (2012b)	-
12	*Luticola dismutica*	Honghe Peatland, NE China (East Asia) 47°47.335′ N 133°37.656′ E	Holocene period (6500 yr BP)	Living in aerophytic habitats	Ma et al. (2018)	Poulickova (2008)
13	*Pinnularia viridis*	Tsuifong Lake, North-eastern Taiwan (East Asia) 24°30′ N 121°36′ E	Holocene period (3500 yr BP)	Dominates mineral-rich water	Wang et al. (2015a)	Wang et al. (2010)
14	*Stauroneis pinnata*	Hoton-Nur Lake, Mongolia (Central Asia) 48°40′ N, 88°18′ E	Holocene period (11,500 yr BP)	Commonly found in disturbed glacial and/or arctic environments	Rudaya et al. (2009)	Haworth (1976); Anderson (2000)

Table 13.11 Summary of diatom taxa insensitive or showing unknown ecological preference condition.

Sr. no	Diatom taxa	Site name	Geological time	Environmental facies inferred from diatom taxa	Study references	Additional references
Insensitive taxa						
1	Achnanthes minutissima	Lake Qinghai, SW China, (East Asia) 25°07′48″–25°08′6″N 98°34′11″–98°34′16″E	Late Pleistocene period (18.5 kyr–till present)	Indifferent taxa	Li et al. (2018)	Patrick and Reimer (1966, 1975); Van Dam et al. (1994); DeNicola (2000); Selby and Brown (2007); Kernan et al. (2009)
2	Cyclotella bodanica	Lake Qinghai, SW China, (East Asia) 25°07′48″–25°08′6″N 98°34′11″–98°34′16″E	Late Pleistocene period (18.5 kyr–till present)	Indifferent, characteristic of lower nutrient concentrations and is interpreted to reflect warm summers with long periods of thermal stratification	Li et al. (2018)	Cholnoky (1968); Beaver (1981); Bracht et al. (2008)
3	Cyclotella bodanica	Lake Qinghai, SW China, (East Asia) 25°07′48″–25°08′6″N 98°34′11″–98°34′16″E	Late Pleistocene period (18.5 kyr–till present)	Indifferent	Li et al. (2018)	Cholnoky (1968); Beaver (1981)
4	Fragilaria construens	Lake Qinghai, SW China, (East Asia) 25°07′48″–25°08′6″N 98°34′11″–98°34′16″E	Late Pleistocene period (18.5 kyr–till present)	Indifferent	Li et al. (2018)	Gasse and Tekaia (1983); Shennan et al. (1993); Kovacs et al. (2006)

Continued

Table 13.11 Summary of diatom taxa insensitive or showing unknown ecological preference condition.—cont'd

Sr. no	Diatom taxa	Site name	Geological time	Environmental facies inferred from diatom taxa	Study references	Additional references
5	*Navicula cryptocephala*	Lake Qinghai, SW China, (East Asia) 25°07'48''–25°08'6'' N 98°34'11''–98°34'16'' E	Late Pleistocene period (18.5 kyr–till present)	Indifferent	Li et al. (2018)	Hustedt (1942); Krasske (1949); Foged (1968, 1980)
6	*Discostella stelligera*	Balkhash Lake, Kazakhstan, (Central Asia) 46°17'–21'' N 74°00'–33'' E	Anthropocene/Holocene period (last 2000 yr)	Euplanktonic, freshwater, Indifferent	Chiba et al. (2016)	Vos and de Wolf (1993); Van Dam et al. (1994); Watanabe et al. (2005)
Taxa showing unknown environmental preference						
7	*Brackysira* sp.	Balkhash Lake, Kazakhstan, (Central Asia) 46°17'–21'' N 74°00'–33'' E	Anthropocene/Holocene period (last 2000 yr)	Benthic, freshwater, unknown	Chiba et al. (2016)	Vos and de Wolf (1993); Van Dam et al. (1994); Watanabe et al. (2005)
8	*Cocconeis* sp.	Balkhash Lake, Kazakhstan, (Central Asia) 46°17'–21'' N 74°00'–33'' E	Anthropocene/Holocene period (last 2000 yr)	Benthic, freshwater, unknown	Chiba et al. (2016)	Vos and de Wolf (1993); Van Dam et al. (1994); Watanabe et al. (2005)
9	*Cymbella* sp.	Balkhash Lake, Kazakhstan, (Central Asia) 46°17'–21'' N 74°00'–33'' E	Anthropocene/Holocene period (last 2000 yr)	Benthic, freshwater, unknown	Chiba et al. (2016)	Vos and de Wolf (1993); Van Dam et al. (1994); Watanabe et al. (2005)
10	*Diploneis* sp.	Balkhash Lake, Kazakhstan, (Central Asia) 46°17'–21'' N 74°00'–33'' E	Anthropocene/Holocene period (last 2000 yr)	Benthic, freshwater, unknown	Chiba et al. (2016)	Vos and de Wolf (1993); Van Dam et al. (1994); Watanabe et al. (2005)

11	*Encyonema* sp.	Balkhash Lake, Kazakhstan, (Central Asia) 46°17′–21″ N 74°00′–33″ E	Anthropocene/Holocene period (last 2000 yr)	Benthic, freshwater, unknown	Chiba et al. (2016)	Vos and de Wolf (1993); Van Dam et al. (1994); Watanabe et al. (2005)
12	*Epithemia* sp.	Balkhash Lake, Kazakhstan, (Central Asia) 46°17′–21″ N 74°00′–33″ E	Anthropocene/Holocene period (last 2000 yr)	Benthic, freshwater, unknown	Chiba et al. (2016)	Vos and de Wolf (1993); Van Dam et al. (1994); Watanabe et al. (2005)
13	*Gyrosigma scalproides*	Balkhash Lake, Kazakhstan, (Central Asia) 46°17′–21″ N 74°00′–33″ E	Anthropocene/Holocene period (last 2000 yr)	Benthic, brackish, unknown	Chiba et al. (2016)	Vos and de Wolf (1993); Van Dam et al. (1994); Watanabe et al. (2005)
14	*Gyrosigma* sp.	Balkhash Lake, Kazakhstan, (Central Asia) 46°17′–21″ N 74°00′–33″ E	Anthropocene/Holocene period (last 2000 yr)	Benthic, brackish, unknown	Chiba et al. (2016)	Vos and de Wolf (1993); Van Dam et al. (1994); Watanabe et al. (2005)
15	*Handmania* sp.	Balkhash Lake, Kazakhstan, (Central Asia) 46°17′–21″ N 74°00′–33″ E	Anthropocene/Holocene period (last 2000 yr)	Euplanktonic, freshwater, unknown	Chiba et al. (2016)	Vos and de Wolf (1993); Van Dam et al. (1994); Watanabe et al. (2005)
16	*Mastogloia* sp.	Balkhash Lake, Kazakhstan, (Central Asia) 46°17′–21″ N 74°00′–33″ E	Anthropocene/Holocene period (last 2000 yr)	Benthic, brackish, unknown	Chiba et al. (2016)	Vos and de Wolf (1993); Van Dam et al. (1994); Watanabe et al. (2005)
17	*Nitzschia* sp.	Balkhash Lake, Kazakhstan, (Central Asia) 46°17′–21″ N 74°00′–33″ E	Anthropocene/Holocene period (last 2000 yr)	Benthic, freshwater, unknown	Chiba et al. (2016)	Vos and de Wolf (1993); Van Dam et al. (1994); Watanabe et al. (2005)

Continued

Table 13.11 Summary of diatom taxa insensitive or showing unknown ecological preference condition.—cont'd

Sr. no	Diatom taxa	Site name	Geological time	Environmental facies inferred from diatom taxa	Study references	Additional references
18	*Parlibellus crucicula*	Balkhash Lake, Kazakhstan, (Central Asia) 46°17′–21″ N 74°00′–33″ E	Anthropocene/Holocene period (last 2000 yr)	Benthic, brackish, unknown	Chiba et al. (2016)	Vos and de Wolf (1993); Van Dam et al. (1994); Watanabe et al. (2005)
19	*Pinnularia* sp.1	Balkhash Lake, Kazakhstan, (Central Asia) 46°17′–21″ N 74°00′–33″ E	Anthropocene/Holocene period (last 2000 yr)	Benthic, freshwater, unknown	Chiba et al. (2016)	Vos and de Wolf (1993); Van Dam et al. (1994); Watanabe et al. (2005)
20	*Pinnularia* sp.2	Balkhash Lake, Kazakhstan, (Central Asia) 46°17′–21″ N 74°00′–33″ E	Anthropocene/Holocene period (last 2000 yr)	Benthic, freshwater, unknown	Chiba et al. (2016)	Vos and de Wolf (1993); Van Dam et al. (1994); Watanabe et al. (2005)
21	*Scoliopleura peisonis*	Balkhash Lake, Kazakhstan, (Central Asia) 46°17′–21″ N 74°00′–33″ E	Anthropocene/Holocene period (last 2000 yr)	Benthic, brackish, unknown	Chiba et al. (2016)	Vos and de Wolf (1993); Van Dam et al. (1994); Watanabe et al. (2005)
22	*Scoliotropis latestriata*	Balkhash Lake, Kazakhstan, (Central Asia) 46°17′–21″ N 74°00′–33″ E	Anthropocene/Holocene period (last 2000 yr)	Benthic, brackish, unknown	Chiba et al. (2016)	Vos and de Wolf (1993); Van Dam et al. (1994); Watanabe et al. (2005)
23	*Surirella* sp.	Balkhash Lake, Kazakhstan, (Central Asia) 46°17′–21″ N 74°00′–33″ E	Anthropocene/Holocene period (last 2000 yr)	Benthic, freshwater, unknown	Chiba et al. (2016)	Vos and de Wolf (1993); Van Dam et al. (1994); Watanabe et al. (2005)

unknown ecological preference conditions. *Achnanthes minutissima, Cyclotella bodanica, Cymbella* sp., *Diploneis* sp., *Gyrosigma* sp., etc., are indifferent or show unknown ecological preference.

Insensitive or unknown ecological preferences of many diatom taxa arise for two reasons.

1. Insensitivity here is associated with taxa such as *Gomphonema gracile, G. parvulum,* and *Discostella stelligera;* these taxa are cosmopolitan, reported across the continents, and there is a huge lumping associated with these groups of taxa.
2. Secondly, many taxa listed in this group are not identified up to species level; hence, there is no conclusion about their environmental inductiveness, so the case of insensitivity/unknown environmental preference here is not associated with diatoms but instead associated with improper taxonomy.

3.2 Brackish environment
3.2.1 Salinity

Even though pH is the most crucial variable in some salt lake systems, especially those dominated by sodium carbonate-bicarbonate chemistry, diatom composition in most salt lakes is usually controlled by the ionic strength of the water measured as salinity or conductivity. Diatoms in salt lakes are predominantly euryhaline and, in many cases, are the same as those in estuarine or coastal waters where brackish water conditions exist. Paleosalinity inferred from diatom records in salt lakes has been used extensively in climate change studies (Battarbee et al., 2001). However, despite their tolerance to salinity fluctuation, different taxa show clearly defined optima along the salinity gradient.

Based on the preference for salinity gradient, diatom taxa have been divided into three different groups.

1. Oligohalobous taxa
2. Mesohalobous taxa
3. Polyhalobous taxa

Huang et al. (2009), Zong et al. (2006), and Schwarz et al. (2017) reported the salinity level preference of the brackish diatom taxa across Central and East Asia, covering (Southern China, SCS, and Kyrgyzstan).

3.2.1.1 Oligohalobous taxa

Taxa such as *Achnanthes minutissima, Aulacoseira granulata, Cocconeis pediculus, Fallacia subhamulata, Gomphonema parvulum,* and *Nitzschia fonticola* indicated less saline environment/oligohalobous, benthic condition (Zong et al., 2006). At the same time, species like *Cyclotella Kuetzingiana, Cyclotella meneghiniana, Cyclotella striata, Cyclotella stylorum,* and *Cyclotella radiosa* indicated less saline environment/oligohalobous, planktonic conditions (Zong et al., 2006).

3.2.1.2 Mesohalobous taxa

Nitszchia sigma indicated moderate saline environment/mesohalobous, benthic, and *Cyclotella striata* indicated moderate saline environment/mesohalobous, planktonic condition. Species like

Anomoeoneis costata, Campylodiscus clypeus, Campylodiscus hibernicus, and *Surirella striatula* indicated periphytic, saline conditions with electrolyte-rich water (Van Campo and Gasse, 1993; Ramrath et al., 1999; Witkowski et al., 2000). Table 13.12 represents a summary of brackish diatom taxa indicating salinity level conditions during the Quaternary period.

3.3 Marine environment
3.3.1 Salinity level condition indicating taxa
The salinity level is one of the critical factors that decide the distribution of marine taxa. Based on the preference for salinity gradient, diatom taxa have been divided into three different groups.

1. Oligohalobous taxa
2. Mesohalobous taxa
3. Polyhalobous taxa

Huang et al. (2009), Zong et al. (2006), and Li et al. (2015c) reported the salinity level preference of the marine diatom taxa across Central and East Asia, covering (Southern China, SCS, and ECS). *Paralia sulcata, Thalassionema frauenfeldii,* and *Diploneis bombus* are cosmopolitan taxa that can tolerate low salinity levels and thrive in an estuarine environment (Zong, 1997). *Coscinodiscus blandus, Coscinodiscus divisus, Cyclotella stylorum,* and *Navicula digitulus* are the planktonic taxa that indicate a preference for a moderate level of salinity (Zong et al., 2006). Species like *Biddulphia rhombus, Chaetoceros radians, Paralia sulcata, Skeletonema costatum, Thalassionema nitschioides,* and *Diploneis bombus* are the taxa that indicate a preference for comparatively higher salinity levels. Table 13.13 summarizes the list of marine diatom taxa showing salinity level conditions across the various parts of Asia.

Huang et al. (2009) and D'Costa and Anil (2010) reported the high-temperature condition inferring taxa across South and SE Asia. *Azpeitia Africana, Azpeitia marina,* and *Rhizosolenia bergonii* were commonly found in warm-water regions and typically oceanic, planktonic species (Hasle and Syvertsen, 1997). Species like *Caloneis westii, Cyclotella caspia, Diploneis crabro, Melosira nummuloides, Navicula distans, Skeletonema costatum,* and *Thalassiosira* spp. indicated moderate to an elevated temperature levels and were found primarily abundant during the postmonsoon period. *Cyclotella stylorum, Navicula elegans,* and *Navicula lyra* indicated high salinity and temperature and were found abundant during the premonsoon period (D'Costa and Anil, 2010). Table 13.14 represents a summary of marine diatom taxa indicating elevated temperature conditions during the Quaternary period.

Chiba et al. (2016) across Central Asia inferred alkaline conditions inside the marine environment using six marine diatom taxa. Table 13.15 lists the marine diatom taxa indicating alkaline conditions during the late Holocene period. Taxa such as *Hippodonta linearis, Campylodiscus clypeus, Navicula digitoradiata, Tryblionella* sp., *Tryblionella levidensis,* and *Cocconeis scutellum* indicated alkaline conditions along with the benthic, marine environment (Vos and de Wolf, 1993; Van Dam et al., 1994; Watanabe, 2005).

Table 13.12 Summary of brackish diatom taxa indicating salinity level condition during the Quaternary period.

Sr. no	Diatom taxa	Site name	Geological time	Environmental facies inferred from diatom taxa	Study references	Additional references
1	Cyclotella striata	South China Sea (SCS) 20°7′ N, 117°23′ E	Late Pleistocene period (15 kyr–till present)	Brackish water species, often abundant in estuaries as planktonic species	Huang et al. (2009)	Hendey (1964)
2	Cyclotella stylorum	South China Sea (SCS) 20°7′ N, 117°23′ E	Late Pleistocene period (15 kyr–till present)	Brackish water species, often abundant in estuaries as planktonic species	Huang et al. (2009)	Hendey (1964)
3	Achnanthes minutissima	Pearl River Estuary, Southern China (East Asia) 26°–22° N	Holocene period (8500 yr. BP)	Oligohalobous (ind.), benthic	Zong et al. (2006)	
4	Aulacoseira granulata	Pearl River Estuary, Southern China (East Asia) 26°–22° N	Holocene period (8500 yr. BP)	Oligohalobous (ind.), benthic	Zong et al. (2006)	
5	Cocconeis pediculus	Pearl River Estuary, Southern China (East Asia) 26°–22° N	Holocene period (8500 yr. BP)	Oligohalobous (hal.), benthic	Zong et al. (2006)	
6	Cyclotella Kuetzingiana	Pearl River Estuary, Southern China (East Asia) 26°–22° N	Holocene period (8500 yr. BP)	Brackish, oligohalobous (ind.), planktonic	Zong et al. (2006)	
7	Cyclotella meneghiniana	Pearl River Estuary, Southern China (East Asia) 26°–22° N	Holocene period (8500 yr. BP)	Oligohalobous (hal.), planktonic	Zong et al. (2006)	
8	Cyclotella radiosa	Pearl River Estuary, Southern China (East Asia) 26°–22° N	Holocene period (8500 yr. BP)	Oligohalobous (ind.), planktonic	Zong et al. (2006)	

Continued

Table 13.12 Summary of brackish diatom taxa indicating salinity level condition during the Quaternary period.—cont'd

Sr. no	Diatom taxa	Site name	Geological time	Environmental facies inferred from diatom taxa	Study references	Additional references
9	*Fallacia subhamulata*	Pearl River Estuary, Southern China (East Asia) 26°–22° N	Holocene period (8500 yr. BP)	Oligohalobous (ind.), benthic	Zong et al. (2006)	
10	*Gomphonema parvulum*	Pearl River Estuary, Southern China (East Asia) 26°–22° N	Holocene period (8500 yr. BP)	Oligohalobous (ind.), benthic	Zong et al. (2006)	
11	*Nitzschia fonticola*	Pearl River Estuary, Southern China (East Asia) 26°–22° N	Holocene period (8500 yr. BP)	Oligohalobous (ind.), benthic	Zong et al. (2006)	
12	*Nitzschia sigma*	Pearl River Estuary, Southern China (East Asia) 26°–22° N	Holocene period (8500 yr. BP)	Brackish, mesohalobous, benthic	Zong et al. (2006)	
13	*Cyclotella striata*	Pearl River Estuary, Southern China (East Asia) 26°–22° N	Holocene period (8500 yr. BP)	Mesohalobous, planktonic	Zong et al. (2006)	
14	*Anomoeoneis costata*	Son Kol Lake, Kyrgyzstan (Central Asia) 41°50′ N, 75°10′ E	Holocene period (6000 yr. BP)	Brackish, prefers saline or electrolyte-rich waters	Schwarz et al. (2017)	
15	*Campylodiscus clypeus*	Son Kol Lake, Kyrgyzstan (Central Asia) 41°50′ N, 75°10′ E	Holocene period (6000 yr. BP)	Periphytic, saline, or electrolyte-rich water-preferring diatoms	Schwarz et al. (2017)	Van Campo and Gasse (1993); Ramrath et al. (1999); Witkowski et al. (2000)
16	*Campylodiscus hibernicus*	Son Kol Lake, Kyrgyzstan (Central Asia) 41°50′ N, 75°10′ E	Holocene period (6000 yr. BP)	Periphytic, saline, or electrolyte-rich water-preferring diatoms	Schwarz et al. (2017)	Van Campo and Gasse (1993); Ramrath et al. (1999); Witkowski et al. (2000)
17	*Surirella striatula*	Son Kol Lake, Kyrgyzstan (Central Asia) 41°50′ N, 75°10′ E	Holocene period (6000 yr. BP)	Periphytic, saline, or electrolyte-rich water-preferring diatoms	Schwarz et al. (2017)	Van Campo and Gasse (1993); Ramrath et al. (1999); Witkowski et al. (2000)

Table 13.13 Summary of marine diatom taxa indicating salinity level condition during the Quaternary period.

Sr. no	Diatom taxa	Site name	Geological time	Environmental facies inferred from diatom taxa	Study references	Additional references
Polyhalobous taxa						
1	*Biddulphia rhombus*	Pearl River Estuary, Southern China (East Asia) 26°–22° N	Holocene period (8500 yr BP)	Polyhalobous, planktonic, marine	Zong et al. (2006)	—
2	*Chaetoceros radians*	Pearl River Estuary, Southern China (East Asia) 26°–22° N	Holocene period (8500 yr BP)	Polyhalobous, planktonic	Zong et al. (2006)	—
3	*Diploneis bombus*	Pearl River Estuary, Southern China (East Asia) 26°–22° N	Holocene period (8500 yr BP)	Polyhalobous, benthic	Zong et al. (2006)	—
4	*Paralia sulcata*	Pearl River Estuary, Southern China (East Asia) 26°–22° N	Holocene period (8500 yr BP)	Polyhalobous, planktonic	Zong et al. (2006)	—
5	*Skeletonema costatum*	Pearl River Estuary, Southern China (East Asia) 26°–22° N	Holocene period (8500 yr BP)	Polyhalobous, planktonic	Zong et al. (2006)	—
6	*Thalassionema nitschioides*	Pearl River Estuary, Southern China (East Asia) 26°–22° N	Holocene period (8500 yr BP)	Polyhalobous, planktonic	Zong et al. (2006)	—

Continued

Table 13.13 Summary of marine diatom taxa indicating salinity level condition during the Quaternary period.—cont'd

Sr. no	Diatom taxa	Site name	Geological time	Environmental facies inferred from diatom taxa	Study references	Additional references
Mesohalobous taxa						
7	*Coscinodiscus blandus*	Pearl River Estuary, Southern China (East Asia) 26°–22° N	Holocene period (8500 yr BP)	Mesohalobous, planktonic	Zong et al. (2006)	—
8	*Coscinodiscus divisus*	Pearl River Estuary, Southern China (East Asia) 26°–22° N	Holocene period (8500 yr BP)	Mesohalobous, planktonic	Zong et al. (2006)	—
9	*Cyclotella stylorum*	Pearl River Estuary, Southern China (East Asia) 26°–22° N	Holocene period (8500 yr BP)	Mesohalobous, planktonic	Zong et al. (2006)	—
10	*Navicula digitulus*	Pearl River Estuary, Southern China (East Asia) 26°–22° N	Holocene period (8500 yr BP)	Mesohalobous, planktonic	Zong et al. (2006)	—
Oligohalobous taxa						
11	*Diploneis bombus*	South China Sea (SCS) 20°7′ N 117°23′ E	Late Pleistocene period (15 kyr–till present)	Tracer of coastal waters and consequently low sea-surface salinity	Huang et al. (2009)	—
12	*Paralia sulcata*	Southern Okinawa Trough, East China Sea 24°48.04′ N 122°29.35′ E	Holocene period (6800 yr BP)	Cosmopolitan species can tolerate reduced salinity in an estuarine environment	Li et al. (2015c)	Zong (1997)
13	*Thalassionema frauenfeldii*	Southern Okinawa Trough, East China Sea 24°48.04′ N 122°29.35′ E	Holocene period (6800 yr BP)	Cosmopolitan species can tolerate reduced salinity in an estuarine environment	Li et al. (2015c)	—

Table 13.14 Summary of marine diatom taxa indicating elevated temperature conditions during the Quaternary period.

Sr. no	Diatom taxa	Site name	Geological time	Environmental facies inferred from diatom taxa	Study references	Additional references
1	*Azpeitia Africana*	South China Sea (SCS) 20°7′ N, 117°23′ E	Late Pleistocene period (15 kyr –till present)	Commonly found in warm-water regions, typical oceanic, planktonic species	Huang et al. (2009)	Hasle and Syvertsen (1997), Jousé et al. (1971); Muhina (1971)
2	*Azpeitia marina*	South China Sea (SCS) 20°7′ N, 117°23′ E	Late Pleistocene period (15 kyr –till present)	Commonly found in warm-water regions, typical oceanic, planktonic species	Huang et al. (2009)	Hasle and Syvertsen (1997), Jousé et al. (1971); Muhina (1971)
3	*Rhizosolenia bergonii*	South China Sea (SCS) 20°7′ N, 117°23′ E	Late Pleistocene period (15 kyr –till present)	Commonly found in warm-water regions, typical oceanic, planktonic species	Huang et al. (2009)	Hasle and Syvertsen (1997), Jousé et al. (1971); Muhina (1971)
4	*Caloneis westii*	West coast of India (South Asia) 18°54′N, 72°40′ E	Anthropocene/Holocene period (last 2000 yr)	Moderate to elevated levels of temperature, abundant in the postmonsoon period	D'Costa and Anil (2010)	
5	*Cyclotella caspia*	West coast of India (South Asia) 18°54′N, 72°40′ E	Anthropocene/Holocene period (last 2000 yr)	Moderate to elevated levels of temperature, abundant in the postmonsoon period	D'Costa and Anil (2010)	
6	*Cyclotella stylorum*	West coast of India (South Asia) 18°54′N, 72°40′ E	Anthropocene/Holocene period (last 2000 yr)	Elevated conditions of temperature, salinity, SPM, and low concentration of nutrients, abundant in premonsoon	D'Costa and Anil (2010)	

Continued

Table 13.14 Summary of marine diatom taxa indicating elevated temperature conditions during the Quaternary period.—cont'd

Sr. no	Diatom taxa	Site name	Geological time	Environmental facies inferred from diatom taxa	Study references	Additional references
7	Diploneis crabro	West coast of India (South Asia) 18°54'N, 72°40'E	Anthropocene/Holocene period (last 2000 yr)	Moderate to elevated levels of temperature, abundant in the postmonsoon period	D'Costa and Anil (2010)	
8	Melosira nummuloides	West coast of India (South Asia) 18°54'N, 72°40'E	Anthropocene/Holocene period (last 2000 yr)	Moderate to elevated levels of temperature, abundant in the postmonsoon period	D'Costa and Anil (2010)	
9	Navicula distans	West coast of India (South Asia) 18°54'N, 72°40'E	Anthropocene/Holocene period (last 2000 yr)	Moderate to elevated levels of temperature, abundant in the postmonsoon period	D'Costa and Anil (2010)	
10	Navicula elegans	West coast of India (South Asia) 18°54'N, 72°40'E	Anthropocene/Holocene period (last 2000 yr)	Elevated conditions of temperature, salinity, SPM, and low concentration of nutrients, abundant in premonsoon	D'Costa and Anil (2010)	
11	Navicula lyra	West coast of India (South Asia) 18°54'N, 72°40'E	Anthropocene/Holocene period (last 2000 yr)	Elevated conditions of temperature, salinity, SPM, and low concentration of nutrients, abundant in premonsoon	D'Costa and Anil (2010)	
12	Skeletonema costatum	West coast of India (South Asia) 18°54'N, 72°40'E	Anthropocene/Holocene period (last 2000 yr)	Prefers warmer temperatures, lower salinity, eutrophic, proliferate after higher precipitation	D'Costa and Anil (2010)	Liu et al. (2005); Mitbavkar and Anil (2000)
13	Thalassiosira spp.	West coast of India (South Asia) 18°54'N, 72°40'E	Anthropocene/Holocene period (last 2000 yr)	Prefers moderate to elevated levels of temperature, abundant in the postmonsoon period	D'Costa and Anil (2010)	

Table 13.15 Summary of marine diatom taxa indicating alkaline conditions during the Quaternary period.

Sr. no	Diatom taxa	Site name	Geological time	Environmental facies inferred from diatom taxa	Study references	Additional references
1	*Hippodonta linearis*	Balkhash Lake, Kazakhstan, (Central Asia) 46°17′–21.2″ N 74°00′–33.6″ E	Anthropocene/Holocene period (last 2000 yr)	Benthic, marine, alkaliphilous	Chiba et al. (2016)	Vos and de Wolf (1993); Van Dam et al. (1994); Watanabe et al. (2005)
2	*Campylodiscus clypeus*	Balkhash Lake, Kazakhstan, (Central Asia) 46°17′–21.2″ N 74°00′–33.6″ E	Anthropocene/Holocene period (last 2000 yr)	Benthic, marine, alkalibiontic	Chiba et al. (2016)	Vos and de Wolf (1993); Van Dam et al. (1994); Watanabe et al. (2005)
3	*Navicula digitoradiata*	Balkhash Lake, Kazakhstan, (Central Asia) 46°17′–21.2″ N 74°00′–33.6″ E	Anthropocene/Holocene period (last 2000 yr)	Benthic, marine, alkaliphilous	Chiba et al. (2016)	Vos and de Wolf (1993); Van Dam et al. (1994); Watanabe et al. (2005)
4	*Tryblionella* sp.	Balkhash Lake, Kazakhstan, (Central Asia) 46°17′–21.2″ N 74°00′–33.6″ E	Anthropocene/Holocene period (last 2000 yr)	Benthic, brackish, alkaliphilous	Chiba et al. (2016)	Vos and de Wolf (1993); Van Dam et al. (1994); Watanabe et al. (2005)
5	*Tryblionella levidensis*	Balkhash Lake, Kazakhstan, (Central Asia) 46°17′–21.2″ N 74°00′–33.6″ E	Anthropocene/Holocene period (last 2000 yr)	Benthic, brackish, alkaliphilous	Chiba et al. (2016)	Vos and de Wolf (1993); Van Dam et al. (1994); Watanabe et al. (2005)
6	*Cocconeis scutellum*	Lake Hamana, Central Japan (East Asia) 34°45.912′ N 137°34.995 E	Holocene period (last 1800 yr)	Brackish to marine (coastal area and lagoon), benthic on aquatic plant	Cho et al. (2019)	Joh (2012)

Continued

Table 13.15 Summary of marine diatom taxa indicating alkaline conditions during the Quaternary period.—cont'd

Sr. no	Diatom taxa	Site name	Geological time	Environmental facies inferred from diatom taxa	Study references	Additional references
Taxa showing unknown environmental preference						
7	*Lyrella lyroides*	Balkhash Lake, Kazakhstan, (Central Asia) 46°17′–21.2″ N 74°00′–33.6″ E	Anthropocene/ Holocene period (last 2000 yr)	Benthic, marine, unknown	Chiba et al. (2016)	Vos and de Wolf (1993); Van Dam et al. (1994); Watanabe et al. (2005)
8	*Tryblionella compressa*	Balkhash Lake, Kazakhstan, (Central Asia) 46°17′–21.2″ N 74°00′–33.6″ E	Anthropocene/ Holocene period (last 2000 yr)	Benthic, marine, unknown	Chiba et al. (2016)	Vos and de Wolf (1993); Van Dam et al. (1994); Watanabe et al. (2005)
9	*Tryblionella granulata*	Balkhash Lake, Kazakhstan, (Central Asia) 46°17′–21.2″ N 74°00′–33.6″ E	Anthropocene/ Holocene period (last 2000 yr)	Benthic, marine, unknown	Chiba et al. (2016)	Vos and de Wolf (1993); Van Dam et al. (1994); Watanabe et al. (2005)

4. Conclusions

Paleoclimate and paleo-monsoonal inference using diatoms are primarily based on the knowledge about the ecological preference and environmental occurrence of those particular diatom taxa derived from the modern diatom community studies. The present compilation includes almost 270 entries with their environmental preference, archived mainly from the monsoon-dominated tropical part of Asia. Despite this vast diversity in the tropics, the environmental choice for many taxa is unknown, which limits their use in paleolimnological reconstruction. Except few, most of the taxa recorded here are cosmopolitan ones; their natural environmental preference in the tropics is unknown; as in most current cases, the environmental preference data comes from other parts of the world. This calls for more studies on the modern diatoms from the tropical region. Even though we have documentation of diatoms from the tropical Indian Ocean and adjoining regions (Desikachary, 1988, 1989), it is only a taxonomic listing with the diatom images, and there is no information on the environmental preference of the taxa listed, whereas paleoclimate and paleo-monsoonal inference using diatoms are primarily based on the knowledge about the ecological importance and environment and habitat-specific occurrence of that particular diatom taxa. So, despite this colossal listing, this cannot be used immediately for paleoecological reconstruction as it requires further confirmation from the modern analogs. Thus, forthcoming studies need to compile information on diatoms' taxonomy and ecology. In the present compilation, most of the entries provide information on the pH-related data, such as preference for acidic and alkaline diatoms in both fresh and marine water. More information from these taxa can add value to the already available ocean acidification data. Although diatoms are known to be sensitive to multiple parameters such as oxygen saturation, nutrient concentration, ionic concentration, water temperature and turbulence, current velocity, and salinity, most present studies essentially list only pH or nutrient-related information, which needs to be remedied in the future with modern diatom community structure along with the environmental information.

Moreover, many studies have emerged here in the tropics in the recent period, which has led to the rise of endemic biodiversity hotspots in regions of "South East Asia". In this case, to catch the signal from the endemic taxa found here, it is a dire need to establish a local or regional database exclusive to tropical countries/parts of the world. Taking this into account, we firmly believe that this catalogue will serve as a starting point for applying diatom-based reconstruction, especially from the Asian tropics and more information needs to be added in the subsequent versions.

Acknowledgments

This research was funded by the Ministry of Earth Sciences (MoES), Government of India, under the paleoclimate program (MoES/CCR/Paleo26/2015). The authors thank the Director Agharkar Research Institute for the support and encouragement.

References

Achyuthan, H., Farooqui, A., Gopal, V., Phartiyal, B., Lone, A., 2016. Late quaternary to Holocene southwest monsoon reconstruction: a review based on lake and wetland systems (studies carried out during 2011-2016). Proc. Indian National. Sci. Acad. 82 (3), 847–868. https://doi.org/10.16943/ptinsa/2016/48489.

Ahmad, S.M., Padmakumari, V.M., Raza, W., Venkatesham, K., Suseela, G., Sagar, N., Chamoli, A., Rajan, R.S., 2011. High-resolution carbon and oxygen isotope records from a scleractinian (Porites) coral of Lakshadweep Archipelago. Quat. Int. 238 (1–2), 107–114. https://doi.org/10.1016/j.quaint.2009.11.020.

An, Z., Kukla, G.J., Porter, S.C., Xiao, J., 1991. Magnetic susceptibility evidence of monsoon variation on the Loess Plateau of central China during the last 130,000 years. Quat. Res. 36 (1), 29–36. https://doi.org/10.1016/0033-5894(91)90015-W.

An, Z., Guoxiong, W., Jianping, L., Youbin, S., Yimin, L., Weijian, Z., Yanjun, C., Anmin, D., Li, L., Jiangyu, M., Hai, C., Zhengguo, S., Liangcheng, T., Hong, Y., Hong, A., Hong, C., Juan, F., 2015. Global monsoon dynamics and climate change. Annu. Rev. Earth Planet Sci. 43, 29–77. https://doi.org/10.1146/annurev-earth-060313-054623.

Anderson, N.J., 1990. The biostratigraphy and taxonomy of small *Stephanodiscus* and *Cyclostephanos* species (Bacillariophyceae) in a eutrophic lake and their ecological implications. Br. Phycol. J. 25, 217–235. https://doi.org/10.1080/00071619000650211.

Anderson, N.J., Rippey, B., Gibson, C.E., 1993. A comparison of sedimentary and diatom-inferred phosphorus profiles: implications for defining pre-disturbance nutrient conditions. In: Boers, P.C.M., Cappenberg, T.E., van Raaphorst, W. (Eds.), Proceedings of the Third International Workshop on Phosphorus in Sediments. Developments in Hydrobiology, vol. 84. Springer, Dordrecht. https://doi.org/10.1007/978-94-011-1598-8_44.

Anderson, N.J., 1997. Reconstructing historical phosphorus concentrations in rural lakes using diatom models. In: Tunney, H., Carton, O.T., Brookes, P.C., Johnston, A.E. (Eds.), Phosphorus Loss from Soil to Water. CAB International, Oxford, pp. 95–118.

Anderson, N.J., 2000. Diatoms, temperature and climatic change. Eur. J. Phycol. 35, 307–314.

Bahls, L., 2010. Stauroneis in the Northern Rockies 50 species of *Stauroneis* sensu stricto from western Montana, northern Idaho, northeastern Washington and southwestern Alberta, including 16 species described as new. Vol. 4. In: Bahls, L. (Ed.), Northwest Diatoms, Montana Diatom Collection, p. 179. Helena.

Band, S., Yadava, M.G., Lone, M.A., Shen, C.C., Sree, K., Ramesh, R., 2018. High-resolution mid-Holocene Indian Summer Monsoon recorded in a stalagmite from the Kotumsar cave, Central India. Quat. Int. 479, 19–24. https://doi.org/10.1016/j.quaint.2018.01.026.

Battarbee, R.W., Mason, J., Renberg, I., Talling, J.F., 1990. Paleolimnology and Lake Acidification. The Royal Society, London, p. 219.

Battarbee, R.W., 2000. Palaeolimnological approaches to climate change, with special regard to the biological record. Quat. Sci. Rev. 19, 124–197. https://doi.org/10.1016/S0277-3791(99)00057-8.

Battarbee, R.W., Jones, V.J., Flower, R.J., Cameron, N.J., Bennion, H., Carvalho, L., Juggins, S., 2001. Diatoms. In: Smol, J.P., Birks, H.J.B., Last, W.M. (Eds.), Tracking Environmental Change Using Lake Sediments, Terrestrial, Algal, and Siliceous Indicators, vol. 3. Kluwer Academic Publisher, Dordrecht, pp. 155–202.

Battarbee, R.W., Thompson, R., Catalan, J., Grytnes, J.A., Birks, H.J.B., 2002. Climate variability and ecosystem dynamics of remote alpine and arctic lakes: the MOLAR project. J. Paleolimnol. 28, 1–6. https://doi.org/10.1023/A:1020342316326.

Battarbee, R.W., Donald, F., Charles, C.B., Cumming, B.F., Renberg, I., 2010. Diatoms as indicators of surface-water acidity. In: second ed.Smol, J.P., Stoermer, E.F. (Eds.), The Diatoms: Applications for the Environmental and Earth Sciences. Cambridge University Press, Cambridge, pp. 98–121.

Beaver, J., 1981. Apparent Ecological Characteristics of Some Common Freshwater Diatoms. Ministry of the Environment, Plankton Taxonomy Unit, Limnology and Taxonomy Section.

Bennion, H., 1994. A diatom-phosphorus transfer function for shallow, eutrophic ponds in southeast England. Hydrobiologia 275 (276), 391–410. https://doi.org/10.1007/978-94-017-2460-9_35.

Bigler, C., Hall, R.I., 2002. Diatoms as indicators of climatic and limnological change in Swedish Lapland: a 100-lake calibration set and its validation for paleoecological reconstructions. J. Paleolimnol. 27, 97–115. https://doi.org/10.1023/A:1013562325326.

Bigler, C., Hall, R.I., 2003. Diatoms as quantitative indicators of July temperature: a validation attempt at century-scale with meteorological data from northern Sweden. Palaeogeogr. Palaeoclimatol. Palaeoecol. 189, 147–160. https://doi.org/10.1016/S0031-0182(02)00638-7.

Birks, H.J.B., ter Braak, C.J.F., Line, J.M., Juggins, S., Stevenson, A.C., 1990. Diatoms and pH reconstruction. Philos. Trans. R. Soc. Lond. B Biol. Sci. 327 (1240), 263–278. https://doi.org/10.1098/rstb.1990.0062.

Bracht, B.B., Stone, J.R., Fritz, S.C., 2008. A diatom record of late Holocene climate variation in the northern range of Yellowstone National Park, USA. Quat. Int. 188, 149–155. https://doi.org/10.1016/j.quaint.2007.08.043.

Bradbury, J.P., 1975. Diatom Stratigraphy and Human Settlement in Minnesota, vol. 172. Geological Society of America, pp. 1–74.

Bradbury, J.P., 1989. Late quaternary lacustrine paleoenvironments in the Cuenca de Mexico. Quat. Sci. Rev. 8, 75–100. https://doi.org/10.1016/0277-3791(89)90022-X.

Bradbury, J.P., Diederich-Rurup, S.K., 1993. Holocene diatom paleolimnology of Elk lake, Minnesota. In: Bradbury, J.P., Dean, W.E. (Eds.), Elk Lake, Minnesota: Evidence for Rapid Climate Change in the North-Central United States: Boulder, Colorado, vol. 276. Geological Society of America Special Paper, pp. 215–238. https://doi.org/10.1130/SPE276-p215.

Bradbury, J.P., Bezrukova, Y.V., Chernyaeva, G.P., Colman, S.M., Khursevich, G., King, J.W., Likoshway, Y.V., 1994. A synthesis of post-glacial diatom records from Lake Baikal. J. Paleolimnol. 10, 213–252. https://doi.org/10.1007/BF00684034.

Bradshaw, E.G., Anderson, N.J., 2003. Environmental factors that control the abundance of *Cyclostephanos dubius* (Bacillariophyceae) in Danish lakes, from seasonal to century scale. Eur. J. Phycol. 38 (3), 265–276. https://doi.org/10.1080/0967026031000136349.

Brugam, R.B., McKeever, K., Kolesa, L., 1998. A diatom-inferred water depth reconstruction for an Upper Peninsula, Michigan. Lake J. Paleolimnol. 20, 267–276. https://doi.org/10.1023/A:1007948616511.

Brugam, R.B., Swain, P., 2000. Diatom indicators of peatland development at pogonia bog pond, Minnesota, USA. Holocene 10, 453–464.

Buczkó, K., Ognjanova-Rumenova, N., Magyari, E., 2010. Taxonomy, morphology and distribution of some *Aulacoseira* taxa in glacial lakes in the south Carpathian region. Pol. Bot. J. 55 (1), 149–163.

Cameron, E.M., Prévost, C.L., McCurdy, M., Hall, G.E.M., Doidge, B., 1998. Recent (1930s) natural acidification and fish kill in a lake that was an important food source for an Inuit community in northern Quebec, Canada. J. Geochem. Explor. 64, 197–213. https://doi.org/10.1016/S0375-6742(98)00033-8.

Cantonati, M., Lange-Bertalot, H., Decet, F., Gabrieli, J., 2011. Diatoms in very-shallow pools of the site of community importance, Danta di Cadore Mires (south-eastern Alps), and the potential contribution of these habitats to diatom biodiversity conservation. Nova Hedwigia 93 (3), 475.

Cao, P., Shi, X., Li, W., Liu, S., Yao, Z., Hu, L., Khokiattiwong, S., Kornkanitnan, N., 2015. Sedimentary responses to the Indian Summer Monsoon variations recorded in the southeastern Andaman Sea slope since 26 ka. J. Asian Earth Sci. 114, 512–525. https://doi.org/10.1016/j.jseaes.2015.06.028.

Cattaneo, A., Couillard, Y., Wunsam, S., Courcelles, M., 2004. Diatom taxonomic and morphological changes as indicators of metal pollution and recovery in Lac Dufault (Québec, Canada). J. Paleolimnol. 32, 163–175. https://doi.org/10.1023/B:JOPL.0000029430.78278.a5.

Chen, S.H., Wu, J.T., 1999. Paleolimnological environment indicated by the diatom and pollen assemblages in an alpine lake in Taiwan. J. Paleolimnol. 22, 149–215. https://doi.org/10.1023/A:1008067928365.

Chen, X., Bu, Z., Yang, X., Wang, S., 2012. Epiphytic diatoms and their relation to moisture and moss composition in two montane mires, Northeast China. Fundam. Appl. Limnol. Archiv fur Hydrobiologie 181 (3), 197.

Chen, X., Li, Y., Metcalfe, S., Xiao, X., Yang, X., Zhang, E., 2014a. Diatom response to Asian monsoon variability during the Late Glacial to Holocene in a small treeline lake, SW China. Holocene 24 (10), 1369–1377.

Chen, X., Qin, Y., Stevenson, M.A., McGowan, S., 2014b. Diatom communities along pH and hydrological gradients in three montane mires, central China. Ecol. Indicat. 45, 123–129. https://doi.org/10.1016/j.ecolind.2014.04.016.

Chevalier, M., Davis, B.A., Heiri, O, Seppä, H., Chase, B.M., Gajewski, K., Lacourse, T., Telford, R.J., Finsinger, W., Guiot, J., Kühl, N., Maezumi, S.Y., Tipton, J.R., Carter, V.A., Brussel, T., Phelps, L.N., Dawson, A., Zanon, M., Vallé, F., Nolan, C., Mauri, A., Vernal, A., Izumi, K., Holmström, L., Marsicek, J., Goring, S., Sommer, P.S., Michelle, C., Kupriyanov, D., 2020. Pollen-based climate reconstruction techniques for late Quaternary studies. Earth Sci. Rev. 210, 103384. https://doi.org/10.1016/j.earscirev.2020.103384.

Chiba, T., Endo, K., Sugai, T., Haraguchi, T., Kondo, R., Kubota, J., 2016. Reconstruction of Lake Balkhash levels and precipitation/evaporation changes during the last 2000 years from fossil diatom assemblages. Quat. Int. 397, 330–341. https://doi.org/10.1016/j.quaint.2015.08.009.

Cho, A., Cheong, D., Kim, J.C., Yang, D.Y., Lee, J.Y., Kashima, K., Katsuki, K., 2018. Holocene climate and environmental changes inferred from sediment characteristics and diatom assemblages in a core from Hwa-jinpo Lagoon, Korea. J. Paleolimnol. 60 (4), 553–570. https://doi.org/10.1007/s10933-018-0040-1.

Cho, A., Kashima, K., Seto, K., Yamada, K., Sato, T., Katsuki, K., 2019. Climate change during the little ice age from the lake hamana sediment record. Estuar. Coast Shelf Sci. 223, 39–49. https://doi.org/10.1016/j.ecss.2019.04.033.

Cholnoky, F.I., 1968. The Ecology of Diatoms in Inland Water. J. Cramer, Hirschberg, Germany, p. 699.

Chung, Y.H., Noh, K.H., 1987. The diatom flora of lowland swamp in Haman County, Korea. Proc. Coll. Nat. Sci. Seoul. Natl. Univ. 12, 75–100.

Colombaroli, D., Marchetto, A., Tinner, W., 2007. Long-term interactions between Mediterranean climate, vegetation and fire regime at Lago di Massaciuccoli (Tuscany, Italy). J. Ecol. 95, 755–770. https://doi.org/10.1111/j.1365-2745.2007.01240.x.

Cook, E.R., Anchukaitis, K.J., Buckley, B.M., D'Arrigo, R.D., Jacoby, G.C., Wright, W.E., 2010. Asian monsoon failure and megadrought during the last millennium. Science 328, 486–489. https://doi.org/10.1126/science.1185188.

Cumming, B.F., Davey, K.A., Smol, J.P., Birks, H.J.B., 1994. When did acid-sensitive Adirondack lakes (New York, U.S.A.) begin to acidify and are they still acidifying? Can. J. Fish. Aquat. Sci. 51, 1550–1568. https://doi.org/10.1139/f94-154.

D'Costa, P.M., Anil, A.C., 2010. Diatom community dynamics in a tropical, monsoon-influenced environment: West coast of India. Continent. Shelf Res. 30 (12), 1324–1337. https://doi.org/10.1016/j.csr.2010.04.015.

DeNicola, D.M., 2000. A review of diatoms found in highly acidic environments. Hydrobiologia 433, 111–122. https://doi.org/10.1023/A:1004066620172.

Desikachary, T.V., 1988. Marine diatoms of the Indian Ocean region. In: Desikachary, T.V. (Ed.), Atlas of Diatoms. Fasc. V. MadrasScience Foundation, Madras, pls, pp. 401–621.

Desikachary, T.V., 1989. Marine diatoms of the Indian Ocean region. In: Desikachary, T.V. (Ed.), Atlas of Diatoms. Madras Science Foundation, Madras [fasc. VI], pp. 1–27 pls 622–809.

Denys, L., 1992. A checklist of the diatoms in the Holocene deposits of the western Belgian coastal plain with a survey of their apparent ecological requirements. I. Introduction, ecological code and complete list. Belsgische. Geolgische. Dienst. 246, 1–41 (Professional Paper).

Dixit, S.S., Dixit, A.S., Evans, R.D., 1988. Sedimentary diatom assemblages and their utility in computing diatom-inferred pH in Sudbury Ontario lakes. Hydrobiologia 169, 135–148. https://doi.org/10.1007/BF00007306.

Dixit, S.S., 1992. Diatoms: Powerful indicators of environmental change. Environ. Sci. Technol. 26 (1), 22–33. https://doi.org/10.1021/es00025a002.

Dixit, S.S., Cumming, B.E., Birks, H.J.B., Stool, S.M., Kingston, J.C., Uutala, A.J., Charles, D.E., Cambrun, K.E., 1993. Diatom assemblages from Adirondack lakes (New York, USA) and the development of inference models for retrospective environmental assessment. J. Paleolimnol. 8, 27—47. https://doi.org/10.1007/BF00210056.

Douglas, M.S.V., Smol, J.P., 1999. Freshwater diatoms as indicators of environmental change in the High Arctic. In: Stoermer, E.F., Smol, J.P. (Eds.), The Diatoms: Applications for the Environmental and Earth Sciences. Cambridge Univ. Press, New York, pp. 227—244.

Edlund, M.B., Stoermer, E.F., 2000. A 200,000-year, high-resolution record of diatom productivity and community makeup from Lake Baikal shows high correspondence to the marine oxygen-isotope record of climate change. Limnol. Oceanogr. 45, 948—962. https://doi.org/10.4319/lo.2000.45.4.0948.

Escobar, J., Serna, Y., Hoyos, N., Velez, M.I., Correa-Metrio, A., 2020. Why we need more paleolimnology studies in the tropics? J. Paleolimnol. 64 (1), 47—53. https://doi.org/10.1007/s10933-020-00120-6.

Fahnenstiel, G.L., Glime, J., 1983. Subsurface chlorophyll maximum and associated Cyclotella Pulse in Lake Superior. Int. Rev. Gesamt. Hydrobiol. Hydrogra. 68 (5), 605—616. https://doi.org/10.1002/iroh.3510680502.

Fishman, D.B., Adlerstein, S.A., Vanderploeg, H.A., Fahnenstiel, G.L., Scavia, D., 2010. Phytoplankton community composition of Saginaw Bay, Lake Huron, during the zebra mussel (*Dreissena polymorpha*) invasion: a multivariate analysis. J. Great Lake. Res. 36, 9—19. https://doi.org/10.1016/j.jglr.2009.10.004.

Fleitmann, D., Burns, S.J., Mangini, A., Mudelsee, M., Kramers, J., Villa, I., Neff, U., Al-Subbary, A.A., Buettner, A., Hippler, D., Matter, A., 2007. Holocene ITCZ and Indian monsoon dynamics recorded in stalagmites from Oman and Yemen (Socotra). Quat. Sci. Rev. 26 (1—2), 170—188. https://doi.org/10.1016/j.quascirev.2006.04.012.

Flower, R.J., Juggins, S., Battarbee, R.W., 1997. Matching diatom assemblages in lake sediment cores and modern surface sediment samples: the implications for conservation and restoration with special reference to acidified systems. Hydrobiologia 344, 27—40. https://doi.org/10.1023/A:1002941908602.

Foged, N., 1968. The freshwater diatom flora of the Varanger peninsula, North Norway. Acta boreal A. Sci. 25, 1—64.

Foged, N., 1980. Diatoms in Oland, Sweden. Bibliotheca Phycologia, vol. 49. Cramer, Vaduz, pp. 1—193.

Fritz, S.C., Juggins, S., Battarbee, R.W., 1993. Diatom assemblages and ionic characterization of lakes of the northern Great Plains, North America: a tool for reconstructing past salinity and climate fluctuations. Can. J. Fish. Aquat. Sci. 50, 1844—1856. https://doi.org/10.1139/f93-207.

Gaiser, E.E., Johansen, J., 2000. Freshwater diatoms from Carolina Bays and other isolated wetlands on the Atlantic coastal plain of South Carolina, USA, with descriptions of seven taxa new to science. Diatom Res. 15 (1), 75—130. https://doi.org/10.1080/0269249X.2000.9705487.

Gasse, F., Tekaia, F., 1983. Transfer functions for estimating paleoecological conditions (pH) from East African diatoms. Hydrobiologia 103, 85—90. https://doi.org/10.1007/978-94-009-7290-2_14.

Genkal, S.I., Kiss, K.T., 1993. Morphological Variability of the Diatom *Cyclotella atomus* Hustedt Var. *Atomus* and *C. Atomus* Var. *Gracilis* Var. Nov. Hydrobiologia 269—270, pp. 39—47. https://doi.org/10.1007/BF00028002.

Gibson, C.E., Anderson, N.J., Haworth, E.Y., 2003. *Aulacoseira subarctica*: Taxonomy, physiology, ecology and palaeoecology. Eur. J. Phycol. 38, 83—101. https://doi.org/10.1080/0967026031000094102.

Ginn, B.K., Cumming, B.F., Smol, J.P., 2007. Diatom-based environmental inferences and model comparisons from 494 northeastern North American lakes. J. Phycol. 43, 647—661. https://doi.org/10.1111/j.1529-8817.2007.00363.x.

Govil, P., Naidu, P.D., 2011. Variations of Indian monsoon precipitation during the last 32 kyr are reflected in the surface hydrography of the Western Bay of Bengal. Quat. Sci. Rev. 30 (27—28), 3871—3879. https://doi.org/10.1016/j.quascirev.2011.10.004.

Gregory, J.M., Oerlemans, J., 1998. Simulated future sea-level rise due to glacier melt bastes on regionally and seasonally resolved temperature changes. Nature 391, 474—476. https://doi.org/10.1038/35119.

Håkansson, H., Regnéll, J., 1993. Diatom succession related to land use during the last 6000 years: a study of a small eutrophic lake in southern Sweden. J. Paleolimnol. 8, 49—69. https://doi.org/10.1007/BF00210057.

Hall, R.I., Smol, J.P., 2010. Diatoms as indicators of lake eutrophication. In: Stoermer, E.F., Smol, J.P. (Eds.), The Diatoms: Applications for the Environmental and Earth Sciences, second ed. Cambridge University Press, Cambridge, pp. 122–153.

Hargan, K.E., Rühland, K.M., Paterson, A.M., Finkelstein, S.A., Holmquist, J.R., MacDonald, G.M., Keller, W., Smol, J.P., 2015. The influence of water-table depth and pH on the spatial distribution of diatom species in peatlands of the Boreal Shield and Hudson Plains, Canada. Botany 93, 57–74. https://doi.org/10.1139/cjb-2014-0138.

Hasle, G.R., Syvertsen, E.E., 1997. Marine diatoms. In: Tomas, R. (Ed.), Identifying Marine Phytoplankton. Academic Press, California, pp. 5–85.

Haworth, E.Y., 1976. Two late-glacial (Late Devensian) diatom assemblage profiles from Northern Scotland. New Phytol. 77, 227–256.

Haworth, E.Y., Hurley, M.A., 1984. Comparison of the stelligeroid taxa of the centric diatom genus Cyclotella. In: Ricard, M. (Ed.), Proceedings of the 8th International Diatom Symposium — Paris. Koeltz Scientific Books, Koenigstein, pp. 43–58.

Haworth, E.Y., Offer, E.H., 1986. An amended description of the diatom *Navicula menda* Carter. Br. Phycol. J. 21, 445–447. https://doi.org/10.1080/00071618600650511.

Hendey, N.I., 1964. An Introductory Account of the Smaller Algae of British Coastal Waters: Part V. Bacillariophyceae (Diatoms), vol. 4. Fishery Investigations, London, p. 317. Series.

Houk, V., 2003. Atlas of freshwater-centric diatoms with a brief key and descriptions. Part I. Melosiraceae, Orthoseiraceae, Paraliaceae and Aulacoseiraceae. Czech Phycol. Suppl. 1, 1–27.

Houk, V., Klee, R., Tanaka, H., 2010. Atlas of freshwater-centric diatoms with a brief key and descriptions. Part III: stephanodiscaceae A. Cyclotella, Tertiarius, Discostella. Fottea 10, 1–498.

Huang, Y., Jiang, H., Sarnthein, M., Knudsen, K.L., Li, D., 2009. Diatom response to changes in palaeoenvironments of the northern South China Sea during the last 15000 years. Mar. Micropaleontol. 72 (1–2), 109. https://doi.org/10.1016/j.marmicro.2009.04.003.

Hustedt, F., 1930. Bacillariophyta. (Diatomeae). In: Pascher, A. (Ed.), Die Süsswasser- Flora Mitteleuropas. Gustar Gustav Fisher, Jena, p. 466.

Hustedt, F., 1942. Süsswasser-Diatomeen des Indomalayischen Archipels und der Hawaii-Inseln. Int. Rev. Ges. Hydrobiol. Hydrogr. 42, 1–252. https://doi.org/10.1002/iroh.19420420102.

Hwang, S., Kim, J., Yoon, S.-O., 2014. Environmental changes and embankment addition of Reservoir Gonggeomji, Sangju City between Late Silla- and Early Goryeo dynasty. J. Kor. Geomorphol. Assoc. 21, 165–180.

Joh, G.J., 2010. Algal Flora of Korea Volume 3 Number1 Freshwater Diatoms I. National Institute of Biological Resource, Incheon, p. 152.

Joh, G., 2011. Algal Flora of Korea. Volume 3, Number 3. Chrysophyta: Bacillariophyceae: Pennales: Raphidineae: Eunotiaceae Freshwater Diatoms III. National Institute of Biological Resources, Ministry of Environment, pp. 1–92.

Joh, G.J., 2012. Algal Flora of Korea Volume3 Number7 Freshwater Diatoms V. National Institute of Biological Resource, Incheon.

Johansson, C., 1982. Attached Algal Vegetation in the Running Waters of Jiimtland, Sweden. Acta Phytogeographica Suecica, Uppsala, p. 71.

Jones, V., Birks, H.J.B., 2004. Lake-sediment records of recent environmental change on svalbard: results of diatom analysis. J. Paleolimnol. 31, 445–466. https://doi.org/10.1023/B:JOPL.0000022544.35526.11.

Jousé, A.P., Kozlova, O.G., Muhina, V.V., 1971. Distribution of diatoms in the surface layer of sediment from the Pacific Ocean. In: Funnell, B.M., Riedel, W.R. (Eds.), The Micropalaeontology of Oceans. Cambridge University Press, London, pp. 263–269.

Julius, M.L., Theriot, E.C., 2010. The diatoms: a primer. In: Smol, J.P., Stoermer Juyal, N., Pant, R.K., Basavaiah, N., Bhushan, R., Jain, M., Saini, N.K., Yadava, M.G., Singhvi, A.K. (Eds.), Reconstruction of Last

Glacial to Early Holocene Monsoon Variability from Relict Lake Sediments of the Higher Central Himalaya, Uttrakhand, India Journal of Asian Earth Science, vol. 34, pp. 437−449. https://doi.org/10.1016/j.jseaes.2008.07.007.

Juyal, N., Pant, R.K., Basavaiah, N., Bhushan, R., Jain, M., Saini, N.K., Yadava, M.G., Singhvi, A.K., 2009. Reconstruction of Last Glacial to early Holocene monsoon variability from relict lake sediments of the Higher Central Himalaya, Uttrakhand, India. J. Asian Earth Sci. 34 (3), 437−449. https://doi.org/10.1016/j.jseaes.2008.07.007.

Kagan, L.,Y., 2001. Human-induced changes in the diatom communities of lake Imandra. Water Resour. 28 (3), 297−306. https://doi.org/10.1023/A:1010452824498.

Kelly, M.G., Whitton, B.A., 1995. The Trophic Diatom Index: A New Index for Monitoring.

Kernan, M., Ventura, M., Bitusik, P., Brancel, A., Clarke, G., Velle, G., Raddum, G., Stuchlik, E., Catalan, J., 2009. Regionalisation of remote European mountain lake ecosystems according to their biota: environmental vs. geographical patterns. Freshw. Biol. 54, 2470−2493. https://doi.org/10.1111/j.1365-2427.2009.02284.x.

Kienel, U., Sigert, C., Hahne, J., 1999. Late quaternary palaeoenvironmental reconstructions from a permafrost sequence (North Siberian Lowland, SE Taymyr Peninsula) − a multidisciplinary case study. Boreas 28, 181−193. https://doi.org/10.1111/j.1502-3885.1999.tb00213.x.

Kilham, S.S., Kilham, P., 1975. *Melosira granulata* (Ehr.) Ralfs: Morphology and ecology of a cosmopolitan freshwater diatom, Verh. Int. Ver. Theor. Angew. Limnol. 19, 2716−2721. https://doi.org/10.1080/03680770.1974.11896368.

Kilham, P., Kilham, S.S., Hecky, R.E., 1986. Hypothesized resource relationships among African planktonic diatoms. Limnol. Oceanogr. 31, 1169−1181. https://doi.org/10.4319/lo.1986.31.6.1169.

Kilham, P., 1990. Ecology of Melosira species in the Great Lakes of Africa. In: Tilzer, M.M., Serruya, C. (Eds.), Large Lakes Ecological Structure and Function. Springer, Berlin, pp. 414−427. https://doi.org/10.1007/978-3-642-84077-7_20.

Kilham, S.S., Theriot, E.C., Fritz, S.C., 1996. Linking planktonic diatoms and climate change in the large lakes of the Yellowstone ecosystem using resource theory. Limnol. Oceanogr. 41 (5), 1052−1062. https://doi.org/10.4319/lo.1996.41.5.1052.

Kim, Y.S., Choi, T.S., Kim, H.S., 2007. Epiphytic diatom communities from two mountain bogs in South Korea. Nova Hedwigia 84, 363−379. https://doi.org/10.1127/0029-5035/2007/0084-0363.

Kirilova, E.P., Cremer, H., Heiri, O., Lotter, A.F., 2010. Eutrophication of moderately deep dutch lakes during the past century: flaws in the expectations of water management? Hydrobiologia 637, 157−171. https://doi.org/10.1007/s10750-009-9993-4.

Kiss, K.T., Acs, E., Szabo, K.E., et al., 2007. Morphological observations on *Cyclotella distinguenda* Hustedt and *C. delicatula* Hustedt from the core sample of a meromictic karstic lake of Spain (Lake La Cruz) with aspects of their ecology. Diatom Res. 22, 287−308. https://doi.org/10.1080/0269249X.2007.9705716.

Kock, A., Taylor, J.C., Malherbe, W., 2019. Diatom community structure and relationship with water quality in Lake Sibaya, KwaZulu-Natal, South Africa. South Afr. J. Bot. 123, 161−169. https://doi.org/10.1016/j.sajb.2019.03.013.

Korsman, T., Birks, H.J.B., 1996. Diatom-based water chemistry reconstructions from northern Sweden: a comparison of reconstruction techniques. J. Paleolimnol. 15, 65−77. https://doi.org/10.1007/BF00176990.

Köster, D., Pienitz, R., 2006. Seasonal diatom variability and paleolimnological inferences −a case study. J. Paleolimnol. 35, 395−416. https://doi.org/10.1007/s10933-005-1334-7.

Kotlia, B.S., Singh, A.K., Joshi, L.M., Dhaila, B.S., 2015. Precipitation variability in the Indian Central Himalaya during last ca. 4,000 years inferred from a speleothem record: Impact of Indian Summer Monsoon (ISM) and Westerlies. Quat. Int. 371, 244−253. https://doi.org/10.1016/j.quaint.2014.10.066.

Kovács, C., Kahlert, M., Padisák, J., 2006. Benthic diatom communities along pH and TP gradients in Hungarian and Swedish streams. J. Appl. Phycol. 18, 105−117. https://doi.org/10.1007/s10811-006-9080-4.

Krammer, K., Lange-Bertalot, H., 1986. Bacillariophyceae 1. Teil: Naviculaceae. In: Ettl, H., Gärtner, G., Gerloff, J., Heynig, H., Mollenhauer, D. (Eds.), Süsswasserflora von Mitteleuropa 2/1. Gustav Fischer Verlag, Jena, Germany, pp. 1−876.

Krammer, K., Lange-Bertalot, H., 1988. Bacillariophyceae 2. Teil: Bacillariaceae, Epithemiaceae, Surirellaceae. In: Ettl, H., Gärtner, G., Gerloff, J., Heynig, H., Mollenhauer, D. (Eds.), Süsswasserflora von Mitteleuropa 2/2. Gustav Fischer Verlag, Jena, Germany, pp. 1−596.

Krammer, K., Lange-Bertalot, H., 1991a. Bacillariophyceae 3. Teil: Centrales, Fragilariaceae, Eunotiaceae. In: Ettl, H., Gärtner, G., Gerloff, J., Heynig, H., Mollenhauer, D. (Eds.), Süsswasserflora von Mitteleuropa 2/3. Gustav Fischer Verlag, Jena, Germany, pp. 1−576.

Krammer, K., Lange-Bertalot, H., 1991b. Bacillariophyceae 4. Teil: Achnanthaceae, Kritische Ergänzungen zu Navicula (Lineolate) und Gomphonema. In: Ettl, H., Gärtner, G., Gerloff, J., Heynig, H., Mollenhauer, D. (Eds.), Süsswasserflora von Mitteleuropa 2/4. Gustav Fischer Verlag, Jena, Germany, pp. 1−437.

Krammer, K., Lange-Bertalot, H., 1999. Suesswasserflora von Mitteleuropa, vols. 1−4. Spektrum Akademischer Verlag Heidelberg, Berlin.

Krammer, K., Lange-Bertalot, H., 2010. Süßwasserflora von Mitteleuropa, 2/1. Naviculaceae Spektrum Akademischer Verlag, Heidelberg.

Krasske, G., 1949. Zur diatomeenflora lapplands. 11. Ann bot. Societas Zoologica-botanica Fennica. Vanamo 23 (5), 1−30.

Krstić, S.S., Zech, W., Obreht, I., Sirčev, Z., Marković, S.B., 2012. Late Quaternary environmental changes in Helambu Himal, Central Nepal, recorded in the diatom flora assemblage composition and geochemistry of Lake PanchPokhari. J. Paleolimnol. 47 (1), 113−124. https://doi.org/10.1007/s10933-011-9563-4.

Laing, T.E., Rülhand, K.M., Smol, J.P., 1999. Past environmental and climatic changes related to tree-line shift inferred from fossil diatoms from a lake near the Lena River Delta, Siberia. Holocene 9 (5), 547−557.

Laing, T.E., Smol, J.P., 2000. Factors influencing diatom distributions in circumpolar treeline lakes of Northern Russia. J. Phycol. 36 (6), 1035−1048. https://doi.org/10.1046/j.1529-8817.2000.99229.x.

Leblanc, K., Arístegui, J., Armand, L., Assmy, P., Beker, B., Bode, A., Breton, E., Cornet, V., Gibson, J., Gosselin, M.-P., Kopczynska, E., Marshall, H., Peloquin, J., Piontkovski, S., Poulton, A.J., Quéguiner, B., Schiebel, R., Shipe, R., Stefels, J., van Leeuwe, M.A., Varela, M., Widdicombe, C., Yallop, M., 2012. A global diatom database − abundance, biovolume and biomass in the world ocean. Earth Syst. Sci. Data 4, 149−165. https://doi.org/10.5194/essd-4-149-2012.

Lee, H., Yun, S.M., Lee, J.Y., Lee, S.D., Lim, J., Cho, P.Y., 2018. Late Holocene climate changes from diatom records in the historical Reservoir Gonggeomji, Korea. J. Appl. Phycol. 30 (6), 3205−3219. https://doi.org/10.1007/s10811-018-1548-5.

Leira, M., Cole, E.E., Mitchell, F.J.G., 2007. Peat erosion and atmospheric deposition impact an oligotrophic lake in eastern Ireland. J. Paleolimnol. 38, 49−71. https://doi.org/10.1007/s10933-006-9060-3.

Li, Y.L., Gong, Z.J., Shen, J., 2012. Effects of eutrophication and temperature on *Cyclotella rhomboideo-elliptica* Skuja. Endem. Diat. China Phycol. Res. 60, 288−296. https://doi.org/10.1111/j.1440-1835.2012.00659.x.

Li, Y., Rioual, P., Shen, J., Xiao, X., 2015a. Diatom response to climatic and tectonic forcing of a palaeolake at the southeastern margin of the Tibetan Plateau during the late Pleistocene, between 140 and 35 ka BP. Palaeogeogr. Palaeoclimatol. Palaeoecol. 436, 123−134. https://doi.org/10.1016/j.palaeo.2015.06.039.

Li, Y., Liu, E., Xiao, X., Zhang, E., Ji, M., 2015b. Diatom response to Asian monsoon variability during the Holocene in a deep lake at the southeastern margin of the Tibetan Plateau. Boreas 44 (4), 785−793. https://doi.org/10.1111/bor.12128.

Li, D., Jiang, H., Knudsen, K.L., Björck, S., Olsen, J., Zhao, M., Li, T., Li, J., 2015c. A diatom record of mid-to-late Holocene palaeoenvironmental changes in the southern Okinawa trough. J. Quat. Sci. 30 (1), 32−43. https://doi.org/10.1002/jqs.2756.

Li, K., Liu, X., Wang, Y., Herzschuh, U., Ni, J., Liao, M., Xiao, X., 2017. Late Holocene vegetation and climate change on the southeastern Tibetan Plateau: Implications for the Indian Summer Monsoon and links to the Indian Ocean Dipole. Quat. Sci. Rev. 177, 235−245. https://doi.org/10.1016/j.quascirev.2017.10.020.

Li, Y., Chen, X., Xiao, X., Zhang, H., Xue, B., Shen, J., Zhang, E., 2018. Diatom-based inference of Asian monsoon precipitation from a volcanic lake in southwest China for the last 18.5 ka. Quat. Sci. Rev. 182, 109−120. https://doi.org/10.1016/j.quascirev.2017.11.021.

Liu, D., Sun, J., Zou, J., Zhang, J., 2005. Phytoplankton succession during a red tide of *Skeletonema costatum* in Jiaozhou Bay of China. Mar. Pollut. Bull. 50, 91−94.

Liu, Y., 2007. Studies on Diatoms of Da'erbin Lake and Swamps Around it in Daxing'anling Mountains. Shanghai Normal University, Shanghai (In Chinese).

Liu, Z., Wen, X., Brady, E.C., Otto-Bliesner, B., Yu, G., Lu, H., Cheng, H., Wang, Y., Zheng, W., Ding, Y., Edwards, R.L., Cheng, J., Liu, W., Yang, H., 2014. Chinese cave records and the East Asia summer monsoon. Quat. Sci. Rev. 83, 115−128. https://doi.org/10.1016/j.quascirev.2013.10.021.

Lotter, A.F., Birks, H.J.B., Hofmann, W., Marchetto, A., 1997. Modern diatom, cladocera, chironomid and chrysophyte cyst assemblages as quantitative indicators for the reconstruction of past environmental conditions in the Alps. I. Climate. J. Paleolimnol. 18, 395−420. https://doi.org/10.1023/A:1007982008956.

Lotter, A.F., Pienitz, R., Schmidt, R., 1999. Diatoms as indicators of environmental change near the Arctic and Alpine treeline. In: Stoermer, E.F., Smol, J.P. (Eds.), The Diatoms: Application for the Environmental and Earth Sciences, 205−226. Cambridge University Press, Cambridge.

Lotter, A.F., Bigler, C., 2000. Do diatoms in the Swiss Alps reflect the length of ice cover? Aquat. Sci. 62, 125−141. https://doi.org/10.1007/s000270050002.

Lotter, A.F., Pienitz, R., Schmidt, R., 2010. Diatoms as indicators of environmental change in subarctic and alpine regions. In: Smol, J.P., Stoermer, E.F. (Eds.), The Diatoms: Applications for the Environmental and Earth Sciences, pp. 231−248.

Lowe, R.L., 1974. Environmental Requirements and Pollution Tolerance of Freshwater Dsiatoms. US Environmental Protection Agency, Cincinnati, OH. EPA-670/4−74-005.

Ma, L., Gao, C., Kattel, G.R., Yu, X., Wang, G., 2018. Evidence of Holocene water level changes inferred from diatoms and the evolution of the Honghe Peatland on the Sanjiang Plain of Northeast China. Quat. Int. 476, 82−94. https://doi.org/10.1016/j.quaint.2018.02.025.

Mann, D.G., Vanormelingen, P., 2013. An inordinate fondness? The number, distributions, and origins of diatom species. J. Eukaryot. Microbiol. 60 (4), 414−420. https://doi.org/10.1111/jeu.12047.

Meriläinen, J.J., Hynynen, J., Palomäki, A., Mäntykoski, K., Witick, A., 2003. Environmental history of an urban lake: a palaeolimnological study of Lake Jyväsjärvi, Finland. J. Paleolimnol. 30, 387−406. https://doi.org/10.1023/B:JOPL.0000007229.46166.59.

Mitbavkar, S., Anil, A.C., 2000. Diatom colonization on stainless steel panels in estuarine waters of Goa, west coast of India. Indian J. Mar. Sci. 29, 273−276.

Moos, M., Laird, K., Cumming, B., 2005. Diatom assemblages and water depth in Lake 239 (Experimental Lakes Area, Ontario): implications for paleoclimatic studies. J. Paleolimnol. 34, 217−227. https://doi.org/10.1007/s10933-005-2382-8.

Morabito, G., Ruggiu, D., Panzani, P., 2002. Recent dynamics (1995−1999) of the phytoplankton assemblages in Lago Maggiore as a basic tool for defining association patterns in the Italian deep lakes. J. Limnol. 61, 129−145.

Muhina, V.V., 1971. Problems of diatom and silicoflagellate Quaternary stratigraphy in the equatorial Pacific Ocean. In: Funnell, B.M., Riedel, W.R. (Eds.), The Micropalaeontology of Oceans. Cambridge University Press, London, pp. 423−431.

Nagayasu, K., Otani, H., Kumon, F., 2017. Climate-deduced diatom changes were elucidated in a sediment core TKN2004 of the late Pleistocene Takano formation in Nagano, central Japan. Quat. Int. 440, 55–63. https://doi.org/10.1016/j.quaint.2016.11.043.

Padisák, J., Crossetti, L.O., Naselli-Flores, L., 2009. Use and misuse in the application of the phytoplankton functional classification: a critical review with updates. Hydrobiologia 621, 1–19. https://doi.org/10.1007/s10750-008-9645-0.

Pajunen, V., Luoto, M., Soininen, J., 2016. Climate is an important driver for stream diatom distributions. Global Ecol. Biogeogr. 25, 198–206. https://doi.org/10.1111/geb.12399.

Pan, Y., Stevenson, R.J., Hill, B.H., Herlihy, A.T., Collins, G.B., 1996. Using diatoms as indicators of ecological conditions in lotic systems: a regional assessment. J. North Am. Benthol. Soc. 15 (4), 481–495.

Pannard, A., Bormans, M., Lagadeuc, Y., 2008. Phytoplankton species turnover controlled by physical forcing at different time scales, Can. J. Fish. Aquat. Sci. 65, 47–60. https://doi.org/10.1139/f07-149.

Park, J., Han, J., Jin, Q., Bahk, J., Yi, S., 2017. The link between ENSO-like forcing and hydroclimate variability of coastal East Asia during the last millennium. Sci. Rep. 7 (1), 1–12. https://doi.org/10.1038/s41598-017-08538-1.

Patrick, R., Reimer, C.W., 1966. The diatoms of the United States 1. Acad. Nat. Sci. Phila. Monograph. 13, 788.

Patrick, R., Reimer, C.W., 1975. The Diatoms of the United States. Exclusive of Alaska and Hawaii, vol. 2. Monographic series of Academy of Natural Sciences of Philadelphia #13, Pennsylvania.

Petrova, N.A., 1986. Seasonality of Melosira-plankton of the great northern lakes. Hydrobiologia 138, 65–73. https://doi.org/10.1007/BF00027232.

Philibert, A., Prairie, Y.T., 2002. Diatom-based transfer functions for western Quebec lakes (Abitibi and Haute Mauricie): the possible role of epilimnetic CO_2 concentration in influencing diatom assemblages. J. Paleolimnol. 27, 465–480. https://doi.org/10.1023/A:1020372724266.

Pienitz, R., Cournoyer, L., 2017. Circumpolar Diatom Database (CDD): a new database for use in paleolimnology and limnology. J. Paleolimnol. 57 (2), 213–219. https://doi.org/10.1007/s10933-016-9932-0.

Pilskaln, C.H., Johnson, T.C., 1991. Seasonal signals in Lake Malawi sediments. Limnol. Oceanogr. 36 (3), 544–557. https://doi.org/10.4319/lo.1991.36.3.0544.

Potapova, M., Charles, D.F., 2007. Diatom metrics for monitoring eutrophication in rivers of the United States. Ecol. Indicat. 7 (1), 48–70. https://doi.org/10.1016/j.ecolind.2005.10.001.

Poulickova, A., Hasler, P., 2007. Aerophytic diatoms from caves in central Moravia (Czech Republic). Preslia 79 (2), 185–204.

Pouličková, A., 2008. Morphology, cytology and sexual reproduction in the aerophytic cave diatom *Luticola dismutica* (Bacillariophyceae). Preslia 80, 87–99.

Ptacnik, R., Diehl, S., Berger, S., 2003. Performance of sinking and non-sinking phytoplankton taxa in a gradient of mixing depths. Limnol. Oceanogr. 48, 1903–1912. https://doi.org/10.4319/lo.2003.48.5.1903.

Quamar, M.F., Bera, S.K., 2020. Pollen records of vegetation dynamics, climate change and ISM variability since the LGM from Chhattisgarh State, central India. Rev. Palaeobot. Palynol. 278, 104237. https://doi.org/10.1016/j.revpalbo.2020.104237.

Racca, J.M.J., Philibert, A., Racca, R., Prairie, Y.T., 2001. A comparison between diatom-based pH inference models using artificial neural networks (ANN), weighted averaging (WA) and weighted averaging partial least squares (WAPLS) regressions. J. Paleolimnol. 26, 411–422. https://doi.org/10.1023/A:1012763829453.

Ramrath, A., Nowaczyk, N., Negendank, J., 1999. Sedimentological evidence for environmental changes since 34 000 years BP from Lago di Mezzano, central Italy. J. Paleolimnol. 21, 423–435. https://doi.org/10.1023/A:1008006424706.

Rimet, F., Chaumeil, P., Keck, F., Kermarrec, L., Vasselon, V., Kahlert, M., Franc, A., Bouchez, A., 2016. R-syst::Diatom: An Open-Access and Curated Barcode Database for Diatoms and Freshwater Monitoring. Database. https://doi.org/10.1093/database/baw016.

Rioual, P., Andrieu-Ponel, V., de Beaulieu, J.L., Reille, M., Svobodova, H., Battarbee, R.W., 2007. Diatom responses to limnological and climatic changes at Ribains maar (French massif Central) during the Eemian and early Würm. Quat. Sci. Rev. 26, 1557—1609. https://doi.org/10.1016/j.quascirev.2007.03.009.

Rioual, P., Chu, G.Q., Li, D., Mingram, J., Han, J., Liu, J., 2009. Climate-induced Shifts in Planktonic Diatoms in Lake Sihailongwan (North-East China): A Study of the Sediment Trap and Palaeolimnological Records. 11th International Paleolimnology Symposium, Guadalajara, Mexico, p. 120.

Rosen, P., Hall1, R., Korsman, T., Renberg, I., 2000. Diatom transfer functions for quantifying past air temperature, pH and total organic carbon concentration from lakes in northern Sweden. J. Paleolimnol. 24, 109—123. https://doi.org/10.1023/A:1008128014721.

Round, F.E., Crawford, R.M., Mann, D.G., 1990. Diatoms: Biology and Morphology of the Genera. Cambridge university press, Cambridge, p. 747.

Rudaya, N., Tarasov, P., Dorofeyuk, N., Solovieva, N., Kalugin, I., Andreev, A., Daryin, A., Diekmann, B., Riedel, F., Tserendash, N., Wagner, M., 2009. Holocene environments and climate in the Mongolian Altai reconstructed from the Hoton-Nur pollen and diatom records: a step towards better understanding climate dynamics in Central Asia. Quat. Sci. Rev. 28 (5—6), 540—554. https://doi.org/10.1016/j.quascirev.2008.10.013.

Rühland, K., Priesnitz, A., Smol, J.P., 2003. Paleolimnological evidence from diatoms for recent environmental changes in 50 lakes across the Canadian arctic treeline. Arctic Antarct. Alpine Res. 35, 110—123.

Rühland, K., Phadtare, N.R., Pant, R.K., Sangode, S.J., Smol, J.P., 2006. Accelerated melting of Himalayan snow and ice triggers pronounced changes in a valley peatland from Northern India. Geophys. Res. Lett. 33, 1—6. https://doi.org/10.1029/2006GL026704.

Rühland, K., Paterson, A.M., Smol, J.P., 2008. Hemispheric-scale patterns of climate-related shifts in planktonic diatoms from North American and European lakes. Global Change Biol. 14, 1—15. https://doi.org/10.1111/j.1365-2486.2008.01670.x.

Rühland, K.M., Paterson, A.M., Smol, J.P., 2015. Lake diatom responses to warming: reviewing the evidence. J. Paleolimnol. 54 (1), 1—35. https://doi.org/10.1007/s10933-015-9837-3.

Rybak, M., Noga, T., Zubel, R., 2018. The Aerophytic diatom assemblages developed on mosses covering the bark of populus alba L. J. Ecol. Eng. 19 (6). https://doi.org/10.12911/22998993/92673.

Saros, J.E., Stone, J.R., Pederson, G.T., Slemmons, K.E.H., Spanbauer, T., Schliep, A., Cahl, D., Williamson, C.E., Engstrom, D.R., 2012. Climate-induced changes in lake ecosystem structure inferred from coupled neo- and paleoecological approaches. Ecology 93, 2155—2164. https://doi.org/10.1890/11-2218.1.

Schelske, C.L., Carrick, H.J., Aldridge, F.J., 1995. Can wind-induced resuspension of meroplankton affect phytoplankton dynamics? J. North Am. Benthol. Soc. 14 (4), 616—630. https://doi.org/10.2307/1467545.

Schmidt, R., Kamenik, C., Lange-Bertalot, H., et al., 2004. *Fragilaria* and *Staurosira* (Bacillariophyceae) from sediment surfaces of 40 lakes in the *Austrian Alps* in relation to environmental variables, and their potential for palaeoclimatology. J. Limnol. 63, 171—189.

Schwarz, A., Turner, F., Lauterbach, S., Plessen, B., Krahn, K.J., Glodniok, S., Mischke, S., Stebich, M., Witt, R., Mingram, J., Schwalb, A., 2017. Mid-to late Holocene climate-driven regime shifts inferred from diatom, ostracod and stable isotope records from Lake Son Kol (Central Tian Shan, Kyrgyzstan). Quat. Sci. Rev. 177, 340—356. https://doi.org/10.1016/j.quascirev.2017.10.009.

Selby, K.A., Brown, A.G., 2007. Holocene development and anthropogenic disturbance of a shallow lake system in Central Ireland recorded by diatoms. J. Paleolimnol. 38, 419—440. https://doi.org/10.1007/s10933-006-9081-y.

Seckbach, J., Kociolek, P. (Eds.), 2011. The Diatom World, vol. 19. Springer Science & Business Media.

Shear, H., Nalewajko, C., Bacchus, H.M., 1976. Some aspects of the ecology of *Melosira* spp. in Ontario lakes. Hydrobiologia 50, 173—176. https://doi.org/10.1007/BF00019821.

Shennan, I., Innes, J.B., Long, A.J., Zong, Y., 1993. Late Devensian and Holocene relative sea-level changes at Rumach, near Arisaig, north west Scotland. Nor. Geol. Tidsskr. 73, 161–174.

Shennan, I., Bruhn, R., Plafker, G., 2009. Multi-segment earthquakes and tsunami potential of the Aleutian megathrust. Quat. Sci. Rev. 28, 7–13. https://doi.org/10.1016/j.quascirev.2008.09.016.

Shukla, S.K., Mohan, R., Sudhakar, M., 2009. Diatoms: a potential tool to understand past oceanographic settings. Curr. Sci. 97 (12), 1726–1734. https://www.jstor.org/stable/24107252.

Smol, J.P., Cumming, B.F., 2000. Tracking long-term changes in climate using algal indicators in lake sediments. J. Phycol. 36 (6), 986–1011. https://doi.org/10.1046/j.1529-8817.2000.00049.x.

Smol, J.P., Stoermer, E.F. (Eds.), 2010. The Diatoms: Applications for the Environmental and Earth Sciences, second ed. Cambridge University Press, Cambridge, p. 686.

Soininen, J., Jamoneau, A., Rosebery, J., Leboucher, T., Wang, J., Kokocinski, M., Passy, S.I., 2019. Stream diatoms exhibit weak niche conservation along global environmental and climatic gradients. Ecography 42, 346–353. https://doi.org/10.1111/ecog.03828.

Sorvari, S., Korhola, A., Thompson, R., 2002. Lake diatom response to recent Arctic warming in Finnish Lapland. Global Change Biol. 8 (2), 153–163. https://doi.org/10.1046/j.1365-2486.2002.00463.x.

Spaulding, S.A., Potapova, M.G., Bishop, I.W., Lee, S.S., Gasperak, T.S., Jovanoska, E., Furey, P.C., Edlund, M.B., 2020. Diatoms of North America. https://diatoms.org/.

Stanek-Tarkowska, J., Noga, T., Kochman-Kedziora, N., Peszek, L., Pajaczek, A., Kozak, E., 2015. The diversity of diatom assemblages developed on fallow soil in Pogórska Wola near Tarnów (southern Poland). Acta Agrobot. 68 (1). https://doi.org/10.5586/aa.2015.011.

Sterrenburg, F.A.S., 1995. Studies on the Genera *Gyrosigma* and *Pleurosigma* (Bacillariophyceae): *Gyrosigma Acuminatum* (Kützing) Rabenhorst, G. Spenceri (Quekett) Griffith, and G. Rautenbachiae Cholnoky. Proceedings of the Academy of Natural Sciences of Philadelphia, pp. 467–480.

Stevenson, A.C., Juggins, S., Birks, H.J.B., Anderson, D.S., Anderson, N.J., Battarbee, R.W., Berge, F., Davis, R.B., Flower, R.J., Haworth, E.Y., Jones, V.J., Kingston, J.C., Kreiser, A., Line, J.M., Munro, M.A.R., Renberg, I., 1991. The Surface Waters Acidification Project Palaeolimnology Programme: Modern Diatom/Lake Water Chemistry Data Set. UCL Environmental Change Research Centre, ENSIS LTD, London.

Stockner, J.G., 1971. Preliminary characterization of lakes of the Experimental Lakes Area, northwestern Ontario; using diatom occurrence in sediments. J. Fish. Board Canada 28, 265–275. https://doi.org/10.1139/f71-037.

Sun, W., Zhang, E., Shulmeister, J., Bird, M.I., Chang, J., Shen, J., 2019. Abrupt changes in Indian summer monsoon strength during the last deglaciation and early Holocene based on stable isotope evidence from Lake Chenghai, southwest China. Quat. Sci. Rev. 218, 1–9. https://doi.org/10.1016/j.quascirev.2019.06.006.

Suokhrie, T., Saalim, S.M., Saraswat, R., Nigam, R., 2018. Indian monsoon variability in the last 2000 years as inferred from benthic foraminifera. Quat. Int. 479, 128–140. https://doi.org/10.1016/j.quaint.2017.05.037.

Talling, J.F., 1957. Photosynthetic characteristics of some freshwater plankton diatoms in relation to underwater radiation. New Phytol. 56, 345–356.

Thoms, M.C., Ogden, R.W., Reid, M.A., 1999. Establishing the condition of lowland floodplain rivers: a palaeo-ecological approach. Freshw. Biol. 41, 407–423. https://doi.org/10.1046/j.1365-2427.1999.00439.x.

Tibby, J., Reid, M.A., 2004. A model for inferring past conductivity in low salinity waters derived from Murray River (Australia) diatom plankton. Mar. Freshw. Res. 55 (6), 597–607. https://doi.org/10.1071/MF04032.

Tiffany, M.A., 2011. Epizoic and epiphytic diatoms. In: Seckbach, J., Kociolek, J.P. (Eds.), The Diatom World. Springer, pp. 195–211.

Tiffany, M.A., Lange, C.B., 2002. Diatoms provide attachment sites for other diatoms: a natural history of epiphytism from southern California. Phycologia 41 (2), 116–124. https://doi.org/10.2216/i0031-8884-41-2-116.1.

Tilman, D., Kilham, S.S., Kilham, P., 1982. Phytoplankton community ecology: the role of limiting nutrients. Annu. Rev. Ecol. Systemat. 13, 349–372.

Trenberth, K.E., Stepaniak, D.P., Caron, J.M., 2000. The global monsoon as seen through the divergent atmospheric circulation. J. Clim. 13 (22), 3969−3993. https://doi.org/10.1175/1520-0442(2000)013%3C3969:TGMAST%3E2.0.CO;2.

Van Campo, E., Gasse, F., 1993. Pollen- and diatom inferred climatic and hydrological changes in Sumxi Co Basin (Western Tibet) since 13,000 yr B. P Quatern. Res. 39, 300−313. https://doi.org/10.1006/qres.1993.1037.

Van Campo, E., Gasse, F., 1994. Pollen-and diatom-inferred climatic and hydrological changes in Sumxi Co Basin (Western Tibet) since 13,000 yr BP. Quat. Res. 39 (3), 300−313. https://doi.org/10.1006/qres.1993.1037.

Van Dam, H., Mertens, A., Sinkeldam, J., 1994. A coded checklist and ecological indicator values of freshwater diatoms from the Neth. J.Aquat. Ecol. 28, 117−133. https://doi.org/10.1007/BF02334251.

Van de Vijver, B.V., Beyens, L., 1997. The epiphytic diatom flora of mosses from Strømness Bay area, South Georgia. Polar Biol. 17 (6), 492−501. https://doi.org/10.1007/s003000050148.

Van de Vijver, B., Beyens, L., 1999. Moss diatom communities from Ile de la Possession (Crozet, Subantarctic) and their relationship with moisture. Polar Biol. 22 (4), 219−231. https://doi.org/10.1007/s003000050414.

Van de Vijver, B., Frenot, Y., Beyens, L., 2002. Freshwater Diatoms from Ile de la Possession (Crozet Archipelago, Subantarctic). Bibl Diatomol, Band 46. J. Cramer, Berlin.

Van de Vijver, B., Beyens, L., Lange-Bertalot, H., 2004. The genus *Stauroneis* in the Arctic and (sub-) Antarctic regions. Bibl. Diatomol. 51, 1−317.

Velez, M.I., Hooghiemstra, H., Metcalfe, S., Martínez, I., Mommersteeg, H., 2003. Pollen- and diatom-based environmental history since the last glacial maximum from the andean core Fúquene-7, Colombia. J. Quat. Sci. 18 (1), 17−30. https://doi.org/10.1002/jqs.730.

Verpoorter, C., Kutser, T., Seekell, D.A., Tranvik, L.J., 2014. A global inventory of lakes based on high-resolution satellite imagery. Geophys. Res. Lett. 41, 6396−6402. https://doi.org/10.1002/2014GL060641.

Verschuren, D., 2003. Lake-based climate reconstruction in Africa: progress and challenges. In: Martens, K. (Ed.), Aquatic Biodiversity. Developments in Hydrobiology, vol. 171. Springer, Dordrecht. https://doi.org/10.1007/978-94-007-1084-9_22.

Vos, P.C., de Wolf, H., 1993. Diatoms as a tool for reconstructing sedimentary environments in coastal wetlands; methodological aspects. Hydrobiologia 285−296. https://doi.org/10.1007/BF00028027.

Vyverman, W., Sabbe, K., 1995. Diatom-temperature transfer functions based on the altitudinal zonation of diatom assemblages in Papua New Guinea: a possible tool in the reconstruction of regional palaeoclimatic changes. J. Paleolimnol. 13, 65−77. https://doi.org/10.1007/BF00678111.

Wang, B., 2006. The Asian Monsoon. Praxis Publishing Ltd. Springer, Chichester.

Wang, L., Lu, H., Liu, J., Gu, Z., Mingram, J., Chu, G., Li, J., Rioual, P., Negendank, J.F.W., Han, J., Liu, T., 2008. Diatom-based inference of variations in the strength of Asian winter monsoon winds between 17,500 and 6000 calendar years BP. J. Geophys. Res. Atmos. 113 (D21). https://doi.org/10.1029/2008JD010145.

Wang, L.-C., Lee, T.-Q., Chen, S.-H., Wu, J.-T., 2010. Diatoms in Liyu Lake, Eastern Taiwan. Taiwania 55, 228−242.

Wang, L., Li, J., Lu, H., Gu, Z., Rioual, P., Hao, Q., Mackey, A.W., Jiang, W., Cai, B., Xu, B., Han, J., Chu, G., 2012a. The East Asian winter monsoon over the last 15,000 years: its links to high-latitudes and tropical climate systems and complex correlation to the summer monsoon. Quat. Sci. Rev. 32, 131−142. https://doi.org/10.1016/j.quascirev.2011.11.003.

Wang, L., Rioual, P., Panizzo, V.N., Lu, H., Gu, Z., Chu, G., Yang, D., Han, J., Liu, J., Mackay, A.W., 2012b. A 1000-yr record of environmental change in NE China is indicated by diatom assemblages from maar lake Erlongwan. Quat. Res. 78 (1), 24−34. https://doi.org/10.1016/j.yqres.2012.03.006.

Wang, Q., Yang, X.D., Hamilton, P.B., Zhang, E., 2012c. Linking spatial distributions of sediment diatom assemblages with hydrological depth profiles in a plateau deep−water lake system of subtropical China. Fottea 12 (1), 59−73. https://doi.org/10.5507/fot.2012.005.

Wang, L.C., Behling, H., Lee, T.Q., Li, H.C., Huh, C.A., Shiau, L.J., Chen, S.H., Wu, J.T., 2013. Increased precipitation during the Little Ice Age in northern Taiwan was inferred from diatoms and geochemistry in a sediment core from a subalpine lake. J. Paleolimnol. 49 (4), 619−631. https://doi.org/10.1007/s10933-013-9679-9.

Wang, Q., Yang, X., Anderson, N.J., Zhang, E., Li, Y., 2014. Diatom response to climate forcing of a deep, alpine lake (Lugu Hu, Yunnan, SW China) during the Last Glacial Maximum and its implications for understanding regional monsoon variability. Quat. Sci. Rev. 86, 1−12. https://doi.org/10.1016/j.quascirev.2013.12.024.

Wang, L.C., Behling, H., Kao, S.J., Li, H.C., Selvaraj, K., Hsieh, M.L., Chang, Y.P., 2015a. Late Holocene environment of subalpine northeastern Taiwan from pollen and diatom analysis of lake sediments. J. Asian Earth Sci. 114, 447−456. https://doi.org/10.1016/j.jseaes.2015.03.037.

Wang, Q., Yang, X., Anderson, N.J., Ji, J.F., 2015b. Diatom seasonality and sedimentation in a sub-tropical alpine lake (Lugu Hu, Yunnan-Sichuan, SW China). Arctic Antarct. Alpine Res. 47, 461−472. https://doi.org/10.1657/AAAR0014-039.

Watanabe, T., 2005. Picture Book and Ecology of the Freshwater Diatoms. Uchida Rokakuho, Tokyo, p. 666.

Webster, P.J., Magaña, V.O., Palmer, T.N., Shukla, J., Tomas, R.A., Yanai, M., Yasunari, T., 1998. Monsoons: Processes, predictability, and the prospects for prediction. J. Geophys. Res. 103 (C7), 14451−14510. https://doi.org/10.1029/97JC02719.

Weckström, J., Korhola, A., Blom, T., 1997. The relationship between diatoms and water temperature in thirty subarctic Fennoscandian lakes. Arctic Antarct. Alpine Res. 29, 75−92.

Weilhoefer, C.L., Pan, Y., 2007. Relationships between diatoms and environmental variables in wetlands in the Willamette Valley, Oregon, USA. Wetlands 27, 668−682. https://doi.org/10.1672/0277-5212(2007)27[668:RBDAEV]2.0.CO;2.

Whitmore, T.J., 1989. Florida diatom assemblages as indicators of trophic state and pH. Limnol. Oceanogr. 34 (5), 882−895. https://doi.org/10.4319/lo.1989.34.5.0882.

Winter, J.G., Duthie, H.C., 2000. Epilithic diatoms as indicators of stream total N and total P concentration. J. North Am. Benthol. Soc. 19 (1), 32−49.

Witkowski, A., Lange-Bertalot, H., Metzeltin, M., 2000. Diatom flora of marine coasts I. In: Iconographica Diatomologia, vol. 7. A. R. G. Ganter Verlag, Ruggell.

Witon, E., Witkowski, A., 2003. Diatom (Bacillariophyceae) flora of early Holocene freshwater sediments from Skalafjord, Faeroe Islands. J. Micropaleontol. 22, 183−208. https://doi.org/10.1144/jm.22.2.183.

Wolin, J.A., Duthie, H.C., 2010. Diatoms as indicators of water level change in freshwater lakes. In: Stoermer, E.F., Smol, J.P. (Eds.), The Diatoms: Applications for the Environmental and Earth Sciences. Cambridge University Press, pp. 174−185.

Wunsam, S., Schmidt, R., Klee, R., 1995. *Cyclotella*-taxa (Bacillariophyceae) in lakes of the Alpine region and their relationship to environmental variables. Aquat. Sci. 57, 4360−4386. https://doi.org/10.1007/BF00878399.

de Wolf, H., 1982. Method of coding of ecological data from diatoms for computer utilization. Mededel Rijks Geol. Dienst. 36 (2), 95−98.

Wolin, J.A., Stone, J.R., 2010. Diatoms as indicators of water-level change in freshwater lakes. In: Stoermer, E.F., Smol, J.P. (Eds.), The Diatoms: Applications for the Environmental and Earth Sciences, 2nd. Cambridge University Press, Cambridge, pp. 174−185.

Wunsam, S., Cattaneo, A., Bourassa, N., 2002. Comparing diatom species, genera and size in biomonitoring: a case study from streams in the Laurentians (Quebec, Canada). Freshw. Biol. 47, 325−340. https://doi.org/10.1046/j.1365-2427.2002.00809.x.

Yang, X.D., Dong, X.H., Gao, G., Pan, H.X., Wu, J.L., 2005. Relationship between surface sediment diatoms and summer water quality in shallow lakes of the middle and lower reaches of the Yangtze River. J. Integr. Plant Biol. 47, 153−164. https://doi.org/10.1111/j.1744-7909.2005.00035.x.

Yang, X.D., Anderson, N.J., Dong, X.H., Shen, J., 2008. Surface sediment diatom assemblages and epilimnetic total phosphorus in large, shallow lakes of the Yangtze floodplain: their relationships and implications for assessing long-term eutrophication. Freshw. Biol. 53, 1273–1290. https://doi.org/10.1111/j.1365-2427.2007.01921.x.

Yoshikawa, S., Yamaguchi, S., Hata, A., 2000. Paleolimnological investigation of recent acidity changes in Sawanoike Pond, Kyoto. Japan. J. Paleolimnol. 23, 285–304. https://doi.org/10.1023/A:1008199830698.

Zhang, X., Zheng, Z., Huang, K., Yang, X., Tian, L., 2020. Sensitivity of altitudinal vegetation in southwest China to the Indian summer monsoon changes during the past 68000 years. Quat. Sci. Rev. 239, 106359. https://doi.org/10.1016/j.quascirev.2020.106359.

Zhao, K., Wang, Y., Edwards, R.L., Cheng, H., Liu, D., Kong, X., 2015. A high-resolved record of the Asian Summer Monsoon from Dongge Cave, China, for the past 1200 years. Quat. Sci. Rev. 122, 250–257. https://doi.org/10.1016/j.quascirev.2015.05.030.

Zong, Y.Q., 1997. Implications of *Paralia sulcata* abundance in Scottish isolation basins. Diatom Res. 12, 125–150. https://doi.org/10.1080/0269249X.1997.9705407.

Zong, Y., Lloyd, J.M., Leng, M.J., Yim, W.S., Huang, G., 2006. Reconstruction of Holocene monsoon history from the Pearl River Estuary, southern China, using diatoms and carbon isotope ratios. Holocene 16 (2), 251–263.

CHAPTER

Environmental disaster in Pithoragarh district, Uttarakhand, India: an analogy to climate change and anthropogenic interference

14

Akhouri Bishwapriya[1], Prakash K. Gajbhiye[2], Milind Vasantrao Dakate[3], Santosh Kumar Tripathi[4] and Abhishek Kumar Chaurasia[1]

[1]*Geological Survey of India, Patna, Bihar, India;* [2]*Geological Survey of India, Pune, Maharashtra, India;* [3]*Geological Survey of India, Nagpur, Maharashtra, India;* [4]*Geological Survey of India, Hyderabad, Telangana, India*

1. Introduction

The fragility of mountain chains in combination with high relief, the adverse inclination of discontinuity planes, intense rain, flash floods, and snowfall/avalanches make the Himalayan ranges highly prone to landslides. Besides, the Himalayas are the product of the collision of Indian and Eurasian Plates, the home of recurring earthquakes and landslides. Excessive human interference has also contributed greatly to the degradation of the Himalayan ecosystem. The gigantic landslides have frequently disrupted human activities and blocked major and minor channels. The menacing effects of these phenomena, though greatly varying in magnitude and space, have caused great miseries to the inhabitants in terms of loss of life and property, disruption of communication lines, disconnecting the communication routes, etc.

Pithoragarh district has witnessed several environmental and environmentally induced geological disasters in the past, accentuated by human interference. A major incidence of a landslide blocking the river Dhauliganga occurred in 1956. The most disastrous landslide was induced by unprecedented heavy rains, killing 221 people occurred during the early hours of August 18, 1998, at Malpa in Pithoragarh district (Sharma, 2009). The landslide on August 8, 2009, near La, Chachna and Rumidoula villages buried houses and took the lives of 43 people and 120 cattle (Sharma, 2009). The earthquake-generated landslides are widespread in the terrain. Severe shaking during 1916 near Dharchula, the Dharchula earthquake (M 6.1) of July 29, 1980, and Indo-Nepal earthquake (M. 5.5) of January 5, 1997, jolted the region severely and induced several landslides (Akhouri et al., 2013).

The climatic aberrations have not been confined to the hilly state only but have been conspicuously experienced in other parts of the subcontinent as well. Some of the recent events are testimony to this belief. In 2022, Rajasthan received 270 mm of rainfall in July, the highest for the month in nearly 7 decades, according to official data. The 270 mm rainfall is 67% more than the average rainfall of

161.4 mm for July, according to the data provided by the meteorological center in Jaipur (India Meteorological Department, 2022). For comparison, Rajasthan received 130.8 mm of rain in July last year. In 1956, Rajasthan recorded 308.7 mm of rainfall in July. The July rainfall in Rajasthan stood at 288 mm in 1908, 281.6 mm in 1943, 270 mm in 2022, 262.3 mm in 2015, and 252.3 mm in 2017. This climatic vagary resulted in flood-like situations in several districts in Rajasthan, a state known for its dry weather and desert topography.

In August 2022, Chaibasa town of West Singhbhum district recorded the highest rainfall of 191.6 mm, followed by 142.2 mm in Ramgarh, 137.2 mm in Jamshedpur, and 54.2 mm in Ranchi. The water level of several rivers has either reached the danger mark or crossed it in the state's Kolhan region comprising West Singhbhum, Saraikela-Kharsawan, and East Singhbhum districts, an official said.

In 2021, the state of Kerela logged its highest annual rainfall in 60 years and the sixth highest rainfall in the past 120 years according to India Meteorological Department (2021). The state recorded an annual rainfall of 3610.2 mm during 12 monthly periods against normal rainfall of 2924.7 mm, an excess of 23.4%. The winter rainfall and northeast rainfall in 2021, were the highest ever recorded in the state, while the premonsoon rainfall was the sixth highest in 120 years. Kerela recorded a percentage departure of 409% above normal in winter rainfall, 108% in premonsoon rainfall, and 109% in the north monsoon rainfall (India Meteorological Department, 2021). On the other hand, the southwest monsoon, which contributes most of the states' annual rainfall quota, was "normal," Although the season's rainfall (June—September) recorded a 16% deficit, percentage departures ranging between -19% and $+19\%$ are classified as normal by IMD.

In comparison, in 2018 and 2019, when the state was subjected to devastating floods and landslips, the surplus in annual rainfall had stood at 20% and 7%, respectively (India Meteorological Department, 2019). In 2020, the annual rainfall exceeded the normal by 2%. The year will be remembered as one of the wettest for Kerela, setting new rainfall records even as the increase in unruly weather systems over the Arabian Sea added fuel to the debate on climate change impacts.

In 2019, parts of Bihar suffered a "short window catastrophe" due to unprecedented rains. As per daily rainfall data from the India Meteorological Department (India Meteorological Department, 2019), the total rainfall in Patna between September 26 and September 29, 2019, was 210.8 mm. The 5 years that come closest to the 210.8 mm rainfall received in the September 26—September 29 periods in 2019 are 1963 (198.1 mm), 1960 (136.1 mm), 2007 (126.7 mm), 1989 (117.8 mm), and 1902 (116.8 mm). What is ironic is the fact that total rainfall in 2019, in both Patna and Bihar was less than the long period average (LPA) rainfall. The total rainfall in Patna and Bihar was only 86% and 98% of LPA between June 1 and September 29. LPA is the average rainfall between 1951 and 2000.

$$Su_1 = Au \quad ; \quad (C-S)u_0 = CV - Au$$

$$T = S\rho u_1(u_1 - V) + (C-S)\rho u_0(u_0 - V) + C(p_1 - p) \tag{14.35}$$

$$\frac{T}{A} = \left(p_1 + \frac{1}{2}\rho u_1^2\right) - \left(p + \frac{1}{2}\rho V^2\right) = \frac{1}{2}\rho(u_1^2 - u_0^2)$$

If the week after September 25, 2019, is excluded, the monsoon rainfall for Patna and Bihar comes to -35% and -18%, respectively. Another metric can be given to highlight the skewed nature of rainfall in Bihar in 2019. Out of the 121 days in the 2019 monsoon—June 1 to September

29—for which daily rainfall data are available for Bihar, 90 days saw lower-than-average rainfall in the state.

These figures indicate the erratic turnaround in moods of precipitation, which can be described as anything but normal and hint toward a more enveloping factor as the reason behind such climatic aberrations. What was worse is that the havoc that resulted from this unexpected phenomenon unsettled the habitation due to a lack of preparedness. The topography of Patna is saucer-shaped with predominant centripetal drainage. The levee portion on which Patna is located is at places commensurate with the bed level of the Ganga river, and when it swells during the flood, the land areas get coagulated with stagnant water. The sorry state of affairs caused by choked and inadequate drainage created a flood-like situation.

Adding to the woes, all new constructions have been made on a raised foundation leaving erstwhile buildings to sustain on a relatively lower foundation thus entrapping a voluminous load of water, which otherwise could have dissipated uniformly. Thus, low-lying areas suffered much more than areas rebuilt on a raised foundation. Some high-rise buildings have been located within the active channel of river Ganga around Patna, a step which out rightly violates the environmental stipulations. The matter may be subjudice but the act is headed for a "climatic surprise," which may have far-reaching implications.

In 2015, a relentless downpour pounded Tamil Nadu's capital city of Chennai on December 1, flooding and submerging one of India's largest cities. The heaviest 1-day rainfall in the region—as much as 494 mm (19.45 inches)—in more than a century left more than three million people without basic services.

To assess the potential link between the heavy rainfall in Chennai on December 1, 2015, and human-caused greenhouse gases in the atmosphere, Climate Central, the Royal Netherlands Meteorological Institute (KNMI), the University of Oxford, and the Indian Institute of Technology, Delhi (IITD)—as part of the World Weather Attribution (WWA) partnership which also includes Red Cross Red Crescent Climate Center and University of Melbourne—conducted independent assessments using multiple peer-reviewed approaches of extreme event attribution methodologies using statistical analysis. The highest observed daily rainfall, 494 mm, was an extreme occurrence meteorologically, with a probability of occurrence in the current climate of one in 600—2500 per year (95% level of confidence). This would have overwhelmed flood prevention measures in any city. Though no direct human intervention was concluded.

The above incidents with the one that happened in the state of Uttarakhand between June 15 and 17, 2013, are not only seen to be occurring frequently in recent times but infested "small time corridors" with their severity. On the contrary, the precipitation regime on a seasonal basis barring this window more or less remained normal or below normal in most of the cases. These erratic climatic behavior unsettling the behavioral aspects of these regions call for awakening to assess them from a wider perspective (Fig. 14.1).

1.1 Study area

The area of study lie between latitude $30°17'00''$ and $29°30'00''$ and longitude $79°40'00''$ and $80°47'00''$ in Survey of India toposheet Nos. 62 B/04, B/08, B/12, 62C/01, C/02, C/05, C/06, C/09, and 62 O/14 in district Pithoragarh, Uttarakhand (Fig. 14.2). The area is well connected by a broad metaled road. The study limits of Pithoragarh district lie within Seraghat in the west to Kali river in the east and

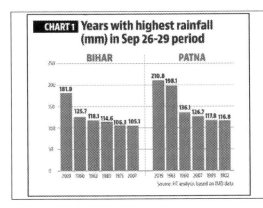
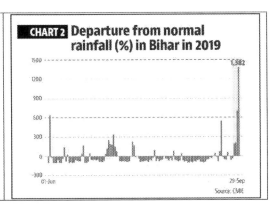

FIGURE 14.1

Rainfall comparison, Bihar.

Source: India Meteorological Department (2019).

Milam in the north to south of Rameshwar Ghat in the south. The National Highway from Champawat district terminates at Pithoragarh headquarters and as such has a total stretch of ~30 km in the district. The other centers of the district are connected by metaled motorable roads/state highways. Physiographically, the area of study is represented by high hills and valleys presenting rugged topography. The prominent drainages are Ramganga River, Goriganga River, Dhauliganga River, and Kali River.

The climate of the area is moderate and tropically characterized by hot and dry summer from March to the middle of June, a humid monsoon or rainy season from the middle of June to September and a cool dry winter spanning between December and February, its severity increases in the higher reaches.

Geology: The area of investigation exposes rocks of central crystalline and Garhwal Group with associated acid intrusives. The central crystalline represent the oldest rocks exposed in Higher Himalayas and consists of gneisses, migmatites, psamitic and mica gneiss, calc gneiss, quartzite, marble, mica schist, and amphibolite. The MCT is a north-northeasterly dipping major tectonic plane in the Kumaon region defining the northern tectonic limit of sedimentation of the Garhwal Group with the Central Crystalline. It marks the position of the MCT north of Dharchula in the Kali Ganga valley and north of Munsyari in the Gori Ganga valley in the Kumaon region. Another major plane of dislocation exposed in Kali Ganga valley near Dar separates the Dar Formation from the Central Crystalline. It is traced to the south of Martoli and named as Dar Martoli fault. It is traceable to Dhauliganga valley at Martoli where it is offset by the Martoli fault (Akhouri et al., 2013) (Fig. 14.3).

The area of study is further subjected to an intense degree of tectonic deformation, folding, and faulting. A transverse roughly east-west trending anticline has its eastern extremity north of Jauljibi in the Gori Ganga valley and its western extremity, east of Tejam in the Ramganga valley. At the confluence of Mandakini and Gori Ganga rivers near Madkot, also lies a small anticlinal axis. A north–south trending anticlinal axis also aligns along the Gori Ganga River immediately north of the confluence of Paina gad and Gori Ganga river, south-west of Golpha and extends southwards up to

1. Introduction

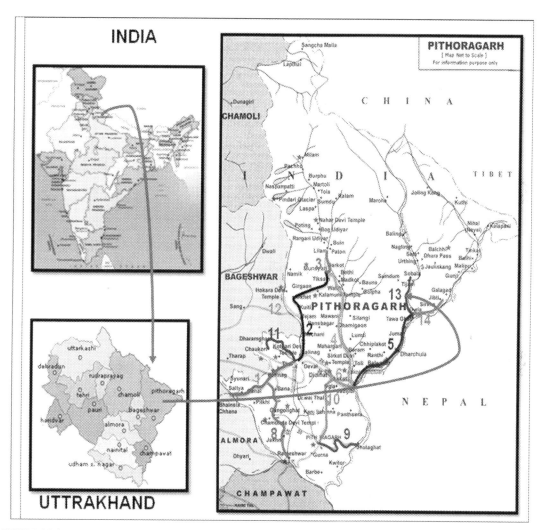

FIGURE 14.2

Study area, Pithoragarh district, Uttarakhand.

Humkumta and Chamoli villages on the right bank of Gori Ganga river. The stretch of the Goriganga river south of Baram up to Jauljibi is also thrown into several anticlines and synclines trending east-west. Similar east-west trending antiforms and synform extend up to Pithoragarh and even south of it. The area north of Tawaghat, south of Nyu, west of Chiplakot, Baram and north of Dharchula represents in a near-circular pattern, a nappe/klippe occupied primarily by acid intrusive, mainly granodiorite (Akhouri et al., 2013).

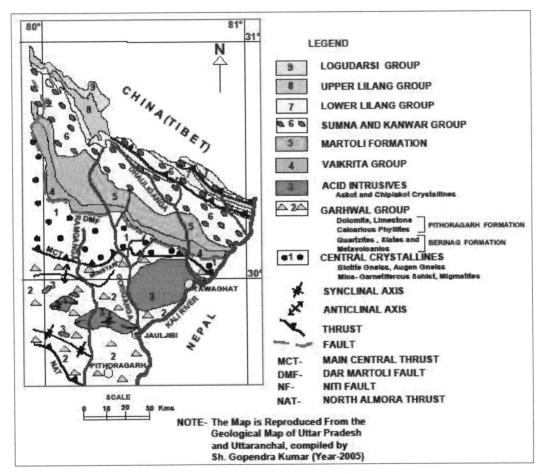

FIGURE 14.3

Regional geological map of study area, Pithoragarh district, Uttarakhand.

1.2 Slope stability assessment

A disastrous event of heavy incessant rains/snowfall occurred from 15th–17th June 2013 in the higher reaches of the region and catchment areas of major rivers like Ramganga, Gori Ganga, Dhauliganga, and Kali. This event caused several landslides, especially in the northern part of the Pithoragarh district, and led to flash floods with huge sediment loads causing unusual swelling of the rivers with high velocity/current, leading to extensive river bank erosion and terrace cutting. This event created havoc and swept away the villages/townships located on river banks, communication routes, bridges, hydropower projects, other infrastructures, etc. that came in the course of the flood.

There are a number of major slide zones in the Ramganga, Gori Ganga, east Dhauliganga, and Kali valley. The stretch from Tejam to Mansuriya Kanth along the right bank of the Ramganga River is also prone to sliding and subsidence. The left bank major tributary of the Ramganga, the Jakula Nadi is particularly prone to a number of slide zones. The most important slide zones are found around Kwiti on the opposite bank and a major one near Sirmola village. There is a major sinking zone just below village Bhandarigaon along the left bank slopes. The nala along Dhapa, La and Jhekla villages is also prone to subsidence and cloud bursts. Similarly, along the Gori Ganga river, the entire stretch up to Jauljibi is prone to landslides on the left bank and bank erosion on both banks. The slope along which Talla Dumar village is located is also prone to long-time subsidence. Similarly, Baikunthi and Kaithi villages are also prone to gradual creep and subsidence along their slopes. Along the Dhauliganga River, from Tawaghat to Sobala-Tijam, several sliding zones are present and river bank erosion/cutting incidences have been recorded. Also, the stretch from Tawaghat to Baluwakot, Jauljibi, and Jhulaghat is affected by river cutting/bank erosion (Fig. 14.4).

The macro-scale mass wasting inventory database of the area (1:50,000 scale) has been prepared for 14 different sectors. The predominant mass wasting process involved are landslides and river bank erosion by flood and land subsidence. Altogether 92 incidences of landslide, bank erosion, and subsidence have been recorded. The predominant mass wasting processes involved are landslide (56.52%, 52 nos.), river bank erosion due to high floods (34.78%, 32 nos.), and land subsidence (8.69%, eight nos.). Out of the total of 60 landslide/subsidence incidences, 40 nos. are debris slides, 19 nos. are rock-cum-debris slides, and one number is a rock slide. Out of the total of 92 incidences, 21 nos. are new landslides, 39 nos. are old reactivated landslides, and all 32 nos. of river bank erosions are new incidences comprising RBM and colluvium debris (Figs. 14.5 and 14.6).

A total of 14 road/river sectors in different parts of the Pithoragarh district have been studied during investigations recording 92 incidences of landslides, subsidence, and river bank erosion (Figs. 14.7 and 14.8, Table 14.1).

The sector-wise details about the location of all the 92 recorded incidences are tabulated as under.

Out of 92 landslide/subsidence/river bank erosion incidences, 64 nos. of locations are recorded along the highways and roads affecting a cumulative stretch of about 21.7 Km. Out of these 64 locations, the road segments at 40 locations have been recommended for realignment to safer locations, probably at higher elevations away from the river course. The total cumulative stretch suggested for realignment corresponds to about 20.80 Km.

Impact on major villages: The disastrous event of June 2013 has not only affected the communication routes but has damaged many villages in different valleys. The damages can primarily be categorized into three types viz. (1) by intense river cutting and bank erosion, (2) subsidence, and (3) landslides. Most pronounced of these has been river cutting wherein sudden gush of sediment-laden high-velocity current took a heavy toll on property. The river current was so high that it washed several terraces encroaching into the river along with settlements thriving on them. The water level also rose significantly (10–15 m) in narrow river stretches and (5–6 m) in wider sections. Interestingly, the river path changed significantly, and the water, which was confined either to the right or left bank before the incident left its path or started flowing along the other bank. The probable reason for this phenomenon was that the existing river path already had a raised river bed profile owing to the slow accumulation of sediments over a long period of time, and thus, this obstruction gave an easy way for the water current to start flowing from lesser sediment/boulder laden zone. This was the only way; an abnormally high discharge could have progressed further. Consequently, it took a heavy toll on everything that came it's way, particularly low-lying river terraces and in some cases, even significantly high terraces. Such was the volume of water that

FIGURE 14.4

Various slides along the study area, Pithoragarh district, Uttarakhand.

1. Introduction

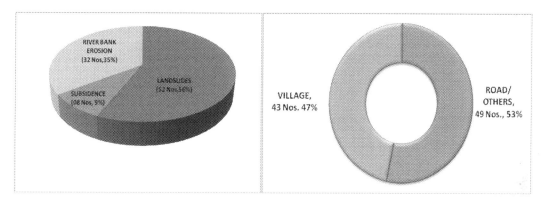

FIGURE 14.5

Distribution of landslide incidences.

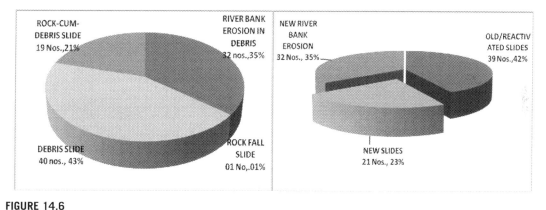

FIGURE 14.6

Distribution of types of slides.

narrow river sections (say 20 m) were widened to as much as 50–70 m as a result of bank erosion. The most affected sectors are along Gori Ganga, Dhauliganga, and Kali rivers, whereas villages in Ram Ganga valley are hardly affected (Fig. 14.9).

Impact on major infrastructures: The sediment-laden water in rivers, particularly Kali, Dhauliganga, and Gori Ganga under high floods and flowing with high velocity and discharge swept away long stretches of communication routes and settlements/houses located in the villages on their banks. Damage was also caused to several infrastructural projects like small hydropower establishments, hydropower schemes/projects, road construction projects, bridges and defense establishments, etc raising serious questions about their safety in the susceptible Himalayan set-up. One Major Power station, 07 small/micro hydel schemes, and 07 steel/suspension bridges have been damaged by the event (Fig. 14.10).

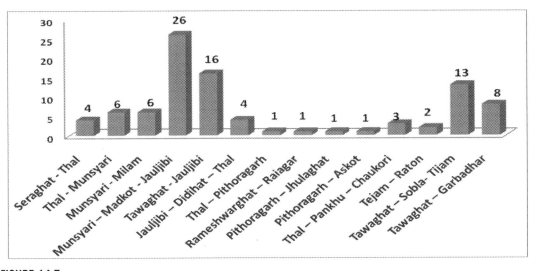

FIGURE 14.7

Sector-wise distribution of incidences.

1.3 Geotechnical assessment

The ground observations indicate that the brunt of devastation is mainly confined along three major rivers draining the area of study, that is, Gori Ganga, Dhauliganga, and Kali. Interestingly, all road sectors away from the proximity of major rivers have barely been affected. It probably suggests that the precipitation on the dates of the natural disaster was mainly confined to very high reaches, and the damages have been caused by probable glacial descent (avalanches), cloud bursts, flash floods, torrent stream conditions, incessant precipitation, and increase in discharge along these rivers in the district. Reportedly, there was not much rainfall recorded in the subdivisions of Dharchula, Munsyari, Didihat, Berinag, or Gangolihat, that is, in the lower reaches of the district.

Glacial melting in peak summer months, their triggering, and supplemented by incessant rainfall could be seen as a pertinent reason for the damage caused along major perennial glacial-fed rivers. However, the information on discharge data, inflow, and daily precipitation especially in the higher reaches is meager and incomplete; therefore, the cause of the devastation is difficult to be quantified.

On the contrary, the areas adjacent to the Ram Ganga river, which is mainly fed by Ram Ganga glacier and which descends from a slightly lower elevation have suffered significantly lesser devastation as compared to Milam and Ralam glaciers feeding Gori Ganga and Dhauliganga rivers. It could be that either the rainfall in the Ramganga basin may have been lesser as compared to Gori Ganga and Dhauliganga river basins or the contribution/discharge from the glaciers was less (Fig. 14.11, Table 14.2).

There is a characteristic deflection from the normal in almost all the districts of Uttarakhand with Dehradun (1436%) recording the maximum departure in rainfall and Pithoragarh district possibly recording the minimum departure (238%) from normal. The normal and actual rainfall in Pithoragarh district is 73 and 246.9 mm, respectively, during the period from 13 to 19 June 2013. The actual rainfall

FIGURE 14.8

Various slides/damages along the study area, Pithoragarh district, Uttarakhand.

Table 14.1 Co-ordinates, toposheet, and ground locations of observed incidences.

Sector	Loc. No	Latitude and longitude	Topo-sheet No	Location
Seraghat–Thal Sector	0001	29°42′33″/79°55′05″	62 O/14	~4.1 km from Seraghat toward Thal
	0002	29°42′50″/79°55′18″	62 O/14	~4.5 km from Seraghat toward Thal
	0003	29°49′35″/80°07′01″	62 C/01	~5 km from Thal toward Seraghat
	0004	29°49′38″/80°07′11″	62 C/01	3.5 km from Thal toward Seraghat
Thal–Munsyari Sector	0005	29°55′57″/80°08′52″ (D/s) 29°56′01″/80°08′46″ (U/s)	62 C/01	~4 km downstream of Tejam village on Thal Munsyari road
	0006	30°00′16″/80°08′46″	62 B/04	Near village Phulibagar on the left bank of river Jakula
	0007	30°00′57″/80°09′02″	62 B/04	~7.7 km from Tejam toward Munsyari
	0008	30°01′20″/80°09′00″	62 B/04	~8.5 km from Tejam toward Munsyari
	0009	30°01′15″/80°09′25″	62 B/04	~8.5 km from Tejam toward Munsyari (Vil-La, Jekhla)
	0010	30°03′15″/80°13′02″	62 B/04	On Thal-Munsyari road, 12 km from Birthi
Munsyari–Milam Sector	0011	30°06′23″/80°14′12″	62 B/04	7 km from Munsyari, toward Milam
	0012	30°06′27″/80°14′11″	62 B/04	~7 km from Munsyari toward Milam
	0013	Senar- 30°06′25″/80°13′47″ Pyangti-30°06′40″/80°13′30″	62 B/04	On the left bank of Senar gad, a tributary of river Goriganga
	0014	30°06′51″/80°14′23″	62 B 04	On the left bank of Suringhata Nadi (Vil-Dhapa)
	0015	30°07′54/80°14′46″	62 B/04	~10 km from Munsyari toward Milam
	0016	30°08′04/80°15′01″	62 B/08	~11 km from Munsyari toward Milam (Vil- Jimi tok, R/b of Goriganga)
Munsyari–Jauljibi Sector	0017	30°07′20″/80°15′22″	62 B/08	Vil-Kultham on the left bank of Goriganga river
	0018	30°06′27″/80°15′04″	62 B/08	

Table 14.1 Co-ordinates, toposheet, and ground locations of observed incidences.—cont'd

Sector	Loc. No	Latitude and longitude	Topo-sheet No	Location
Munsyari–Jauljibi Sector	0019	30°05′09″/80°15′10″	62 B/08	Vil- Talla Dumar on the right bank of river Goriganga ±8 Km from Munsyari toward Madkot near the steel bridge
	0020	30°04′51″/80°16′16″-Kainthi 30°04′04/80°16′00″–Paikuti	62 B/08	Village Kainthi, Paikuti, ±6 km from Madkot toward Munsyari
	0021	30°04′27″/80°16′46″	62 B/08	Village Bhadeli, Munsyari-Jauljibi Rd
	0022	30°04′07″/80°17′20″	62 B/08	3 km from Madkot toward Munsyari
	0023	30°03′27″/80°17′32″-u/s 30°03′16″/80°17′37″ - In d/s near school	62 B/08	Madkot village (proper), 22 Km from Munsyari on Munsyari - Jauljibi road
	0024	30°02′53″/80°18′15″	62 B/08	Village Devibagar (walthi), 1 Km from Madkot toward Bangapani
	0025	30°00′50″/80°19′8″	62 B/08	1 km from Seraghat toward Madkot
	0026	30°0′33″/80°19′13″	62 B/08	Village Seraghat, 7 km from Madkot toward Bangapani
	0027	29°59′08″/80°19′06″	62 C/05	11 km from Madkot toward Bangapani
	0028	29°58′40″/80°18′54″ (U/s) 29°58′04″/80°18′25″ (D/s)	62 C/05	12 km from Madkot toward Bangapani
	0029	29°58′11″/80°18′44″	62 C/05	2 Km from Bangapani toward Madkot
	0030	29°57′54″/80°18′50″	62 C/05	5 km from Seraghat toward Bangapani
	0031	30°57′41″/80°18′22″	62 C/05	Vill- Mawani Dawani, U/s of Bangapani
	0032	29°57′30″/80°18′17″	62 C/05	100m from Bangapani toward Madkot
	0033	29°57′30″/80°18′17″	62 C/05	Village Bangapani, 15 Km from Madkot toward Jauljibi

Continued

Table 14.1 Co-ordinates, toposheet, and ground locations of observed incidences.—cont'd

Sector	Loc. No	Latitude and longitude	Topo-sheet No	Location
	0034	29°56'54"/80°18'08"	62 C/05	Village- Chhoribagar, 2 km from Bangapani toward Jauljibi
	0035	29° 55'32"/80° 18'16"	62 C/05	12 km from Baram toward Bangapani
	0036	29°54'19"/80°18'46"	62 C/05	Village- Mori Talla, 6.5 km from Bangapani toward Jauljibi
	0037	29°54'12"/80°18'55"	62 C/05	9 km from Baram toward Bangapani
	0038	29°53'33"/80°19'05"	62 C/05	Village-Lumti (Talla), 8 km from Bangapani toward Jauljibi
	0039	29°52'50"/80°19'32"	62 C/05	6 km from Baram toward Bangapani
	0040	29,°49'45"/80,°21'58"	62 C/05	Village: Ghatta bagar (Bander khet), 3.2 Km from Baram toward Jauljibi
	0041	29°49'05"/80°21'54"	62 C/05	10 km from Jauljibi toward Bangapani
	0042	29°48'34"/80°22'13"	62 C/05	7.5 km from Jauljibi toward Bangapani
Tawaghat—Jauljibi Sector	0043	29°57'29"/80°36'03"	62 C/09	Tawaghat village, 16 Km u/s of Dharchula
	0044	29°57'08"/80°35'40"	62 C/09	About 2 km downstream of Tawaghat
	0045	29°55'22"/80°34'11"	62 C/09	Vil-Ela gad, about 6 km downstream of Tawaghat
	0046	29°54'06"/80°34'12" (u/s) 29°54'01/80°34'12" (d/s)	62 C/09	About 2 km from Ela gad toward Dharchula
	0047	29°53'06"/80°34'02"	62 C/09	About 8 km from Tawaghat toward Dharchula
	0048	29°52'37"/80°33'45"	62 C/09	Vil-Dobat, about 9 km from Tawaghat toward Dharchula
	0049	29°52'20"/80°33'40"	62 C/09	About 700 m from Dobat village toward Dharchula

Table 14.1 Co-ordinates, toposheet, and ground locations of observed incidences.—cont'd

Sector	Loc. No	Latitude and longitude	Topo-sheet No	Location
	0050	29°52′13″/80°33′21″	62 C/09	Vil-Tapovan, about 10 km from Tawaghat toward Dharchula
	0051	29°51′15″/80°32′58″ (from road level)	62 C/09	Vil-Khotala, about 3.5 km from Dharchula toward Tawaghat
	0052	29°50′48″/80°32′11″ (from road level)	62 C/09	Dharchula, about 500 m d/s of a suspension bridge across Kali
	0053	29°49′25″/80°30′26″ (Nigalpani) & 29°49′02″/80°29′59″	62 C/09	3 Km from Dharchula toward Jauljibi
	0054	29°48′18″/80°29′41″	62 C/05	Vil-Joshikhet, 1 Km from Kalika village toward Jauljibi
	0055	29°47′37″/80°29′32″	62 C/05	4 Km from Kalika village toward Jauljibi
	0056	29°47′48″/80°28′56″	62 C/05	Vil-Nayee Basti-sipu, 12 Km from Dharchula toward Jauljibi
	0057	29°48′05″/80°25′47″	62 C/05	Vil- Baluwakot, about 16 km from Dharchula toward Jauljibi
	0058	29°45′03″/80°22′53″	62 C/05	Vil- Jauljibi
Jauljibi-Didihat-Thal sector	0059	29°45′30″/80°20′51″	62 C/05	±5 km from Askot toward Jauljibi
	0060	29°45′50″/80°19′53″	62 C/05	±1 km from Askot toward Jauljibi
	0061	29°45′14″/80°14′07″	62 C/01	Village- Jaurasi in Didihat Tehsil
	0062	29°47′37″/80°14′41″	62 C/01	Village- Urgaon in Didihat Tehsil
Thal-Pithoragarh Sector	0063	29°42′33″/80°09′44″	62 C/02	On Thal –Pithoragarh road
Rameshwarghat–Raiagar sector	0064	29°34′50″/80°04′50″	62 C/02	±22 km from Gangolihat toward Rameshwar Ghat
Pithoragarh –Jhulaghat Sector	0065	29°34′17″/80°22′58″	62 C/06	Village-Jhulaghat
	0066	29°45′40″/80°18′58″	62 C/05	

Continued

Table 14.1 Co-ordinates, toposheet, and ground locations of observed incidences.—cont'd

Sector	Loc. No	Latitude and longitude	Topo-sheet No	Location
Pithoragarh—Askot Sector				~5 km from Askot toward Pithoragarh
Thal-Pankhu-Chaukori sector	0067	29°49′21″/80°08′10″	62 C/01	Village-Garjia tok (Jyoguda-Thal)
	0068	29°53′03″/80°02′28″	62 C/01	Village-Haliyadob
	0069	29°53′14″/80°02′18″	62 C/01	Village-Gelthi (Dasholi)
Tejam—Raton sector	0070	29°57′29″/80°07′10″	62 C/01	~3.5 km from Tejam toward Masuriya Kanth
	0071	29°59′04″/80°05′11″	62 C/01	~12.5 km from Tejam toward Masuriya Kanth
Tawaghat-Sobla-Tejam sector	0072	29°58′37″/80°35′23″	62 C/09	3 km u/s of Tawaghat
	0073	29°58′41″/80°35′03″	62 C/09	4.5 Km u/s of Tawaghat
	0074	29°58′41″/80°35′03″	62 C/09	5 Km u/s of Tawaghat
	0075	29°59′32″/80°33′56″	62 C/09	6 km U/s of Tawaghat between Jamkhu and Khet villages
	0076	29°59′50″/80°33′58″	62 C/09	Village-Khet, 7 km u/s of Tawaghat
	0077	30°01′06″/80°34′18″	62 B/12	Village-Kanchauti and Khimtalla
	0078	30°01′17″/80°34′23″	62 B/12	10.5 km u/s of Tawaghat
	0079	30°02′02″/80°34′58″	62 B/12	12 km u/s of Tawaghat, just d/s of Suwa gad
	0080	30°02′22″/80°35′09″	62 B/12	Village: Jhimargaon
	0081	30°03′12″/80°35′16″	62 B/12	Village: Nyu (Suwa)
	0082	30°03′34″/80°35′14″	62 B/12	Village: Sobala and Thari
	0083	30°04′17″/80°36′12″	62 B/12	Village: Dar
	0084	30°03′52″/80°34′14″	62 B/12	Village-Tijam
Tawaghat—Garbadhar sector	0085	29°57′38″/80°36′53″	62 C/09	About 1.7 km u/s of Tawaghat
	0086	29°57′55″/80°37′58″	62 C/09	About 3 km u/s of Tawaghat
	0087	29°57′51″/80°38′09″	62 C/09	About 3.6 Km u/s of Tawaghat
	0088	29°57′34″/80°38′58″	62 C/09	About 4.7 Km u/s of Tawaghat
	0089	29°57′30″/80°40′07″	62 C/09	7 km u/s of Tawaghat
	0090	29°57′41″/80°40′45″	62 C/09	8 km u/s of Tawaghat
	0091	29°59′51″/80°43′00″	62 C/09	15 km u/s of Tawaghat, 1 km d/s of Mangti bridge
	0092	30°00′09″/80°43′09″	62 B/12	16 km u/s of Tawaghat, 150 m d/s of Mangti bridge

 a. Location of the earlier suspension bridge (Swept away) connecting with Nepal at Jauljibi	 b. Location of earlier SSB check post (Swept away) on the right bank of river Kali at Jauljibi
 c. Rock slide due to road widening on Jauljibi Didihat road	 d. Debris slide (scars) on Jauljibi-Didihat road Near Askot village, Pithoragarh
 e. Tilting of the house at Village Jaurasi due to subsidence	 f. House at risk in Jaurasi village due to subsidence near nala edges

FIGURE 14.9

Various slides/damages along the study area, Pithoragarh district, Uttarakhand.

a. The weir gates of the Himalayan Hydro project on Paina gad badly damaged

b. Another view of the damaged gates of the weir, removal of excess silt in progress

c. Water diversion concrete structure badly damaged on the left bank of Paina gad

d. Huge boulders brought along by the water current responsible for major destruction to the scheme

e. Downstream view of the weir site severely affected by the incessant discharge and Huge boulders

f. Damages as observed in the open desilting tank of the Himalayan Hydro project on Paina Gad on its left bank

FIGURE 14.10

Various slides/damages along the study area, Pithoragarh district, Uttarakhand.

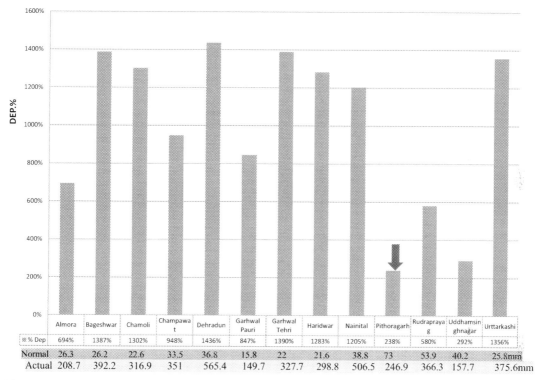

FIGURE 14.11
Percentage of rainfall departure in all districts of Uttarakhand.

Table 14.2 Deflection in rainfall departure data of all the districts of Uttarakhand.

District	Actual (mm)	Normal (mm)	% Dep
Almora	208.7	26.3	694%
Bageshwar	391.2	26.3	1387%
Chamoli	316.9	22.6	1302%
Champawat	351	33.5	948%
Dehradun	565.4	36.8	1436%
Garhwal Pauri	149.7	15.8	847%
Garhwal Tehri	327.7	22	1390%
Haridwar	298.8	21.6	1283%
Nainital	506.5	38.8	1205%
Pithoragarh	246.9	73	238%
Rudraprayag	366.3	53.9	580%
Uddhamsinghnagar	157.7	40.2	292%
Uttarkashi	375.6	25.8	1356%

Source: India Meteorological Department (2013).

in the week was up to 1436% of normal rainfall and was excessive in all districts of Uttarakhand. The state received 322 mm of rainfall in the week, which was 847% higher than the normal rainfall of the week at 34 mm. It is rather strange to see such a vast area facing simultaneous high-intensity rainfall, an event not recorded in the last 100 years or so. The district despite showing minimum deflection among the rest of the districts also witnessed an increase of 238% which is phenomenal.

Another cause for heavy flooding may be that in the premonsoon season, there has been heavy snowfall in the Himalayas, and by June, the snow started melting thus increasing the water level in the glacial-fed rivers. Moreover, the intense rainfall in the higher reaches helped in causing the melting of the snow much faster, as water which has the highest heat capacity than air, helps in melting of snow or ice much faster when it comes in contact, even when both air and water have the same temperature. This phenomenon and climate change need to be understood in this context.

Apart from this, the river morphology, geological set-up adjacent to it, land encroachment along the riverside, river gradient, width of the river and carrying potential, river-borne vis-a-vis rocky terraces, and, in turn, change in river path could well have acted as a catalyst to the natural disaster.

River profiles: A study of the river profile along the three major rivers is also attempted to draw a comparison in the river gradient and other factors, which may have been responsible for destruction in typical sectors (Table 14.3).

River Gori Ganga: It is one of the important high-order drainages in the area of study and has witnessed significant devastation, particularly along its banks. The river descends from an approximate elevation of 4570 m above msl to its confluence with river Kali at EL 580 m near Jauljibi. It flows initially in the north-to-south direction (62 B/03) and shifts toward the southeast up to its confluence with River Kali. It is fed prominently by Milam, Shakram, and other minor glaciers, which originate at a height of about 6300 m. Along its course, it is joined by tributaries like Ralam River, Pyunsani

Table 14.3 Details of the morphology and damage pattern of the major rivers.

River	Sector	Gradient in %	Geology/material involved	Damage pattern
Gori Ganga	Kultham—Josha	1.92	Overburden debris, Colluvium/Hill wash and river borne material (RBM)	Slide/River bank erosion
	Josha—Sheraghat	2.27		River bank erosion
	Sheraghat—Kalikabagar	1.78		
	Kalikabagar—Jauljibi	1.23		
Kali	Garbadhar—tawaghat	3.07	Overburden debris, Colluvium/Hill wash	Slides/River bank erosion
	Tawaghat—Dharchula	1.70	Colluvium/Hill wash and river borne material (RBM)	River bank erosion
	Dharchula—Baluwakot	1.00		
	Baluwakot—Jauljibi	0.90		
Dhauliganga	Tijam—Sobla	12.12	Glacial debris/RBM	River bank erosion/slides
	Sobla—Khet	3.68	Colluvium/Hill wash and (RBM)	
	Khet—tawaghat	4.26		

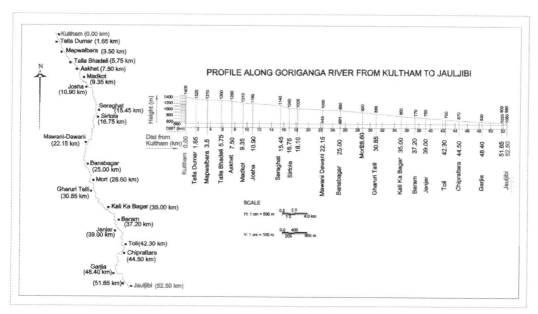

FIGURE 14.12

Profile along Gori Ganga river from Kultham to Jauljibi.

gadbhera, Mandakini River, Palna *gad*, Ghosi *gad*, and Rauntis *gad*, which significantly contribute to its discharge (Fig. 14.12).

River Kali: The river forms the international boundary between India and Nepal. In the Garbadhar—Jauljibi sector, the river initially flows in a south—west to east—west direction, taking a conspicuous ENE—WSW trend near Tawaghat up to Salvari, flowing almost westerly up to Dhunga before flowing southward up to Jauljibi. There are prominent bends at Tawaghat, Salvari, and Dhunga and several other small kinks all along the course of the river (Figs. 14.13 and 14.14).

River Dhauliganga: The sector from Tijam to Tawaghat along river Dhauliganga suffered the maximum damages due to the natural disaster. The river flows from north to south taking a conspicuous turn toward east near Duk-Ki-Gad and finally bending toward south at Gurguwa village. The river in the sector has a high gradient (4.77%) along a length of 16.75 km. Between Tijam and Sobala (along Thari gad), the gradient is very high at 12.12%, between Sobala and Khet, it is 3.68%, and between Khet and Tawaghat, it is 4.26%. Understandably, the entire stretch has suffered widespread damages with Tawaghat, Gurguwa, Khet, Kanchuti, Khimtala, Suwa, Sobala, and Tijam being the worst affected centers. Besides this, the motorable road connecting Dar (immediately northeast of Sobala) and Tawaghat (~22 km) mostly along river Dhauliganga has been badly damaged and left unoperational. Barring a few patches which have escaped the effect of flooding and bank erosion, the transportation route has either been damaged or completely washed away. The infrastructural set-up has also been badly affected (Figs. 14.15 and 14.16).

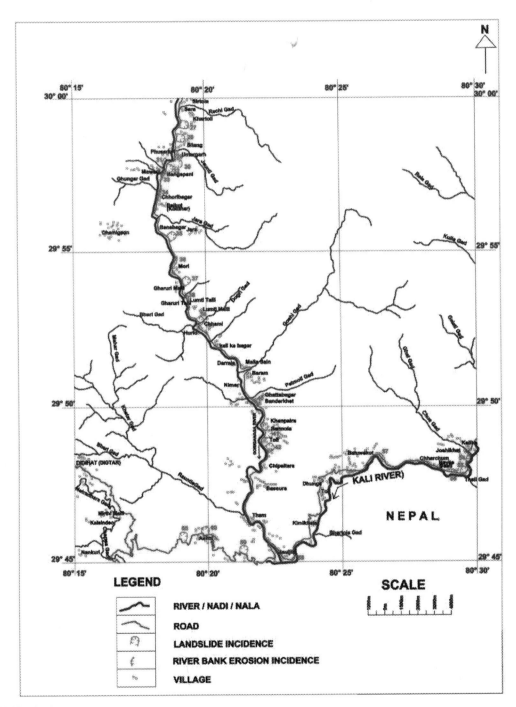

FIGURE 14.13

Kali river track showing events in the study area, Pithoragarh, Uttarakhand.

1. Introduction   399

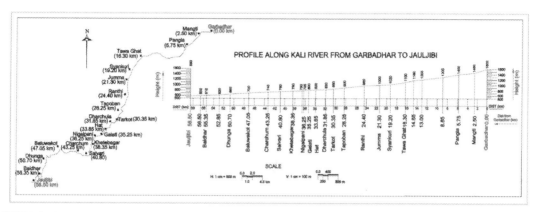

FIGURE 14.14

Profile along Kali river from Garbadhar to Jauljibi.

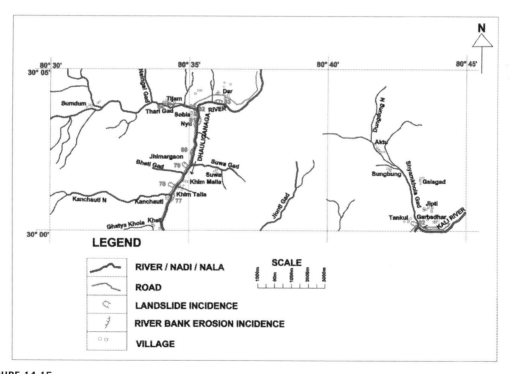

FIGURE 14.15

Dhauliganga river track showing events in the study area, Pithoragarh, Uttarakhand.

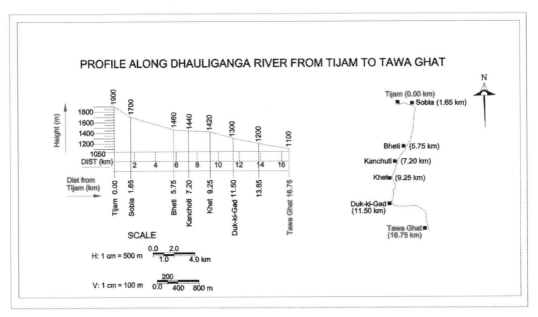

FIGURE 14.16

Profile along Dhauliganga river from Tijam to Tawa Ghat.

2. Conclusions

The climatic turnaround leading to environmental disaster and several induced effects in the hilly state gives credence to the "small time corridor" catastrophe, which has taken many provinces by surprise. Their recurrence with unfailing regularity in different parts of the country and sometimes unfamiliar territory drives the point that such abnormal precipitation cannot be delinked to a climatic aberration and could perhaps be a testimony to short-term climate change and a precursor to a change that will be irreversible.

In the case of the hilly district under reference, the local population which has significantly increased over the years was rather oblivious of a climatic turnaround and its potential to cast such extensive damage to life and property. The lack of preparedness of people in the hills locating themselves on the river-borne terraces of major rivers found themselves sitting on a catastrophe in the wake of the event. The fragility of the mountain chain with high relief, the adverse inclination of the discontinuity planes, intense rain, flash flood, and avalanches made the Himalayan terrain vulnerable to an event like this. Arguably, excessive human interference with a lack of awareness and preparedness contributed to the degradation of the Himalayan ecosystem.

What surprises me is, that except for the vagaries of climate dominating these corridors of uncertainty, the precipitation regime on a seasonal basis barring this window more or less remained normal or below normal in most of the cases. This erratic climatic disposition unsettling the behavioral aspects of these regions calls for an awakening to assess them from a wider perspective.

Prudent steps were taken by the district administration to provide relief and rehabilitation for those who lost their lives or adobe. But once normalcy would have regained, would the people and administration take lessons and proactive steps from this and similar disasters not to repeat their earlier mistakes and yet again find themselves on the receiving ends of the uncertainties of nature?

Detailed susceptibility mapping, risk zoning, adopting systematic and scientific planning with a backup arrangement and giving space to reforms which make conducive but stringent guidelines are probably what is required for better preparedness for the future to combat perceptible shifts in the climatic regime.

Acknowledgments

The authors express their sincere gratitude to the Director General, Geological Survey of India, CHQ, Kolkata for allowing the carrying out of this work of great societal relevance. The authors are grateful to the Deputy Director General and HOD, GSI, NR, Lucknow and Dy. Director General, (Mission IV activities), CHQ, GSI, Kolkata for their moral support and constant encouragement. The authors are thankful to the district administration, Pithoragarh and Sub-divisional Magistrates of Munsyari, Dharchula and Didihat Tehsils for providing infrastructural support and desired information in regard to the natural disaster.

References

Akhouri, B., Gajbhiye, P.K., Dhakate, M.V., 2013. Preliminary assessment of disaster-affected areas of Pithoragarh District, Uttarakhand (Post landslide/flood disaster of June 2013). Indian Journal of Geosciences 67 (3 and 4)), 329–336.
India Meteorological Department, 2022. Hydromet Division, New Delhi, District-Wise Rainfall Distribution.
SAE, 1996. Closed-test-section wind tunnel blockage corrections for road vehicles. SAE SP-1176.
India Meteorological Department, 2021. Hydromet Division, New Delhi, District-Wise Rainfall Distribution.
India Meteorological Department, 2019. Hydromet Division, New Delhi, District-Wise Rainfall Distribution.
India Meteorological Department, 2013. Hydromet Division, New Delhi, District-Wise Rainfall Distribution.
Sharma, V.K, Geological Assessment of Landslide Disaster of 8th August 2009, District Pithoragarh, Uttarakhand (Unpublished GSI Report), F. S. 2008-09.

Further reading

Akhouri, B., Gajbhiye, P.K, Dhakate, M.V, Interim Report on Preliminary Slope Stability Assessment of Recent Disaster Affected Areas of Pithoragarh District, Uttarakhand (Unpublished GSI report, F.S, 2013-14).

CHAPTER 15

Geological/paleontological applications in marine archeology: few examples from Indian waters

Rajiv Nigam

CSIR-National Institute of Oceanography, Dona Paula, Goa, India

The instinct of all human beings is to know his/her roots and get attached to the same. Archeology helps them to connect to their cultural heritage and makes them proud of their roots. The dividing line between history and archeology is a little diffused. History is generally considered as written records of the developments and happenings of the different rulers and dynasties, their rise and fall, and the living conditions of the man during their regime with some information on the climatic fluctuations. Whereas artifact-based reconstructions of similar information, beyond the written records, is archeology.

According to Dr. Graham Clark, Professor of Archeology at the University of Cambridge, "archeology may be subtly defined as the systematic study of antiquities as a means of reconstructing the past." Archeology is not only of academic importance to cultural heritage but also has large financial implications as well due to Archeo-tourism (Table 15.1).

Table 15.1 Earning in Rs. During 2013–14 [Hampi 2016–17] from few popular archeological/historical sites exhibiting scope for archeotourism in India as we have much older spectacular sites but lesser known like Lothal and Dholavira.

Tajmahal	21,84,88,950
Kutubminar	10,16,05,890
Agra fort	10,22,56,790
Humanu tomb	7,12,88,110
Red fort	6,15,89,750
Fatehpur sikri	5,62,14,640
Mahabalipuram	2,72,93,480
Konark	2,43,52,060
Khajuraho	2,24,47,030
Ellora	2,06,72,820
Hampi [2016–17]	3,40,00,000

Therefore, archeological studies are gaining a lot of attention. To enhance the value and reliability of archeological interpretations, inputs from the different branches of sciences are the need of the hour. Here, we attempted to exhibit the usefulness of the geological/paleontological tools in the field of marine archeological investigations. Marine archeology is the branch of archeology where the sea has played a role in shaping the destiny of the ancient coastal towns and other maritime activities (Nigam, 2005). The study of shipwrecks also comes under marine archeology. But, we decided to concentrate here on the role of geology in explaining or making new findings related to ancient coastal towns with examples from the Indian waters.

Based on the artifacts, a number of climatic changes in the past are documented in archeological investigations, which are reported to have been the cause of the rise and fall of human civilizations. The unique contribution of the marine sediments has been for deciphering the changes in oceanographic conditions due to climatic variations in the past. Since the common aim of oceanography and archeology lies in the illumination of the past, it is apparent to bring coherence between the two.

There are various geological, geophysical, and paleontological tools have applications in the field of marine archeological studies (Nigam, 2005). However, in recent years, studies of microfossils (mainly foraminifera) and ground penetrating radar (GPR) gain special significance and are discussed here in detail with examples.

1. Ground penetrating radar for subsurface information

Much information about the past is hidden under the cover of sediments in the coastal areas like Dholavira, Mahabalipuram, etc. Archaeologists go for excavations in any potential archeological sites based on their experience based on preliminary information available on the surface. Excavation is a very expensive and time-consuming process and sometimes does not ravel the expected structures and cause disappointment and discouragement. The excavation of every nook and corner of the big archeological sites is not possible due to logistic reasons, and thus, sometimes, the important information remained buried forever. This necessitated the need for an instrument, which can give clues about buried structures without excavation.

Ground penetrating radar (GPR) is such an instrument, which can assist the archeologist in very effective ways. GPR is a geophysical tool designed to generate sub-surface details. Initially, this instrument was extensively used by civil engineers for locating buried pipelines, cables, etc. GPR is a tool archeologist which is used more and more for excavations in the 21st century.

GPR essentially consists of a controlling unit, recording, and storage unit and antenna (Fig. 15.1).

The guiding principle, the antenna, transmits pulses of high-frequency (radio waves) energy into materials. The other unit records the strength and the time required for the return of any reflected energy. The reflection phenomenon is mainly governed by the electrical properties of the ground. A GPR profile is generated when the antenna is moved along the surface. This movement can be done by hand, by vehicle, or even by air. The radar unit emits and receives reflected signals up to a 1000 times per second. As a result, not only do the relative depths and "strengths" of the targets appear, but the image or shape of the target is "seen" on the monitor. GPR data are presented as a Radargram (Fig. 15.2).

This instrument is used to study the subsurface sediments in Dholavira and Mahabalipuram areas, and results are discussed later.

FIGURE 15.1

Ground penetrating radar (GPR). (A) different components of the instruments, (B) survey for a small area by dragging the antenna with a walk, and (C) survey for the large area by car by dragging the antenna.

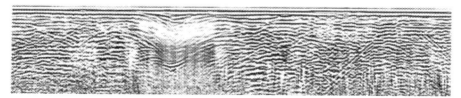

FIGURE 15.2

The output of ground penetrating radar (GPR) as radargram showing subsurface features.

2. Marine sediments as a source of reconstruction of the past

There is a continuous supply of sediments from land to the sea, either by river or wind. These sediments contained a lot of information about the climate and keep getting deposited year by year and layer by layer. These layers also contained the fossils of the organisms that lived in equilibrium with seawater conditions. Therefore, these layers served as pages of climatic changes on Earth at various time scales. With help of research vessels, through coring equipment, these layered sediments can be collected (Fig. 15.3). With the help of microfossils like foraminifera and dating techniques like radiocarbon dates, we can reconstruct the past climate on any time scale from year to thousands of years.

3. Microfossils with special reference to foraminifera

As discussed above, microfossils embedded in layered marine sediments are the prime source for the reconstruction of past climates. The study of microfossils is known as micropaleontology, which is an important subdivision of marine geology and oceanography. Based on composition, morphology, and ecology, microfossils are classified as Calcareous (Foraminifera, Calcareous Nannoplankton, Ostraoda, Pteropods, Calcareous Algae, Bryozoa), Silicieous (Radiolaria, Marine diatoms,

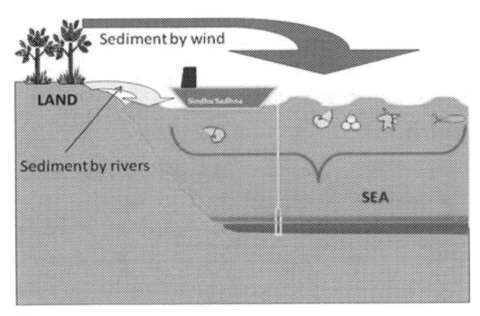

FIGURE 15.3

Coring through layered sediments (embedded with fossils of microorganisms) deposited at the bottom of the sea.

Silicoflagellates), Phosphatic (Conodonts), and organic-walled (Dinoflagellates, Acritarchs and tismanida; spores and pollen) microfossils (Haq and Borsema, 1998). Among these, foraminifera has been used traditionally for biostratigraphy with applications in oil industries. Besides this, foraminifera is useful (Fig. 15.4) for studying sea-level changes, paleomonsoons, sediment transport, paleotsunamis, fisheries, pollution monitoring, etc (Nigam, 2005). The usefulness of foraminifera in reconstructing paleomonsoon reconstructions and application of the same to support the two great floods in 2000 and 1500BCE reported in archeological literature (Rao et al., 1963) has already been published (Nigam and Khare, 1992). The main objective of this article is to create more awareness about their use as a tool to address the marine archeological applications where sea level fluctuations played a great role by providing examples from the Indian waters. In this communication, examples are given where either the presence/absence of foraminifera as an indicator of marine and nonmarine past environments (Lothal and Dholavira) or the sea-level curves (largely using foraminifera-based information) are used to explain marine archeological findings like Lothal, Neolithic settlements off Surat, Dholavira, Submerged Dwarka and Ramsetu, etc. Therefore, it is essential to know some basic aspects of foraminifera before discussing their applications.

4. Foraminifera

Foraminifera (except for a few species without good fossilization potentials) are exclusively marine organisms (Boltovskoy and Wright, 1976). They have widespread geographical (horizontal) and bathymetric (vertical) distribution in ocean including marginal marine bodies like estuaries, lagoons,

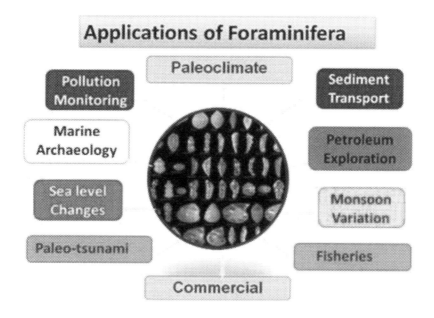

FIGURE 15.4

Applications of foraminifera (microfossil) in different fields including marine archeology.

bays, etc. With reference to habitat, they are classified as benthic (or benthonic) and planktic (or planktonic) (Fig. 15.5).

They are very sensitive to all the physico-chemical environmental parameters. During their lifetime, their hard covering shell (known as a test), which is made up of calcium carbonate, incorporates the effect of prevailing parameters. After their death, this test gets buried in sediments as fossils and served as the key to reconstructing past climatic conditions.

5. Reconstruction of past sea-level changes

It is important to understand sea level fluctuations in the recent past to assess accurately the magnitude of the suspected future rise in sea level due to global warming. At the same time, knowledge of sea-level fluctuations helps to understand the rise and fall of several important cities in coastal areas.

Sea-level fluctuations studied all over the world by many workers (Nigam and Henriques, 1992) cannot be applied as such to our region. Due to the possible role of neotectonics and other factors, curves exhibiting relative sea-level changes are expected to be region specific. In numerous international and national conferences, it has been emphasized that specific areas should have their Holocene sea-level curves rather than following or adopting the generalized curves generated for other regions. The final report of the International Geological Correlation Program (sea-level changes during the last hemicycle) also contains recommendations for these effects (Pirazzoli, 1991).

In India, some of the earliest attempts to study variations within the Holocene sea level on the western Indian shelf were made by Nair (1974), kale and Rajaguru (1985), Merh (1987, 1992). Later,

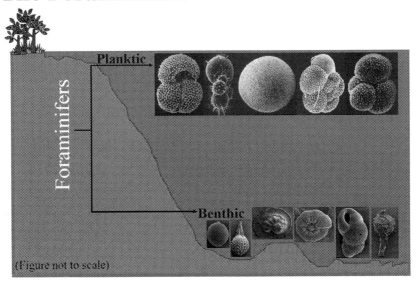

FIGURE 15.5

The types of foraminifera as per habitat, benthic and planktic.

based on the foraminifera, many publications appeared which had enhanced our understanding of sea-level fluctuations. Nigam et al. (1990) collected evidence of paleo-sea level fluctuations in religious and archeological records and supplemented them with inferences from oceanographic studies. Based on the planktic percentage of foraminifera in the surface sediment of the Arabian Sea, a regional model was developed for paleodepth determination. One of the main advantages of the proposed model is that it requires no detailed taxonomic study of the fauna. It is sufficient to separate the fauna into two groups, that is, planktic and benthic. Preliminary sea-level fluctuation curves can be prepared by using paleobathymetric information through the proposed model. Such information is of immense use to develop predictive models (Nigam et al., 1992). During investigations, it was discovered that barnacle fouling on relict foraminiferal tests could be used as an additional tool for identifying paleoshore line changes (Nigam et al., 1993). During the last 10,000 years, three major episodes of sea-level fluctuations were also reported (Nigam, 1990). To produce an updated sea-level curve for the Arabian Sea, Hashimi et al. (1995) summarized a large number of published records. The studies successfully demonstrated that, as compared to the present, sea level was lower by 100 m about 14,500 years BP and 60 m by about 10,000 years BP. Later publications (Nigam and Hashimi, 2002) used this information to further enhance our understanding of the sea level curve (Nigam, 2012) and its use in marine archeology. Fig. 15.6 exhibits a comprehensive account of the fluctuations in sea levels in the Arabian Sea.

By adopting a similar approach, another comprehensive sea-level curve was published for the Bay of Bengal (Loveson and Nigam, 2016). Many workers, explain a number of paleoclimatic events like the destruction of ancient coral reefs, intensification of monsoons, etc. Here, some examples are given below to show how this curve helps explain some of the important archeological discoveries like Lothal, Dholavira, submerged Dwarka, Neolithic settlement off Surat, Ramsetu, etc.

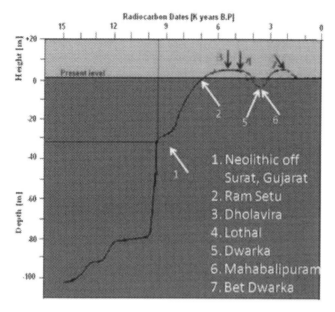

FIGURE 15.6

Holocene sea-level curve showing sea level fluctuations off the West Coast of India (Arabian Sea) and the position of various ancient Coastal towns.

6. Sea-level fluctuations and marine archeology

The archeological monuments which were shaped by fluctuating sea are studied under the heading Marine Archeology. The necessity to study Marine Archeology was felt about 4 decades ago, and Marine Archeology emerged as a new arena of scientific pursuit in India. Dr. S.R.Rao, a retired archeologist from the Archeological Survey of India (ASI), came to NIO in 1981 and started this new branch of Marine Archeology as a project. Later, encouraged by the initial findings, officially in 1990, a unit of Marine Archeology was established at NIO, Goa. Subsequently, both geological evidence and archeological explorations underneath sea waters have worked in tandem but with a synergy between them. Such integrated scientific investigations have yielded good results, and some very interesting long pending unanswered archeological riddles could be solved with the aid of geological interpretations. Some of the examples are discussed below.

6.1 Lothal

Lothal is an important and famous name in Indian cultural heritage, presented as the oldest dockyard (ever discovered in the world as claimed by archeologists) and evidence of advanced maritime activities around 4500 Years B.P. (before present). Lothal was discovered as a result of a systematic survey undertaken by S.R. Rao in the year 1954 as a part of the program locating Harappan settlements within the present-day boarders of India (Fig. 15.7). Out of several important structures excavated, a large basin-like structure (Fig. 15.8A) became the most important and disputed one. Several arguments were given for its possible use. Using almost similar artifacts like triangular and/or rectangular stones (Fig. 15.8B) to interpret the purpose of the structures to different opinions was advanced by galaxies of Indian and foreign archeologists. One school (Rao, 1979–1985; Wheeler, 1973) proposed the

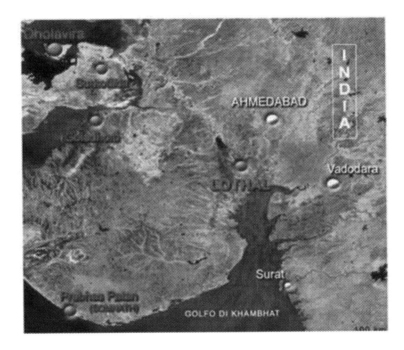

FIGURE 15.7

Location of Lothal (Gujarat), India where the oldest dockyard was discovered. The location of Dholavira is also shown.

possible use as a "Dockyard" for maritime activities (Fig. 15.8C) whereas, other schools (Shah. 1960; Sankalia, 1974; Lesnik, 1968; Fairservice, 1971) opined in favor of "freshwater storage tank" mostly for irrigation purpose (Fig. 15.8D) and bathing. This controversy was continuing for almost 4 decades and was finally solved with the help of foraminiferal studies (Nigam, 1988).

Since fossils of foraminifera are exclusively marine, their presence and absence could be a decisive factor in interpreting whether any ancient water body was filled with fresh or marine (brackish) waters. Therefore, with this intention, samples were collected from the sediments deposited at the bottom of the rectangular body. The study ravels the presence of well-preserved in situ foraminiferal assemblages (Fig. 15.8E) comparable with marginal marine environments (Nigam, 1988). This helped to summarize that the rectangular structure was a dockyard, connected to the sea through the estuarine channel and high tidal range and thus settled the old archeological controversy.

However, one question remained to be answered. If this was a dockyard, how it is on land? This question is reasonably answered through the sea-level curve for this region (Fig. 15.6). Around 4500 years B.P. (time of Lothal), the sea level was higher than today. When the sea level has gone down, this connection was lost and the dock became out of use. These results were further supported by Khadkaikar et al. (2004), who by using remote sensing techniques, reported ancient estuarine channels through which this dockyard was connected to the sea (Fig. 15.8F). However, some limited roles played by neotectonic cannot be ruled out.

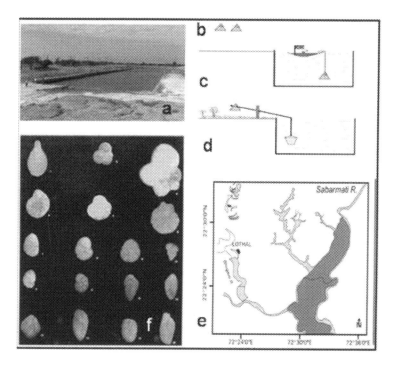

FIGURE 15.8

(A) Tank-like body deciphered as dockyard, (B) triangular stones, (C) showing stones as anchor thus a dockyard, (D) stone as a counter weight to lift the water thus an irrigation tank, (E) remote sensing data showing the connection with the sea, and (F) foraminifera in sediments from the bottom of the rectangular body confirming the dockyard.

Considering the good conditions in which various ancient structures, especially the world's oldest dockyard at Lothal, are there and also the tremendous scope for archaeo-tourisms, under consideration for the tag of the world"s cultural heritage by UNESCO and the Government of India decided to have a National Maritime Museum at Lothal.

6.2 Unknown city off Surat, Gujarat

During 2000–02, the discovery of remnants of an ancient city in an offshore area off Surat in the Gulf of Khambhat (Gupta, 2002) was real excitement when the famous magazine India Today brought out a cover story (Chengappa, 2002) as this was a milestone in marine archeological studies (Fig. 15.9).

As described by Gupta (2002) "The material collected at the site include artifacts, possible construction elements with holes and studs, pot shreds, beads, fossil bones etc. which provide significant evidence of human activity in the area. A detailed examination of the area has revealed riverine conglomerate at a water depth of 30–40 m between 20 and 40 km west of Hazira near Surat (Gujarat)." Based on the radiocarbon dating of wooden pieces recovered from the site, age of 7500 BCE (~9500

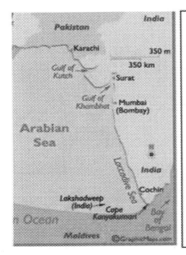 

FIGURE 15.9

Location of Gulf of Khambhat (Combay), timing and special issues of India today on the discovery of ancient submerged city off Surat (Gujarat).

BP) is assigned to the human settlement on the bank of the river that was present at that time. The remnant of this time, "push back the hitherto held view of first human civilization from around 3500BCE (C.3500BC-valley of Sumer; c.3000 BCE, - Egypt and c. 2500 BC-Harappa) to 7500 BCE, thus making the present find the oldest known to man" (Gupta, 2002). Because of its linkage to cultural heritage, this discovery was hotly debated and contested for archeological significance (Bavadam, 2002; kathiroli, 2004 etc.). Here, discussion on artifacts is beyond the scope and may be left to archaeologists. But, the role of sea-level fluctuations regarding this settlement was largely ignored (Nigam and Hashimi, 2002).

If the age of discovered settlement (Gupta, 2002) is ∼7500 B.C. (i.e., ∼9500 years old) as plotted in Fig. 15.6, it gives a depth of 30–40 m water depth. This matches with the depth zone in which this settlement is reported. If the isobaths of 40 m are considered as paleo shoreline, the view of Prof. Rajguru (in Gupta, 2002) comes true that "Bhavnagar and Hazira were probably connected at 7000B.C." This also implies that the river (as suggested) must have been passing through the area (now under the sea) before joining the sea of that time and thus, supporting the postulation of Gupta (2002) "Further, acoustic images present channel-like features indicating the presence of the river in the region." Further observation of the curve (Fig. 15.6) indicates that the sea came to stand still (or rise very slowly) at the said level for some time, thus providing time for civilization to flourish before being engulfed by the sea. This is similar to the happening at submerged Dwarka, which is discussed below.

6.3 Submerged Dwarka

Explorations off Dwarka (Fig. 15.10) by the Marine Archeology Center of the National Institute of Oceanography were initiated by Dr. S.R. Rao in 1983 with small funding from Indian National Science Academy (INSA) and have been continued with other agencies like the Archeological Survey

6. Sea-level fluctuations and marine archeology

FIGURE 15.10

Location of Dwarka and Bet Dwarka (Gujarat).

of India (ASI), Department of Science and Technology (DST), Council and Scientific and Industrial Research (CSIR), Department of Ocean Development (DOS, now Ministry of Earth Sciences) etc (NIO, Report, 2001).

Exploration of Dwarka has created great interest among historians, archaeologists, and oceanographers. Rao (1997) stated, "the existence of a port town more than 3500 years ago and its submergence by the sea are now proved beyond doubt as results of scientific excavation in the Arabian Sea off Dwarka and in the Gulf of Kutch off Bet Dvaraka island both of which are considered holy because they are associated with Sri Krishna's activities" (Rao, 1997). He (Rao, 2006) further stated "… Dwarka on the mainland was built by Krishna, which was contemporary to Bet Dwaraka (Kussthali). It can be dated to the 17th century BC. This date is also confirmed by Gaur and Sundaresh (2004) for the Belapur Fishhook." Alok Tripathi (2015) presents an account of excavations and explorations done on land and in areas of Dwarka led by Prof. H.D. Sankalia of Deccan College and Research Institute, Pune; Dr. S.R Rao and Prof. A. Tripathi of the Archeological Survey of India (ASI) and 2 decades of Marine surveys off Dwarka by the National Institute of Oceanography, Goa. He postulated that " …the structural remains found scattered on the bed of the sea are not in situ but transported by waves and currents."

However, the underwater photographs of the area presented in Fig. 15.11 show stones one above another one exhibiting it as part of an ancient wall and origin as autochthonous. Thus remnants exhibit the existence of an ancient city at that time which was later engulfed by sea. However, the submergence of the Ancient city off Dwarka remained debatable. There are three possibilities suggested for the destruction/submergence of Dwarka. First, the cause could be erosion by the sea. Of course, out of several geological works of the sea, one is to erode the coastal rocks and deposit the eroded material in form of beaches. If beach sediments are eroded and deposited somewhere else, overlaying structures may collapse but that will not cause submergence. Second, the cause could be a tsunami. Yes, a tsunami can submerge a large area on land and destroy many structures due to very high waves. But submergence by the tsunami is a temporary phenomenon. Once the tsunami is over, the water returned

FIGURE 15.11

Photographer showing submerged remnant of ancient wall exhibiting stones one above another indicating in-situ structure.

to its original level and thus cannot be permanently submerged (more than 6 m water depth where structures are found at Dwarka) in any area. Therefore, the last and most pliable explanation is the sea-level fluctuations as the cause of submergence. The relative sea-level curve presented in Fig. 15.6 shows that after the mid-Holocene higher than today's sea levels, a gradual regression of level started and leaving behind the large land area, which could be used to build the city of Dwarka (as the port town). At the time of Dwarka (1500–1700 BCE), the sea level reached to lowest and further transgressions lead to the submergence of Dwarka.

6.4 Dholavira

The recent announcement of declaring Dholavira as India's 40th World Heritage Site by UNESCO created tremendous interest in the common man in knowing more about Archeology in India and abroad. This site was first noticed by Dr. J. P. Joshi of ASI (Fig. 15.7). Later, the archeological excavations in Gujarat from 1989–90 to 2004-05 led to the discovery of Dholavira (Fig. 15.12), which is located on Khadir island between the two minor rivers Mahar and Mansa and comes under district Kutch, Gujarat. It covers an area of about 50 ha (Bisht, 2015). Dholavira is a fortified city and world-famous for its excellent water conservation system. Dholavira is the second-largest Harappan settlement known in India. To quote the Chief excavator, Dr. R.S. Bisht, "it was, perhaps, the best planned Harappan city with several divisions and many new features hitherto unknown."

This well-planned urban settlement flourished for about 1500 years from about 5000 to 3450 years before the present. Archeological excavations show that the township comprised three parts—the castle, middle town and the lower town. Thick walls (12–18 m) around the city and castle was built as a protective measure. The real purpose of the Dholavira thick wall has been a topic of considerable debate. Intriguingly, walls of such thickness are not found even in historic times when conflicts have been more common, and weapons have become increasingly destructive. Even the third century BC China wall is 4.6–9.1 m thick at the base and tapers to 3.7 m at the top. Similarly, protection against floods is not tenable as this is a water-starved area, which is evident from the extensive water conservation system developed at Dholavira. Dholavira being a port town could have been vulnerable to

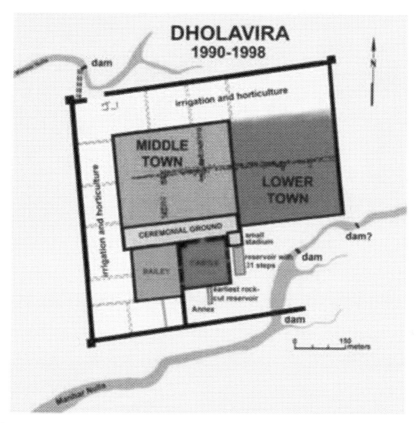

FIGURE 15.12

Site plan of Dhoavira (Gujarat).

oceanic calamities. Are the unusually thick walls an answer to this? Being part of the Makran coast, the area is prone to tsunami-like events Mahmood et al. (2012). Evidence of paleotsunami at 8000-7000 years BP is known from the region (Nigam and Chaturvedi, 2006). Therefore, it can be postulated that ancient Settlers at Dholavira were aware of the vulnerability of the area due to tsunamis but due to trade-related strategic location to build this city around 5 to 6 1000 years back. However, the thick wall was built to protect the town from extreme oceanic events such as tsunamis. Similar protective walls come under coastal hazard management as adopted in Japan and New Orleans in recent times. Therefore, it can be concluded that ancient Dholavirians were aware of tsunami protective measures and that is probably the oldest evidence in the world (Nigam et al., 2016).

To study the possibility of the area impacted by tsunami post-construction, CSIR-NIO has carried out additional work at this site. A team of paleoclimatologists, archaeologists, and geophysicists from the institute surveyed a hitherto unexcavated area using GPR and systematically collected soil samples in the middle town area (Fig. 15.13). The GPR records show a 2.5–3.5 m thick homogenous soil layer (without any layering) below the surface, which suggests its episodic deposition, possible due to an extreme event.

416 Chapter 15 Geological/paleontological applications

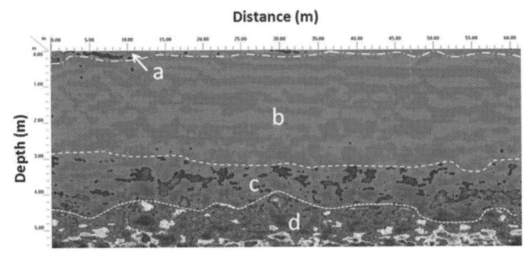

FIGURE 15.13

Ground penetrating radar (GPR) record at middle town of Fig. 15.12 showing thick subsurface sediments without any layering (~3m).

A 2.5 × 2.5 m trench was dug in the north-western corner of the Middle Town to a depth of 3.65 m (Fig. 15.14A). The soil samples have been found to contain fossils of *"foraminifera,"* microscopic organisms that build calcareous shells and live only in seawater. The presence of shells of marine

FIGURE 15.14

(A) Trench at the location of ground penetrating radar (GPR) record showing sediments without layering in middle town and (B) foraminifera encountered in sediments.

organisms foraminifera (Fig. 15.14B) in the soil strongly suggests an episodic deposition of marine sediments in the area. This deposition could have occurred as a result of a massive tsunami. The exact timing of the sediments deposited in Dholavira is yet to be established. The presence of exclusively marine organisms (foraminifera) in subsurface sediments outside the citadel (Middle town) indicated the possibility that Dholavira, at least in part, could have been destroyed by the tsunami. In addition to this, the thick wall in Dholavira shows that the Harappans were not only aware of the potential threats from tsunamis, but they were also pioneers in coastal disaster management.

Studies indicate a need for a long-term detailed survey of the entire area by GPR to decipher the buried structures, geological analysis of the subsurface sediments, dating of the datable material like radiocarbon dating, and/or TL/OSL dating to ascertain the time of the event/s. The deciphered structures in unexcavated area/s may be excavated by the Archaeological Survey of India to expose the full grandiose of this most important archaeological site. These will enlighten the golden pages of the rich cultural heritage of Gujarat, India. A fully exposed site will also enhance tourism potential.

6.5 Ramsetu

Ramsetu (or Adam's bridge) at the southern tip of India also became a very lively controversy in India due to its religious connection with lord Ram and thus a very emotional issue. It is believed that the army of lord Ram constructed the bridge over the sea between India and Sri Lanka. The supposed to be pictured by NASA (Fig. 15.15) also generated a big interest in this area. However, the controversy became very intense when Setu Samdurum Project was conceived by the government to dredge the area so that the ships can pass through this region while going from the Arabian Sea to the Bay of Bengal. Otherwise, the water depth in this area is approximately only 3 m, and therefore, ships have to take more time to encircle Sri Lanka. But the belief of its association with Lord Ram created a big debate due to the response of the government challenging any such structure ever existed/constructed.

However, most of the arguments given in support of the possible existence of Ramsetu came from the religious literature and for a more conclusive settlement, a scientific understanding of the area is the need of the hour. This required the answer to the 3 questions.

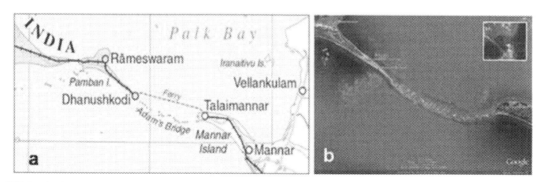

FIGURE 15.15

(A) Location of ramsetu (Adam's bridge) at rameswaram and (B) NASA picture showing the possibility of ramsetu.

i. what was the timing of Ram and thus the construction of Ramsetu?
ii. What is the water depth between India and Sri Lanka now?
iii. Is it possible to know the fluctuations of the sea levels at that time and is it possible to cross over to Sri Lanka at that time?

The answer to the first question came from following a very novel approach considering the positions of stars at the time of Rama as mentioned in ancient literature (mostly in Sanskrit) and planetarium software (Pushkar Bhatnagar *In* Bala, 2019 and Bala and Mishra, 2012). The timing concluded was 5100 BCE (i.e., ~7100 years old). The second question is easy to answer through the hydrographic (bathymetric) charts of this region. The sea between India and Sri Lanka is very shallow, and the average depth is about 3 m. The answer to the third question and the most relevant information was provided by the understanding of the Holocene sea-level fluctuations (Fig. 15.6) in published literature. By adopting various scientific equipment and methods (Hashimi et al., 1995; Nigam, 2012; Loveson and Nigam, 2016) relative sea-level fluctuations were inferred from the sea-level curves published for the Arabian Sea and the Bay of Bengal. Both the curves show that at the time of Rama ~7100 BP, the sea level was 2–3 m lower than today. This employs that during this time sea between India and Si Lanka was very shallow (almost connected) and it was possible to connect the two regions with some efforts to fill the depressions in between. The correct word used in ancient literature is "Setubandh Rameshwaram" bandh means the causeway and not the bridge (which give the impression of a flyover) and thus supports our conclusion. This was the only time when construction of a causeway (Bandh) was possible. Because, if the timing is older, there was no need to construct Ramsetu as the water depth level was much older (30–40 m at ~9.5 K and ~100 m at 14.5 k) which automatically connect India and Sri Lanka without any sea in between. Whereas if the timing is Younger than 7.5 k, the sea level was higher than today (when Dholavira and Lothal were port towns due to higher than today sea level) and thus it was not possible to make a bandh or causeway.

In view of the foregoing, it is summarized that the scientific evidence indicates that it was possible ~7100 BP to construct Ramsetu as a causeway between India and Sri Lanka.

6.6 Other areas

Some of the other important sites of marine Archeological interests where sea level must have played a leading role to shape the destiny of these cities are Pumphar, Mahabalipuram on the east coast and Somnath, Gopikapatnam (Goa) and a few areas of Kerala on the west coast of India. Investigations are going on in these areas by various agencies like the National Institute of Oceanography, Archeological Survey of India, Bhartidasan University, various IITs etc. are in progress and results are expected to be published in due course.

7. Conclusions

Marine Archeology is an upcoming branch of Science which needs a multidisciplinary approach for more acceptable conclusions. In this, the tools used by Geologists and Paleontologists can play significant roles as exhibited by the few examples from The Indian waters. India has more than a 7000 km long cost line and a large number of researchers are required with inclinations to apply their expertise in the field of marine archeology. This need requires more training programs and starting of specialized courses in universities and research organizations. Marine archeological sites especially places like Lothal (oldest dockyard) and Dholavira (oldest evidence of tsunami protection measures and water conservation knowledge) also provide an excellent opportunity to enhance international tourism and thus increase national earning and job opportunities.

Acknowledgements

The author would like to acknowledge the help received from Drs. N.H. Hashimi, Rajeev Saraswat, Rupal Dubey, Thejasino Suokhrie, V. J. Loveson and M.C. Pathak along with members of survey team from the Geological Oceanography Division, and the colleagues from Marine Archaeology Division, especially, Dr. A.S. Gaur, Sundaresh and Dr. Sila Tripati. Dr. Neloy Khare is acknowledged for motivating to write this review paper.

References

Bala, S., 2019. Ramayan Retold with Scientific Evidences. Prabhat Prakashan, New Delhi, p. 475.

Bala, S., Mishra, K., 2012. Historicity of Vedic and Ramayan Era: Scientific Evidences from the Depth of Ocean to the Heights of Skies. Institute of Scientific Research on Vedas (ISERVE), New Delhi, p. 288.

Bavadam, L., 2002. Questionable claims: archaeologists debunk the claim that underwater structures in the Gulf of Khambat point to the existence of a pre-Harappan civilisation. Frontline 2–15 March 2002.

Bisht, R.S., 2015. Excavation at Dholavira (1989–90 to 2004–05). Archaeological Survey of India, New Delhi, India.

Boltovskoy, E., Wright, 1976. Recent Foraminifera. The Hauge, p. 515.

Chengappa, R., 2002. The lost civilization. India Today 27 (6), 38–46.

Fairservice Jr., W.A., 1971. The Roots of Ancient India: The Archaeology of Early Indian Civilization, New York, p. 512

Gaur, A.S., Sunderesh, 2004. A late bronze age Copper fish-hook from Bet Dwarka, Gujarat: an evidence on the advance fishing technology. Curr. Sci. 86 ($0), 512–514.

Gupta, H.K., 2002. Oldest neolithic settlements discovered in Gulf of Cambay. J. Geol. Soc. India 59 (3), 277–278.

Haq, B.U., Borsma, A., 1998. Introduction to Marine Micropaleontology. Elsevier, p. 376.

Hashimi, N.H., Nigam, R., Nair, R.R., Rajagopalan, G., 1995. Holocene sea-level fluctuations on western Indian continental margin: an update. J. Geol. Soc. India 46, 157–162.

Kale, V.S., Rajguru, S.N., 1985. Neogene and quaternary transgressions and regressional history of the West Coast of India: an overview. Bull. Deccan College Res. Inst. 44, 153–165.

Kathiroli, S., 2004. Recent marine archaeological finds in Khambhat, Gujarat. Journal of Indian Ocean Archaeology 141–149 (Online).

Khadkikar, A.S., Rajshekhar, C., Kumaran, K.P.N., 2004. Palaeogeography around the Harappan port of Lothal, Gujarat, Western India. Antiquity 78 (302), 896–903.

Leshnik, L.S., 1968. The Harappan "port" at lothal: another view. American Anthropologist, New Series 70 (5), 911–922.

Loveson, V.J., Nigam, R., 2019. Reconstruction of late pleistocene and holocene sea level curve for the east coast of India. J. Geol. Soc. India 93, 507–514.

Mahmood, N., Khan, K., Rafi, Z., Lovholt, F., 2012. Mapping of tsunami hazard along Makran coast of Pakistan. Technical Report No. PMD-20 32.

Merh, S.S., 1987. Quaternary sea level changes: the present status vis-a- Vis records along coasts of India. Indian J. Earth Sci. 14, 235–251.

Merh, S.S., 1992. Quaternary sea level changes along Indian coasts. Proc. Ind. Natl. Sci. Acad. 58, 461–472.

Nair, R.R., 1974. Holocene sea levels on the Western continental shelf of India. Proc. Indiana Acad. Sci. 79 b, 197–203.

National Institute of Ocenography, 2001. Marine Archaeological Exploration off Dwarka, Bet Dwarka and Somnath (Gujarat)- Phase II, Final Report, p. 87.

Nigam, R., 1988. Was the large rectangular structure at lothal (Harrapan settlement) a "dockyard" or an "irrigation tank". Proc. 1st Ind. Confer. on Marine Archaeology of Indian Ocean Countries 20–22.

Nigam, R., 1990. Paleoclimatic implications of size variation of *Orbulina universa* in a core from North Indian Ocean. Curr. Sci. 59, 46–47.

Nigam, R., 1996. Potentiality of Foraminifera in deciphering paleo sea levels. In: Qasim, S.Z., Roonwal, G.S. (Eds.), India's Exclusive Economic Zone. Omega Scientific Publisher, New Delhi, pp. 225–232.

Nigam, R., 2005. Addressing the environmental issues through foraminifera- Case studies from the Arabian Sea. J. Paleontol. Soc. India 50, 25–36.

Nigam, R., 2006. Foraminifera [marine microfossil] as an additional tool for archaeologists; examples from the Arabian Sea. In: Vora, et al. (Eds.), Glimpses of Marine Archaeology in India. National Institute of Oceanography, Dona Paula, Goa, pp. 94–99.

Nigam, R., 2012. Sea Level Fluctuations during the last 15000 years and their impact on human settlements. In: Bala, S., Mishra, K. (Eds.), Historicity of Vedic and Ramayan Eras. Institute of Scientific Research on Vedas (ISERVE), pp. 193–210.

Nigam, R., Chaturvedi, S.K., 2006. Do inverted depositional sequences and allochothonous foraminifers in sediments along the coast of Kachchh, NW India, indicate palaeostorm and/or tsunami effects? Geo Mar. Lett. 26, 40–52.

Nigam, R., Hashimi, N.H., 2002. Has sea level fluctuations modulated human settlements in Gulf of Khambat (Cambay)? J. Geol. Soc. India 59, 583–584.

Nigam, R., Henriques, P.J., 1992. Planktonic percentage of foraminiferal fauna in surface sediments of the Arabian Sea (Indian Ocean) and a regional model for paleodepth determination. Palaeogeogr. Palaeoclimatol. Palaeoecol. 91, 89–98.

Nigam, R., Henriques, P.J., Wagh, A.B., 1993. Barnacle fouling on relict foraminiferal specimens from the western continental margin of India: An indicator to paleo-sea level. Cont. Shelf Res. 13, 279–286.

Nigam, R., Khare, N., 1992. In: Ghose, N.C., Nayak, B.U. (Eds.), Oceanographic Evidences of the Great Floods in 2000 and 1500 BC Is Documented in Archaeological Records, vol. 2. New trends in Ind. Arts. and Archaeology, Dharwar, pp. 517–522.

Nigam, R., Hashimi, N.H., Pathak, M.C., 1990. Sea level fluctuations: inferences from religious and archaeological records and their Oceanographic evidences. J. Marit. Archaeol. 1, 16–18.

Nigam, R., Hashimi, N.H., Menezes, E.T., Wagh, A.B., 1992. Fluctuating sea level off Bombay (India) between 14,500 to 10,000 years, before present. Curr. Sci. 62, 309–311.

Nigam, R., Pathak, M.C., Hashimi, N.H., 1991. In: Rao, S.R. (Ed.), Consequences of Sea Level Rise Due to Greenhouse Effect for Coastal Archaeological Monuments. Recent Advances in Marine Archaeology, Bangalore, pp. 116–121.

Nigam, R., Dubey, R., Saraswat, R., Sundaresh, Gaur, A.S., Loveson, V.J., 2016. Ancient Indians (Harappan settlement) were aware of tsunami/storm protection measures: a new interpretation of thick walls at Dholavira, Gujarat. Curr. Sci. 111, 2040–2243.

Pirazzoli, P.A., 1991. World Atlas of Sea Level Changes. Elsevier, p. 300.

Rao, S.R., 1979 and 1985. Lothal—A Harappan Port Town (1955–62), Memoir of the Archaeological Survey of India, No.78, vols. I and II (New Delhi).

Rao, S.R., 1997. Keynote address. In: An Integrated Approach to Marine Archaeology, Proc. 4th Indian Conference on Marine Archaeology of Indian Ocean Countries, Vishakhapatnam-1994. Published by National Institute of Oceanography, Goa, pp. 1—5.

Rao, S.R., 2006. Keynote address. Seven national conference of marine archaeology. In: Glimpses of Marine Archaeology in India. Published by Society of Marine Archaeology, Goa, pp. 1—3.

Ro, S.R., Lal, B.B., Nath, B., Ghosh, S.S., Lal, K., 1963. Excavation at Rangpur and other explorations in Gujarat. In: Ancient India: Bull. Of Archaeological Survey of India, No. 18 and 19, pp. 5—207.

Sankalia, H.D., 1974. The Prehistory and Protohistory of India and Pakistan, Poona, p. 592.

Shah, U.P., 1960. Lothal—A Port. Jour, vol. 9. Oriental Institute, pp. 310—320.

Tripathy, A., 2015. Excavation at Dwarka: critical analysis of archaeological remains. In: Gaur, A.S., Sundaresh (Eds.), Recent Researches on Indus Civilization and Marine Archaeology in India, pp. 261—270.

Wheeler, R.E.M., 1973. Forward. In: Rao, S.R. (Ed.), Lothal and Indus Civilization. London, p. 234.

CHAPTER 16

Introspecting contribution and preparedness of tropical agriculture against climate change: an Indian perspective

S. Suresh Ramanan[1], M. Prabhakar[2], Mohammed Osman[2] and A. Arunachalam[1]

[1]*ICAR-Central Agroforestry Research Institute, Jhansi, Uttar Pradesh, India;* [2]*ICAR-Central Research Institute for Dryland Agriculture, Hyderabad, Telangana, India*

1. Background

Climate change is a simple phrase that has been the talk of the 21st century which implies every sector, especially agriculture. From the tropical climate change perspective, agriculture is both the contributor as well as highly vulnerable to climate change, that is, both accused and victim. Food is one of the essential needs of human life—food, cloth, and shelter. Any statistics projecting the food demand for the future is a simple attempt to justify the indispensability of food available to us. Notwithstanding, a few studies have predicted the impact of climate change on food production. The agricultural sector gains its significance not only in terms of contribution to GDP but also to the percentage of people it supports in developing countries (Gulati et al., 2020). Ever since World War II, countries have been working together under the embodiment of the Food and Agriculture Organization (FAO) to ensure food security. Among many programs of FAO, Climate-Smart Agriculture (CSA) started as an idea in the report titled "Food Security and Agricultural Mitigation in Developing Countries: Options for Capturing Synergies" (Lipper and Zilberman, 2018). The idea got implemented at the Global Conference on Agriculture, Food Security and Climate Change at Hague in 2010. In brief, CSA means adopting agricultural activities and interventions that aim at developing a sustainable agricultural landscape. Technically, it is defined "as a strategy to address the challenges of climate change and food security through enhancing productivity, bolstering resilience, reducing GHG emissions, for achieving the national food security and development goals" (Chandra et al., 2018). The three essential pillars of CSA are (i) sustainably increasing agricultural productivity and incomes, (ii) adapting and building resistance to climate change, and (iii) reducing and/or removing greenhouse gases emission. In simple terms, food security, adaptation, and mitigation are the pillars of preparedness (van Wijk et al., 2020). Elementally, the adoption of climate-smart agriculture at the farm level means replacing or switching to varieties tolerant for abiotic stress; adjusting of cropping calendar; nutrient management like micro-dosing and fertigation; improving livestock quality and their stress tolerance level; improving breed, feed, and shelter management, *etc*. Overall, CSA aims at crop diversification, integration of crops and livestock through agroforestry, and integrated pest and nutrient management approach, along with soil and water conservation efforts.

Prior to the mooting of CSA at the FAO conference in 2010, there were many works on food security, sustainable agriculture, and climate change aspects. Starting from 1996, the World Food Summit led to the adoption of the Rome Declaration on World Food Security. Subsequently, followed the "World Food Summit after Five years" in Rome in 2002, which led to the drafting of "Right to Food" guidelines. After many high-level meetings and consultations at the country and global level, it all culminated at the 2010 FAO conference for initiating climate-smart agriculture (IISD, 2010). During these meetings, ensuring food security in developing nations is considered the top priority agenda of FAO. India is an apt example in this context with the increasing population and limited natural resources, the impact of climate change on Indian agriculture could be a disaster.

Incidentally, agricultural research has always been proactive, as it has transformed India from a food-deficit nation to a food-surplus nation in the past 60 years. The Indian Council of Agricultural Research (ICAR) and Agricultural Universities (AU) are the key contributors. India accounted for approximately 8% of global agricultural research publications and ranked second, following the United States between 1993 and 2012 as per the Web of Science database (Sagar et al., 2013). Climate-dependent agricultural research also began before the 2010 FAO conference. For instance, ICAR started a Network Project on Climate Change (NPCC) in 2004 to identify vulnerable regions and develop methodologies to assess the impacts of climate change. The researchers predicted yield loss in major crops like 3.9 million tonnes of wheat due to climate change by 2020. The predictions altered the thought process of policymakers to have a larger research program on climate-resilient agriculture. The announcement was made in the budget speech 2010-11 by then the Finance Minister as "The gains already made in the green revolution areas have to be sustained through conservation farming, which involves concurrent attention to soil health, water conservation and preservation of biodiversity. I propose an allocation of Rs. 200 crores for launching this climate resilient agriculture initiative." It marked the beginning of the National Initiative for Climate Resilient Agriculture (NICRA)—a flagship program of the ICAR in 2011 with a larger mandate to develop climate-resilient technologies for sustainable agriculture. Later on, in 2014—15, the name of the program was tweaked to National Innovations in Climate Resilient Agriculture.

Numerous factors influence the successful transfer of lab findings to the farmer's field. The findings of the research works need not necessarily end up creating utility or commodity. Sometimes findings of the scientists can enrich existing scientific knowledge and enable scientists to do basic research on focused areas. To be more elaborate: Scientific research can be categorized as basic, strategic, and applied research. Intriguingly, there is no hard-fast distinction between these aspects of research, especially in agricultural research. The agricultural scientists toil both in the field and labs, and they do contribute toward ensuring food security through the development of high-yielding varieties and technologies. All the findings of the agricultural scientists reach the global scientific fraternity first in the form of high-end publications. In this context, research articles and other scholarly publications become the best means of communicating their research findings. So, one can probably agree that the worth of scientific research can be assessed through scholarly publications. The Indian scientific fraternity has begun to debate the usage of scholarly publications and citations as the sole criteria for assessing the productivity of an individual scientist. One can argue that number of scientific publications and their citations cannot be the only true means for assessing an individual scientist. However, the usage of several publications and citations to evaluate the veracity of a network project or research program is more valid. Therefore, we assessed the veracity of scholarly publications to understand climate change research through the NICRA program vis-à-vis their contribution before an assessment on the farmers' field.

Another specific need to understand the research contributions under NICRA is also to elucidate its fundamental differences from Climate-Smart Agriculture (CSA). However, most of the literature does consider CSA and climate-resilient agriculture as similar. Meanwhile, the Indian Government has decided to focus more on adaptation against climate change in the agriculture sector (Richards et al., 2015). Thus, there is a need for introspection of the research efforts of the Indian Agricultural Research system toward climate change per se.

2. Methodology

To assess the veracity of the NICRA research program, we decided to analyze the research papers and communications which are published in journals indexed in the Web of Science (WoS). WoS database is considered to provide broad and holistic coverage of reputed journals. In this context, the articles related to the NICRA program were retrieved from the WoS database using the keyword search "NICRA" (All fields) with a time limit between 2010 and 20 (on December 25, 2020). After manual checking through the search results, a total number of 457 scholarly publications were screened out. Out of which, 432 research articles and proceeding papers were included for the analysis. The metadata of these publications was downloaded in the plaintext format and analyzed using bibliometrix R-package (Ver.1.3.959) (http://www.bibliometrix.org) (Aria and Cuccurullo, 2017). To understand the important thematic areas of NICRA research, the science mapping tool—Science Mapping Analysis Software Tool (SciMAT) (v 1.1.04) was used (Cobo et al., 2012a). It is open-source software and has some advantages compared to other science mapping tools. This can perform data retrieval, preprocessing, network extraction, normalization, mapping, analysis, visualization, and interpretation (Cobo et al., 2011, 2012b; Zambrano-Gonzalez et al., 2017). The analysis was designed with the following parameters: Unit of analysis, Words (authorRole = true, sourceRole = true, addedRole = true); Kind of network: Co-occurrence; Normalization measure: Equivalence Index; Cluster algorithm: Simple Centers (Max cluster size: 3, Min cluster size: 1); Evolution measure: Jaccard Index; Overlapping measure: Inclusion Strength. Moreover, to have a clear perspective on a different theme of NICRA research, the scholarly publications were sorted out manually into different subject matter divisions of ICAR based on the corresponding author affiliations and two-way ANOVA was performed in SPSS (ver. 22.0) to assess the difference in publication numbers among the divisions.

The grouping of publications into different SMDs of ICAR was based on the affiliation of the corresponding authors alone. Though there were authors from other institutions, we assumed that the corresponding authors in most cases will be directly linked with the NICRA project either as an investigator of that institute or received a grant from the NICRA project. While agricultural research mostly tends to address the issues of local relevance, many research papers might have been published also in journals/bulletins that are not listed in WoS. The National Academy of Agricultural Sciences (NAAS) has developed a journal rating system, which is widely recognized in India at least in agricultural sciences but not listed in WoS (Xia, 2019). However, there is no dedicated database for NAAS-rated journals. Eventually, the contributions in those journals are excluded from this study.

3. Results

The results showed that the number of publications increased at the rate of 20.87% per annum. The total number of scholarly publications and their citations is given in Fig. 16.1. There were 424 research articles and seven proceeding papers published between 2012 and 20 by 1336 authors. Most of the research articles were multi-authored publications with 3.09 authors per document yielding 0.324 documents per author.

The strategic diagram was developed in the SciMAT tool using the co-occurrence analysis, after grouping out author keywords and keyword-plus (Fig. 16.2). It provides an overview of different thematic areas sorted into four quadrants based on density and centrality. Anticlockwise, the first quadrant indicates the clusters that are core or mainstream. The second quadrant indicates developed but isolated themes. The clusters in the third quadrant represent emerging and/or declining themes. In other words, it represents the themes in flux. Lastly, the fourth quadrant represents general and broadly researched themes.

There are many sub-disciplines in the broad spectrum of agricultural research such as plant breeding, biochemistry, pathology, veterinary, etc. The WoS database categorizes the documents into different subjects. However, agricultural research under ICAR is usually categorized under six subject matter divisions (SMDs). Fig. 16.3 depicts the research output under different SMDs of ICAR during different years. About 131 research articles were published by institutes under the crop science (Crp Sci.) division followed by the animal science (Ani. Sci.) division with 108 publications. In 9 years, 47 research publications were from researchers affiliated with more than 20 institutions outside the ICAR domain. Scientists affiliated with 53 ICAR research institutes contributed to the scientific output. The collaboration index among the authors was 3.1 and there were 6.17 co-authors per scholarly publication.

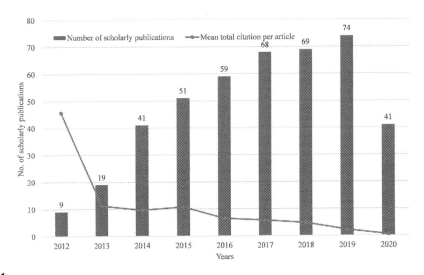

FIGURE 16.1

The total number of papers published between 2012 and 20 under National Innovations in Climate Resilient Agriculture (NICRA) in the Web of Science (WoS) database.

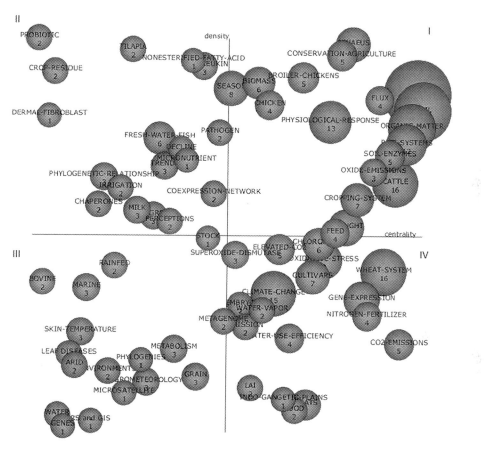

FIGURE 16.2

Prominent research themes of the National Innovations in Climate Resilient Agriculture (NICRA) according to the strategic diagram.

Among the total 1336 authors, Fig. 16.4 depicts the top 10 authors based on the number of publications. Most of the authors belong to the Animal Science and Crop Science Divisions. A list of actual research papers based on the citation count is given in Table 16.1. It is evident from the data that most of the papers deal with rice or wheat (crop science) with an exception of one paper on buffalo (animal science).

4. Discussion

Technically, the CRA initiative of ICAR is much different from the FAO's CSA concept. Owing to this difference, it is envisaged that the present introspection will provide useful insights about this flagship program. Overall, the number of articles increased temporally with a drop in 2020 although the mean total citation per article declined. Predominately, the research findings are attributed to the team effort

428　Chapter 16 An Indian perspective on tropical agriculture's role in climate change

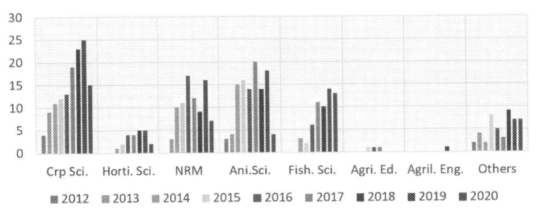

FIGURE 16.3

The research output of the National Innovations in Climate Resilient Agriculture (NICRA) in terms of scholarly publications across the time.

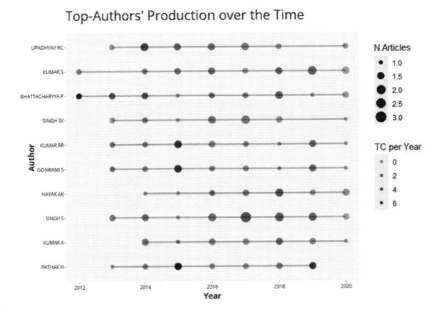

FIGURE 16.4

Top 10 authors under the National Innovations in Climate Resilient Agriculture (NICRA) in terms of publication numbers.

4. Discussion

Table 16.1 Top 10 papers published in NICRA research program based on citation counts.

Author, year and journal name	Title	Total citations
Katiyar et al. (2012), Bmc Genomics	Genome-wide classification and expression analysis of MYB transcription factor families in rice and arabidopsis	200
Bhattacharyya et al. (2012), Soil Tillage res	Effects of rice straw and nitrogen fertilization on greenhouse gas emissions and carbon storage in tropical flooded soil planted with rice	142
Purakayastha et al. (2015), Geoderma	Characterization, stability, and microbial effects of four biochars produced from crop residues	58
Kumar et al. (2015), Funct Integr Genomics	Novel and conserved heat-responsive microRNAs in wheat (*Triticum aestivum* L.)	46
Yadav et al. (2017), J Clean Prod	Energy budgeting for designing sustainable and environmentally clean/safer cropping systems for rainfed rice fallow lands in India	44
Bhattacharyya et al. (2014), Agric Ecosyst Environ	Tropical low-land rice ecosystem is a net carbon sink	40
Bhattacharyya et al. (2013), Sci Total Environ	Impact of elevated CO_2 and temperature on soil C and N dynamics in relation to CH_4 and N_2O emissions from tropical flooded rice (*Oryza sativa* L.)	39
Kumar et al. (2015), J Therm Biol	Expression profiling of major heat shock protein genes during different seasons in cattle (*Bos indicus*) and buffalo (*Bubalus bubalis*) under tropical climatic condition	37

Continued

Table 16.1 Top 10 papers published in NICRA research program based on citation counts.—cont'd

Author, year and journal name	Title	Total citations
Mangrauthia et al. (2017), J Exp Bot	Genome-wide changes in microRNA expression during short and prolonged heat stress and recovery in contrasting rice cultivars	37
Yadav et al. (2018), J Clean Prod	Energy budget and carbon footprint in a no-till and mulch based rice—mustard cropping system	37

which indirectly reflected on the extensive cooperation and collaboration level among scientists in the NICRA program only except for two single-authored publications. The collaboration index was 3.1, which supports the former contention. Earlier studies, however, have reported >70% of publications to tend to be co-authored in agricultural research (Tripathi and Garg, 2014). The decline may be attributed to either the delay in the publication of the submitted paper or the reduction in publication number because of the COVID pandemic.

Despite these, one of the major concerns will be the key thematic research areas. The different thematic areas of climate change research are depicted using a strategic diagram (Fig. 16.3). The figure produced in the SciMAT tool is based on the co-occurrence of author keywords and keywords plus. Studies have proven that the author keywords and keyword-plus as appropriate means to visualize the knowledge structure of scientific fields (Zhang et al., 2016). In the SciMAT tool, both author keywords and keywords plus are used for science mapping. Thus, it can be considered an appropriate representation of research happening under NICRA. The strategic diagram presents the keywords as clusters on density (y-axis) and centrality (x-axis). The density measures the internal binding force, while the centrality measures the strength of the binding of a cluster to other clusters. The mainstream theme in NICRA research was represented by clusters "HSPs" (heat shock proteins), "Organic-matter," "Cattle," "Soil," "Methane Emission," "Soil Enzymes," "Penaeus," "Conservation Agriculture," "Drought," "feed," "Physiological Response," "Chicken," and "Season." There is also an overlap of clusters—"HSP," "Soil," "Organic-matter," "Rice-systems," and clusters—"Feed" and "Drought." This means that these clusters are strongly connected to other clusters and have a high density of work. For instance, the HSP-related works were predominant (Kumar et al., 2017, 2020; Rajkumar et al., 2018; Soren et al., 2018). Another apt example will be the "Penaeus" cluster. In the fisheries discipline under NICRA, researchers worked on brackish water aquaculture exclusively in the last 5 years (Jannathulla et al., 2020; Kathyayani et al., 2019; Saraswathy et al., 2020) compared to freshwater fish which is represented as "Fresh-Water Fish" cluster in quadrant II, that is, developed but isolated. The third quadrant represents emerging and declining themes of research. Here, the clusters like "Abiotic stress," "Salinity," and "Sequencing" can be regarded as an emerging issue as there are many recent works on these areas. However, clusters like "Emission" and "elevated CO_2" are declining. Overall, it can be stated that there is less focus on the clusters present in the third quadrant where researchers are to plan their future programs. The last and final quadrant represented the general and broad theme of research, which very well reflected, via clusters like "Wheat-system," "Water-use-efficiency," "Metagenome," "Cultivars," and many others.

Irrespective of the overview of different research themes, we grouped the scholarly publications based on the ICAR classification of agricultural disciplines. The analysis revealed that there was a significant difference in terms of publication numbers across the disciplines ($P < .00$) (Table 16.2). The crop science division (n = 131) was on par with the animal science division (n = 108). This trend is also reflected in the top 10 authors as well as the 10 most cited papers too are visibly from crop and animal scientists (Fig. 16.4 and Table 16.1), as crop and animal sciences divisions have been given more emphasis in NICRA because of their critical and applied importance. Rice and wheat-based cropping systems are predominant in India. Eventually, they became the crops of popular choice for research work globally. In India too, 40% of papers between 1965 and 2010 were on rice, followed by wheat ($\sim$20%) (Tripathi and Garg, 2016). Thus, increased research interest in rice and wheat is usual among agricultural researchers. However, other countries like China and the United States are global leaders in terms of most cited papers (Yuan and Sun, 2020). In a global agricultural publication analysis, it was reported that India was the second leading nation in dairy and animal sciences in terms of publication numbers (Sagar et al., 2013).

Researchers were intentionally focusing on it as any production implication will affect both livelihoods of the majority of Indian farmers and the food security of the nation. Similarly, livestock like cows, buffalo, goats, and sheep play a vital role in the farm income, which falls under the animal science division. The Natural Resource Division (NRM) focuses research on areas like soil and water conservation, integrated farming systems, abiotic stress management, agroforestry, and problematic land reclamation. The number of publications under NRM (n = 85) was slightly higher than that of the Fisheries Science Division (n = 59) (Fish. Sci.).

As compared to the budget allocated to different divisions *vis-a-vis* outcomes, no correlation could be achieved mainly because of the nature of work carried out among divisions. For example, research

Table 16.2 Total number of publications and total budget allocated under NICRA (2011–20).

S.No.	Division	Number of scholarly publications (2012–20)	Total budget allocated under NICRA (2011–20) (Rupees in lakhs)
1.	Crop science	131[a]	13,962
2.	Animal science	108[ab]	3,799
3.	Natural resource management	85[bc]	19,105
4.	Fisheries science	59[c]	5,258
5.	Horticulture science	23[d]	5,238
6.	Agricultural education	3[d]	71
7.	Agricultural engineering	1[d]	882

The same letters indicate statistically insignificant differences ($P > .05$).

on natural resource management will take more time for designing, and planning of the project and high execution cost. Typically, the benefits from NRM-based research exceed the total projects. It is also reported that the Internal Rate of Return for NRM research will be less when compared to crop-based research because of the difficulty in the quantification of environmental benefits and services. This ultimately makes NRM-based research appear less lucrative and productive (Pal, 2011; Science Council Secretariat, 2006). Further, the time lag between the initiation of the project and publication release will be long. For example, soil carbon improvement needs 10 years of field observation for publishing in reputed journals (Bhardwaj et al., 2019). Thus, the comparison between the divisions is not for criticism but to project and uphold the works of researchers collectively toward climate resilient agriculture.

5. Conclusions

The periodical assessment of the research performance of any organization or research program is necessary to know about its present status and future scope. In recent times, the need for evaluation of the research program is growing owing to the need for greater accountability and effectiveness. Bibliometrics and science mapping are some of the reliable tools for research performance evaluation. It plays a crucial role in decision-making on national research policies and funding. It potentially analyzes secondary data on scholarly outputs and give meaningful information on the essence of scientific work at individual researcher level, research group level, organization level, and country level (Bornmann and Leydesdorff, 2014). Although multiple bibliometric databases like PubMed, Google Scholar, Scopus, etc., are available, WoS not only provides reliable and comprehensive information but also provides several analyzing tools (Ekundayo and Okoh, 2018; Sweileh et al., 2016; Zyoud et al., 2017). To understand the veracity and evolution of research, thematic maps are used. We can coherently visualize the scientific contribution and development of climate change research in India under NICRA through the bibliometric and science mapping approach.

Acknowledgments

The authors like to thank the entire team of NICRA and the NICRA Implementing Agency at ICAR-CRIDA, Hyderabad for translating agricultural research outputs and enabling climate resilience in Indian agriculture.

References

Aria, M., Cuccurullo, C., 2017. bibliometrix: an R-tool for comprehensive science mapping analysis. J. Informetr. 11, 959–975. https://doi.org/10.1016/j.joi.2017.08.007.

Bhardwaj, A.K., Rajwar, D., Mandal, U.K., Ahamad, S., Kaphaliya, B., Minhas, P.S., Prabhakar, M., Banyal, R., Singh, R., Chaudhari, S.K., Sharma, P.C., 2019. Impact of carbon inputs on soil carbon fractionation, sequestration and biological responses under major nutrient management practices for rice-wheat cropping systems. Sci. Rep. 9, 9114. https://doi.org/10.1038/s41598-019-45534-z.

Bhattacharyya, P., Neogi, S., Roy, K.S., Dash, P.K., Nayak, A.K., Mohapatra, T., 2014. Tropical low land rice ecosystem is a net carbon sink. Agric. Ecosyst. Environ. 189, 127–135.

Bhattacharyya, P., Roy, K.S., Neogi, S., Adhya, T.K., Rao, K.S., Manna, M.C., 2012. Effects of rice straw and nitrogen fertilization on greenhouse gas emissions and carbon storage in tropical flooded soil planted with rice. Soil Tillage Res. 124, 119−130.

Bhattacharyya, P., Roy, K.S., Neogi, S., Dash, P.K., Nayak, A.K., Mohanty, S., Baig, M.J., Sarkar, R.K., Rao, K.S., 2013. Impact of elevated CO_2 and temperature on soil C and N dynamics in relation to CH_4 and N_2O emissions from tropical flooded rice (*Oryza sativa* L.). Sci. Total Environ. 461, 601−611.

Bornmann, L., Leydesdorff, L., 2014. Scientometrics in a changing research landscape. EMBO Rep. 15, 1228−1232. https://doi.org/10.15252/embr.201439608.

Chandra, A., McNamara, K.E., Dargusch, P., 2018. Climate-smart agriculture: perspectives and framings. Clim. Pol. 18, 526−541. https://doi.org/10.1080/14693062.2017.1316968.

Cobo, M.J., López-Herrera, A.G., Herrera-Viedma, E., Herrera, F., 2012a. SciMAT: a new science mapping analysis software tool. J. Am. Soc. Inf. Sci. Technol. 63, 1609−1630. https://doi.org/10.1002/asi.22688.

Cobo, M.J., López-Herrera, A.G., Herrera-Viedma, E., Herrera, F., 2011. Science mapping software tools: review, analysis, and cooperative study among tools. J. Am. Soc. Inf. Sci. Technol. 62, 1382−1402. https://doi.org/10.1002/asi.21525.

Cobo, M.J., Lopez-Herrera, A.G., Herrera, F., Herrera-Viedma, E., 2012b. A note on the ITS topic evolution in the period 2000−2009 at T-ITS. IEEE Trans. Intell. Transport. Syst. 13, 413−420. https://doi.org/10.1109/TITS.2011.2167968.

Ekundayo, T.C., Okoh, A.I., 2018. A global bibliometric analysis of Plesiomonas-related research (1990 − 2017). PLoS One 13, 1−17. https://doi.org/10.1371/journal.pone.0207655.

Gulati, A., Kapur, D., Bouton, M.M., 2020. Reforming Indian agriculture. Econ. Polit. Wkly. 55, 35−42.

IISD, 2010. Summary of the Global Conference on Agriculture, Food Security and Climate Change: 31 October−5 November 2010. Global conference on agriculture, food security and climate change bulletin 184 (6). https://enb.iisd.org/events/global-conference-agriculture-food-security-and-climate-change/summary-report-31-october-5.

Jannathulla, R., Dayal, J.S., Ambasankar, K., Chitra, V., Muralidhar, M., 2020. Effect of water salinity on tissue mineralisation in penaeus vannamei (Boone, 1931). Indian J. Fish. 67, 138−145. https://doi.org/10.21077/ijf.2019.67.1.74661-19.

Kathyayani, S.A., Poornima, M., Sukumaran, S., Nagavel, A., Muralidhar, M., 2019. Effect of ammonia stress on immune variables of Pacific white shrimp Penaeus vannamei under varying levels of pH and susceptibility to white spot syndrome virus. Ecotoxicol. Environ. Saf. 184. https://doi.org/10.1016/j.ecoenv.2019.109626.

Kumar, R., Gupta, I.D., Verma, A., Kumari, R., Verma, N., 2017. Molecular characterization and SNP identification in HSPB6 gene in Karan Fries (*Bos taurus* x *Bos indicus*) cattle. Trop. Anim. Health Prod. 49, 1059−1063. https://doi.org/10.1007/s11250-017-1235-6.

Katiyar, A., Smita, S., Lenka, S.K., Rajwanshi, R., Chinnusamy, V., Bansal, K.C., 2012. Genome-wide classification and expression analysis of MYB transcription factor families in rice and Arabidopsis. BMC Genomics 13 (1), 1−19.

Kumar, A., Ashraf, S., Goud, T.S., Grewal, A., Singh, S.V., Yadav, B.R., Upadhyay, R.C., 2015. Expression profiling of major heat shock protein genes during different seasons in cattle (*Bos indicus*) and buffalo (*Bubalus bubalis*) under tropical climatic condition. J. Therm. Biol. 51, 55−64.

Kumar, R.R., Goswami, S., Rai, G.K., Jain, N., Singh, P.K., Mishra, D., Chaturvedi, K.K., Kumar, S., Singh, B., Singh, G.P., Rai, A.K., Chinnusamy, V., Praveen, S., 2020. Protection from terminal heat stress: a trade-off between heat-responsive transcription factors (HSFs) and stress-associated genes (SAGs) under changing environment. Cereal Res. Commun. https://doi.org/10.1007/s42976-020-00097-y.

Kumar, R.R., Pathak, H., Sharma, S.K., Kala, Y.K., Nirjal, M.K., Singh, G.P., Goswami, S., Rai, R.D., 2015. Novel and conserved heat-responsive microRNAs in wheat (*Triticum aestivum* L.). Funct. Integr. Genomics 15, 323−348.

Lipper, L., Zilberman, D., 2018. A Short History of the Evolution of the Climate Smart Agriculture Approach and its Links to Climate Change and Sustainable Agriculture Debates, pp. 13–30. https://doi.org/10.1007/978-3-319-61194-5_2.

Mangrauthia, S.K., Bhogireddy, S., Agarwal, S., Prasanth, V.V., Voleti, S.R., Neelamraju, S., Subrahmanyam, D., 2017. Genome-wide changes in microRNA expression during short and prolonged heat stress and recovery in contrasting rice cultivars. J. Exp. Bot. 68 (9), 2399–2412.

Pal, S., 2011. Impacts of CGIAR crop improvement and natural resource management research: a review of evidence. Agric. Econ. Res. Rev. 24, 185–200.

Purakayastha, T.J., Kumari, S., Pathak, H., 2015. Characterisation, stability, and microbial effects of four biochars produced from crop residues. Geoderma 239, 293–303.

Rajkumar, U., Vinoth, A., Reddy, E.P.K., Shanmugam, M., Rao, S.V.R., 2018. Effect of supplemental trace minerals on Hsp-70 mRNA expression in commercial broiler chicken. Anim. Biotechnol. 29, 20–25. https://doi.org/10.1080/10495398.2017.1287712.

Richards, M., Bruun, T.B., Campbell, B.M., Gregersen, L.E., Huyer, S., et al., 2015. How Countries Plan to Address Agricultural Adaptation and Mitigation, How Countries Plan to Address Agricultural Adaptation and Mitigation. Copenhagen, Denmark.

Sagar, A., Kademani, B., Bhanumurthy, K., 2013. Research trends in agricultural science: a global perspective. J. Scientometr. Res. 2, 185. https://doi.org/10.4103/2320-0057.135409.

Saraswathy, R., Muralidhar, M., Balasubramanian, C.P., Rajesh, R., Sukumaran, S., Kumararaja, P., Dayal, J.S., Avunje, S., Nagavel, A., Vijayan, K.K., 2020. Osmo-ionic regulation in white leg shrimp, Penaeus vannamei, exposed to climate change-induced low salinities. Aquacult. Res. https://doi.org/10.1111/are.14933.

Science Council Secretariat, 2006. Natural Resources Management Research Impacts: Evidence from the CGIAR.

Soren, S., Vir Singh, S., Singh, P., 2018. Seasonal variation of mitochondria activity related and heat shock protein genes in spermatozoa of Karan Fries bulls in the tropical climate. Biol. Rhythm. Res. 49, 366–381. https://doi.org/10.1080/09291016.2017.1361584.

Sweileh, W.M., Al-Jabi, S.W., Sawalha, A.F., AbuTaha, A.S., Zyoud, S.H., 2016. Bibliometric analysis of publications on Campylobacter: (2000–2015). J. Health Popul. Nutr. 35, 39. https://doi.org/10.1186/s41043-016-0076-7.

Tripathi, H.K., Garg, K.C., 2016. Scientometrics of cereal crop science research in India as seen through SCOPUS database during 1965-2010. Ann. Libr. Inf. Stud. 63, 222–231.

Tripathi, H.K., Garg, K.C., 2014. Scientometrics of Indian crop science research as reflected by the coverage in Scopus, CABI and ISA databases during 2008–2010. Ann. Libr. Inf. Stud. 61, 41–48.

van Wijk, M.T., Merbold, L., Hammond, J., Butterbach-Bahl, K., 2020. Improving assessments of the three pillars of climate smart agriculture: current achievements and ideas for the future. Front. Sustain. Food Syst. 4. https://doi.org/10.3389/fsufs.2020.558483.

Xia, J., 2019. A preliminary study of alternative open access journal indexes. Publish. Res. Q. 35, 274–284. https://doi.org/10.1007/s12109-019-09642-y.

Yadav, G.S., Das, A., Lal, R., Babu, S., Meena, R.S., Saha, P., Singh, R., Datta, M., 2018. Energy budget and carbon footprint in a no-till and mulch based rice–mustard cropping system. J. Clean. Prod. 191, 144–157.

Yadav, G.S., Lal, R., Meena, R.S., Datta, M., Babu, S., Das, A., Layek, J., Saha, P., 2017. Energy budgeting for designing sustainable and environmentally clean/safer cropping systems for rainfed rice fallow lands in India. J. Clean. Prod. 158, 29–37.

Yuan, B.Z., Sun, J., 2020. Mapping the scientific research on maize or corn: a bibliometric analysis of top papers during 2008–2018. Maydica 65, 1–9.

Zambrano-Gonzalez, G., Ramirez-Gonzalez, G., Almanza-P, M., 2017. The evolution of knowledge in sericultural research as observed through a science mapping approach. F1000 Res. 6, 2075. https://doi.org/10.12688/f1000research.12649.1.

Zhang, J., Yu, Q., Zheng, F., Long, C., Lu, Z., Duan, Z., 2016. Comparing keywords plus of WOS and author keywords: a case study of patient adherence research. J. Assoc. Inf. Sci. Technol. 67, 967—972. https://doi.org/10.1002/asi.23437.

Zyoud, S.H., Waring, W.S., Al-Jabi, S.W., Sweileh, W.M., 2017. Global cocaine intoxication research trends during 1975—2015: a bibliometric analysis of Web of Science publications. Subst. Abuse Treat. Prev. Pol. 12, 6. https://doi.org/10.1186/s13011-017-0090-9.

CHAPTER 17

Mechanisms and proxies of solar forcing on climate and a peek into Indian paleoclimatic records

Rajani Panchang[1], Mugdha Ambokar[1], Kalyani Panchamwar[1] and Neloy Khare[2]

[1]*Marine Bio-Geo Research-Lab for Climate and Environment (MBReCE), Department of Environmental Science, Savitribai Phule Pune University, Pune, Maharashtra, India;* [2]*Ministry of Earth Sciences, New Delhi, India*

1. Introduction

The International Monetary Fund considers climate change as a potential threat to global economy, finance, and development. Climate change is all set to increase insurance premiums, impact agricultural produce and food security, threaten human health, impact infrastructure to extreme climatic events, and alter water cycles and energy demands. Thus, prediction of climate change has a direct bearing on resource management. An understanding of the past climate is necessary to understand the natural climatic variability over different timescales as well as identify the magnitude, process, and causes of such climatic changes. Such studies help compute rates of climate change and to identify how sensitive it is to the changes that humans are making to the energy balance (by changing the atmospheric structure and composition); the concern being "Has the climatic threshold or tipping point been reached?"

2. Sun as the eternal driver of the Earth's climate: an historical perspective

The Earth's climate is modulated from time to time due to the change in Earth's energy budget, that is, the amount of energy the Earth receives from the Sun and loses to outer space. This budget is dependent on the composition of the atmospheric layer that surrounds the Earth. The Earth's atmosphere has not been the same, since its origin (Boutaud, 2018). The concentration of greenhouse gases (GHG) such as water vapor, methane, carbon dioxide, particulate matter, sulfur dioxide, etc. have varied throughout the 4.5 billion years history of evolution of the Earth, which involves plate tectonics, volcanism, evolution of continental crust, formation of early atmosphere, precipitation, and continental weathering as sinks of carbon-di-oxide. The earth's atmosphere has evolved throughout its history and has been controlled as well as shaped by the relative abundances of three very critical gases, namely methane, carbon-di-oxide, and oxygen (*See Fact Box*). Whenever, the methane (the GHG which is 30 times more potent than CO_2) concentrations have dropped, the Earth has witnessed planet-wide glaciation, a state better known as "Snow-ball Earth" (approximately 2500 mya. and 650 mya). It is thus very evident that atmospheric interaction in the earth's radiative budget is the most important internal forcing on its climate.

438 Chapter 17 Mechanisms and proxies of solar forcing

Gilles Ramstein, Climatologist and CEA senior researcher at the Laboratoire des Sciences du Climat et de l'Environnement states that, "From the very beginning, the Earth's atmosphere has behaved like a heating blanket!" For most of the Earth's history, it has been a greenhouse keeping the Earth free of ice (i.e., it did not house any of the icecaps existent today, even at the poles! (*See Fact Box*). As per Ramstein, even though 4.5 billion years ago, the sun radiated 30% lesser heat than it does today, the Earth was much warmer than it is at present! (Boutaud, 2018).

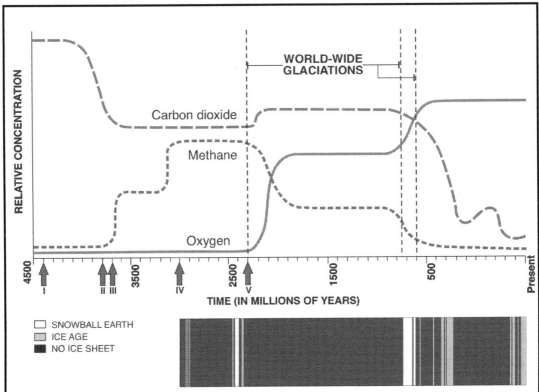

The history of Earth's Ice Ages since its origination 4500 million years ago has been depicted in the illustration above. (Modfied after Hoffmann et al., 2017 and Boutaud, 2018). Fossil evidence for past 3000 million years suggests Earth has behaved like a green house and has been free of icecaps on its surface as indicated on the banded panel. The 5 red arrows [dark gray arrows in print] mark significant thresholds of change in Earth's atmospheric composition, that changed the solar budget. Event I: Warm Earth despite 30% lower solar luminosity due to high volcanic carbon-dioxide emissions; Event II: Emergence of Methanogens, that consumed CO_2 and released methane; Event III: Rapid rise in methane; Event IV: Emergence of Cyanobacteria, oxygen producing bacteria which oxidized methane to CO_2; Event V: Great Oxygenation Event, rapid increase in atmospheric oxygen. Note the correlation between absence of ice sheets on earth and the peaked concentrations of methane in the Earth's history.

Studying deep sea paleoclimatic records over the past 65 million years provide alternative views for climate forcing mechanisms. Zachos et al. (2001) suggested that the 5° to 6°C rise in deep ocean temperature within 10 ky (Fig. 17.1), 55 million years ago during the Late Paleocene Thermal

2. Sun as the eternal driver of the Earth's climate: an historical perspective

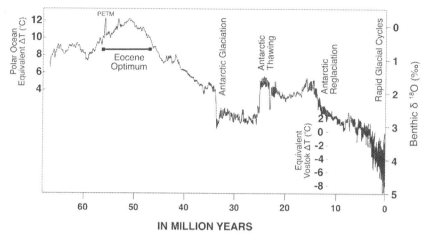

FIGURE 17.1

Reconstruction of paleotemperatures for the past 65 million years using benthic foraminiferal stable oxygen isotopic concentrations, reveal a prominent, but brief excursion attributed to rapid warming during the Paleocene-Eocene Thermal Maxima (PETM). It was a time period marked by 5°–8°C rise in average global temperature across the event (Haynes and Honische, 2020). The associated period of massive carbon release into the atmosphere has been estimated to have lasted from 20,000 to 50,000 years. The entire warm period lasted for about 200,000 years. Zachos et al. (2001) identified low-frequency glacial interglacial cyclicity in these data coinciding with sun's orbital cyclicity.

Maxima, may not have been due to greenhouse forcing, but instead GHG may only have been the trigger or the amplifying factor. They found low-frequency 400 ky cycles in glacial-interglacials records in the Cenozoic which coincided with eccentricity minimas but could be explained by the superimposition of the obliquity of the sun's orbital variations.

Henley and Abram (2017) compiled data for Antarctic temperature changes across the ice ages using data from Parrenin et al. (2020), Snyder (2016) and Bereiter et al. (2015) (Fig. 17.2) and found that were very similar to globally averaged temperatures. Though these studies revealed a tight connection between CO_2 and temperature over the past 800,000 years, what was a striking observation was the positive correlation between the onset of a glaciation every 100,000 years that matched very well with the cyclicity in the orbital relationship that the Sun shares with the Earth! The CO_2 concentrations amplified the effect of these natural 100,000-year orbital positions.

Such studies underscore the role of the solar forcing as the sole external forcing of the Earth's climate that contributes to natural climatic variability on long- and short-term time scales. Climatic shifts due to orbital variations, if computed right can be predicted and their effects can be teased apart from those of isolated events of high magnitude. For example, methane outgassing (Zachos et al., 2001) or burning of fossil fuels and such internal forcing. Identifying and testing mechanisms help reduce the number of potential variables influencing the dynamics of rapidly changing climates and abrupt events.

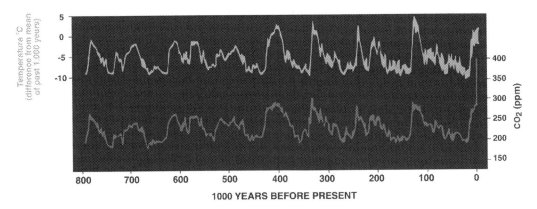

FIGURE 17.2

Reconstruction of global average surface temperature (GAST) over the past 800,000 years using stack of deep-sea oxygen isotopes $\delta^{18}O$ in blue [light gray in print] (Leisiecki and Raymo, 2005), in comparison with stacked reconstruction of atmospheric CO_2 concentrations (p.p.m.) in red [dark gray in print] (Bereiter et al., 2015). Note the correlation of events at a periodicity of 100,000 years.

3. Sun: the external driver of Earth's climate

The Sun, which has existed for about 4.603 billion years, has been the prime source of heat and light, to the Earth since its origin (4.52 billion years), enabling life. Every second, the Sun's core fuses about 600 million tons of hydrogen into helium and in the process, converts 4 million tons of matter into energy. This energy which is the source of the Sun's light and heat takes about 10,000 and 170,000 years to escape the core and is radiated throughout the solar system in the form of visible, infrared, and ultraviolet radiations. The energy, that is, power per unit area received from the sun, in the form of electromagnetic radiation is called "solar irradiance" and the intensity of solar irradiance including all wavelengths is known as the "Total Solar Irradiance" (TSI). Given the limited understanding of the solar physics, the radiative and climate forcings arising from changes in the Sun's insolation are expected to continue to be small (Jones et al., 2012). The solar insolation, that is, the TSI received by the Earth is estimated to have increased by about 7% every billion years. Earth's TSI varies with both, solar activity and planetary orbital dynamics.

Ultraviolet irradiance (EUV) varies by approximately 1.5% from solar maxima to minima, for 200—300 nm UV (Hathway et al., 2004) thereby modulating the production and loss of ozone. Increase in UV irradiance can cause higher ozone production, leading to stratospheric heating and poleward displacements in the stratospheric and tropospheric wind systems. This demonstrates how solar radiations interact with the Earth's atmospheric chemistry, modifying it, and modulating Earth's climate.

The original chemical composition of the Sun is derived to have contained about 71.1% hydrogen and 27.4% helium (Lodders, 2003a, 2003b). The fusion of hydrogen into helium over the past 4.6 billion years has increased the proportion of helium to 60% and caused it to settle from the photosphere toward the center of the Sun because of gravity. In the current photosphere, the helium fraction is thus reduced. Heat is transferred outward from the Sun's core by radiation and not by convection, and thus the fusion products are not lifted outward by heat; rather they remain in the core (Hansen et al., 2012). Effectively, helium has gradually started to form the inner core because presently the Sun's core is not hot or dense enough to fuse helium. The Sun today is roughly halfway through the most stable part of its life.

Solar flares, coronal mass ejections, high-speed solar wind, and solar energetic particles are all forms of solar activity. All solar activity is driven by the solar magnetic field. Solar flares are made of photons. They travel out directly from the flare site and only impact the side of the Earth facing them. Coronal mass ejections (CME) are large clouds of plasma and magnetic field that erupt from the sun which can erupt in any direction and continue to flow in that direction through the solar wind. However, only when the clouds are directed toward the Earth can they hit and impact it. High-speed solar wind streams originate from coronal holes that can be formed anywhere on the surface of the Sun. However, they impact the Earth only when they are close to the solar equator. All these phenomena produce high-energy charged particles which follow magnetic field lines that pervade the space between the Sun and the Earth. Only those particles that follow magnetic field lines intersecting those of the Earth have the potential to impact it. Sunspot activity which is discussed in detail in the following sections of the chapter provides deep insights into the timescales and magnitudes at which the sun continues to work as an external climate forcing. As discussed before, the TSI is estimated to continue to slowly increase as the sun continues to age at a rate of about 1% each 100 million years. Such a rate of change is far too small to be detectable within measurements and is considered insignificant on human timescales.

The orbital dynamics that the Earth shares with the Sun includes annual cycles and Milkovich shifts. The Earth follows an elliptical orbit around the Sun which causes a fluctuation in the TSI received from season to season. In July, when the Earth is at aphelion, it receives about 1321 W m^{-2}, while in early January at perihelion, it receives about 1412 W m^{-2} (Lewis, 2021). These changes in irradiance could appear minor influences resulting from the annual cycling in Earth's relative tilt direction. However, this causes change in seasons and latitudinal climatic zones which have major socio-economic bearing. Further, superimposition of these minor annual shifts in these cycles are amplified and manifested in decadal climatic variations.

4. The climatic manifestations of solar activity: obvious correlations of the inquisitive

Climate-related changes like droughts, floods, glaciation, and interglaciation periods, to name a few cannot be termed chance events; rather these oscillate with the solar cycle. Distinct proxies for such climatic phenomenon all over the world are in phase with the periodicity of solar cycles. The link between long-term changes in solar activity and earth's climate was first suggested by Eddy (1976). He reported the interconnection between the solar cycles and climate by comparing the Schwabe and Wolf's sunspot data and ^{14}C data generated from tree rings. On the basis of similarities between globally averaged sea surface temperature data and the solar activity data, Reid (1987) strengthened the idea that solar activity and climate variation were linked.

Hoyt and Schatten (1997) suggested crucial inter-relationships between the Sun and climate over long periods of time. They ascertained that the changes in solar rotation rate is a compelling proof that the convection at deeper levels are varying and its effect is seen in the variation in solar luminosity and irradiance. The Sun exhibits a differential solar rotation period; unlike the Earth, it takes 25 days to complete one rotation at the equator and 35 days at its poles. With increase in latitude, the rotational period keeps on increasing; at mid latitudes, it is around 28–29 days. Convection is the main mechanism of energy transmission inside the Sun. Packets of gas that contain energy rise during this convection and eventually disintegrate after going a certain distance. Theoretically, as convection is typically thought to be what drives rotation, there should be a significant link between the two.

The Sun was believed to be constant until the late nineteenth century; the truth is farther from this belief. Rather the varying nature of the Sun is constant. Spectral and temporal variations from equatorial to polar region and from billion years to a few seconds are the fundamental characteristics of the Sun (Beer et al., 2000). Solar magnetic field plays a crucial role in the instability that occurs on the Sun. The realignment of the magnetic field releases vast amount of energy into the space in the form of solar flares. Along with it, the solar material consisting of hot gas of electrically charged hydrogen and helium is expelled out of the Sun in form of CME. The CMEs take approximately 3 days to reach the Earth, whereas solar flares require only 8 minutes, since they travel with the speed of light. SEPs or high-energy charged particles consisting of protons, electrons, and heavy ions are associated with the CME and solar flare explosions. The mechanism and causes of this complex nature of the magnetic field still need the explanation, to predict them even before they occur. The fluctuation in the solar irradiance results from these different solar activities. In turn, this aspect of the Sun plays an important role in the periodic variation of the Earth's climate. The solar activity as received by the Earth can be affected by.

1. Actual changes occurring within the Sun
2. Orbital variations of the Earth

Between 1761 and 1776, Christian Horrebow observed sunspots from the Rundentaarn observatory in Copenhagen, Denmark. Through his observations, he hypothesized the cyclical nature of the solar cycle. One of the main observations he mentioned was *"It appears that after the course of a certain number of years, the appearance of the sun repeats itself with respect to the number and size of the spots."* Thus, started the journey of studying solar cycles!

5. Solar cycles

Lockyer (1874) wrote *"Surely in Meteorology, as in Astronomy, the thing to hunt down is a cycle and if that is not to be found in the temperate zone, then go to the frigid zones or to the torrid zone to look for it, and if found, then above all things, and in whatever manner, lay hold of, study it, record it and see what it means."*

5.1 Sunspots: structure and nature

Sunspots, viz. strong concentrations of magnetic flux on the solar surface are considered to be the longest-studied direct tracers of solar activity (Solanki et al., 2004). Sunspots are dark spots observed on the surface of the sun, which have a dark region called the Umbra and a surrounding lighter region called Penumbra. The Umbra is much cooler (3700°C.) than the surrounding photosphere (5500°C.) and thus appear as dark spots on the bright surface of the Sun. However, the magnetic fields within a sunspot could be about 0.4 T, which is about 1000 times stronger than that in the other parts of the sun (which is 2500 times than the earth!). The stronger, concentrated magnetic field prevents the movement of hot gas from the Sun's interior to the surface, producing sunspots.

A monochromatic photograph acquired by spectroheliograph helped to identify a vortical structure in the center of the sunspots. Similar vortical structure was found in other observed spots establishing it as a general characteristic of sunspots. Moreover, photographs showed that the masses of hydrogen from a great distance were being sucked into the vortex at the center of the sunspots. This observation led to the theory that the whirling of electrified particles at high speed may give rise to magnetic field in sunspots. Subsequently, abundant proofs were found reinforcing the existence of a magnetic field in sunspots (Hale, 1908).

The average size of the sunspot is same as the size of the Earth. Sunspots mostly occur in pairs or groups though single spots do also appear. In one hemisphere, all the leading sunspots have the same polarity, and all the trailing spots share the same polarity, while the polarities of sunspots in the other hemisphere are oppositely aligned. The magnetic field of the sun flips around every 11 years, turning north into south and south into north. The level of activity on the solar surface is influenced by changes in the magnetic field of the sun (Hale, 1919).

Hale et al. (1919) classified sunspots based primarily upon the determination of their magnetic polarities. Three classes of spots are included in the scheme: (a) unipolar, (b) bipolar, and (c) multipolar.

a. Unipolar spots—Single spots or groups of small spots, having the same magnetic polarity.
b. Bipolar spots—Bipolarity of the sunspots was well established by the fact that nearly 90% percent of observed sunspots were bipolar. The simplest and most characteristic group of sunspots consisting of two spots of opposite polarity.
c. Multipolar spots—Groups of this character hardly constitute >1% of the total number of spots observed. They contain sunspots of both polarities so irregularly distributed, that we cannot classify them as bipolar.

It should be noted however that unipolar spots, and the chief components of bipolar groups, may occasionally have small companions of opposite polarity, which play such a minor and sporadic part in the group that they are disregarded in the classification.

Though astronomers were observing sunspots through telescopes since the 17th century, it was Rudolf Wolf in 1848 in Zurich, who first thought of compiling the information about sunspot numbers, generated by various observers (Vaquero and Vazquez, 2009). Thus, Wolf number measures the total number of sunspots and groups of sunspots that are present on the surface of the Sun. It is also called as relative sunspot number or Zürich number. The compiled sunspot series initially had Wolf's name, but now it is more commonly referred to as the international sunspot number series which is still being produced at the Observatory of Brussels. The annual data value for the international sunspot number is available from 1700, while the daily values extend back to 1818. However, this series was revised and updated on July 1, 2015 making modern counts closer to their raw values. The count of number of sunspots were reduced marginally after 1947 in order to compensate for bias introduced by a new counting method adopted that same year; in this method, sunspots are weighted according to their size (Clette et al., 2014).

5.2 Grand Minima and Grand Maxima

One of the very fascinating features that astonished and excited the early sunspot observer was the sudden disappearance of the sunspots from the entire surface of the sun. Curiosity led to the further explanation of the phenomenon which was within the realm of the studies that existed during that time.

Historical accounts during A.D. 1645–1715 recorded very few sunspots numbers, anomalous cosmic ray modulation, and reduced auroral activities (Eddy, 1976). Eddy, in 1977, recorded the similar events in the long-term data of radiocarbon in tree-rings (Stuiver and Quay, 1980; Stuiver and Grootes, 1980) and in auroral histories by Siscoe (1980).

This period of solar inactivity was called "Maunder Minima." Total solar irradiance (TSI) is predicted to be 0.25% lower than present. Lean et al. (1992) proposed that a 0.25% drop in TSI might result in a 0.2–0.6°C drop in the global equilibrium temperature. Since it occurred during the time when the telescope existed, the data obtained during this period are considered more reliable, and this event became a prototype for all minimas that would have occurred before.

Reconstruction of solar activity for past few millennia was possible due to the advanced development of the technologies using different proxies. Usoskin et al. (2016) identified 20 grand solar minimas and 14 grand maximas in the past 9000 years (Table 17.1). Grand minima and maxima are time periods when Sun exhibits lesser than the averaged solar activity and the enhanced solar activity respectively for a minimum of 2–3 decades. The Sun's magnetic activity is moderate for around 3/4 of

Table 17.1 Table showing the grand solar minimas and maximas complied from (Usoskin et al. 2007; Inceoglu et al., 2015; Usoskin et al., 2016; Wagner et al., 2005; Martin-Puertas et al., 2012).

Solar events	Center of cycle	Duration of solar event
Modern maxima	1970 AD	80 years
Dalton minima	1805 AD	30 years
Maunder minima	1680 AD	80 years
Spörer minima	1470 AD	160 years
Wolf minima	1310 AD	80 years
Oort minima	1030 AD	80 years
U. min	690 AD	80 years
U. max	505 AD	50 years
U. max	305 AD	30 years
U. max	245 BCE	70 years
U. min	360 BCE	80 years
U. max	435 BCE	50 years
U. min	750 BCE	120 years
Homeric minimum	875 BCE	150 years
U. min	1385 BCE	70 years
U. max	2065 BCE	50 years
U. min	2450 BCE	40 years
U. min	2855 BCE	90 years
U. max	2955 BCE	30 years
U. max	3170 BCE	100 years
U. min	3325 BCE	90 years
U. max	3405 BCE	50 years
U. min	3495 BCE	50 years

Table 17.1 Table showing the grand solar minimas and maximas complied from (Usoskin et al. 2007; Inceoglu et al., 2015; Usoskin et al., 2016; Wagner et al., 2005; Martin-Puertas et al., 2012).—cont'd

Solar events	Center of cycle	Duration of solar event
U. min	3620 BCE	50 years
U. max	3860 BCE	50 years
U. min	4220 BCE	30 years
U. min	4315 BCE	50 years
U. min	5195 BCE	50 years
U. min	5300 BCE	50 years
U. min	5460 BCE	40 years
U. min	5610 BCE	40 years
U. max	6120 BCE	40 years
U. max	6280 BCE	40 years
U. min	6385 BCE	130 years
U. max	6515 BCE	70 years
U. max	6710 BCE	40 years

the time (averaged over 10 years). The remaining time is either a grand minimum (approximately 17%) or a grand maximum (about 9%). The grand maxima condition of solar activity is consistent with the current solar activity. Grand minima and maxima do not indicate a dominating periodic behavior. Instead, grand minima only have a tendency to cluster over a quasi-period of roughly 2400 years, while grand maxima do not appear to occur in any perceptible pattern (Fig. 17.3).

Solar cycles continue to occur during these great solar minimum periods, although at a lower intensity. When long-term solar activity declines, mid-latitude glaciers advance and the temperature cools; when long-term solar activity rises, glaciers recede and the climate warms.

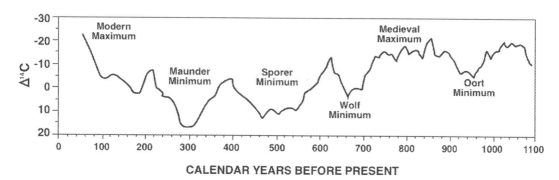

FIGURE 17.3

Grand solar maxima and minima reconstructed from the carbon-14 data from tree rings.

Modified after Damon and Sonnet, 1991a.

5.3 ENSO and Solar cycles

Quinn et al. (1987) compiled records of occurrences and intensity of El Nino events since 1525 from different sources such as ship logs of pirates, privateers, and explorers. When compared it with the Southern Oscillation Index (SOI) Quinn et al. (1987) revealed inconsistency in ranking the strength of El Nino Events. Later, Anderson (1990) correlated sunspot numbers and El Nino occurrences (Fig. 17.4) between 1800 and 1980. The period during 1880–1930 was marked by weaker 11-year solar cycles but frequent El Nino events, while post-1930, the 11-year cycles were stronger, but the El Nino events were less frequent. This suggested an inverse relationship between sunspots numbers and frequency of El Nino.

Dunbar (1983) corelated the delta^{18}O in *Globigerina bulloides* data from the Santa Barbara Basin for the changes in the temperature in the near-surface waters with the changes in El Nino frequency from 1750 to 1950 (Fig. 17.5) reported by Quinn et al. (1987). He noticed a close correlation between isotopic delta ^{18}O changes with the changes in the El Nino frequency and suggested that when the surface layer in the basin is cooler El Nino events are more frequent.

The Multivariate ENSO Index (MEI) was constructed based on the sea level pressure, surface zonal and meridional wind components, sea surface temperature, surface air temperature, and cloudiness over the tropical Pacific (Wolter and Timlin, 1998) (Fig. 17.6). El Nino was predominant from 1976 to 1996, while the Hale cycle and the La Nina prevailed during 1954–1976. Such an alternating pattern dates back to 1900, as far as the Southern Oscillation Index is available (Landscheidt, 2000; Höhler, 2017).

The nonrecurrence of ENSO at the same periodicity makes it a complex weather phenomenon, suggesting it is not entirely triggered by the solar cycle. Though paleoclimatic data (Moy et al., 2002),

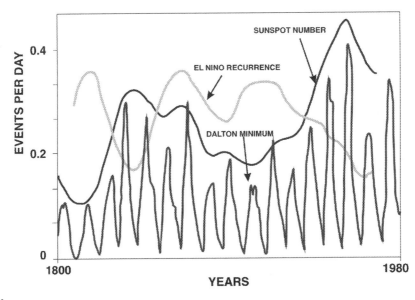

FIGURE 17.4

Graph representing correlation between sunspot number and El Nino occurrence (Anderson, 1990).

5. Solar cycles 447

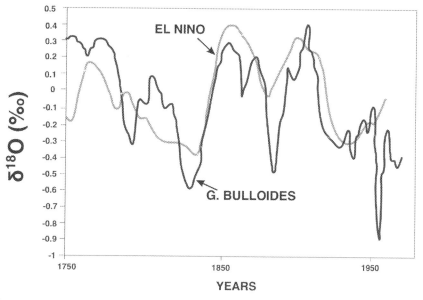

FIGURE 17.5

Graph corelating δ18O (G. Bulloides) with El Nino (Dunbar, 1983).

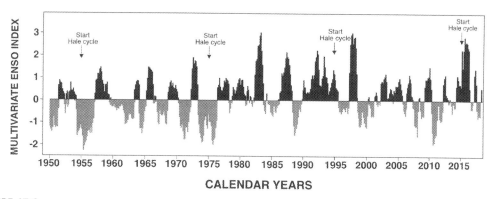

FIGURE 17.6

Alternating La Nina and El Nino events with the hale solar cycle.

Modified after Landscheidt (2000) and Höhler (2017).

solar physics (Leamon and McIntosh, 2017), modeling and reanalysis (van Loon and Meehl, 2008), and frequency analysis (White and Liu, 2008) suggest a strong influence of the Sun on El Nino Southern Oscillations (ENSO), there exist few initiatives explaining the Sun's influence.

5.4 Schwabe cycle (the 11-year cycle)

The Schwabe Cycle was discovered by a German pharmacist and an enthusiast Astronomer Heinrich Schwabe by accident. He believed the presence of the planet named Vulcan inside the orbit of Mercury, and its presence in the close vicinity of the Sun made it difficult to observe. In his search for this planet, Schwabe (1843) observed the Sun for 17 years from 1826 to 1843 (Fig. 17.7). He did not discover the planet Vulcan, but rather discovered a 11-year pattern in the number of sunspots. He also noted that the activity on the Sun increases and decreases over approximately 11-year cycle.

This 11-year cyclicity in the occurrence and variability of sunspots is known as the "Schwabe Solar cycle" or the "solar magnetic activity cycle" or "The Sunspot Cycle." Period and amplitude variation are one of the most common characteristics of solar cycles. Their irregular nature can be explained through the Schwabe cycle. The standard period of Schwabe cycle is 11 years, but it can vary anywhere in between 8 and 15 years and the effects of its climatic modulations have been recorded to occur between 9 to 14 years, from year 1700–2012 (UCAR). Out of 24 solar cycles recorded till 2012, 20 cycles have varied between 10 and 11 years. Similarly, variations can be found in the amplitude specially when the sunspot numbers reduce due to very weak cycles.

Sun's activity cycle occurs in phases, starting from minimum (when sunspots maybe absent for several weeks) to maximum (when 20 or more groups may be present at one time) and in the next phase, the number will decrease from maximum to minimum again. The time duration of appearance of both the phases is unequal; from minimum to maximum, the average time required is 4 years, while for maximum to minimum, the average time is 7 years. One of the examples of Schwabe cycle spans from 1986 to 1996. In 1986 at the minimum phase, only 13 sunspots were observed, whereas during solar maximum in the year 1989, more than 157 sunspots were spotted and then number kept on decreasing to fewer than nine till the year 1986, which was the next solar minima year.

During a solar cycle, the Sun's polar magnetic fields weaken, go to zero, and then emerge again with the opposite polarity. This is a regular part of the solar cycle (NASA). Apart from the variation in numbers and duration of the cycle, the sunspots also vary in latitudinal location. In the starting phase, that is, at solar minima, sunspots start to form around latitudes of 30–45° North and South of

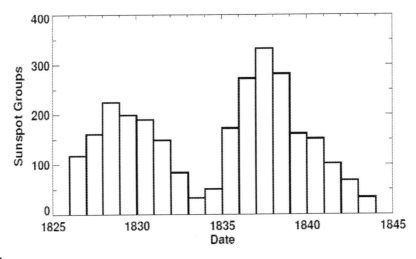

FIGURE 17.7

Number of sunspots group observed by Heinrich Schwabe (1844) between year 1826 and 1843.

the equator of the Sun. Gradually with time, sunspots start migrating toward the equator. As the cycle reaches its maximum, sunspots appear around the 15-degree latitudes. The migration continues toward the equator, and upon reaching solar minima in the end, sunspots tend to form closer to the 7° North and South latitudes. This latitudinal migratory pattern frequently overlaps with the solar minima when sunspots from the previous cycle develop at low latitudes and sunspots from the next cycle form at high latitudes. The magnetic polarity of the leading spot in the northern hemisphere is diametrically opposed to that of the leading spot in the southern hemisphere, and it reverses at the start of each 11-year cycle.

In the early 1860s, German astronomer Gustav Sporer and the Englishman Richard Christopher Carrington independently observed gradual equatorward migration of sunspots with the progression of sunspot cycle. The place of occurrence of the sunspots serves as a criterion to determine the phase of the solar cycle.

Changes in the Sun's activity during the entire cycle have different terrestrial effects. The finding that the quantity and intensity of aurora borealis sightings were highest during sunspot maxima, when the sun was most active (active sun), were at their lowest during sunspot minima (quiet sun), was one of the earliest associations between sunspots and terrestrial phenomena. It changes the Earth's magnetic field. During the solar maximum phase, there are numerous sunspots, solar flares and coronal mass ejections. These high energy particles are carried by the solar winds to the earth and disrupts the communication, power system, navigation, and weather.

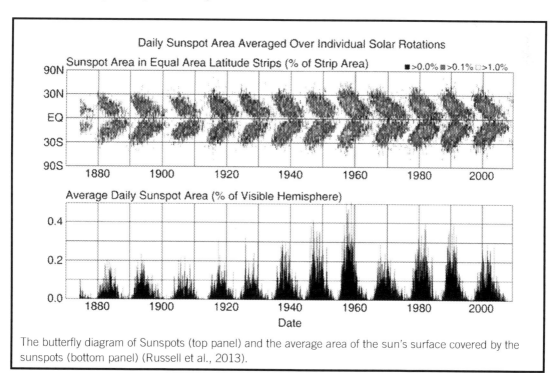

The butterfly diagram of Sunspots (top panel) and the average area of the sun's surface covered by the sunspots (bottom panel) (Russell et al., 2013).

> **—CONT'D**
> **The butterfly diagram: graphical plot of the sunspot migration trend**
> In 1904, E.W. Maunder constructed a 'butterfly' diagram to represent the gradual migration of sunspots throughout a cycle. He began with drawing a coordinate grid like the Earth's geographic coordinates and rotational axis trending North-South. He then split the visible solar disc into latitudinal strips with constant projected area. In each such strip, he computed and color-coded the proportion of the area occupied by active areas and/or sunspots. The average sunspot coverage at any given time was described by a one-dimensional (vertical) array. Considering time as a second parameter, a 2D array and repeating the same steps used before, he stacked the arrays one beside the other at regular intervals to make the butterfly diagram.
>
> Few important characteristics of this diagram are that no sunspots are spotted at higher latitudes probably > 40 degrees, at any time during the cycle. Sunspots are practically never seen within a few degrees of latitude of the equator, and at solar minima, spots from each new cycle start to show at mid-latitudes while spots from the previous cycle are still visible near the equator. Sunspot maximum (1991, 1980, 1969, ...) occurs about midway along each butterfly, when sunspot coverage is maximal at about 15 degrees latitude. One of the striking features is the equatorward migration of the sunspots with progression in cycle.

5.5 Hale cycle (the 22-year cycle)

The Hale cycle is described in terms of magnetic fields and is considered complete only when the field has the polarity similar to what it was at the start of the cycle. This duration takes 22 years (Hale et al., 1919). In the Schwabe cycle, polarity of the magnetic field switches every 11 years. Two such cycles make the 22-year cycle. Thus, the Hale cycle is also known as the "Double Sunspot Cycle." The Hale Cycle starts with the odd number sunspot cycle and ends with the even numbered solar cycle allowing for the possibility of forecasting the amount of activity of the latter from the former. As per the Gnevyshev-Ohl (G-O) empirical rule, the total number of sunspots occurring in the odd number sunspot is more than its preceding even-numbered sunspot cycle (Gneyshev and Ohl, 1948).

Terminator events mark the end of the Hale Magnetic Cycle and start of the new cycle (Dikpati et al., 2019; Hurd and Cameron, 1984; McIntosh et al., 2019). The cancellation of oppositely polarized magnetic bands causes sunspots to be formed at the equator (Fig. 17.8) in both the hemispheres at the end stages. In this transition period between two cycles; old cycle flux is completely removed, while new sunspots forming at the mid-solar latitude modify the magnetic flux to emerge.

Indian evidence of the Hale Cycle: Between the 89-year period of 1891–1979, Bhalme and Mooley (1981) attributed the fluctuation in flood area indices over India to the 22-year Hale Cycle. Their study was further reinforced by Ananthakrishnan and Parthasarathy (1984) who identified excessive rainfall years around the peak phase of the alternate sunspot cycle.

Bhalme and Mooley (1981) interpreted that the floods are clustered in the positive sunspot cycle suggesting a connection between large-scale flood repetition over India and the Hale sunspot cycle (Fig. 17.9). Such a correlation did not exist for droughts! The drought season can occur in the positive and negative sunspot cycles, suggesting no relation between Hale cycle and drought occurrences. This study, however, established that the solar activity related to its magnetic field variability, control large-scale flood occurrences in India, to some extent. Even though the physical mechanism behind it is presently unknown, continued and directed study of this relation can have a practical applicability in managing water resources and planning flood situations in India.

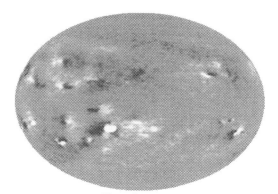
Cycle 22
1989 August 02

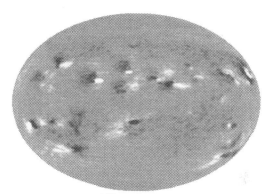
Cycle 23
2000 June 26

FIGURE 17.8

Magnetogram of sunspot cycle 22 and 23 is shown, with yellow (gray in print) and blue (dark gray in print) depicting positive and negative polarity, respectively. Leading sunspots of the same cycle have opposite polarity, in both the hemispheres. When shifting from one sunspot cycle to another, the polarity flips.

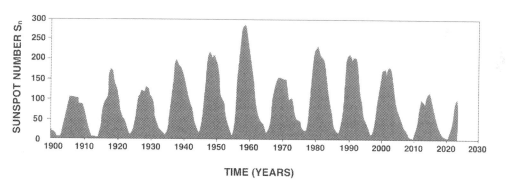

FIGURE 17.9

The Sunspot numbers follow the Hale Cycle. Note the alternating heights; the shorter cyles are the negative (minor) Hale cycles, whereas the taller are positive (major) Hale cycles.

Modified after Bhalme and Jadhav (1984). Graph updated till 1st June 2023 with data from the Royal Observatory of Belgium.

5.6 Brückner-Egeson-Lockyer Cycle (the 30- to 40-year cycle)

The Brückner-Egeson-Lockyer Cycle or the BEL Cycle is a 30- to 40-year climate cycle. Charles Egeson, a map compiler at the Sydney Observatory, conducted some of the early studies on climatic cycles in the late 19th century. He first defined it on the basis of frequent flooding episodes in southern Australia that coincided with sunspot activity and foresaw a terrible drought striking Australia at the turn of the 20th century (Egeson 1889). He studied shorter period of "sunspot-induced data" and agreed with Bacon (1597) that the character of weather recurs every 35 years. Soon, Bruckner (1890) who made the most thorough observations, referred to it as a secular cycle. Lockyer (1901) compared

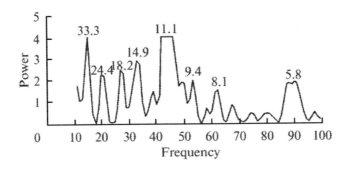

FIGURE 17.10

Time series of monthly aurora data for New England from 1500 to 1948, in terms of power spectrum. Over the peaks are displayed the values of the periods in years (Silverman, 1992).

the 35-year cycle that he discovered in the period of variation of sunspots with the Brückner's climatic cycle, and hence, the cycle was names after all the three scientists.

The outcomes of a spectral examination of the aurora time series support the BEL cycle's actual existence. According to Silverman's analysis of a nearly 500-year dataset of 45,000 auroral sightings based on monthly averages from 1500 to 1948 (Fig. 17.10), the power spectrum has a high peak around 33.3 years. The BEL or para-tri-decadal cycles have also been able to explain various other long-term cycles including those affecting economics, human physiology, and historical events (Halberg et al., 2010).

5.7 Gleissberg cycle (70- to 100-year cycle)

Several secondary aspects of the sunspot cycle have long been noted (Eddy, 1977) like the time required to reach maxima is shorter than the time required to reach the minima and the alternating height of maxima from cycle to cycle. Long-term amplitudes of maxima seem to follow a longer period of seven or eight 11-year cycles, as first noted by Wolf (1862) and later elaborated by Gleissberg (1944). Gleissberg (1939) observed the sunspots number since the year 1750 and noted periodic modulations of sunspot and mean displacement of sunspots from the equator over a 90-year cycle. Thus, this cycle is called as the "Gleissberg Cycle" after the famous German Astronomer Wolfgang Gleissberg and is considered to be the amplitude modulation of the solar cycle with a period of 70–100 years or seven or eight solar cycles.

Between 1879 and 2002, Javaraiah (2006) analyzed the sunspot data and determined the 90-year variation in their differential rotation rates, thus suggesting a connection between differential rotation rate and the Wolf–Gleissberg Cycle and relate it to terrestrial phenomena. Spectral analysis of decennial frequency of aurora records since the 689 B.C. to 1519 A.D. revealed an 88-year cycle (Attolini et al., 1988, 1990). According to Reid and Gage (1988), quasi sinusoidal Wolf-Gleissberg Cycle and variance in the solar luminosity are associated. They correlated the sea surface temperatures (SST) to solar luminosity and concluded that the globally averaged direct sea-surface temperature observations were in accordance with a modulation of the Sun's luminance over a period of approximately 80 years and an amplitude of roughly 0.5%. As the sunspot observation began in 17th century; hence, only three Wolf-Gleissberg cycles were recorded from the historical data. Recently, new techniques using cosmogenic isotopes have been able to reconstruct past solar activity for longer time. Raisbeck et al. (1990) measured concentration of cosmogenic isotope ^{10}Be at the polar observation station at Antarctica for the last 1100 calendar years. Spectral analysis of the data showed two dominant cycles; one for 93 years and another for 202 years.

Influence of Wolf-Gleissberg Cycle is evident from the sharp changes in herring fisheries that Baltic and North Sea are undergoing. About a thousand years ago herring fishery commenced in the Atlanto-Scandian region. The number of adult herring catches during the winter, off the western Norwegian coast closely follow two Wolf-Gleissberg cycles. The number of catches maximised during the W-G cycle maxima. From 70,000 metric tons in 1830 they weighed 100,000 metric tons during the 1860 (i.e. the maxima of W-G cycle) and then declined to just 13000 metric tonnes in 1880s and 1890s. With the advent of next cycle, at the peak of the cycle, catches were 10 million metric tons in 1957 (again the maxima of the W-G cycle). As the cycle progressed, only 3 million metric tons catches were recorded in 1963 (Krovnin and Rodionov, 1992). This trend of declining herring fishery was also noted in USSR and Iceland (Krovnin and Rodionov, 1992). There has been a visible correlation between density of sea weeds off the Scotland shore and sunspot cycles from 1946 to 1955 (Hoyt and Schatten, 1997 and references therein). Increased abundances of herring also indirectly suggest abundance of food necessary to sustain the fish (Picture Credits: picture-alliance/dpa).

5.8 Suess Cycle or de Vries Cycle (∼210-year cycle)

Suess Cycle (or the de Vries Cycle) refers to the approximately 210-year cycle that occurs in the radiocarbon proxies of solar activity, namely ^{14}C and ^{10}Be. It is named after Hans Eduard Suess (1980) and de Vries (1958), who discovered this cycle. In 1984, they suggested that the 208-year cycle in solar activity proxies may be connected to variations in tree-ring width that occur every 200 years. In some parts of the world, tree rings, which record variations in precipitation or temperature, have supported this discovery (Anchukaitis et al., 2017; Schmidt et al., 2012). Breitenmoser et al. (2012) used seventeen tree chronologies (with near global distribution) to analyse 200-year periodicity over the last two millenia. They discovered significant periodicities in the 208-year frequency band (Fig. 17.11), which corresponds to the DeVries Cycle of solar activity, indicating a solar contribution to the temperature and precipitation.

Sunspots were observed to a great extent even in the absence of telescopes. As stated before, such naked eye observations are documented in many historical records. Naked-eye sunspot series was formulated from 200 BCE to 1918 AD by Vaquero et al. (2002). A strong peak at nearly 250 years was observed after performing spectral analysis of the series using the multitaper method (MTM) and later confirmed using singular spectrum analysis (SSA). Later, Ma and Vaquero (2009) used a modified Lomb-Scargle periodogram analysis, to discover a spectral peak of around 229 years in the annual number of naked-eye observations of sunspots. Based on outcomes of the weighted wavelet Z-transform of the naked eye observations of the sunspots, it was validated that the Suess/de Vries Cycle is due to solar activity. The fluctuation exhibits substantial time variation. The periodic signals are manifested from 200 BCE to 400 AD, 800 AD to 1340 AD, and 1610 AD to 1918 AD. The signal's equivalent periods are 211, 195, and 235 years, respectively, and the signal is strongest between 800 and 1340 AD (Ma and Vaquero, 2020).

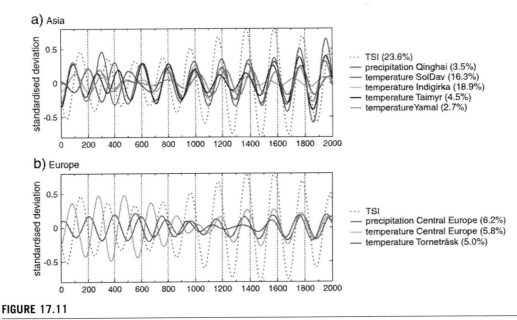

FIGURE 17.11

Dendrochronological climate response to the de Vries solar cycle for the last 2000 years (Breitenmoser et al., 2012).

5.9 Eddy Cycle (~1000-year cycle)

One of the most consistent solar periodicities observed is the 976-year Eddy Cycle, often known as the "millennial cycle." The 980-year solar cycle was named by Abreu et al. (2010). Studies conducted from the lake sediment core from the Vienna basin revealed a periodicity of 1000 years since Miocene to Holocene. This periodicity was reported from the wavelet analysis on the magnetic susceptibility data and the ostracod records (Kern et al., 2012). According to wavelet analysis, the 1000-year periodicity has a high signal between 11,500 and 4000 years ago and between 2000 and 0 years ago but a very low signal between 4000 and 2000 years ago (Ma, 2007; Kern et al., 2012). Every frequency study of Holocene solar activity reconstructions reveals a significant peak around 1000 years ago (Fig. 17.12) (Darby et al., 2012; Kern et al., 2012).

Solanki et al. (2004) reconstructed the sunspot numbers using cosmogenic ^{10}Be and ^{14}C (solid black line in Panel a). Kern et al. (2012) identified a 1000-yr cycle by comparing Ostracods data and the magnetic susceptibility data with the Solanki et al. (2004) (represented in orange arches in Panel a). Ma (2007) generated wavelet analysis using SSN data of Solanki et al. (2004). From the wavelet analysis Ma (2007) identified Eddy cycle (1000 years) and Bray cycle (2400 years) (Panel b and c).

The analysis of the link between solar variability and climate change over the Holocene reveals that the longer the solar cycle, the greater the severity of the climatic change experienced during its lows.

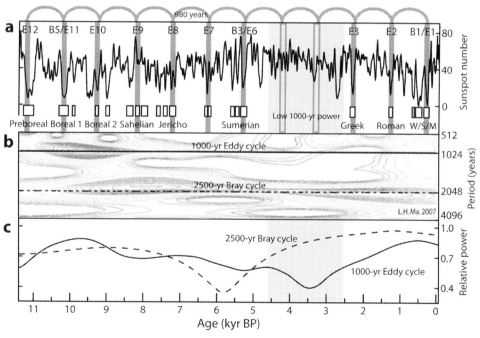

FIGURE 17.12

Compilation of SSN (Solanki et al., 2004) and wavelet analysis (Ma et al., 2007) showing 980 years periodicity in the sunspot number data.

Source Javier (2017).

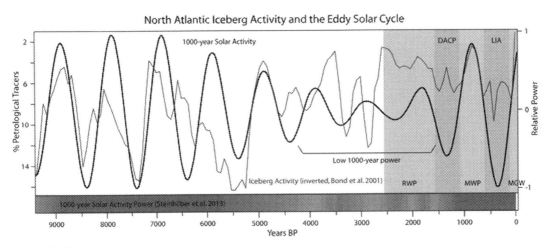

FIGURE 17.13

Corelation between North Atlantic iceberg activity and the Eddy solar cycle. North Atlantic iceberg activities derived from petrological tracers are represented by blue curves, whereas black curves are for Eddy cycle periodicity for the same time. Pink and Blue bars are representing warm and cold period, respectively. *DACD*, Dark Ages Cold Period; *LIA*, Little Ice Age; *MGW*, Modern Global Warming; *MWP*, Medieval Warm Period; *RWP*, Roman Warm Period. Periods with stronger correlation between both curves correlate to periods with high signal amplitude, whereas periods with lower correlation correspond to periods with low signal amplitude. For interpretation of the references to color in this figure legend, please refer online version of this title.

Source Javier (2017).

Lows in the 1000-year Eddy Cycle are associated with significantly higher iceberg releases in the North Atlantic. The most visible climate correlation of the Eddy Cycle is with many maxima in the Bond series of ice rafted debris that represent cold periods in the North Atlantic region (Bond et al., 2001) (Fig. 17.13).

5.10 Hallstatt/Bray cycle (the 2500-year cycle)

In 1968, J. Roger Bray identified a solar-driven climate cycle based on the advances and retreats of the global glaciers of 2600-year cyclicity from Holocene period. Bray studied the Glacier fields in North America, Greenland, Eurasia, New Zealand, and South America and only took into consideration the maximum glacier advances or large readvances that approached the maximum.

A maximum ice advance reached the farthest from the ice field during the post-Pleistocene and a major readvance came close to maximum advance, occasionally leaving moraine without coming in contact (Bray, 1968).

These advances were dated with the help of lichenometry, tree rings, and radiocarbon dating. The glacial episodes over the past 13,700 years revealed a 2600-year cyclicity. Bray in 1967 constructed a solar activity index (Bray, 1967), starting in 527 BCE wherein he combined telescopic sunspot observations with naked-eye sunspot (Schove, 1955) and auroral observations (Schove, 1962). He also constructed an index for postglacial major ice readvances from glaciers all over the world. He correlated these two indices and found that there existed a high degree of correlation between Icelandic sea-ice and ^{14}C production variations (Bray, 1968). He also observed that the 2600-year cyclicity was perfectly correlating with vegetation transitions (Atlantic/Sub-Boreal, or the Sub-Boreal/Sub-Atlantic) and glacier expansion/retreat from the past (Table 17.2).

By 1970s, the existence of a 2500-year climatic cycle that caused glacier advances and recessions and that separated significantly different vegetation stages and cultural phases became evident. Though Bray originally classified it as a 2500-year cycle, it is also estimated as low as 2200 years from other records (Steinhilber et al., 2012; Thamban et al., 2007). O'Brien et al. (1995) too report that the glacial events occurring in the Holocene are separated by 2500 years. The Little Ice Age (1500–1800 years AD), the Hallstattzeit cold epoch (750–400 years BC), and the previous cold epoch (3200–2800 years BC) are all separated by 2200–2500 years, according to Middle Europe Oak dendroclimatology. The timing and periodicity of these epochs are connected to the times when there are significant ^{14}C fluctuations (Damon and Sonett, 1992).

The earliest work of correlating fluctuations in ^{14}C and climate change was described by Hessel de Vries (1958) by noticing a large increase in Carbon-14 during Little Ice Age marked by reduced solar activity (Maunder Minima) and climate deterioration. Later in the 1990s Damon and Sonnet (1991)

Table 17.2 The high and low solar activity phases computed by Bray (1968) by applying the 2600-year periodicity he had derived from his solar activity index. He simply back-calculated 2600-year intervals since little ice age to late pleistocene and demonstrated how well the solar activity phases correlated with climate-induced change in landscape.

Years before present	General climate	Solar activity
0–700	Little ice age, lowered sea levels, increased high altitude glaciation	Low
700–2,000	Subatlantic climatic recovery culminating in little climatic optimum, which was relatively mild, warm, dry climate, posthypsithermal temperature maximum	High
2,000–3,300	Temperatures dropped to near posthypsithermal minimum, increased mountain glaciation, regeneration of bogs	–
3,300–4,600	Drier and slightly cooler (by about 0.1°C) than in the preceding Atlantic but still warmer than today. Maximum sea level, reduced mountain glaciation	High
4,600–5,900	Atlantic/sub-boreal transition, brief temperature decline, minor sea level fall, increased mountain glaciation, cool phase	Low
5,900–7,200	Peakhypsithermal-Atlantic, post-Wisconsin temperature maximum, disappearance of almost all continental glaciers, reduced mountain glaciation	High
7,200–8,500	Brief temperature decline, minor sea level decline, increased mountain glaciation	Low
8,500–9,800	Postglacial-early hypsithermal pre-boreal, temperature increase, major regression of continental glacier	High
9,800–11,100	Minor continental readvances in North Atlantic, temperature decline, post-allerod climatic recession	Low
1,1,100–12,400	Minor continental glacier recession, temperature increase	High
1,2,400–13,700	Major continental glacier readvance	Low

Modified after Bray (1968).

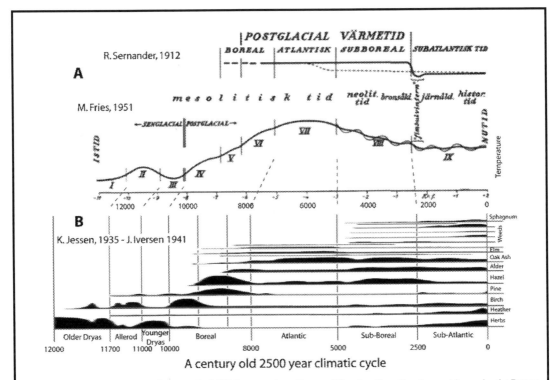

In an attempt to reconstruct the post-glacial Holocene palaeoclimate of Sweden, Scandinavian peat bog scientist Rutger Sernander (1912) used peat-bog stratigraphy to identify wet/dry and cold/warm periods (Top of panel A). Post-1930, advanced palynological investigations (Jessen, 1935; Iversen, 1941) allowed improved biozonation which led to the development of a summer vegetation-based temperature scale (Panel B). Frics (1951) identified nine climate-driven vegetation cycles of 2500-year periodicity, wherein transitions were abrupt and not gradual.

performed the spectral analysis on ^{14}C data to confirm the cycle and named it Hallstattzei (later Hallstatt) based on the late bronze-early iron cultural transition in an Australian archeological site during the cycle's last minimum, 2800 years ago. Usoskin et al. (2016) reconstructed the sunspot numbers using cosmogenic nuclides ^{14}C and ^{10}Be. Singular spectrum analysis on the data revealed a 2400-year periodicity. The synchronous graphs of the two cosmogenic isotopic reconstruction implies that the 2400-year cycle is the product of long-term solar activity (Fig. 17.14). Based on similar data, he proposed a list of Grand Solar Minimas and Maximas (Table 17.1).

What causes a ~2400-yr cycle in solar variability is not known. However, while some attribute it to solar inertial motion (Charvátová and Hejda, 2014), others defend a "planetary hypothesis" where the giant planets in the Solar System would cause these cycles by affecting the movements on the Sun around the Baricenter of the Solar System, through changes in planetary torque (Abreu et al., 2012). It has been suggested that the cycle could also be modulated by the orbits of the larger planets that have a repeating pattern of 2318 years (Scafetta et al., 2016). Due to this cycle's irregularity and the challenge in distinguishing its signal from the 1000-year Eddy cycle, its average length has not been estimated with great clarity.

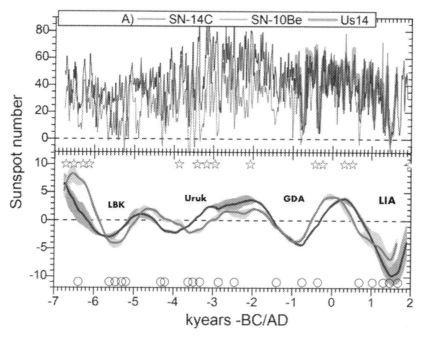

FIGURE 17.14

The top panel is a compilation of sunspot numbers reconstructed using ^{14}C (blue curve) and ^{10}B (red curve) (Usokin et al., 2016) and past 3 kyr. ^{14}C (green curve) (Usoskin et al., 2014). The lower panel shows the results of spectral analysis done on the same data wherein blue hollow circles represent grand minima and red stars represent grand maxima. The grand minima and grand maxima have occurred at irregular intervals for the past 9000 years, and the likelihood of clustering in the Bray's high and low solar activity is visible in the figure. For interpretation of the references to color in this figure legend, please refer online version of this title.

5.11 Milankovitch cycles—Earth's orbital variations

Milankovitch cycles describe the collective effects of changes in the Earth's movements on its climate over thousands of years. The term is named for Serbian geophysicist and astronomer Milutin Milanković who was able to explain the relationship between the changes in the Earth's climate and its orbital configurations.

James Croll in 1864 almost 60 years before Milankovitch proposed that the astronomical aspects are responsible for the glacial-interglacial climate cycles experienced by the Earth. But the lack of the then available scientific knowledge and his own limitations in mathematics and astronomy led to the rejection of his theory. Later, in 1920, Milutin Milankovitch, with great mathematical details computed solar insolation at different latitudes and related it with the heat budget of the Earth. The Milankovitch theory explains the changes occurring in the Earth's orbital configurations which in turn redistributes the solar energy (insolation) received by the Earth and modulating its climate. The theory, to some extent, was able to explain the glacial—interglacial cycles occurring during the Quaternary period. According to him, for the onset and prevalence of the glacial periods, the high

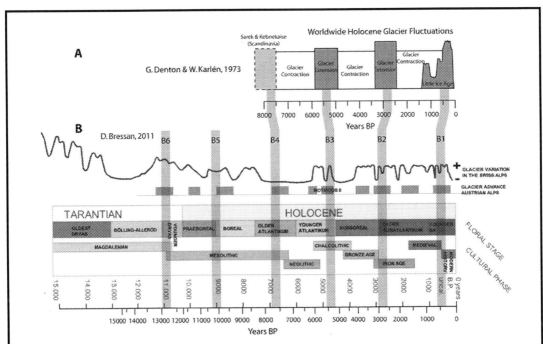

The Bray Cycle's lows correspond with the coldest periods of the Holocene, which are defined by massive glacial readvances (panel A). (Panel B) The low points of the approximately 2400-year Bray cycle are indicated by blue bars. The uncalibrated radiocarbon dates along with the corresponding calibrated scale in calendar years B.P. Holocene glacier advances and retreats in the European Alps not always corresponding with the Australian and Swiss Alps.

Source (A) Denton and Karlén (1973); (B) Bressan (2011).

The Greek Dark Ages, or GDA, was the time during which the Late Bronze Age came to an end and the Early Iron Age began, following a period of civilizational collapse. The "Uruk" Bray low event takes place at the same time as the Uruk civilization's rise and some of the earliest cities in the planet. The Middle East passes from the Copper Age to the Early Bronze Age at the end of the Uruk Bray low, and cuneiform writing begins to be used. The earliest Bray low recorded correlates to the start of the "LBK", or Linear Pottery Culture, along Europe's Danube River. This time period represents the end of the hunter-gatherer civilization in Europe and the beginning of the emergence of an agricultural economy. It is not sure that the LBK and Uruk historical events were caused by Bray lows; it is only mentioned, to place the lows in the context of human history. The more recent Greek Dark Ages and Little Ice Age, on the other hand, are well recognised colder times with significant historical climate crises. Each Bray low corresponds to a major cultural transition. The LBK is roughly the end of the Early Neolithic in Europe which is when agriculture started to spread. The Uruk period marks the transition of the Middle East from the Copper Age to the Early Bronze Age. The GDA occurs as the Middle East moves from the Bronze Age to the Iron Age. Humans transitioned from the pre-industrial era to the Industrial era and that's when LIA occurs.

latitudes of the northern hemisphere must be cold enough to prevent the melting of the winter snow during summers.

The effective increase in annual snow and ice cover would increase the albedo of the Earth, inducing positive feedback, causing an ice-age. Conversely, high intensity of summer insolation beyond the 65° N latitude along with a 100 kyr. eccentricity cycle was proposed to induce a nonlinear response to the Earth's climate, making it the main determinant of a glacial period termination cycle and the pacing of interglacials. Thus, the intensity of the climatic aberrations varies depending upon the combination of the orbital configurations in operation at any particular moment in time. The Earth undergoes three main orbital variations.

- Eccentricity: changes in the shape of the Earth's orbit every 100,000 yrs.
- Obliquity: changes in the tilt of the Earth's rotational axis every 41,000 yrs.
- Precession: wobbles in the Earth's rotational axis every 26,000 yrs.

5.11.1 Eccentricity—the 100,000-year cycle

The Earth's orbit is not circular but elliptical or eccentric. Eccentricity means the measure of how much the shape of Earth's orbit departs from a perfect circle. This shape of the Earth's orbit around the Sun is constantly fluctuating; the orbital shape ranges between more and less elliptical (0%–5% ellipticity) on a cycle of about 100,000 years (Fig. 17.15) These oscillations, from more elliptic to less elliptic, are of prime importance to climatic variability because it alters the distance of the Earth from the Sun resulting in alteration in the amount of solar heat received.

The difference in the distance between Earth's closest approach to the Sun known as "perihelion" and its farthest departure from the Sun known as "aphelion" is currently about 5.1 million kilometers. This causes 6.8% more incoming solar radiation in January than in July. Currently, the orbital eccentricity is nearly at the minimum of its cycle. When the Earth's orbit is most elliptical the amount of solar energy received at the perihelion would be in the range of 20–30% more than at aphelion. This cyclic change in the eccentricity of the Earth's orbit has orchestrated wide spread glaciation on the Earth approximately every 100,000 years as evidenced from long-term proxy data (Gupta et al., 2001; Liebrand et al., 2016; Mitchell and Karr, 1998). Some workers have also identified and attributed 400 ky. cycles to eccentricity (Gupta et al., 2001; Rodriguez-Tovar et al., 2010).

5.11.2 Obliquity—the 41,000-year cycle

Obliquity refers to is the inclination of the Earth's axis in relation to its plane of orbit around the Sun. It is the reason for the Northern and Southern hemispheres of the Earth having disjointed seasons during the same time of the year. The axial tilt varies cyclically between 22.1 degree and 24.3 degrees over the course of 41,000 years. Currently, the tilt is 23.44 degrees and is continuously decreasing. With less axial tilt the Sun's solar radiation is more evenly distributed between winter and summer and also decreases the difference in radiation received between the equatorial and polar regions. With the increase in the Earth's axial tilt, intensity of seasons increases.

High obliquity results in interglacial, while lower obliquity is associated with glacial periods. One hypothesis for Earth's reaction to less axial tilt is that it would promote the growth of ice sheets. This response would be due to a warmer winter, in which warmer air would be able to hold more moisture and subsequently produce a greater amount of snowfall. In addition, summer temperatures would be cooler, resulting in less melting of the winter's accumulation. Obliquity does not change

Eccentricity: 100,000 yr

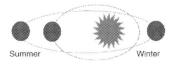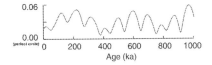

Obliquity: 41,000 yr

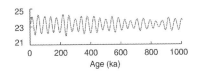

Axial precession: 26,000 yr

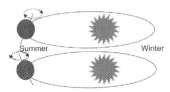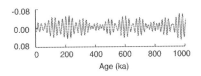

FIGURE 17.15

Illustration of Milankovitch cycle.

Modified from Zachos et al. (2001).

the amount of insolation the Earth receives, but it does change the amount of insolation each latitude receives which is large at high latitudes. The 41 kyr cyclicity has been observed in multiple long-term paleoclimatic records (Lourens et al., 2010; Wunch, 2004; Kutterolf et al., 2013; Liu et al., 1999).

5.11.3 Precession—the 26,000-year cycle

Precession is the clockwise circular motion of the Earth's axis, as it rotates around itself. As the Earth spins around itself, it slowly wobbles; this wobbling of the rotational axis, can be compared to a top running down, and beginning to wobble back and forth on its axis. Tidal forces caused by the gravitational influences of the Sun and Moon result in the Earth to bulge at the equator, affecting its rotation, which in turn cause the wobble.

As the Earth wobbles, its axis traces a circular path, and points to different stars which have a fixed position in space. The axis moves slowly, at a rate of a little more than a half-degree per century. Currently, the axis points to a star called "Polaris," which we call the "North Star." Approximately, 13,000 years ago, the axis pointed to another star called the "Vega" (the then North Star!). So, Polaris has not always been and will not always be, the pole star. For example, when the pyramids were built, around 2500 BCE, the pole was near the star Thuban (Alpha Draconis). The path that the axis follows from Polaris to Vega and back to Polaris has a periodicity of 26,000 years.

Axial precession makes seasonal contrasts in one hemisphere more extreme and less extreme in the other. It can also affect the timing of the season of a place. Because the direction of the Earth's axis of rotation determines at which point in the Earth's orbit the seasons will occur, precession will cause a particular season (for example, northern hemisphere winter) to occur at a slightly different place from year to year. The 26,000 years precessions cyclicity has been reported form the paleoclimatic records (Cole, 1998; Fischer et al., 1991; Stansell, 2009).

5.11.4 Application of the Milankovitch theory

Milankovitch linked the cycles to develop a thorough mathematical model for estimating variations in solar radiation and surface temperatures at various latitudes. The model functions similarly to a climate time machine; it may be run backward and forward to investigate past and future climate conditions. According to the Milankovitch Theory, glacial origination occurs when summer insolation at 65°N permits more ice to survive the summer each year, causing formation of the Laurentide, Fennoscandian, and Siberian ice. This mechanism is supported by ice-albedo and other feedbacks which gradually cool the Earth with a simultaneous drop in sea level. The glacial period endures many cycles of increased summer insolation at 65°N, as it gradually grows colder and the sea level falls. The following eccentricity cycle, between 95 and 125 kyr later, generates a nonlinear reaction on precession, triggering a glacial termination with the next spike in 65°N summer insolation. This is considerably a quicker process than glaciation caused by feedback effects such as reduced ice albedo and/or accumulation of greenhouse gases.

5.11.5 Limitations of the Milankovitch cycle

The Milankovitch theory has been unable to explain the change in the periodicity in the occurrences of glacial periods between Pliocene and Pleistocene (Fig. 17.16). Until about a million years ago, glaciations occurred at 41-kyr intervals, indicating that obliquity was the primary cause. However, since then, glaciations have occurred at 100-kyr intervals. This transition from early Pleistocene glaciations to late Pleistocene glaciations seems to have occurred without any change in insolation.

6. Proxies and paleoclimatic records of solar cycles
6.1 Short-term solar activity

In the 16th century, before the telescope was invented, the sunspots were observed by naked eyes. After the discovery of the telescope and the camera obscura, the solar activity was recorded via direct measurements of sunspots. The primary observer would measure the sunspots, and the relative sunspot number was calculated each day. It is interesting to note that there exists a legacy of European primary observers; it appears as if the baton of sunspot observation was passed on from one to the other. Historical records document the contributors as Staacher (observed from 1749 to 1787), Flaugergues (observed from 1788 to 1825), Schwabe (observed from 1826 to 1847), Wolf (observed from 1848 to 1893), Wolfer (observed from 1893 to 1928), Brunner (observed from 1929 to 1944), Waldmeier (observed from 1945 to 1980), and Koeckelenbergh (observed from 1980).

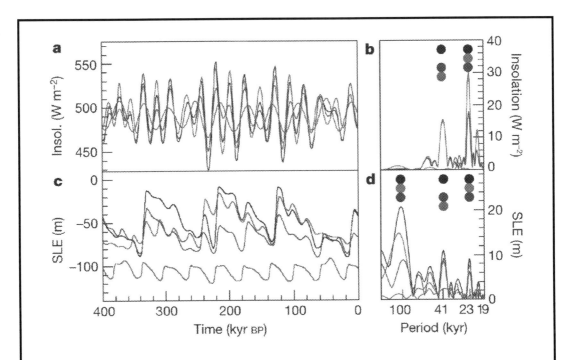

Abe-Ouchi et al. (2013) demonstrated the role of Role of eccentricity, obliquity and precession in the 100-kyr cycle. Time series of the model experiments with one of eccentricity, obliquity or precession fixed for a constant atmospheric CO_2 concentration of 220 p.p.m. (A) Insolation forcing (insolation at latitude 65° N on 21 June) with variations in eccentricity, obliquity and precession (*black lines*); with obliquity fixed at 23.5° (*red lines*); with eccentricity fixed at 0.02 (*blue lines*); and with perihelion passage fixed at the spring equinox and no precession (*green lines*). (B) Corresponding spectra of insolation change in a; (C) Calculated ice-volume change, expressed as sea-level equivalent (colours same as in a); (D) Corresponding spectra of calculated ice-volume change in c. For interpretation of the references to color in this figure legend, please refer online version of this title.

Recent work with sophisticated computer models of climate and ice sheet growth has led to a better understanding of how the effects of precession and eccentricity might work together to produce the ice ages as seen in figure above (Abe-Ouchi et al., 2013).

The short-term variation in the solar activity is also recorded by measuring the total solar irradiance (TSI). The TSI measurement has only been possible because of the instruments designed to measure it are fitted in the satellites. These instruments measured the TSI most accurately. However, they provide records for short durations, that is, only since there deployment and thus ineffective in studying long-term solar irradiance changes.

In the Indian subcontinent, various studies related to solar activities have been conducted. Dumka et al. (2021) studied the behavior of total electron content in the ionosphere to indicate sunspot activity and demarcated the 24th solar cycle. The recent changes in the solar activity have been studied from the Indian subcontinent on the basis of available multiple spectral analysis on Indian rainfall data, coral

6. Proxies and paleoclimatic records of solar cycles

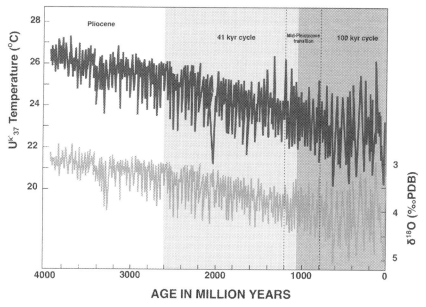

FIGURE 17.16

Reconstruction of the mid-Pleistocene transition using temperature proxies, alkenone (UK'37 in green) and benthic delta ^{18}O (orange) in marine sediments. The light grey panel indicates the duration when the influence of Milankovich's Obliquity if recorded and that in dark grey was influenced by Eccentricity. For interpretation of the references to color in this figure legend, please refer online version of this title.

Modified after Lawrence et al. (2006).

growth rate data, and satellite temperature records suggested the multidecadal (66–70 years) cycle, weak signal of 13- and 22-year cycle, and 2.5–7.5 years solar cycle associated with ENSO (Tiwari and Rao, 2004).

6.2 Long-term solar activity

It was only in 1976 when people started realizing the effects of solar cycles in nature and started correlating solar cycles as a forcing of climate (Eddy, 1976). After the advent of instruments, direct measurements of solar activity were maintained. A statistical analysis of these records provided insights to periodicity in solar activity.

In the Indian subcontinent the primary climate phenomenon, that is, the Indian monsoon system is controlled by the solar insolation (Sengupta, 1960). The strength of solar activity is directly linked with the strength of monsoonal winds. Hence, proxies that help decipher rainfall pattern, viz. speleothems, marine sediments, and tree rings, can also help demarcate variation in solar activities. The long-term changes in the solar activity are recorded as paleoclimatic signatures in natural archives such as marine sediments, lake sediments, tree rings, speleothems, ice cores, etc. These paleoclimatic records are deciphered indirectly through bio-geo-chemical components of the Earth System, which have been affected due to solar forcing. Such components are called "proxies" and vary in their sensitivities.

6.2.1 Cosmogenic isotopes

In the upper atmosphere, several radioactive isotopes are produced when cosmic rays collide with atmospheric molecules at high speed. These isotopes are known as cosmogenic isotopes. The production rate of the cosmogenic isotopes depends on the strength of the cosmic radiation, which in turn varies with the strength of the Earth's magnetic field and solar activity. Hence, records of cosmogenic isotope production rates are invaluable for understanding the relation between past climate change, the Earth magnetic field, and variations in the solar activity. The Earth magnetic field shields the Earth from charged cosmic particles, such that a relatively strong magnetic field reduces the production of radiogenic isotopes (Fig. 17.17). The solar wind is a stream of charged particles emitted from the Sun, which varies with the solar activity. The Earth reacts to the solar wind by increasing the strength of its shielding magnetic field. Therefore, higher solar activity results in stronger shielding and thus lower production of cosmogenic isotopes. The abundance of cosmogenic isotopes in depositional archives reflects past variations in solar activity as well as the strength of the Earth magnetic field (Svensson, 2022).

There are a wide variety of useful cosmogenic nuclides, which can be measured in soil, rocks, groundwater, and the atmosphere (Schaefer et al., 2022). All these nuclides share the common feature of being absent in the host material at the time of formation. Some of these nuclides have a short half-life, where-in they decay since nucleosynthesis. But the half-lives are long enough to allow measurement of their accumulated concentrations. For example, ^{10}Be (half-life 13,87,000 years) and ^{14}C (half-life 5730 years) are both used for dating sediments, ice cores, and tree rings. They are also used to

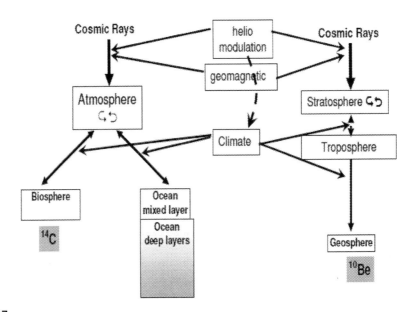

FIGURE 17.17

The schematic represents the production chain of ^{14}C and ^{10}Be. Showing the incoming cosmic ray flux being affected by geomagnetic field change and heliospheric modulation. The dashed line represents influence of solar activity on climate (Usoskin, 2013).

understand the past variations in the solar activity because the source of these isotopes are cosmic rays, which enter the upper atmosphere and produce cosmogenic isotopes (Lal and Peter, 1967). Their concentrations are measured using accelerated mass spectrometers.

6.2.1.1 Beryllium-10 (^{10}Be)

Beryllium-10 (^{10}Be) is a radioactive isotope of beryllium. It is formed in the Earth's atmosphere mainly by cosmic ray spallation of nitrogen and oxygen (Beer et al., 2012). Beryllium occurs in solutions below about pH 5.5 and thus dissolves in rainwater (precipitation) and reaches the surface of the Earth. As the pH rises, ^{10}Be precipitates out and gets accumulated in the sediments. The measurement of ^{10}Be is complicated and expensive as it is measured using accelerator mass spectrometer. However, as ^{10}Be has long half-life, it occurs in very low concentrations, making it difficult to measure (Beer, 2000).

6.2.1.2 Carbon-14 (^{14}C)

Carbon-14 (^{14}C) is produced in the upper layers of the troposphere and the stratosphere by thermal neutrons absorbed by nitrogen atoms. The highest rate of carbon-14 production takes place at altitudes of 9–15 km (30,000 to 49,000 ft) and at high geomagnetic latitudes. When cosmic rays enter the atmosphere, they undergo various transformations, including the production of neutrons (Fig. 17.18).

Carbon-14 may also be produced by lightning. About 78% of the Earth's atmosphere is made up of nitrogen gas, which itself is a diatomic molecule made of two nitrogen atoms. Each time a neutron collides with a nitrogen nucleus, which consists of seven protons and seven neutrons, there is a finite probability that it will react with that nucleus, replacing one of the protons. As a result, a nitrogen-14 atom (and a neutron) transforms into a carbon-14 atom (and a proton). Once you produce that carbon-14, it behaves just like any other atom of carbon. It readily forms carbon dioxide in our atmosphere and mixes throughout the atmosphere and the oceans. It gets incorporated into plants, consumed by animals, and readily makes its way into living organisms. The ^{14}C incorporates in the plant and animal tissue through photosynthesis and respiration, respectively. The ^{14}C amount in the tissue is always in equilibrium with that of the atmosphere, because as the old cell dies, they are replaced with new ^{14}C. However, the exchange and replacement of ^{14}C ends, when the organism dies. As the organism dies, ^{14}C content starts decreasing due to decay of ^{14}C to nitrogen. The ^{14}C decreases at a negative exponential rate after the organism is dead. Therefore, if the 10 half-lives are passed (i.e., 57,300 years), the sample will contain less than 0.01% of the original ^{14}C content of the organism. Hence, the specimens older than 50,000 years cannot be dated using radiocarbon dating.

Nuclear weapons testing in the 1950s and 1960s brought about a reaction that simulated atmospheric production of carbon-14 in unnatural quantities. The huge thermal neutron flux produced by nuclear bombs reacted with the nitrogen atoms present in the atmosphere to form carbon-14. This carbon-14 produced is known as "bomb carbon" or "artificial radiocarbon." It nearly doubled the atmospheric

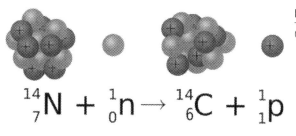

FIGURE 17.18

Creation of ^{14}C is.

carbon 14-content as measured in around 1965. The level of bomb carbon was about 100% above normal levels between 1963 and 1965. The level of bomb carbon in the northern hemisphere reached a peak in 1963 and in the southern hemisphere around 1965. Even after nuclear weapon testing was banned, the bomb effect persists. Though the excess carbon-14 produced during nuclear weapons testing has already decreased partly due to the global carbon exchange cycle, it is still high enough to mask the true measurements of natural fluxes of C-14. Radiocarbon scientists used this knowledge to test their theories regarding the mixing rates of Carbon-14 through various carbon reservoirs.

Due to the magnitude of the carbon reservoirs, the amplitudes of short-term production fluctuations are substantially attenuated, making it challenging to detect the Schwabe cycle in the ^{14}C data (atmosphere, biosphere, and ocean). ^{10}Be records have an 11-year Schwabe cycle, according to Steig et al. (1998) and Beer et al. (1990).

6.2.1.3 Chlorine-36 (^{36}Cl)

Chlorine-36 (^{36}Cl) is produced in-situ from the lithosphere and atmospheric, both. ^{36}Cl in the atmosphere is produced when high-energy particles of Argon-40 are bombarded violently. The interaction between solar winds and galactic cosmic rays (GCR) with Earth's magnetic field controls the production of atmospheric ^{36}Cl (Bard, 1998). The amount of ^{36}Cl on the earth surface is not only controlled by its rate of production but also its transportation and deposition. The two processes by which the ^{36}Cl arrives on earth surface are by ^{36}Cl flux, which is atmospheric source, and by ^{36}Cl fallout, by the means of transportation and deposition.

As the ^{36}Cl production is dependent on the latitude of the Earth, Blinov et al. (2000) created a model to measure ^{36}Cl flux and fallout for 10° latitude. He observed that the mid-latitudinal regions which showed high ^{36}Cl fallout provide higher resolution of data rather than the low latitudinal or polar regions (Fig. 17.20). However, the ^{36}Cl behavior is difficult to understand as studies about its depositional fallout was obstructed due to substantial amount of ^{36}Cl production during nuclear weapon testing in 1950s, which stayed in the environment for years (Elmore et al., 1982).

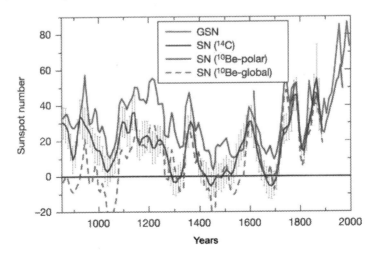

FIGURE 17.19

Graph corelating sunspot numbers reconstruction by ^{10}Be and ^{14}C (Solanki et al., 2004).

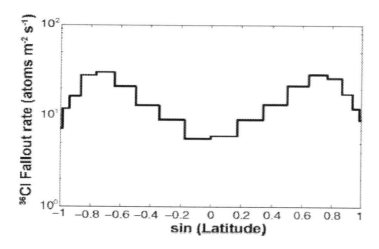

FIGURE 17.20

Schematic represents the latitudinal variation of ^{36}Cl fallout (Blinov et al., 2000).

Ideally, ^{36}Cl is preferred over the ^{10}Be for speleothems studies, as the ^{10}Be shows high particle active behavior, causing ^{10}Be to be retained in soil, longer than ^{36}Cl, leading to uncertainties in interpretation. These uncertainties could occur due to increased retention time or mixing and buffering in soil. ^{10}Be exhibits complex relation with pH changes in the soil, due to which the sorption capacity of ^{10}Be varies. This could cause flushing of ^{10}Be and variation in ^{10}Be recorded in the speleothems that are unrelated to solar activity. As compared to ^{10}Be, ^{36}Cl has hydrophilic nature due to which it rapidly transfers from atmosphere through soil and into speleothem deposits with negligible mixing or retention (Johnston, 2010). Hence, coupling ^{36}Cl isotopic studies with speleothems could give an excellent proxy for paleoclimatic reconstruction.

Cosmogenic isotopic concentrations in ice cores provide very high-resolution records of not only past solar activity but also that of the Earth's magnetic field. However, abundance of ^{14}C being very low, ^{10}Be and ^{36}Cl provide more data in ice cores, as they occur in measurable quantities.

6.2.1.4 Factors that influence cosmic ray fluxes

The shape and geomorphology of the Earth are complex and not a uniform smooth spheroid. Thus, bulge at the Earth's Equator, deep oceanic trenches, and tall mountains cause uneven bombardment of cosmic rays based on the latitude and altitude. Thus, accurate cosmic flux determination demands several geographic and geologic considerations. Even, altitudinal change in atmospheric pressure influences the production rate of nuclides by a factor of 30 between mean sea level and the top of a 5 km high mountain. Even variations in the slope of the ground can affect penetration of nuclides into the subsurface (Dunne et al., 1999). Geomagnetic field strength which varies over time is one of the most influential factors affecting the production rate of cosmogenic nuclides. However, models which take account of these variations assume that the strength gets averaged out over geological time scales and thus rule it out.

6.2.1.5 Paleoclimatic records from employing cosmogenic proxies

During the solar maxima, intense electromagnetic field is generated, causing less penetration of cosmic rays to the earth's atmosphere (Bard et al., 2000). Hence, high solar activity leads to decreased production rate of cosmogenic isotopes. The solar activity is inversely proportional to the cosmogenic isotopes. High production rate of ^{10}Be and ^{14}C signifies low solar activity (Fig. 17.19) and decreased temperature on the Earth's surface. The production rate of the ^{14}C (Stuiver and Quay, 1980) and ^{10}Be (Beer et al., 1990) is also controlled by the variations in the solar wind magnetic properties. Beer et al. (1985) reconstructed the solar activity by measuring the ^{10}Be from the polar ice, and later even ^{14}C was included as proxy for solar activity reconstruction. Usoskin et al. (2004) reconstructed the sunspot activity of the last millennium by measuring ^{10}Be data from Greenland ice core as well as Antarctica ice core. Hence, reproducing the Maunder Minima, Sporer Minima, Wolf Minima, and Oort Minima from the data. Steinhilber et al. (2009) was the first scientist to reconstruct the TSI records for the entire Holocene period, using ^{10}Be from the ice core. The longest continuous record of solar activity based on radioactive isotopes, documents almost 12,000 years of solar history (Solanki et al., 2004) derived on the basis of comparison between ^{10}Be and ^{14}C from mid-latitude tree rings. Periods and quasi-periods of solar activity were previously detected in these data also by Yin et al. (2007), who also performed a wavelet analysis.

Agnihotri et al. (2002) seems to be the only one to have studied solar activity in the Indian subcontinent using cosmogenic nuclides. They used a sediment core from the Arabian sea and reconstructed the paleo-TSI data for 1200 years using ^{10}Be and ^{14}C to understand the influence of solar activity on the Indian Summer Monsoon (ISM). They suggested that the Schawbe (11 years) cycle may not be controlling the ISM as the TSI, and rainfall data showed no coherence among themselves. Though, they did identify control of multidecadal cycles such as ∼60, ∼100, and ∼200 years cycles, on the ISM.

6.3 Strontium isotopic ratios

Strontium is an element with chemical properties similar to calcium (Ca); it has the same ionic radius as that of Calcium and thus Sr^{2+} ion substitutes for Ca^{2+} in minerals, rocks, and tissues of living organisms. This makes strontium a very efficient tool for tracing paleoclimatic processes involving calcium. Strontium is present in natural and anthropogenic reservoirs in amounts that can be isotopically analyzed by mass spectrometric methods. Strontium locked up in rocks is rarely bioavailable. Only weathering or leaching can release it in aquatic or marine systems, making it available for organisms to construct their calcium-carbonate shells, for example, Molluscs, Foraminifera, and Corals.

The isotopes of Sr exhibit mass-dependent fractionation due processes such as evaporation and precipitation. Generally, given the long residence time of strontium in the oceans (∼3−5 million years) in comparison to the effective mixing time of the oceans (∼1000 years), the differences observed in their temporal concentrations are usually small. The $^{87}Sr/^{86}Sr$ isotope ratios of marine sediments reach 0.707 to 0.709; recent seawater values are 0.7092. Significant deviation from this value in paleoclimatic records are suggestive of large magnitude events that could have contributed to the change.

Rahaman et al. (2022) measured of $^{87}Sr/^{86}Sr$ ratios on corals from Lakshadweep in south-eastern Arabian Sea to determine past ENSO events from the Indian subcontinent, in order to trace its impacts on regional hydrology. In an attempt to develop $^{87}Sr/^{86}Sr$ ratios as a proxy for ENSO events, they

6. Proxies and paleoclimatic records of solar cycles

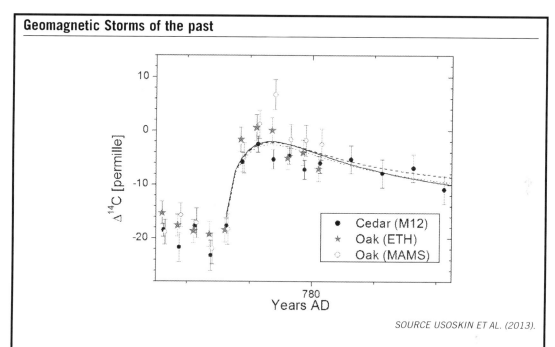

Geomagnetic Storms of the past

SOURCE USOSKIN ET AL. (2013).

The carbon-14 levels remained approximately constant on the Earth, throughout the last millennia. Because of nuclear weapon detonation in the early 2000s caused fluctuations in the carbon-14 pattern. So, the scientists were surprised to find that when two separate cedar trees in Japan were analysed for carbon-14, they showed a 20 times larger spike than natural readings in 8th century. Even though the excess was about 1.2% more than the normal value, it was still lot more than the normal variations. This spike in the carbon-14 value has been reported in tree rings from the multiple regions of the world (Germany, Russia, New Zealand, USA). Later the tree ring data was compared with the Beryllium-10 and Chlorine-36 data from the Antarctica Ice core and similar spike was seen in those data as well. The only explanation that they could come up with, during that period was, that extreme bombardment of cosmic rays on the earth increased the carbon-14 value, caused due to geomagnetic storm. A similar geomagnetic storm like a Carrington event of 1859 was experienced during the 8th century. Though the Carrington event is known to have been largest solar storm in the recent past.

validated their results against instrumental data from 1990 to 2014. The $^{87}Sr/^{86}Sr$ ratios showed an inverse relationship with ENSO events; higher values coincided with La Nina years while the lower values coincided with the El Nino years (Fig. 17.21). They attributed the changes in strontium isotopic ratios to ENSO driven changes in precipitation and its impact on sub-marine groundwater discharge. Significant studies have been able to correlate ENSO events to 11-year and 22-year sunspot cycles; thus, strontium isotopes (rather all proxies) detecting ENSO events could be used effectively as indicators of past solar activity.

6.4 Magnetic susceptibility

Magnetic susceptibility (MS) is the degree to which a material can be magnetized in an external magnetic field. MS is used primarily as a relative proxy indicator for changes in composition that can be linked to paleoclimate-controlled depositional processes. The physical link of MS to particular

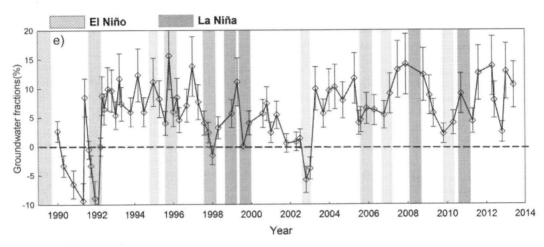

FIGURE 17.21

Groundwater fractionation graph showing El Nino and La Nina events (Rahaman et al., 2022).

sediment components, ocean or wind current strength and direction, or provenance usually requires more detailed magnetic properties studies.

Joint studies by the Solar Observatory of the Carnegie Institution of Washington at Mt. Wilson, California, and the Department of Research in Terrestrial Magnetism under the direction of Professor George E. Hale probed the connection between solar activity and the Earth's magnetic storms. Early around the late 1800s to 1900s, observation of phenomenon, like sudden swinging of magnetic needles out of their usual position or direction with no obvious cause, generation of strong Earth electric currents interfering with telegraph and cable lines, sudden and momentary suspension of electric car line traffic, and occurrence of wide-spread auroras sparked correlations between the solar activity affecting the Earth's magnetic field. Though, the correlation seemed unreasonable, given the vast distance between the Sun and the Earth, interestingly such incidences coincided with sun-spot activity, reinforcing the connection between solar activity and terrestrial phenomenon. Results from actual observations found that the electrical currents in the atmosphere not only follow the general fixed directions but also have a magnetic field whose axis is displaced from that of the Earth through an angle of 32 degrees to the west and south (Bauer, 1910).

The magnetic fields of the earth and the atmosphere differ. Though both their north magnetic poles lie in the northern hemisphere, it is below the surface for the Earth and above the surface for atmosphere. Hence, while the effects of the compass needle are same for both, their effects on the dip needle are opposite. The earth's field makes the north-seeking end of the needle dip below the horizon while that of the atmosphere would cause the end of the needle to point above the horizon. Hence, the effects of the two fields on the vertical magnetic component are opposite in the same hemisphere (Bauer, 1910). This observation led to the development of magnetic susceptibility in sediments as a proxy for past solar activity.

The magnetic susceptibility of soils has been linked with climate, mainly through rainfall, by numerous investigators. MS increases with increasing rainfall from about 200 mm/yr to 1000–1200 mm/yr. Above 1200 mm/yr, MS decreases as rainfall increases up to about 2000 mm/yr. Under arid and semi-arid conditions, below about 200 mm/yr of rainfall, MS and rainfall exhibit no relationship and has been attributed to limited pedogenic activity (Balsam et al., 2011).

MS of most materials is a temperature-dependent property, and thus, sediments are first equilibrated to room temperature. The temperature dependence of paramagnetic minerals such as clays is described by the Curie–Weiss law, $k = c/T$, where c is the Curie constant and T is the temperature in Kelvin. It theorizes that if the room temperature is 20°C, the MS of a pure paramagnetic material that is 5°C below room temperature will be 1.7%; 3.5% if it is 10°C lower, 7.1% if it is 20°C lower, and so on. The temperature dependence of other materials in the temperature range between 0°C and 20°C is less significant.

6.4.1 Paleoclimatic records Indian Subcontinent employing magnetic susceptibility as a proxy

Marine and Lake sediments have been used as reliable archives for paleoclimatic records for reconstructing long-term changes in climate at different resolutions. Hence, excellent proxy to study changes in the solar forcings. As discussed before, based on the depositional process, the physical, chemical, and mineralogical composition as well as magnetic susceptibility of sediments varies.

Sandeep et al. (2017) studied the magnetic variations in the lake sediments of 3100 years from Pookot Lake, Kerala, southern India. He observed that increased monsoonal strength is caused due to high total solar irradiance and vice versa. He also observed that the low total solar irradiance during little ice age and high total solar irradiance during medieval warm period. He reported the periodicity of 210, 128, and 96 years cycle in the magnetic susceptibility data, suggesting that these solar cycles affected the paleoenvironment of Indian subcontinent.

Thamban et al. (2007) studied magnetic susceptibility over the Holocene period in Arabian Sea using sediment core and reported the periodicities of 2200, 1350, 950, 750, 470, 320, 220, 156, 126, 113, 104, and 92 years. These periodicities were observed by wavelet analysis on the magnetic susceptibility data.

Warrier et al. (2017) found periodicities of 906, 232, 128, 96, 61, 54 and 44 years in magnetic susceptibility data from lake sediments in Karnataka and attributed it to solar forcing on the ISM. The magnetic susceptibility of the sediments of the Chilika lake, along the northern east coast of India for the Holocene period, revealed periodicities of 2523, 573, 147, 93, 59, and 55 years cycles (Dash et al., 2022).

6.5 Foraminiferal signatures

Foraminifera are the most widely used proxies of past climate. They are unicellular, calcium carbonate shelled, marine microorganisms, that occur at the base of the food-chain as zoobenthos or zooplankton. Thus, foraminifera are extremely sensitive to the slightest change in their environment. Foraminifera respond in different ways to changes to their environment; they either multiply in numbers under favorable conditions or drastically reduce in abundance or diversity in response to stress (Nigam et al., 2006; Nigam, 2005). Foraminifera construct their shells by utilizing calcium carbonate from their surrounding waters, resulting in their shell geochemistry being in equilibrium with the ambient water. Trace-elemental analysis, stable oxygen and carbon isotopes, and boron isotopes measured on foraminiferal shells hold clues for paleotemperatures, global ice volume, and past ocean pH Emiliani (1954); Duplessy et al. (1981); Lea (1999); Foster et al. (2016). Foraminiferal reproductive behavior is also an extremely sensitive attribute used as a reliable indicator of climate. Favorable climates promote sexual reproduction, producing benthic foraminiferal specimen with larger mean proloculus sizes (MPS) (Boltovoskoy and Wright, 1976). Different solar cycles, right from the Milankovich Cycles to the shorter sun-spot cycles, have been reported from the periodicity in the different attributes of foraminiferal records from marine sediments.

6.5.1 Paleoclimatic records from the Indian subcontinent using foraminifera as proxy

Nigam et al. (1995) was the first to report the Gleissberg Cycle from the Central West Coast of India. He recorded a 77-year cyclicity in the MPS of benthic foraminiferal species *Rotalidium annectens* and suggested that the western Indian region experienced a drought every 77 years. He suggested that the Gleissberg cycle caused reduced freshwater influx from the Kali River Estuary, increased salinities, resulting in asexual reproduction in *R. annectans* (indicated by smaller MPS).

Similar studies from the Irrawaddy (Ayeyarwaddy) Delta, Myanmar (Burma) also suggested the influence of Gleissberg cycle on the ISM (Panchang and Nigam, 2012). Spectral analysis of the past 500-year record of MPS of benthic foraminiferal species, *Asterorotalia trispinosa* revealed a 93-year cyclicity in the ISM, which is in the purview of the Gleissberg Cycle. Additionally, the foraminiferal abundance, low-salinity assemblage, as well as $\delta^{18}O$ evidenced three fresh water pulses at ~1675, 1765, and 1850 AD. The authors attributed them to the Sporer, Maunder, and Dalton Minimas and suggested that the Little Ice Age of Europe was manifested as a wet period, over the Indian Sub-continent.

Rana and Nigam (2009) studied the past 700-year paleoclimatic record from the Central East Coast of India and reported 200 years of periodicity from the benthic foraminiferal data. Spectral analysis of the downcore distribution of benthic foraminiferal abundance as well as that of angular asymmetrical foraminifera revealed a 200 ± 50-year solar cyclicity influencing the monsoon in the region. Das et al. (2017) studied the distribution of the *Bolivina* spp. of benthic foraminifera from the sediment core off Oman margin in the Arabian Sea and reported 208 years of cyclicity and the 57–60 years cyclicity from the 14,000 years of data, as they observed the cyclicity in this oxic benthic foraminiferal species, they suggested that the OMZ intensity in the Arabian Sea is controlled by these solar cycles. Based on the planktonic/benthic foraminifera ratio and $\delta^{18}O$ data, the periodicities of ~512, ~388, ~304, ~250, ~235, ~217, ~152, ~139, and ~135 years have been reported from the sediment cores off Saurashtra, Arabian Sea (Azharuddin et al., 2019). Gupta et al. (2013) observed Suess (208) cycle in the planktonic foraminiferal record from the sediment core, representing the high-resolution data of 14,000 to 8000 years. The sediment core was taken off Oman margin, Arabian Sea. Hence, suggesting that the changes in the ISM winds during the Allerod period was caused due to changes in solar irradiance. Govil et al. (2022) studied the variability of ISM over the mid- to late Holocene using sediment core taken from Bay of Bengal. He observed the periodicities of ~418, ~165, ~139, ~109, ~97, and ~86 years from spectral analysis on the planktonic foraminiferal record. Suokhrie et al. (2022) studied the changes in the productivity over the Holocene from the Bay of Bengal and reported the 153, 183, and 210 years periodicity from *G. bulloides* and 87, 150, and 255 years periodicities from the sedimentary and isotopic datasets. Naidu and Malmgren (1995) reported the 2200 years cycle from the planktonic foraminifera (*G. bulloides*). As *G. bulloides is* an upwelling indicator, they suggested that the upwelling and the ocean circulation changes in the Arabian sea are affected by the ISM. This process of upwelling due to ISM has the 2200 years periodicity. Almogi-Labin et al. (2000) studied the variation in the productivity using benthic and planktonic foraminifera in the Gulf of Aden over the 530 years. They observed that benthic foraminiferal productivity indicated 100 (eccentricity) and 41 ka (obliquity) Milankovitch cyclicity. As the productivity in the Arabian Sea is affected by the monsoonal system; hence, the NE winter monsoon enhanced during the glacial period leading to the increased productivity in the Gulf of Aden. This process is regulated by the 23 years (precession) Milankovitch cyclicity. Shimmield et al. (1990) also observed that the productivity in the Arabian Sea is controlled by the 23 years (precession) Milankovitch cyclicity using planktonic foraminifera as the proxy. Gupta et al. (2001) reported the −400 kyr and −100 kyr Milankovitch cycles in the benthic foraminiferal species from the eastern Indian Ocean.

6.6 Speleothem records

Speleothems, commonly known as stalactite and stalagmites, are mineral deposits formed in limestone caves. The layers in speleothems are formed due to percolation of precipitated groundwater from the carbonate rocks. The percolated waters evaporate or degassing of CO_2 takes place from the water droplets causing seasonal growth of speleothems, making them high-resolution archives of paleoclimate. While evaporation process occurs only near the cave entrance, speleothems in the deep caves are presumed to be formed only due to degassing (Fairchild and Baker, 2012). When water percolating through soils comes in contact with decaying organic matter, it accumulates the partial pressure of carbon dioxide over time. The partial pressure of carbon dioxide of water exceeds that of caves atmosphere. Hence, when the water enters the cave, it is supersaturated with calcite, causing degassing of carbon dioxide and thus precipitating calcium carbonate. Hence, speleothems are primarily composed of calcium carbonate (Fig. 17.22).

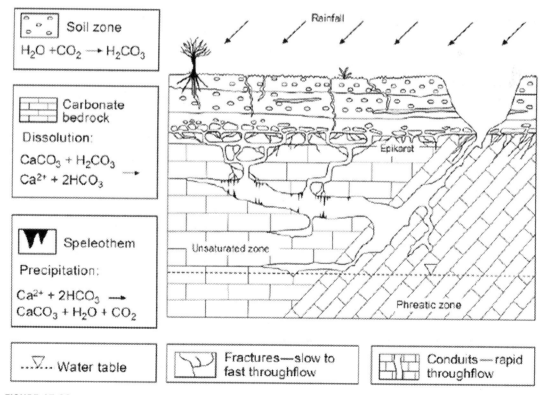

FIGURE 17.22

Carbonate dissolution and deposition in cave system (Fairchild et al., 2006).

The slow movement of water and air in the cave establishes the thermal equilibrium between the air in the cave and temperature within the bedrock, leading to approximation of mean annual surface temperature. During calcite deposition from the percolated water, CO_2 is lost, and oxygen isotope fractionates at the rate which is temperature dependent for deposition. Therefore, oxygen isotope variations in the speleothems provide a proxy for temperature variations through time. Hence, stable oxygen isotopic changes in the speleothems are used to reconstruct the paleoclimatic records. The stable oxygen isotopic ratios of different layers of a speleothem serve as indicators of monsoon strength and frequency. The Indian subcontinent is also affected by annual reversal of monsoonal winds, which are also recorded in the speleothems of the Himalayas and East India.

When water percolates through the host rock, it also accumulates certain amount of trace elements (uranium, thorium etc). These trace elements help date speleothems. High precision uranium series dating can be done using inductively coupled plasma mass spectrometer (ICP-MS) or thermal ionization mass spectrometer (TIMS). These methods of dating are only useful for time range of 500,000–100 years BP.

6.6.1 Paleoclimatic records over the Indian subcontinent using speleothem as proxies

Speleothems found in the Himalayan caves and other parts of Indian subcontinent extensively records the paleomonsoon, along with tool of stable isotope ($\delta^{18}O$ and $\delta^{13}C$) analysis and statistical analysis such as spectral and wavelet analysis, the influence of solar activities on the long-term monsoonal changes is studied.

While studying speleothems from Karnataka, Yadava and Ramesh (2007) reported 132 years periodicity in the Indian Monsoons on bases of $\delta^{18}O$ and $\delta^{13}C$. They observed a strong 21-year cycle and a weak 11-year cycle in rainfalls and suggested that low rainfall in the Indian subcontinent corresponds to low solar activity. Jaglan et al. (2020) observed strong 190-year solar cyclicity in the speleothems from Baratang and Mahadev Caves in Andaman Island. Lone et al. (2014) reconstructed the 1000 years of monsoonal variability using $\delta^{18}O$ on speleothems found in southern India. He observed the periodicity of 12 and 33 years cycle. Berkelhammer et al. (2010) observed the 89 years cyclicity in the ISM, by reconstructing the $\delta^{18}O$ data from the speleothems in east central India, suggesting that the ISM is weakly controlled by Schwabe cycle and strongly controlled by Hale Cycle, Gleissberg cycle, and Seuss Cycle.

6.7 Tree rings

The tree stems/trunks contain alternate light and dark bands, representing seasonal growth. The light bands contain sequence of large, thinly walled cells known as "earlywood," surrounded by dark bands called "latewood." They grow annually in combination and are called tree rings (Fritts, 1976). In wet and warm years, trees grow wider rings and in cold and dry years, they grow narrower rings. The factors controlling the mean width of the tree rings are tree age, tree species, soil nutrients, availability of stored food in the tree and climate (temperature, precipitation, humidity and sunlight) (Vaganov et al., 2011), and are thus used as high-resolution proxies of climate change. Douglass (1867–1962) worked extensively on this subject and established a relation between the solar activity and terrestrial climate. From his extensive research on the trees in arid southwest USA, he observed that rainfall was most important factor in the tree growth. He perceived that width of annual growth rings of tree will reflect the rainfall variations. He assumed that rainfall pattern is affected by solar heat, as the ocean evaporation due to solar heating created winds that brought evaporated seawater to the continents as rain. Therefore, he concluded that solar variations caused rainfall variations, which are reflected in the ring pattern of trees. Apart from studying the structure of the rings, variations in their isotopic signatures (namely those of ^{14}C, $\delta^{13}C$, $\delta^{18}O$, $\delta^{2}H$) are used for paleoclimatic reconstruction.

The source of carbon in tree rings is atmospheric carbon dioxide to which the tree canopy is exposed. This carbon dioxide is converted into carbohydrates in leaves of plant through photosynthesis. The radiocarbon and the stable carbon isotopic concentrations in plants are assumed to be in equilibrium with the atmosphere. Whereas when the carbon dioxide is converted to carbohydrate in plant cells, the isotopic fractionation of ^{14}C decreases more than 5% than atmospheric level. This level of ^{14}C depletion is inconsistent among other organisms (Olsson, 1974). Tree rings do not exchange radiocarbon with other tree rings. This fact has supported the use of dendrochronology in radiocarbon dating, particularly in constructing radiocarbon calibration curves.

The $\delta^{13}C$ is $-8\%_0$ in the atmosphere today, as compared to the preindustrial era when the $\delta^{13}C$ was $\sim 1.68\%_0$, this is because of fossil fuel combustion which led to input of isotopically light carbon dioxide. During photosynthesis, the plants discriminate against $\delta^{13}C$, resulting in $\delta^{13}C$ value to be in the range of -20 to -30, causing variation in the $\delta^{13}C$ values of tree rings from year to year. The most important parameter in the $\delta^{13}C$ fractionation is the concentration of carbon dioxide in the intercellular space within leaves. If the intercellular carbon dioxide level is high and the rate of photosynthesis is low, the resultant photosynthates will be ^{13}C depleted (Schleser et al., 1999).

The $\delta^{13}C$ value in the tree rings is negatively correlated with temperature and sunlight in the tropical region and positively correlated in the mid- and high-latitude region During the drought conditions, the plant stomata closes to conserve the water; as a result, low concentration in carbon dioxide is observed in the leaves. Whereas, in both dry and warm conditions, $\delta^{13}C$ value will increase in the wood.

The oxygen isotope ratio ($\delta^{18}O$) analysis of tree rings is a promising tool for reconstruction of paleoclimatic variations in the terrestrial environment. The $\delta^{18}O$ reflects the air temperature, relative humidity, transpiration, leaf temperature, and precipitation during the tree growth period.

The water in the soil is taken up by the roots; this water is originated from the local precipitation, this is reflected in the $\delta^{18}O$ value.

The tree rings $\delta^{2}H$ represents the isotopic composition of precipitation and atmospheric water. The hydrogen isotope ratio ($\delta^{2}H$) reflects the water content in the plant organ used during photosynthesis. The $\delta^{2}H$ of precipitation is controlled by the temperature during condensation and air humidity during evaporation. After the evaporation from the vegetation or surfaced water, the water vapors are depleted in ^{2}H (low $\delta^{2}H$) compared to its original composition. This is caused due to fractionation during the transition from liquid phase to gaseous phase, leading to preferential evaporation of lighter isotopes ($^{1}H^{16}O$) over the heavier isotopes ($^{2}H^{18}O$). Whereas, when the water condenses in the cloud, precipitation is enriched in ^{2}H (high $\delta^{2}H$) than the water vapors in the air. Hence, the precipitation becomes depleted in ^{2}H with rainfall, as the air mass condenses and precipitates while moving inland or at higher elevations. This process strongly depends on temperature conditions during condensation. Hence, $\delta^{2}H$ value shows strong dependency on temperature of precipitation. The $\delta^{2}H$ values change in precipitation due to factors such as temperature, latitude, elevation, rainfall, atmospheric circulation factor, distance from the coast, and weather. The Vienna Standard Mean Ocean Water (VSMOV) is used as the international reference point of ocean water for isotopic measurements, with $\delta^{2}H$ value of 0%.

The maximum amount of work on paleoclimatic reconstruction using dendrochronology has been carried out on the trees from mid- to high-latitude region because of the strong seasonality in the temperature which forms well developed tree rings. Whereas in the tropical region, minimum seasonal temperature changes lead to no annual tree ring production. However, tropical trees respond strongly to seasonal change in rainfall and evaporation rate and have a significant impact in physiology of tree. Hence, the hydrogen and oxygen isotope compositions show positive correlation between the temperature and precipitation in the mid- and high-latitude region. In the tropics, it has no correlation with the temperature but the strong negative correlation with the precipitation.

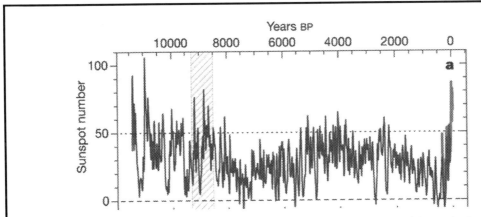

Sunspot numbers over past 11,000 years reconstructed using 14C activity in decadal samples from mid-latitude from tree-rings

Solanki et al. (2004) used ^{14}C concentrations in trees from the mid-latitudes to determine the chronology and reconstruct decadal history of sunspot activity for the past 11,000 years. The study concluded that the level of solar activity since 1940 has been exceptionally high and that, such a period of equally high activity occurred over 8,000 years ago. They found with an extremely high level of confidence (95%) that in the past 11,000 years, the sun has been in high activity state (defined as periods marked by sunspot number >50) only for about 780 to 1,060 years distributed across 31 periods! Additionally, they also stated that the earlier high magnetic activity periods were short lived (on an average 30 years) as compared to the past 70 years; the longest such duration of 90 years span was identified to have occurred about 9,000 years ago. On the basis of the distribution of such events in the past, Solanki et al. (2004) suggested that the number of high-activity periods decreases exponentially with increasing duration. However, the solar activity in past 70 years has not only been unusually high, but has also lasted unusually long. However, Solanki (2003) compared the sunspot activity with the solar flux since 1850 to 2000 and suggested that despite the rarity of the current episode of high average sunspot numbers, solar variability is not the dominant cause of the strong warming during the past three decades.

The oxygen and hydrogen isotopic content in the tree rings preserves the records of rainfall amount or past temperatures in this region, through the uptake of water from the soil with a specific isotopic signature.

While a lot of work has been done to understand the Indian Summer Monsson from tree ring records in India, unfortunately no work has attempted to decipher solar activity in the data.

6.7.1 Limitations of using tree rings as proxies

There are many factors which control the resultant oxygen and hydrogen isotopic content of the tree rings. Such as, while synthesis of woody material, isotopic fractionation occurs in the tree rings. These kinetic and biological fractionation are dependent on factors such as temperature, relative humidity, and evapotranspiration. The duration and amount of evapotranspiration control the isotopic content of the soil and hence also affect the evaporative enrichment occurring within the leaves. Another limitation of isotopic signatures in dendrochronology is, when the tree has deep roots, it taps into isotopically homogenous groundwater. Hence making it difficult to determine changes in the tree rings isotopic ratios.

6.8 Ice cores

The ice cores are formed from the accumulation of past snowfall in the ice sheets and polar ice caps. They provide valuable record of paleoclimatic conditions by chemical and physical analyses of the ice and firn. Firn is the remnant snow that sustained after the summer ablation season. They are found on the ice surface from the high elevations. The snow crystals convert to firn and then to ice due to weight of overlying material, causing crystal to settle, deform, and recrystallize. The firn buried underneath snow accumulation results in increase in density as the air spaces between snow crystals decrease. The interconnected air spaces between grains are locked into individual air bubbles (Herron and Langway, 1980). The snowmelt and sublimation are extremely less in the dry snow zones of polar ice sheets leading to continuous snow accumulation over a million years. Hence, the ice cores provide an insight into past air temperature, atmospheric composition, and solar activity (Jouzel and Masson-Delmotte, 2010). Paleo-temperature is studied by stable isotopic analysis such as deuterium, hydrogen, oxygen, and other elements such as argon and nitrogen. The past solar activity and geomagnetic field changes are studied using cosmogenic nuclide such as ^{10}Be, ^{14}C, and ^{36}Cl. Thamban et al. (2011) studied the two ice cores from the Antarctica and reconstructed the past temperature from 1500 AD to 2000 AD using stable oxygen isotope and stable deuterium isotope. He shows decreased temperature and precipitation during the Maunder minima and Dalton minima.

Apart from polar ice cores, Himalayan ice cores are also used for paleoclimatic studies. Though the there are multiple studies on Himalayan ice cores for atmospheric composition and monsoonal variability, there are no studies about solar activity.

7. Prediction of solar cycles

A U.S. Weather Bureau expert once said "The problem of predicting the future climate of Planet Earth would seem to depend on predicting the future energy output of the Sun." Modelers depend on assumptions on magnetic fields within the Sun based on what is observed on the photosphere, which accounts for some of the limitations. Many meteorologists as well as astrophysicists attempt to predict the forthcoming duration and amplitude of solar cycle and or activity for theoretical and societal causes. However, solar cycles are not quite perfect and tend to, by and large, occur before or after estimates, if not completely go absent! In short, solar activities cannot be predicted using classical extrapolation techniques (Courtillot, 2021).

Effect of solar activity on human health

MOTHER → Embryo/FETUS → INFANT → GROWTH & DEVELOPMENT → PUBERTY

decreased UVR 1ST trimester, 50% more MS in children*

decreased vitamin D increases UVR receptors

Embryo (very sensitive to environment)

Low receptor sensitivity requires *more environmental UVR* (UV-A) to suppress the immune system setting the stage for more autoimmune disorders in later life; flexibility in system if more UVR supplied at this stage which could mitigate autoimmune disorders.

reflex increase in the fetal UVR receptors

Fixed UVR *receptor* activity, If low→ increased incidence of multiple sclerosis or other autoimmune diseases.
If high→ more repair of soma, less inflammatory activity and longer lifespan.

*Staples J, Ponsonby A-L BMJ 2010;340: c1640

If low UVR *at birth*, UVR receptors stay high (from maternal settings) and upregulate the epigenome. *If UVR is elevated,* UVR receptors downregulate.

It has been observed that the solar activity affects the human life span. Increased solar activity during the birth, reduces the lifespan of Humans by an average of 8 years (Davis and Lowell, 2018). They compared the death records of 63 million natural deaths with sunspot number (SSN) and observed, Ultraviolet radiations (UVR) which are estimated from sunspot number were inversely proportional to the lifespan. When the SSN was less than 90, the average lifespan was ~74 for male and ~78 for female. When the SSN was more than 90 average lifespans of was ~66 for male and ~70 for female. Davis and Lowell (2018) also observed that the people born with the autoimmune disorder such as multiple sclerosis had born during the high solar activity, as high UVR during the birth might damage the DNA and membrane in developing foetus.

Several predictive approaches have been adopted by scientists, while most use satellite data that are available only for the past few decades, some use ground-based observations spanning hundreds of years. In recent years, researchers depend on predictive machine outputs by running some of their input data through algorithms. These predictive methods are also called "machine learning," which can be supervised on unsupervised. Supervised learning trains a model on known input and output data so that it can predict future outputs, and unsupervised learning, finds hidden patterns or intrinsic structures in input data.

Scientists can choose to focus on different precursors as the cause or links to the solar activity they are trying to predict. The Earth's magnetic field, which responds to the Sun and the strength of the magnetic field at the Sun's poles are most common phenomenon considered. In recent years, the polar magnetic field model has proven to be the most successful (NASA, 2020). It uses measurements of the magnetic field at the two poles of the Sun. The presumption made in this case is that "the magnetic field at the Sun's poles acts like a seed for the next cycle." If it is strong during solar minimum, the next solar cycle will be strong; if it is diminished, the next cycle would be weak.

Petrovay's (2020) comprehensive review distinguishes three main groups of solar cycle prediction methods: precursor methods that rely on some (often magnetic) measure of solar activity, model-based methods based on dynamo models, and extrapolation methods based on the premise that, "the physical process giving rise to the sunspot number record is statistically homogeneous." He concluded that "precursor methods have clearly been superior to extrapolation methods; nevertheless, some extrapolation methods may still be worth further study."

Food security, political instability and solar cycles

SOURCE CENTRE FOR DISEASE CONTROL AND PREVENTION.

Impact Team (1977) in their book "The Weather Conspiracy: the coming of the new ice age" suggested that one fifth of the world population will face issue of food security, if the Europe's temperature is dropped by one degree. It will lead to drop in number of people sustained by one hectare of land from three individuals to two. It has been observed that good climatic conditions from 1930 to 1960 was a ideal time for world food production, food security in Green revolution. In 1970's CIA encouraged the research in solar activity because of its influence on food production and political instability, as there was possibility that the climatic conditions of the 19th century will return, marking the drop in the global temperature. It was suggested that the decrease in temperature will lead to widespread drought and flood conditions, because of which there will be increased food insecurity causing many revolutions and political instability. As same implications were observed in the past years due to decreased solar activity.

---

8. Conclusions

Proxy records stand testimony to the influence of solar cycles on the climate of the Earth throughout its history. The Sun remains the eternal and primary external forcing of the Earth's climate, and its activity has varied over the past 4.6 billion years. Not only has it aged, but its evolution causes its activity to vary in cycles. In certain instances, the solar activity can induce climate change, and in some other instances, it can combine with other internal forcings (like GHGs) and amplify a climatic phenomenon. Thus, understanding solar activity and their modeling/prediction can prove crucial in sustainable development.

While exhaustive paleoclimatic records exist in and around the Indian subcontinent, those exploring the role of solar activity in modulating climate are far and few. Many researchers through wavelet analysis do identify different lengths of cycles in their proxy data but refrain from correlating them with solar cycles. If attributed to solar activity, most studies neither attempt to identify the process that was modulated nor discuss how these processes were modulated/influenced.

When there is no denying that climate change is for real, Eg. The Hockey Stick Curve (Mann, 2021) and is human induced, some credit is due to natural climatic variability which has been there even before the advent of man. While extreme climatic events, urban heat island effects, and such events have obvious anthropogenic linkages, it is worth exploring if the shifting of climatic zones and their timing could be related to external forcing. Given the knowledge that obliquity has the potential to modulate latitudinal insolation and precession can cause shifting of place and time of seasons, there is a need to recognize the role of the sun in continuous and gradual climate change.

References

Abe-Ouchi, A., Saito, F., Kawamura, K., Raymo, M.E., Okuno, J.I., Takahashi, K., Blatter, H., 2013. Insolation-driven 100,000-year glacial cycles and hysteresis of ice-sheet volume. Nature 500 (7461), 190–193.

Abreu, J.A., Beer, J., Ferriz-Mas, A., June 2010. Past and future solar activity from cosmogenic radionuclides. In: Soho-23: Understanding A Peculiar Solar Minimum, Vol. 428, p. 287.

Abreu, J.A., Beer, J., Ferriz-Mas, A., Mccracken, K.G., Steinhilber, F., 2012. Is there a planetary influence on solar activity? Astron. Astrophys. 548, A88.

Agnihotri, R., Dutta, K., Bhushan, R., Somayajulu, B.L.K., 2002. Evidence for solar forcing on the Indian monsoon during the last millennium. Earth Planet Sci. Lett. 198 (3–4), 521–527.

Almogi-Labin, A., Schmiedl, G., Hemleben, C., Siman-Tov, R., Segl, M., Meischner, D., 2000. The influence of the NE winter monsoon on productivity changes in the Gulf of aden, NW Arabian Sea, during the last 530 ka as recorded by foraminifera. Mar. Micropaleontol. 40 (3), 295–319.

Ananthakrishnan, R., Parthasarathy, B., 1984. Indian rainfall in relation to the sunspot cycle: 1871–1978. J. Climatol. 4 (2), 149–169.

Anchukaitis, K.J., Wilson, R., Briffa, K.R., Büntgen, U., Cook, E.R., D'Arrigo, R., Davi, N., Esper, J., Frank, D., Gunnarson, B.E., Hegerl, G., 2017. Last millennium Northern Hemisphere summer temperatures from tree rings: Part II, spatially resolved reconstructions. Quaternary Science Reviews 163, 1–22.

Anderson, R.Y., 1990. Solar-cycle modulations of ENSO: a possible source of climatic change. In: Betacourt, J.L., MacKay, A.M. (Eds.), Proceedings of the Sixth Annual Pacific Climate Workshop. California Department of Water Resources, Interagency Ecological Studies Program Technical Report.

Attolini, M.R., Galli, M., Nanni, T., 1988. Long and short cycles in solar activity during the last millennia. In: Secular Solar and Geomagnetic Variations in the Last 10,000 Years. Springer, Dordrecht, pp. 49–68.

Attolini, M.R., Cecchini, S., Galli, M., Nanni, T., 1990. On the persistence of the 22yr solar cycle. Sol. Phys. 125 (2), 389–398.

Azharuddin, S., Govil, P., Singh, A.D., Mishra, R., Shekhar, M., 2019. Solar insolation driven periodicities in southwest monsoon and its impact on NE Arabian Sea paleoceanography. Geosci. Front. 10 (6), 2251–2263.

Bacon, F., 1957. Vicissitude of Things. Essays, Civil and Moral.

Balsam, W.L., Ellwood, B.B., Ji, J., Williams, E.R., Long, X., El Hassani, A., 2011. Magnetic susceptibility as a proxy for rainfall: worldwide data from tropical and temperate climate. Quat. Sci. Rev. 30 (19–20), 2732–2744.

Bard, E., 1998. Geochemical and geophysical implications of the radiocarbon calibration. Geochem. Cosmochim. Acta 62 (12), 2025–2038.

Bard, E., Raisbeck, G., Yiou, F., Jouzel, J., 2000. Solar irradiance during the last 1200 Years based on cosmogenic nuclides. Tellus B 52 (3), 985–992.

Bauer, L.A., 1910. Solar activity and terrestrial magnetic disturbances. Proc. Am. Phil. Soc. 49 (194), 130–144.

Beer, J., 2000. Long-term indirect indices of solar variability. Space Sci. Rev. 94 (1), 53–66.

Beer, J., Andree, M., Oeschger, H., Stauffer, B., Balzer, R., Bonani, G., Stoller, C., Suter, M., Wolfli, W., Finkel, R.C., Langway, C.C., 1985. 10Be variations in polar ice cores. Greenl. Ice Core: Geophys. Geochem. Environ. 33, 66–70.

Beer, J., Blinov, A., Bonani, G., Finkel, R.C., Hofmann, H.J., Lehmann, B., Oeschger, H., Sigg, A., Schwander, J., Staffelbach, T., Stauffer, B., 1990. Use of 10Be in polar ice to trace the 11-year cycle of solar activity. Nature 347 (6289), 164–166.

Beer, J., Mende, W., Stellmacher, R., 2000. The role of the sun in climate forcing. Quat. Sci. Rev. 19 (1–5), 403–415.

Beer, J., Mccracken, K., Steiger, R., 2012. Cosmogenic Radionuclides: Theory and Applications in the Terrestrial and Space Environments. Springer Science and Business Media.

Bereiter, B., Eggleston, S., Schmitt, J., Nehrbass-Ahles, C., Stocker, T.F., Fischer, H., Kipfstuhl, S., Chappellaz, J., 2015. Revision of the epica dome C CO_2 record from 800 to 600 kyr before present. Geophys. Res. Lett. 42 (2), 542–549.

Berkelhammer, M., Sinha, A., Mudelsee, M., Cheng, H., Edwards, R.L., Cannariato, K., 2010. Persistent multidecadal power of the Indian summer monsoon. Earth Planet Sci. Lett. 290 (1–2), 166–172.

Bhalme, H.N., Jadhav, S.K., 1984. The double (Hale) sunspot cycle and floods and droughts in India. Weather 39 (4), 112–116.

Bhalme, H.N., Mooley, D.A., 1981. Cyclic fluctuations in the flood area and relationship with the double (Hale) sunspot cycle. J. Appl. Meteorol. (1962–1982), 1041–1048.

Blinov, A., Massonet, S., Sachsenhauser, H., Stan-Sion, C., Lazarev, V., Beer, J., Synal, H.A., Kaba, M., Masarik, J., Nolte, E., 2000. An excess of 36 Cl in modern atmospheric precipitation. Nucl. Instrum. Methods Phys. Res. Sect. B Beam Interact. Mater. Atoms 172 (1–4), 537–544.

Boltovskoy, E., Wright, R., 1976. Recent Foraminifera. Springer, Dordrecht.

Bond, G., Kromer, B., Beer, J., Muscheler, R., Evans, M.N., Showers, W., Hoffmann, S., Lotti-Bond, R., Hajdas, I., Bonani, G., 2001. Persistent solar influence on North Atlantic climate during the Holocene. Science 294 (5549), 2130–2136.

Bray, J.R., 1967. Variation in Atmospheric $ Sup 14$ C Activity Relative to a Sunspot-Auroral Solar Index. Grasslands Division Department of Scientific and Industrial Research, Palmerston North, New Zealand.

Bray, J.R., 1968. Glaciation and solar activity since the fifth century BC and the solar cycle. Nature 220 (5168), 672–674.

Breitenmoser, P., Beer, J., Brönnimann, S., Frank, D., Steinhilber, F., Wanner, H., 2012. Solar and volcanic fingerprints in tree-ring chronologies over the past 2000 years. Palaeogeogr. Palaeoclimatol. Palaeoecol. 313, 127–139.

Brückner, E., 1890. Klima-schwankungen Seit 1700: Nebst Bemerkungen Über die Klimaschwankungen der diluvialzeit. E. Hölzel 4 (2).

Cameron, J., Hurd, G.A., 1984. Terminator. Orion.

Charvátová, I., Hejda, P., 2014. Responses of the basic cycles of 1787 and 2402 Yr in solar–terrestrial phenomena during the Holocene. Pattern Recogni. Phys. 2 (1), 21–26.

Clette, F., Svalgaard, L., Vaquero, J.M., Cliver, E.W., 2014. Revisiting the sunspot number. Space Sci. Rev. 186 (1), 35–103.

Cole, R.D., 1998. Possible Milankovitch Cycles in the Lower Parachute Creek Member of Green River Formation (Eocene), North-Central Piceance Creek Basin, Colorado: An Analysis.

Courtillot, V., Lopes, F., Le Mouël, J.L., 2021. On the prediction of solar cycles. Sol. Phys. 296 (1), 1−23 (Current Science).
Damon, P.E., Jirikowic, I., 1992. Radiocarbon evidence for low frequency solar oscillations. In: Rare Nuclear Processes.
Damon, P.E., Sonett, C.P., 1991a. Solar and terrestrial components of the atmospheric ^{14}C variation spectrum. Sun. Times 360.
Damon, P., Sonett, C., 1991b. In: Sonett, C., Giampapa, M., Matthews, M. (Eds.), The Sun in Time. University of Arizona Press), Tucson, USA, p. 360.
Darby, D.A., Ortiz, J.D., Grosch, C.E., Lund, S.P., 2012. 1,500-Year cycle in the arctic oscillation identified in Holocene arctic sea-ice drift. Nat. Geosci. 5 (12), 897−900.
Davis Jr., G.E., Lowell, W.E., 2018. Solar energy at birth and human lifespan. J. Photochem. Photobiol. B Biol. 186, 59−68.
Das, M., Singh, R.K., Gupta, A.K., Bhaumik, A.K., 2017. Holocene strengthening of the oxygen minimum zone in the Northwestern Arabian Sea linked to changes in intermediate water circulation or Indian monsoon intensity? Palaeogeogr. Palaeoclimatol. Palaeoecol. 483, 125−135.
Dash, C., Shankar, R., Pati, P., Jose, J., Seong, Y.B., Dhal, S.P., Manjunatha, B.R., Sandeep, K., 2022. An environmental magnetic record of Holocene climatic variability from the Chilika Lagoon, southern mahanadi delta, East Coast of India. J. Asian Earth Sci. 230, 105190.
Denton, G.H., Karlén, W., 1973. Holocene climatic variations—their pattern and possible cause. Quat. Res. 3 (2), 155−205.
De Vries, H., 1958. Variation in concentration of radiocarbon with time and Location on earth. Proc. Koninklijke Nederl. Akademie Wetenschappen B 61, 1−9.
Dikpati, M., Mcintosh, S.W., Chatterjee, S., Banerjee, D., Yellin-Bergovoy, R., Srivastava, A., 2019. Triggering the birth of new cycle's sunspots by solar tsunami. Sci. Rep. 9 (1), 1−8.
Dumka, R.K., Suribabu, D., Taloor, A.K., Prajapati, S., Kotlia, B.S., 2021. Demarcation of solar cycle 24 and characterization of ionospheric GPS-tec towards the western part of India. Quat. Int. 575, 111−119.
Dunbar, R.B., 1983. Stable isotope record of upwelling and climate from Santa Barbara Basin, California. In: Coastal Upwelling, its Sediment Record, Part B. Sedimentary Records of Ancient Coastal Upwelling. Plenum), New York, pp. 217−246.
Dunne, J., Elmore, D., Muzikar, P., 1999. Scaling factors for the rates of production of cosmogenic nuclides for geometric shielding and attenuation at depth on sloped surfaces. Geomorphology 27 (1−2), 3−11.
Duplessy, J.C., Bé, A.W.H., Blanc, P.L., 1981. Oxygen and carbon isotopic composition and Biogeographic distribution of planktonic foraminifera in the Indian Ocean. Palaeogeogr. Palaeoclimatol. Palaeoecol. 33 (1−3), 9−46.
Eddy, J.A., 1976. The maunder minimum: the reign of Louis XIY appears to have been a time of real anomaly in the behavior of the sun. Science 192 (4245), 1189−1202.
Eddy, J.A., 1977. Climate and the changing sun. Climatic Change 1 (2), 173−190.
Egeson, C., 1889. Egeson's Weather System of Sunspot Causality Being Original Researches in Solar and Terrestrial Meteorology.
Elmore, D., Tubbs, L.E., Newman, D., Ma, X.Z., Finkel, R., Nishiizumi, K., Beer, J., Oeschger, H., Andree, M., 1982. 36Cl bomb pulse measured in A shallow ice core from dye 3, Greenland. Nature 300 (5894), 735−737.
Emiliani, C., 1954. Depth habitats of some species of pelagic foraminifera as indicated by oxygen isotope ratios. Am. J. Sci. 252 (3), 149−158.
Fairchild, I.J., Baker, A., 2012. Speleothem Science: From Process to Past Environments. John Wiley and Sons.
Fairchild, I.J., Smith, C.L., Baker, A., Fuller, L., Spötl, C., Mattey, D., Mcdermott, F., 2006. Modification and preservation of environmental signals in speleothems. Earth Sci. Rev. 75 (1−4), 105−153.
Fischer, A.G., Bottjer, D.J., 1991. Orbital forcing and sedimentary sequences. J. Sediment. Res. 61 (7).

Foster, G.L., Lécuyer, C., Marschall, H.R., 2016. Boron stable isotopes. In: Encyclopedia of Geochemistry, Encyclopedia Earth Science Series. Springer, Berlin, pp. 1–6.

Frics, M., 1951. Pollcnanalytiska Vittnesbord Om Senkv Altar." Vegetntionsutveckling, Siirskilt Skogshistoria, Ivry/Ogeogr. Sire(2y, 1-220. I Nord,Ii\Tra Gotaland. (Zusanimenfn\\Ung.) Acfn.

Fritts, H.C., 1976. Tree Rings and Climate. Academic Press, New York.

Gleissberg, W., 1939. A long-periodic fluctuation of the sunspot numbers. Observatory 62, 158.

Gleissberg, W., 1944. A table of secular variations of the solar cycle. Terr. Magnetism Atmos. Electr. 49 (4), 243–244.

Gnevyshev, M.N., Ohl, A.I., 1948. On the 22-year cycle of solar activity. Astron. Zh. 25 (1), 18.

Govil, P., Mazumder, A., Agrawal, S., Azharuddin, S., Mishra, R., Khan, H., Kumar, B., Verma, D., 2022. Abrupt changes in the Southwest monsoon during mid-late Holocene in the Western Bay of Bengal. J. Asian Earth Sci. 227, 105100.

Gupta, A.K., Dhingra, H., Mélice, J.L., Anderson, D.M., 2001. Earth's eccentricity cycles and Indian summer monsoon variability over the past 2 million years: evidence from deep-sea benthic foraminifer. Geophys. Res. Lett. 28 (21), 4131–4134.

Gupta, A.K., Mohan, K., Das, M., Singh, R.K., 2013. Solar forcing of the Indian summer monsoon variability during the Allerød period. Sci. Rep. 3 (1), 1–5.

Halberg, F., Cornelissen, G., Sothern, R.B., Czaplicki, J., Schwartzkopff, O., 2010. Thirty-five-year climatic cycle in heliogeophysics, psychophysiology, military politics, and economics. Izvestiya Atmos. Ocean. Phys. 46 (7), 844–864.

Hale, G.E., 1908. Solar vortices and the Zeeman effect. Publ. Astron. Soc. Pac. 20 (121), 220–224.

Hale, G.E., Ellerman, F., Nicholson, S.B., Joy, A.H., 1919. The magnetic polarity of sun-spots. Astrophys. J. 49, 153.

Herron, M.M., Langway, C.C., 1980. Firn densification: an empirical model. J. Glaciol. 25 (93), 373–385.

Hansen, C.J., Kawaler, S.D., Trimble, V., 2012. Stellar Interiors: Physical Principles, Structure, and Evolution. Springer Science and Business Media.

Hathaway, D.H., Wilson, R.M., 2004. What the sunspot record tells us about space climate. Sol. Phys. 224 (1), 5–19.

Haynes, L.L., Hönisch, B., 2020. The seawater carbon inventory at the paleocene–eocene thermal maximum. Proc. Natl. Acad. Sci. USA 117 (39), 24088–24095.

Henley, B., Abram, N., 2017. The three-minute story of 800,000 Years of climate change with A sting in the tail. The Conversation 13.

Hoffman, P.F., Abbot, D.S., Ashkenazy, Y., Benn, D.I., Brocks, J.J., Cohen, P.A., Cox, G.M., Creveling, J.R., Donnadieu, Y., Erwin, D.H., Fairchild, I.J., 2017. Snowball earth climate dynamics and cryogenian geology-geobiology. Sci. Adv. 3 (11), E1600983.

Höhler, S., 2017. Local disruption or global condition? El Niño as weather and as climate phenomenon. Geo. Environ. 4 (1), e00034.

Hoyt, D.V., Schatten, K.H., 1997. The Role of the Sun in Climate Change. Oxford University Press on Demand.

Impact Team, 1977. The Weather Conspiracy: the Coming of the New Ice Age: a Report. Ballantine Books.

Inceoglu, F., Simoniello, R., Knudsen, M.F., Karoff, C., Olsen, J., Turck-Chiéze, S., Jacobsen, B.H., 2015. Grand solar minima and maxima deduced from 10be and 14c: magnetic dynamo configuration and polarity reversal. Astron. Astrophys. 577, A20.

Iversen, J., 1941. Land Occupation in Denmark's Stone Age: A Pollen-Analytical Study of the Influence of Farmer Culture on the Vegetational Development.

Jessen, K., 1935. Archaeological Dating in the History of North Jutland's Vegetation. Munksgaard.

Krovnin, A.S., Rodionov, S.N., 1992. Atlanto-Scandian Herring: A Case Study. In: Glantz, M.H. (Ed.), Climate Variability, Climate Change and Fisheries, 260. Cambridge University Press, p. 231.

Jaglan, S., Gupta, A.K., Cheng, H., Clemens, S.C., Dutt, S., Balaji, S., 2020. Indian summer monsoon variability during the last millennium as recorded in stalagmite from Baratang Mahadev cave, andaman islands. Palaeogeogr. Palaeoclimatol. Palaeoecol. 557, 109908.

Javaraiah, J., Ulrich, R.K., 2006. Solar-cycle-related variations in the solar differential rotation and meridional flow: a comparison. Sol. Phys. 237 (2), 245–265.

Javier, 2017. https://Judithcurry.Com/2017/07/11/Nature-Unbound-Iv-The-2400-Year-Bray-Cycle-Part-A/.

Johnston, V.E., 2010. The Distribution and Systematics of 36 Cl in Meteoric Waters and Cave Materials with an Assessment of its Use as A Solar Proxy in Speleothems. Doctoral Dissertation, University College Dublin.

Jones, G.S., Lockwood, M., Stott, P.A., 2012. What influence will future solar activity changes over the 21st century have on projected global near-surface temperature changes? J. Geophys. Res. Atmos. 117, D05103. https://doi.org/10.1029/2011JD017013.

Jouzel, J., Masson-Delmotte, V., 2010. Paleoclimates: what do we Learn from deep ice cores? Wiley Interdisciplinary Reviews: Clim. Change 1 (5), 654–669.

Kern, A.K., Harzhauser, M., Piller, W.E., Mandic, O., Soliman, A., 2012. Strong evidence for the influence of solar cycles on A late Miocene lake system revealed by Biotic and abiotic proxies. Palaeogeogr. Palaeoclimatol. Palaeoecol. 329, 124–136.

Kutterolf, S., Jegen, M., Mitrovica, J.X., Kwasnitschka, T., Freundt, A., Huybers, P.J., 2013. A detection of Milankovitch frequencies in global volcanic activity. Geology 41 (2), 227–230.

Lal, D., Peters, B., 1967. Cosmic ray produced radioactivity on the earth. In: Kosmische Strahlung Ii/Cosmic Rays Ii. Springer, Berlin, Heidelberg, pp. 551–612.

Landscheidt, T., 2000. Solar forcing of El Niño and La Niña. In: The Solar Cycle and Terrestrial Climate, Solar and Space Weather, vol 463, p. 135.

Lawrence, K.T., Liu, Z., Herbert, T.D., 2006. Evolution of the eastern tropical pacific through plio-pleistocene glaciation. Science 312 (5770), 79–83.

Lea, D.W., Mashiotta, T.A., Spero, H.J., 1999. Controls on magnesium and strontium uptake in planktonic foraminifera determined by live culturing. Geochem. Cosmochim. Acta 63 (16), 2369–2379.

Leamon, R.J., Mcintosh, S.W., December 2017. Predicting the La Niña of 2020-21: termination of solar cycles and correlated variance in solar and atmospheric variability. In: Agu Fall Meeting Abstracts, vol 2017. Sh42a-05.

Lean, J., Skumanich, A., White, O., 1992. Estimating the sun's radiative output during the maunder minimum. Geophys. Res. Lett. 19 (15), 1591–1594.

Liebrand, D., Beddow, H.M., Lourens, L.J., Pälike, H., Raffi, I., Bohaty, S.M., Hilgen, F.J., Saes, M.J., Wilson, P.A., Van Dijk, A.E., Hodell, D.A., 2016. Cyclostratigraphy and eccentricity tuning of the early oligocene through early Miocene (30.1–17.1 Ma): cibicides mundulus stable oxygen and carbon isotope records from walvis ridge site 1264. Earth Planet Sci. Lett. 450, 392–405.

Lisiecki, L.E., Raymo, M.E., 2005. A pliocene-pleistocene stack of 57 globally distributed benthic $\Delta 18o$ records. Paleoceanography 20 (1).

Liu, T., Ding, Z., Rutter, N., 1999. Comparison of Milankovitch periods between continental loess and deep sea records over the last 2.5 Ma. Quat. Sci. Rev. 18 (10–11), 1205–1212.

Lockyer, N., 1874. Contributions to Solar Physics: I. A Popular Account of Inquiries into the Physical Constitution of the Sun, with Special Reference to Recent Spectroscopic Researches; Ii. Communications to the Royal Society of London, and the French Academy of Sciences, With Notes. Macmillan and Company.

Lockyer, W.J., 1901. The solar activity 1833–1900. Proc. Roy. Soc. Lond. 68 (442–450), 285–300.

Lodders, K., 2003a. Solar system abundances and condensation temperatures of the elements. Astrophys. J. 591 (2), 1220.

Lodders, K., 2003b. Abundances and condensation temperatures of the elements. Meteorit. Planet. Sci. Suppl. 38, 5272.

Lone, M.A., Ahmad, S.M., Dung, N.C., Shen, C.C., Raza, W., Kumar, A., 2014. Speleothem based 1000-year high resolution record of Indian monsoon variability during the last deglaciation. Palaeogeogr. Palaeoclimatol. Palaeoecol. 395, 1–8.

Lourens, L.J., Becker, J., Bintanja, R., Hilgen, F.J., Tuenter, E., Van De Wal, R.S., Ziegler, M., 2010. Linear and non-linear response of late Neogene glacial cycles to obliquity forcing and implications for the Milankovitch theory. Quat. Sci. Rev. 29 (1–2), 352–365.

Ma, L.H., 2007. Thousand-year cycle signals in solar activity. Sol. Phys. 245 (2), 411–414.

Ma, L.H., Vaquero, J.M., 2009. Is the suess cycle present in historical naked-eye observations of sunspots? N. Astron. 14 (3), 307–310.

Ma, L., Vaquero, J.M., 2020. New evidence of the Suess/de Vries cycle existing in historical naked-eye observations of sunspots. Open Astron. 29 (1), 28–31.

Mann, M.E., 2021. Beyond the hockey stick: climate Lessons from the common era. Proc. Natl. Acad. Sci. USA 118 (39). E2112797118.

Martin-Puertas, C., Matthes, K., Brauer, A., Muscheler, R., Hansen, F., Petrick, C., Aldahan, A., Possnert, G., Van Geel, B., 2012. Regional atmospheric circulation shifts induced by A grand solar minimum. Nat. Geosci. 5 (6), 397–401.

Mcintosh, S.W., Leamon, R.J., Egeland, R., Dikpati, M., Fan, Y., Rempel, M., 2019. What the sudden death of solar cycles can tell us about the nature of the solar interior. Sol. Phys. 294 (7), 1–24.

Mitchell, S.F., Carr, I.T., 1998. Foraminiferal response to mid-cenomanian (upper cretaceous) palaeoceanographic events in the anglo-paris basin (Northwest Europe). Palaeogeogr. Palaeoclimatol. Palaeoecol. 137 (1–2), 103–125.

Moy, C.M., Seltzer, G.O., Rodbell, D.T., Anderson, D.M., 2002. Variability of El Niño/southern oscillation activity at millennial timescales during the Holocene epoch. Nature 420 (6912), 162–165.

Naidu, P.D., Malmgren, B.A., 1995. Do benthic foraminifer records represent a productivity index in oxygen minimum zone areas? An evaluation from the Oman margin, Arabian Sea. Mar. Micropaleontol. 26 (1–4), 49–55.

NASA, 2020. https://www.nasa.gov/content/goddard/the-suns-magnetic-field-is-about-to-flip/#:~:text=The%20sun's%20magnetic%20field%20changes,magnetic%20dynamo%20re%2Dorganizes%20itself.

Nigam, R., 2005. Addressing Environmental Issues through Foraminifera–Case Studies from the Arabian Sea.

Nigam, R., Khare, N., Nair, R.R., 1995. Foraminiferal evidence for 77-year cycles of droughts in India and its possible modulation by the Gleissberg solar cycle. J. Coast Res. 1099–1107.

Nigam, R., Saraswat, R., Kurtarkar, S.R., 2006. Laboratory Experiment to Study the Effect of Salinity Variations on Benthic Foraminiferal Species-Pararotalia Nipponica (Asano).

O'brien, S.R., Mayewski, P.A., Meeker, L.D., Meese, D.A., Twickler, M.S., Whitlow, S.I., 1995. Complexity of Holocene climate as reconstructed from A Greenland ice core. Science 270 (5244), 1962–1964.

Olsson, I.U., 1974. Some problems in connection with the evaluation of C14 dates. Geol. Foren. Stockh. Forh. 96 (4), 311–320.

Parrenin, F., Masson-Delmotte, V., Köhler, P., Raynaud, D., Paillard, D., Schwander, J., Barbante, C., Landais, Petrovay, K., 2020. Solar cycle prediction. Living Rev. Sol. Phys. 17 (1), 1–93.

Panchang, R., Nigam, R., 2012. High resolution climatic records of the Past ~ 489 years from Central Asia as derived from benthic foraminiferal species, Asterorotalia trispinosa. Mar. Geol. 307, 88–104.

Petrovay, K., 2020. Solar cycle prediction. Living Rev. Sol. Phys. 17 (1), 1–93. https://doi.org/10.1007/s41116-020-0022-z.

Quinn, W.H., Neal, V.T., Antunez De Mayolo, S.E., 1987. El Niño occurrences over the past four and A half centuries. J. Geophys. Res.: Oceans 92 (C13), 14449–14461.

Rahaman, W., Tarique, M., Fousiya, A.A., Prabhat, P., Achyuthan, H., 2022. Tracing impact of El Niño southern oscillation on coastal hydrology using coral 87sr/86sr record from Lakshadweep, south-eastern Arabian Sea. Sci. Total Environ. 843, 157035.

Raisbeck, G.M., Yiou, F., Jouzel, J., Petit, J.R., 1990. 10be and Δ2h in polar ice cores as A probe of the solar variability's influence on climate. Phil. Trans. Roy. Soc. Lond. Math. Phys. Sci. 330 (1615), 463–470.

Rana, S.S., Nigam, R., 2009. Cyclicity in the Late Holocene Monsoonal Changes from the Western Bay of Bengal. Foraminiferal Approach.

Reid, G.C., 1987. Influence of solar variability on global sea surface temperatures. Nature 329 (6135), 142–143.

Reid, G.C., Gage, K.S., 1988. The climatic impact of secular variations in solar irradiance. In: Secular Solar and Geomagnetic Variations in the Last 10,000 Years. Springer, Dordrecht, pp. 225–243.

Rodríguez-Tovar, F.J., Reolid, M., Pardo-Igúzquiza, E., 2010. Planktonic versus benthic foraminifera response to Milankovitch forcing (late Jurassic, betic cordillera): testing methods for cyclostratigraphic analysis. Facies 56 (3), 459–470.

Russell, C.T., Jian, L.K., Luhmann, J.G., 2013. How unprecedented a solar minimum was it? J. Adv. Res. 4 (3), 253–258.

Sandeep, K., Shankar, R., Warrier, A.K., Yadava, M.G., Ramesh, R., Jani, R.A., Weijian, Z., Xuefeng, L., 2017. A multi-proxy lake sediment record of Indian summer monsoon variability during the Holocene in southern India. Palaeogeogr. Palaeoclimatol. Palaeoecol. 476, 1–14.

Sengupta, P.K., 1960. Sunspot influence on movement of storms over the Bay of Bengal and associated atmospheric variations. Weather 15 (2), 52–58.

Scafetta, N., Milani, F., Bianchini, A., Ortolani, S., 2016. On the astronomical origin of the Hallstatt oscillation found in radiocarbon and climate records throughout the Holocene. Earth Sci. Rev. 162, 24–43.

H. Schwabe, 1843 "Sonnen Beobachtungen im Jahre 1843," astron? Omische Nachrichten, Vol. 21, Pp. 233–236.

Schaefer, J.M., Codilean, A.T., Willenbring, J.K., Lu, Z.T., Keisling, B., Fülöp, R.H., Val, P., 2022. Cosmogenic nuclide techniques. Nature Reviews Methods Primers 2 (1), 1–22.

Schleser, G.H., Helle, G., Lücke, A., Vos, H., 1999. Isotope signals as climate proxies: the role of transfer functions in the study of terrestrial archives. Quat. Sci. Rev. 18 (7), 927–943.

Schmidt, M.W., Weinlein, W.A., Marcantonio, F., Lynch-Stieglitz, J., 2012. Solar forcing of Florida straits surface salinity during the early Holocene. Paleoceanography 27 (3).

Schove, D.J., 1955. The sunspot cycle, 649 Bc to ad 2000. J. Geophys. Res. 60 (2), 127–146.

Schove, D.J., 1962. The reduction of annual winds in Northwestern Europe, ad 1635–1960. Geogr. Ann. 44 (3–4), 303–327.

Schwabe, H., 1844. Sonnenbeobachtungen im Jahre 1843. Von herrn hofrath Schwabe in dessau. Astron. Nachr. 21, 233.

Sernander, R., 1912. Die geologische entwicklung des Nordens Nach der eiszeit in ihrem verhältnis Zu den archäologischen perioden. In: Kossina, G. (Ed.), Mannus, Der Erste Baltische Archäologen-Kongress Zu Stockholm, vol. 13.

Shimmield, G.B., Mowbray, S.R., Weedon, G.P., 1990. A 350 ka history of the Indian southwest monsoon—evidence from deep-sea cores, Northwest Arabian Sea. Earth Environ. Sci. Trans. Royal Soc. Edinburgh 81 (4), 289–299.

Silverman, S.M., 1992. Secular variation of the aurora for the past 500 years. Rev. Geophys. 30 (4), 333–351.

Siscoe, G.L., 1980. Evidence in the auroral record for secular solar variability. Rev. Geophys. 18 (3), 647–658.

Snyder, C.W., 2016. Evolution of global temperature over the past two million years. Nature 538 (7624), 226–228.

Solanki, S.K., 2003. Sunspots: an overview. Astron. AstroPhys. Rev. 11 (2), 153–286.

Solanki, S.K., Usoskin, I.G., Kromer, B., Schüssler, M., Beer, J., 2004. Unusual activity of the sun during recent decades compared to the previous 11,000 years. Nature 431 (7012), 1084–1087.

Stansell, N.D., 2009. Rapid Climate Change in the Tropical Americas during the Late-Glacial Interval and the Holocene. Doctoral Dissertation, University Of Pittsburgh.

Steig, E.J., Brook, E.J., White, J.W.C., Sucher, C.M., Bender, M.L., Lehman, S.J., Morse, D.L., Waddington, E.D., Clow, G.D., 1998. Synchronous climate changes in Antarctica and The North Atlantic. Science 282 (5386), 92–95.

Steinhilber, F., Beer, J., Fröhlich, C., 2009. Total solar irradiance during the Holocene. Geophys. Res. Lett. 36 (19).

Steinhilber, F., Abreu, J.A., Beer, J., Brunner, I., Christl, M., Fischer, H., Heikkilä, U., Kubik, P.W., Mann, M., Mccracken, K.G., Miller, H., 2012. 9,400 years of cosmic radiation and solar activity from ice cores and tree rings. Proc. Natl. Acad. Sci. USA 109 (16), 5967–5971.

Stuiver, M., Grootes, P.M., 1980. The Ancient Sun.

Stuiver, M., Quay, P.D., 1980. Changes in atmospheric carbon-14 attributed to A variable sun. Science 207 (4426), 11–19.

Suess, H.E., 1980. The radiocarbon record in tree rings of the last 8000 years. Radiocarbon 22 (2), 200–209.

Suokhrie, T., Saraswat, R., Saju, S., 2022. Strong solar influence on multi-decadal periodic productivity changes in the central-western Bay of Bengal. Quat. Int. 629, 16–26.

Svensson, 2022. Iceandclimate www.Iceandclimate.Nbi.Ku.Dk/Research/Past_Atmos/Cosmogenic_Isotopes/#:~:Text=The%20abundance%20of%20cosmogenic%20isotopes,Widely%20applied%20for%20radiometric%20dating.

Thamban, M., Kawahata, H., Rao, V.P., 2007. Indian summer monsoon variability during the Holocene as recorded in sediments of the Arabian Sea: timing and implications. J. Oceanogr. 63 (6), 1009–1020.

Thamban, M., Laluraj, C.M., Naik, S.S., Chaturvedi, A., 2011. Reconstruction of antarctic climate change using ice core proxy records from the coastal dronning maud land, East Antarctica. J. Geol. Soc. India 78 (1), 19–29.

Tiwari, R.K., Rao, K.N.N., 2004. Signature of enso signals in the coral growth rate record of Arabian Sea and Indian monsoons. Pure Appl. Geophys. 161 (2), 413–427.

Usoskin, I.G., Mursula, K., Solanki, S., Schüssler, M., Alanko, K., 2004. Reconstruction of solar activity for the last millennium using Be data. Astron. Astrophys. 413 (2), 745–751.

Usoskin, I.G., Solanki, S.K., Kovaltsov, G.A., 2007. Grand minima and maxima of solar activity: new observational constraints. Astron. Astrophys. 471 (1), 301–309.

Usoskin, I.G., Kromer, B., Ludlow, F., Beer, J., Friedrich, M., Kovaltsov, G.A., Solanki, S.K., Wacker, L., 2013. The Ad775 cosmic event revisited: the sun is to blame. Astron. Astrophys. 552, L3.

Usoskin, I.G., Hulot, G., Gallet, Y., Roth, R., Licht, A., Joos, F., Kovaltsov, G.A., Thébault, E., Khokhlov, A., 2014. Evidence for distinct modes of solar activity. Astron. Astrophys. 562, L10.

Usoskin, I.G., Gallet, Y., Lopes, F., Kovaltsov, G.A., Hulot, G., 2016. Solar activity during the Holocene: the Hallstatt cycle and its consequence for grand minima and maxima. Astron. Astrophys. 587, A150.

Vaganov, E.A., Anchukaitis, K.J., Evans, M.N., 2011. How Well Understood Are the Processes that Create Dendroclimatic Records? A Mechanistic Model of the Climatic Control on Conifer Tree-Ring Growth Dynamics. Springer, Dordrecht, pp. 37–75.

Van Loon, H., Meehl, G.A., 2008. The response in the pacific to the sun's decadal peaks and contrasts to cold events in the southern oscillation. J. Atmos. Sol. Terr. Phys. 70 (7), 1046–1055.

Vaquero, J.M., Vázquez, M., 2009. The Sun Recorded through History, vol. 361. Springer Science and Business Media.

Vaquero, J.M., Gallego, M.C., García, J.A., 2002. A 250-year cycle in naked-eye observations of sunspots. Geophys. Res. Lett. 29 (20), 58–61. https://doi.org/10.1029/2002GL014782.

Wagner, S., Zorita, E., 2005. The influence of volcanic, solar and Co2 forcing on the temperatures in the Dalton minimum (1790–1830): a model study. Clim. Dynam. 25 (2), 205–218.

Warrier, A.K., Sandeep, K., Shankar, R., 2017. Climatic periodicities recorded in lake sediment magnetic susceptibility data: further evidence for solar forcing on Indian summer monsoon. Geosci. Front. 8 (6), 1349–1355.
White, W.B., Liu, Z., 2008. Non-linear alignment of El Nino to the 11-yr solar cycle. Geophys. Res. Lett. 35 (19).
Wolf, R., 1862. Mittheilungen Über die Sonnenflecken Xiv. Astronomische Mitteilungen Der Eidgenössischen Sternwarte Zurich 2, 119–131.
Wolter, K., Timlin, M.S., 1998. Measuring the strength of enso events: how does 1997/98 rank? Weather 53 (9), 315–324.
Wunsch, C., 2004. Quantitative estimate of the milankovitch-forced contribution to observed quaternary climate change. Quat. Sci. Rev. 23 (9–10), 1001–1012.
Yadava, M.G., Ramesh, R., 2007. Significant longer-term periodicities in the proxy record of the Indian monsoon rainfall. N. Astron. 12 (7), 544–555.
Yin, Z.Q., Han, Y.B., Ma, L.H., Le, G.M., Han, Y.G., 2007. Short-term period variation of relative sunspot numbers. Chin. J. Astron. Astrophys. 7 (6), 823.
Zachos, J., Pagani, M., Sloan, L., Thomas, E., Billups, K., 2001. Trends, rhythms, and aberrations in global climate 65 Ma to present. Science 292 (5517), 686–693.

Website references

Boutaud, A.S., 2018. When Earth Was a Snowball. https://News.Cnrs.Fr/Articles/When-Earth-Was-A-Snowball.
Bressan, D., 2011. September 19, 1991: The Iceman Natural History. Scientific American. https://Blogs.Scientificamerican.Com/History-Of-Geology/September-19-1991-The-Icemannatural-History/.
Cdc https://www.Cdc.Gov/Climateandhealth/Effects/Food_Security.Html.
https://Judithcurry.Com/2017/12/02/Nature-Unbound-Vi-Centennial-To-Millennial-Solar-Cycles/.
Honesthistory https://Honesthistory.Net.Au/Wp/Henley-Ben-Nerilie-Abram-The-Three-Minute-Story-Of-800000-Years-Of-Climate-Change-With-A-Sting-In-The-Tail/.
https://Judithcurry.Com/2017/07/11/Nature-Unbound-Iv-The-2400-Year-Bray-Cycle-Part-A/.
https://Www.Nasa.Gov/Feature/Goddard/2020/What-Will-Solar-Cycle-25-Look-Like-Sun-Prediction-Model.
Lewis, S., 2021. Earth Reaches Perihelion, Closer to the Sun than Any Other Day. Cbs News. Archived From the Original On 2021-05-24. Retrieved 2021-05-24.
https://www.Nasa.Gov/Content/Goddard/The-Suns-Magnetic-Field-Is-About-To-Flip/.
Odp http://www.Odp.Tamu.Edu/Publications/Tnotes/Tn34/Tn34_4.Htm#:~:Text=Magnetic%20susceptibility%20is%20the%20degree,In%20an%20external%20magnetic%20field.
Pubs https://Pubs.Usgs.Gov/Fs/Fs-0095-00/Fs-0095-00.Pdf.
Research www.Iceandclimate.Nbi.Ku.Dk/Research/Past_Atmos/Cosmogenic_Isotopes/#:~:Text=The%20abundance%20of%20cosmogenic%20isotopes,Widely%20applied%20for%20radiometric%20dating.
scied https://Scied.Ucar.Edu/Learning-Zone/Sun-Space-Weather/Sunspot-Cycle#:~:Text=The%20duration%20of%20the%20sunspot,As%20long%20as%20fourteen%20years.
Siegel, E., 2020. Carbon-14 Spiked Worldwide over 1200 Years Ago, and the Sun Is to Blame. Forbes. https://www.Forbes.Com/Sites/Startswithabang/2020/04/02/Carbon-14-Spiked-Worldwide-Over-1200-Years-Ago-And-The-Sun-Is-To-Blame/?Sh=6369ceef3e81.

Glossary

Afro-Alpine The afro-alpine region comprises the high mountains of Ethiopia and tropical East Africa, which represent biological "sky islands" with a high level of endemism. However, some primarily arctic-alpine plants also occur in the afro-alpine mountains.

Alluvial Sediments Clay or silt or gravel carried by rushing streams and deposited where the stream slows down. Synonyms: alluvial deposit, alluvion, and alluvium. Types: delta. A low triangular area of alluvial deposits where a river divides before entering a larger body of water is called Alluvial Sediments.

Archaeology It is the study of the human past using material remains. These remains can be any objects that people created, modified, or used. Portable remains are usually called artifacts. Artifacts include tools, clothing, and decorations. Nonportable remains, such as pyramids or postholes, are called features.

Atmospheric General Circulation The general circulation of the atmosphere takes place when hot air rises in the tropics and moves towards north or south, descends and returns to the equatorial "Hadley cells." Its path along the surface is bent into the trade winds by the Earth's rotation (Coriolis effect).

Bibliometrics Bibliometrics, or research impact, is the quantitative method of citation and content analysis for scholarly journals, books, and researchers. The quantitative impact of a given publication is appraised by measuring the amount of times a certain work is cited by other resources.

Boreal Summer Intra-Seasonal Oscillation (BSISO) It profoundly impacts the Northern Hemisphere monsoon onsets and breaks, tropical cyclones, and many climate extremes. BSISO exhibits more complex propagation patterns than the dominant eastward propagation of the Madden–Julian Oscillation.

Central Crystalline The crystalline sheet of the Higher Himalaya, referred to as the Central Crystallines, is a continuous lithotectonic unit which can be traced from the River Kali of eastern Kumaun in the east to Sankoo in the Suru River valley of Kashmir in the west.

Climate Change Impact Assessment It seeks to characterize, diagnose, and project risks or impacts of environmental change on people, communities, economic activities, infrastructure, ecosystems, or valued natural resources.

Climate Change Climate change encompasses global warming but refers to the broader range of changes that are happening to our planet, including rising sea levels; shrinking mountain glaciers; accelerating ice melt in Greenland, Antarctica, and the Arctic; and shifts in flower/plant blooming times.

Climate Data Tools (CDTs) The Climate Data Tool (CDT) is a free, open-source R package created specifically for national meteorological services. It is a free and open-source R package. It was developed as part of the Enhancing National Climate Services (ENACTS) initiative by the International Research Institute for Climate and Society (IRI), part of the Earth Institute at Columbia University.

Climate Migration Takes place when people leave their homes and livelihood behind because of climate stressors such as changing rainfall, heavy flooding, and sea level rise. It makes their homes uninhabitable.

Climate Variability Climate variability is the way aspects of climate (such as temperature and precipitation) differ from an average. Climate variability occurs due to natural and sometimes periodic changes in the circulation of the air and ocean, volcanic eruptions, and other factors.

Climate-resilient agriculture It is an approach that includes sustainably using existing natural resources through crop and livestock production systems to achieve long-term higher productivity and farm incomes under climate variabilities.

Climate-smart agriculture It is an approach that helps guide actions to transform agri-food systems toward green and climate resilient practices. CSA supports reaching internationally agreed goals such as the SDGs and the Paris Agreement. It aims to tackle three main objectives: sustainably increasing agricultural productivity and incomes; adapting and building resilience to climate change; and reducing and/or removing greenhouse gas emissions, where possible.

Climate-stressed There are three main types of climate-related risks that should be considered in a climate change stress scenario—physical risk, transition risk, and liability risk. Physical risk is the risk associated with the physical effects of climate change, for example flood damage caused by extreme weather events.

Climatic Fluctuation Large-scale annual and decade fluctuations in climate and weather are caused by changes in patterns of ocean circulation and atmospheric pressures.

Cloudburst It is a sudden, very heavy rainfall, usually local and of brief duration. Most so-called cloudbursts occur in connection with thunderstorms. In these storms, there are violent uprushes of air, which at times prevent the condensing raindrops from falling to the ground.

Continental Crust It is the outermost layer of Earth's lithosphere that makes up the planet's continents and continental shelves and is formed near the subduction zones at plate boundaries between continental and oceanic tectonic plates. The continental crust forms nearly all of Earth's land surface.

Coronal Mass Ejections (CMEs) These are large expulsions of plasma and magnetic field from the Sun's corona. They can eject billions of tons of coronal material and carry an embedded magnetic field (frozen in flux) that is stronger than the background solar wind interplanetary magnetic field (IMF) strength.

Deltas Deltas are the unique result of the interaction of rivers and tidal processes resulting in the largest sedimentary deposits in the world. Although comprising only 5% of the land area, deltas have up to 10 times higher than the average population—a number, which is increasing rapidly, especially for deltas in Asia.

Desertification Desertification is land degradation in arid, semiarid, and dry subhumid areas, collectively known as drylands, resulting from many factors, including human activities and climatic variations.

Diatoms It refers to any member of a large group comprising several genera of algae, specifically microalgae, found in the oceans, waterways, and soils of the world.

Diurnal Temperature Range (DTR) It is defined as the difference between daily maximum and minimum temperature, which describes the within-day temperature variability and reflects weather stability.

Diurnal Variation On the other hand, refers to the fluctuations that happen during the day and the variations in the day–night cycle that are not regulated by intrinsic or endogenous mechanisms but rather by extraneous factors.

El Nino-Southern Oscillation (ENSO) It is a recurring climate pattern involving changes in the temperature of waters in the central and eastern tropical Pacific oceans.

Environment-Induced Conflicts Environmental conflicts have emerged as key issues challenging local, regional, national, and global security. Environmental crises and problems throughout the world are widespread and increasing rapidly.

Foraminiferal studies Foraminiferal tests serve to protect the organism within. Owing to their generally hard and durable construction (compared to other protists), the tests of foraminifera are a major source of scientific knowledge about the group. Openings in the test that allow the cytoplasm to extend outside are called apertures.

Glacial Lake Outburst Flood (GLOF) It is a sudden release of water from a lake fed by glacier melt that has formed at the side, in front, within, beneath, or on the surface of a glacier.

Glacial melting The melting of the glaciers, a phenomenon that intensified in the 20th century, is leaving our planet iceless. Human activity is the main culprit in the emission of carbon dioxide and other greenhouse gases. The sea level and global stability depend on how these great masses of recrystallized snow evolve.

Global Circulation Model (GCM) It is a model that simulates the general circulation of the planetary atmosphere or oceans. The term general circulation is used to indicate large-scale atmospheric or oceanic motions with their persistent as well as transient features on various scales.

Global Earth System The earth system is itself an integrated system, but it can be subdivided into four main components, subsystems, or spheres: the geosphere, atmosphere, hydrosphere, and biosphere. These components are also systems and they are tightly interconnected.

Greenhouse Gases Emission Greenhouse gases are measured in "carbon dioxide-equivalents" (CO_2e). Today, we collectively emit around 50 billion tonnes of CO_2e each year. This is more than 40% higher than emissions in 1990, which were around 35 billion tonnes.

Ground penetrating radar (GPR) It is a geophysical method that uses radar pulses to image the subsurface. It is a nonintrusive method of surveying the subsurface to investigate underground utilities such as concrete, asphalt, metals, pipes, cables, or masonry. This nondestructive method uses electromagnetic radiation in the microwave band (UHF/VHF frequencies) of the radio spectrum, and detects the reflected signals from subsurface structures. GPR can have applications in a variety of media, including rock, soil, ice, fresh water, pavements, and structures. In the right conditions, practitioners can use GPR to detect subsurface objects, changes in material properties, and voids and cracks.

High-Resolution Climate Models To simulate the Earth's climate, a model divides up the Earth's surface and overlying the atmosphere into a giant grid. The model then calculates climate variables, such as temperature, humidity, and rainfall, for each grid cell.

Himalayan Ecosystem The Himalayan ecosystem is fragile and diverse. It includes over 51 million people who practice hill agriculture and remains vulnerable. The Himalayan ecosystem is vital to the ecological security of the Indian landmass, through providing forest cover, feeding perennial rivers that are the source of drinking water, irrigation, and hydropower, conserving biodiversity, providing a rich base for high value agriculture, and spectacular landscapes for sustainable tourism.

Himalayan Frontal Thrust The Main Frontal Thrust (MFT), also known as the Himalayan Frontal Thrust (HFT), is a geological fault in the Himalayas that defines the boundary between the Himalayan foothills and Indo-Gangetic Plain. The fault is well expressed on the surface and thus could be seen via satellite imagery. It is the youngest and southernmost thrust structure in the Himalayas deformation front. It is a splay branch of the Main Himalayan Thrust (MHT) as the root décollement. It runs parallel to other major splays of the MHT, Main Boundary Thrust (MBT), and Main Central Thrust (MCT). The Sunda Megathrust, which extends from the Banda Islands to Myanmar, is joined with the MFT. The fault strikes in an NW-SE direction and dips at an angle of 20−30 degrees in the north.

Holocene Period The Holocene is the current geological epoch. It began approximately 11,650 cal years before the present (c. 9700 BCE), after the last glacial period, which concluded with the Holocene glacial retreat. The Holocene and the preceding Pleistocene together form the Quaternary period. The Holocene has been identified with the current warm period, known as MIS 1. It is considered by some to be an interglacial period within the Pleistocene Epoch, called the Flandrian interglacial.

Hydro-Meteorological Hazards It is the process or phenomenon of atmospheric, hydrological, or oceanographic nature that may cause loss of life, injury or other health impacts, property damage, loss of livelihoods and services, social and economic disruption, or environmental damage. The hazards include tropical cyclones (typhoons and hurricanes), thunderstorms, hailstorms, tornados, blizzards, heavy snowfall, avalanches, coastal storm surges, floods including flash floods, drought, heatwaves, and cold spells.

Indian Ocean Dipole (IOD) It is defined by the difference in sea surface temperature between two areas (or poles, hence a dipole)—a western pole in the Arabian Sea (western Indian Ocean) and an eastern pole in the eastern Indian Ocean south of Indonesia. The IOD affects the climate of Australia and other countries that surround the Indian Ocean basin and is a significant contributor to rainfall variability in this region.

Interdecadal Pacific Oscillation (IPO) It is a large-scale, long-period oscillation that influences climate variability over the Pacific basin. The IPO operates at a multidecadal scale, with phases lasting around 20−30 years.

Inter-Tropical Convergence Zone (ITCZ) It is known by sailors as the doldrums or the calms because of its monotonous windless weather; it is the area where the northeast and the southeast trade winds converge. It encircles Earth near the thermal equator though its specific position varies seasonally. When it lies near the geographic equator, it is called the near-equatorial trough. Where the ITCZ is drawn into and merges with a monsoonal circulation, it is sometimes referred to as a monsoon trough, a usage that is more common in Australia and parts of Asia.

Kuroshio Current The Kuroshio Current, also known as the Black or Japan Current (Nihon Kairyū) or the Black Stream, is a north-flowing, warm ocean current on the west side of the North Pacific Ocean basin. It was named for the deep blue appearance of its waters. Similar to the Gulf Stream in the North Atlantic, the Kuroshio is a powerful western boundary current that transports warm equatorial water poleward and forms the western limb of the North Pacific Subtropical Gyre. Off the East Coast of Japan, it merges with the Oyashio Current to form the North Pacific Current.

Lake sediments These are comprised mainly of clastic material (sediment of clay, silt, and sand sizes), organic debris, chemical precipitates, or combinations of these. The relative abundance of each depends upon the nature of the local drainage basin, the climate, and the relative age of a lake.

Limestone Cave Limestone cave or cavern is a natural cavity that is formed underneath the Earth's surface that can range from a few meters to many kilometers in length and depth. Most of the world's caves, including those at the Cradle of Humankind, are formed in porous limestone.

Lithofacies A mappable subdivision of a designated stratigraphic unit, distinguished from adjacent subdivisions based on lithology; facies characterized by lithologic features.

Marine Archaeology It is a discipline within archaeology as a whole that specifically studies human interaction with the sea, lakes, and rivers through the study of associated physical remains, be they vessels, shore-side facilities, port-related structures, cargoes, human remains, and submerged landscapes.

Marine Environment It is an essential component of the global life-support system. Oceans cover 71 per cent of the Earth's surface and provide us with food, oxygen, and jobs. But they are probably the least understood, most biologically diverse, and most undervalued of all ecosystems.

Marine Heatwaves Heatwaves can also occur in the ocean and these are known as marine heatwaves or MHWs. These marine heatwaves, when ocean temperatures are extremely warm for an extended period, can have significant impacts on marine ecosystems and industries. Marine heatwaves can occur in summer or winter—they are defined based on differences in expected temperatures for the location and time of year.

Marine Sediments Any deposit of insoluble material, primarily rock, and soil particles, transported from land areas to the ocean by wind, ice, and rivers, as well as the remains of marine organisms, products of submarine volcanism, chemical precipitates from seawater, and materials from outer space is called marine sediment.

Micro-dosing Microdosing, or micro-dosing, is a technique for studying the behavior of drugs in humans through the administration of doses so low ("subtherapeutic") they are unlikely to produce whole-body effects, but high enough to allow the cellular response to be studied. This is called a "Phase 0 study" and is usually conducted before clinical Phase I to predict whether a drug is viable for the next phase of testing. Human microdosing aims to reduce the resources spent on nonviable drugs and the amount of testing done on animals.

Microfossils These are the tiny remains of bacteria, protists, fungi, animals, and plants. Microfossils are a heterogeneous bunch of fossil remains studied as a single discipline because rock samples must be processed in certain ways to remove them and microscopes must be used to study them.

Milankovitch Theory A century ago, Serbian scientist Milutin Milankovitch hypothesized the long-term, collective effects of changes in Earth's position relative to the Sun are a strong driver of Earth's long-term climate and are responsible for triggering the beginning and end of glaciation periods (Ice Ages).

Natural Disasters Also referred to as natural hazards, they are extreme, sudden events caused by environmental factors such as storms, floods, droughts, fires, and heatwaves. Natural disasters are now occurring with increasing severity, scope, and impact.

Oligocene Period Oligocene Epoch, the third and last major worldwide division of the Paleogene Period (65.5 million to 23 million years ago), spanning the interval between 33.9 million to 23 million years ago. The Oligocene Epoch is subdivided into two ages and their corresponding rock stages: the Rupelian and the Chatti.

Orbital Relationship It was proposed early in the 20th century. In molecular orbital theory, electrons in a molecule are not assigned to individual chemical bonds between atoms but are treated as moving under the influence of the atomic nuclei in the whole molecule.

Ozone Depletion When chlorine and bromine atoms come into contact with ozone in the stratosphere, they destroy ozone molecules. One chlorine atom can destroy over 100,000 ozone molecules before it is removed from the stratosphere. Ozone can be destroyed more quickly than it is naturally created.

Paleolithic Site The Paleolithic or Palaeolithic or Palæolithic, also called Old Stone Age (from Greek palaios—old, lithos—stone), is a period in prehistory distinguished by the original development of stone tools that covers 99% of the period of human technological prehistory. It extends from the earliest known use of stone tools by hominins c. 3.3 million years ago, to the end of the Pleistocene c. 11,650 cal BP.

Paleontological It also spelled palaeontology or palæontology; it is the scientific study of life that existed prior to, and sometimes including, the start of the Holocene epoch. It includes the study of fossils to classify organisms and study their interactions with each other and their environments.

Participatory Rural Appraisal (PRA) It is an approach used by nongovernmental organizations (NGOs) and other agencies involved in international development. The approach aims to incorporate the knowledge and opinions of rural people in the planning and management of development projects and programs.

Permafrost Degradation It refers to a naturally or artificially caused decrease in the thickness and/or areal extent of permafrost. Evidence of change in the southern boundary of the discontinuous permafrost zone in the past decades has been reported.

Phytoplankton These are the autotrophic components of the plankton community and a key part of ocean and freshwater ecosystems. The name comes from the Greek words φυτόν, meaning "plant," and πλαγκτός, meaning "wanderer" or "drifter."

Planetary Hypothesis The planetary hypothesis is also known as the nebular hypothesis. It is a catastrophic hypothesis proposed in 1905. In this, the planets of the Solar system are seen to arise from an encounter between the Sun and another star.

Pleistocene Period The Pleistocene (often referred to as the Ice Age) is the geological epoch that lasted from about 2,580,000 to 11,700 years ago, spanning the Earth's most recent period of repeated glaciations. Before a change was finally confirmed in 2009 by the International Union of Geological Sciences, the cut-off of the Pleistocene and the preceding Pliocene was regarded as being 1.806 million years Before the Present (BP).

Precipitation Concentration Index (PCI) It is defined by Oliver and is also a powerful indicator for temporal precipitation distribution. Like CI, PCI is generally used for evaluating seasonal precipitation changes to investigate the heterogeneity of monthly rainfall data.

RCP4.5 Representative Concentration Pathway (RCP) 4.5 is a scenario of long-term, global emissions of greenhouse gases, short-lived species, and land-use-land-cover which stabilizes radiative forcing at 4.5 Watts per meter squared (W/m^{-2}, approximately 650 ppm CO_2-equivalent) in the year 2100 without ever.

RCP8.5 RCP8.5, generally taken as the basis for worst-case climate change scenarios, was based on what proved to be an overestimation of projected coal outputs. It is still used for predicting mid-century (and earlier) emissions based on current and stated policies.

Riverine Flooding Riverine flooding is when streams and rivers exceed the capacity of their natural or constructed channels to accommodate water flow and water overflows the banks, spilling out into adjacent low-lying, dry land.

Salinity Intrusion Saltwater intrusion is the movement of saline water into freshwater aquifers, which can lead to groundwater quality degradation, including drinking water sources, and other consequences. Saltwater intrusion can naturally occur in coastal aquifers, owing to the hydraulic connection between groundwater and seawater. Because saline water has a higher mineral content than freshwater, it is denser and has a higher water pressure. As a result, saltwater can push inland beneath the freshwater. In other topologies, submarine groundwater discharge can push fresh water into saltwater.

Schwabe Cycle The solar cycle, also known as the solar magnetic activity cycle, sunspot cycle, or Schwabe cycle, is a nearly periodic 11-year change in the Sun's activity measured in terms of variations in the number of observed sunspots on the Sun's surface.

SciMAT Tool It is an open-source (GPLv3) software tool developed to perform a science mapping analysis under a longitudinal framework. SciMAT provides different modules that help the analyst to carry out the steps of the science mapping workflow.

Sea Surface Temperature The temperature of the water at the ocean surface is an important physical attribute of the world's oceans. The surface temperature of the world's oceans varies mainly with latitude, with the warmest waters generally near the equator and the coldest waters in the Arctic and Antarctic regions. As the oceans absorb more heat, sea surface temperature increases, and the ocean circulation patterns that transport warm and cold water around the globe change.

Sea Level Rise Sea level rise is caused primarily by two factors related to global warming: the added water from melting ice sheets and glaciers, and the expansion of seawater as it warms. The first graph tracks the change in global sea level since 1993, as observed by satellites. The second graph, which is from coastal tide gauge and satellite data, shows how much the sea level changed from about 1900 to 2018. Items with pluses (+) are factors that cause the global sea level to increase, while minuses (−) are what cause the sea level to decrease. These items are displayed at the time they were affecting sea level.

Siwalik Rocks The Siwalik Group is a thick sedimentary sequence forming the youngest mountain belt, extending throughout the East−West of the foothills of the Himalayas, and is separated from the Lesser Himalaya to the north by the Main Boundary Thrust, and the Indo-Gangetic Plain to the south by the Himalayan Frontal Thrust.

Slope Stability Assessment Slope stability analysis is performed to assess the safe design of a human-made or natural slope (e.g. embankments, road cuts, open-pit mining, excavations, landfills, etc.) and the equilibrium conditions. Slope stability is the resistance of inclined surface to failure by sliding or collapsing.

Steep Rivers The steepest gradient in the long profile of a river is found in the upper course near the source.

Supraglacial Lakes A supraglacial lake is any pond of liquid water on the top of a glacier. Although these pools are ephemeral, they may reach kilometers in diameter and be several meters deep. They may last for months or even decades at a time but can empty in hours.

Sustainable Development Goals (SDGs) The Sustainable Development Goals are the blueprint to achieve a better and more sustainable future for all. They address the global challenges we face, including poverty, inequality, climate change, environmental degradation, peace, and justice. Learn more and act.

Thematic maps A thematic map shows the spatial distribution of one or more specific data themes for selected geographic areas. The map may be qualitative in nature (e.g., predominant farm types) or quantitative (e.g., percentage population change).

Total Solar Irradiance The Total Solar Irradiance (TSI) Climate Data Record (CDR) measures the spectrally integrated energy input to the top of the Earth's atmosphere at a base mean distance from the Sun (i.e., one Astronomical Unit), and its unit is W/m^2.

Tropical Cyclones Tropical cyclones, also known as typhoons or hurricanes, are among the most destructive weather phenomena. They are intense circular storms that originate over warm tropical oceans and have maximum sustained wind speeds exceeding 119 kilometers per hour and heavy rains.

Water Cycles There are four main parts to the water cycle: Evaporation, Convection, Precipitation, and Collection. Evaporation is when the Sun heats up water in rivers or lakes or the ocean and turns it into vapor or steam.

Author index

Note: 'Page numbers followed by "f" indicate figures, "t" indicates tables, and "b" indicate boxes.'

A

Abbate, E., 181–183
Abbot, D. S., 438b
Abdel-Monem, A. A., 106, 112–113
Abdul Razak, M. A., 106
Abdullah, I., 112–113
Abe-Ouchi, A., 464f
Abel, A., 177–178, 195
Abram, N., 439
Abreu, J. A., 457–458
Abu, M., 124
AbuTaha, A. S., 432
Achyuthan, H., 112–113, 296, 470–471, 472f
Acs, E., 327
Adachi, S. A., 146
Adams, H., 51
Adger, W.N., 51
Adnan, M., 82–84
Afroz, S., 37
Afroz, T., 53
Agarwal, A., 22, 27, 29, 137
Ageta, Y., 124, 126–127
Agnihotri, R., 470
Agrawal, S., 209, 474–475
Ahamad, S., 431–432
Ahmad, S. M., 296, 476
Ahmed, A. U., 82–84
Ahmed, B., 34
Ahmed, F., 41t–42t
Ahmed, S., 53
Ahsan, R., 53
Aide, M., 131
Aiello, L. C., 108–110
Ajey, L., 13–14
Alahacoon, N., 84
Alanko, K., 470
Aldahan, A., 444t–445t
Aldrian, E., 77
Alexander, L. V., 84–86
Alier, J.M., 53–54
Allan, A., 51
Allen, J. R., 282, 289–290
Allen, R. J., 2, 4–5
Almanza-P, M., 425
Almeida, F., 245, 251
Althor, G., 51

Amadi-Mgbenka, C., 52
Ambasankar, K., 430–431
Ambrose, S. H., 108–110, 113–114
An, S. I., 175
An, Z., 296, 299–300
Ananda, M. B., 227
Ananthakrishnan, R., 450
Anchukaitis, K. J., 296, 476–477
Anderson, D. M., 208–209, 259–260, 446–447, 461
Anderson, D. S., 317
Anderson, N. J., 305, 317, 327, 337
Anderson, R. Y., 446f
Andree, M., 468, 470
Andreev, A., 337
Andrews, C. W., 105–106, 110–111
Andrieu-Ponel, V., 305, 337
Anguelovski, I., 53–54
Anmin, D., 299–300
Annaka, T., 51
Antunez De Mayolo, S. E., 446
Anuchitworawong, C., 27–28
Anupama, T. S., 227
Anwar, R., 35
Aparicio, R., 134
Apel, H., 25
Arakawa, O., 146, 161–166
Archer, C. L., 4–5
Aria, M., 425
Aristegui, J., 296–297
Ariyaratne, T. R., 71
Armand, L., 296–297
Arndt, N., 181–183
Arnold, L., 113–114
Arora, A., 52
Artemi, C., 37
Arthur Chen, C. -T., 172
Arunrat, N., 137
Aryal, J. P., 125
Aschauer, S., 124–125
Ashkenazy, Y., 438b
Ashley, G. M., 289–290
Ashraf, J., 82–84
Ashrafuzzaman, M., 34, 36–37
Asrat, A., 177–178
Assmy, P., 296–297
Atapattu, S., 35

Attavanich, W., 27
Attolini, M. R., 452–453
Attri, S. D., 7
Avunje, S., 430–431
Awulachew, S. B., 181, 188
Ayalew, L., 178–180
Aye, K. S., 216
Aye, M. M., 216
Ayelew, D., 181–183
Azharuddin, S., 209, 474–475
Azzaroli, A., 181–183

B

Babel, M., 22, 27, 29, 137
Babel, S. M., 137
Babernits, S., 124–125
Bachmann, G. H., 275
Bacon, F., 451–452
Bagnold, R. A., 201–202
Bahls, L., 327
Baidya, S. K., 82–84
Baig, N., 208–209
Bajgai, Y., 125
Bajracharya, S. R., 124–125
Bak, R. P. M., 244
Baker, A., 475, 475f
Bakhtiari, F., 54
Bakry, N., 105–110, 113–114
Bala, S., 418
Balaji, S., 476
Balasubramanian, C. P., 430–431
Balbon, E., 247
Balsam, W. L., 472
Balzer, R., 470
Band, S., 296
Banerjee, D., 450
Banerjee, P. K., 251
Bansok, R., 23
Banyal, R., 431–432
Barai, K.R., 51
Barbante, C., 439
Bard, E., 468
Barreto, L., 209
Barros, V., 134, 136
Basnayake, B., 79, 79f–82f
Basnayake, B. R. S. B., 77, 81–84, 83f–84f
Bastien, F., 181–183
Batchelor, B., 112
Battarbee, R. W., 295–296, 305, 317, 337, 351
Bauer, L. A., 472
Bavadam, L., 411

Bayissa, Y., 132
Bazile, D., 125
Beaver, J., 296–297
Becker, J., 461–462
Beckman, S., 105–106
Beddow, H. M., 461
Beer, J., 442, 454f–455f, 455–458, 467–468, 469f, 470, 471b, 478f
Behera, S. K., 77
Behling, H., 171–173, 296–297, 317, 327–337
Behrens, D. W., 245
Bejranonda, S., 27
Beker, B., 296–297
Bender, F., 247
Bender, M. L., 468
Benn, D. I., 438b
Bennion, H., 295, 317, 351
Benny, H. -J., 108–110
Bera, S. K., 296
Bereiter, B., 439, 440f
Berge, F., 317
Berhane, A., 132
Berhanu, A., 132
Berkelhammer, M., 476
Bernzen, A., 51
Beyens, L., 300, 317, 327–337
Bhalla, S. N., 208–209, 237
Bhalme, H. N., 450, 451f
Bhanumurthy, K., 424, 431
Bharathi, V., 191–192
Bhardwaj, A. K., 431–432
Bhate, J., 79
Bhatia, S. B., 209
Bhatnagar, A. K., 7
Bhattacharya, S. K., 208–209
Bhaumik, A. K., 474–475
Bhowmick, A., 177–178, 191–192, 195
Bhuiyan, S., 53
Bhushan, R., 470
Bianchini, A., 458
Bianchini, N., 124–125
Bigler, C., 317, 327, 337
Billa, L., 20
Billah, M., 33–34
Billups, K., 438–439, 439f, 462f
binAriffin, Z., 105–110, 113–114
Bintanja, R., 461–462
Bird, M. I., 216, 296
Birks, H. J. B., 295, 300, 317, 337
Bisht, M., 121–124, 126–127
Bisht, R. S., 414
Blackett, P., 24–25

Black, R., 33—34
Blahutová, K.K., 51
Blair, T. C., 287—289
Blatter, H., 464f
Blinov, A., 468, 469f
Bode, A., 296—297
Bohaty, S. M., 461
Bohra-Mishra, P., 51
Boivin, N., 113—114
Bolch, T., 124—125
Boltovskoy, E., 237, 406—407, 473
Bonani, G., 455—456, 468, 470
Bond, G., 455—456
Boonwichai, S., 27, 137
Bora, D. S., 275, 289—291
Bora, J., 113—114
Borg, F.H., 52
Borgaonkar, H. P., 82—84
Bornmann, L., 432
Borsma, A., 405—406
Bosch, D., 181—183
Bouchez, A., 296—297
Bousetta, H., 36
Boutaud, A. S., 438b
Bouton, M. M., 423—424
Boyer, T., 214f
Bracht, B. B., 305
Bradbury, J.P., 337
Bradbury, J. P., 305
Bradford, E. F., 110—111
Bradshaw, J. S., 256
Brady, E. C., 296
Brand, D., 124—125
Brauer, A., 444t—445t
Brauer, G., 108—110
Braun, B., 51
Bray, J. R., 456—457, 457t
Brody, S., 34
Brown, O., 35
Bryman, A., 39
Breitenmoser, P., 454f
Bremer, U. F., 124—125
Bressan, D., 460f
Breton, E., 296—297
Bridge, J. S., 277—281
Brocks, J. J., 438b
Broecker, W. S., 260
Brönnimann, S., 454f
Brook, E. J., 468
Broqua, R. J., 134
Brückner, E., 451—452
Bruckner, H., 251

Brugam, R. B., 300
Bruni, P., 181—183
Brunner, I., 457
Bruun, T. B., 425
Brzoska, M., 33—34
Bu, Z., 300
Buckley, B. M., 296
Buerkert, A., 51
Buettner, A., 296
Bureau, S., 145
Burgemann, R.., 227
Burgener, L. K., 124
Burns, S. J., 296
Burton, K. W., 209
Butterbach-Bahl, K., 423—424

C

Caesar, J., 84—86
Cai, B., 317, 337
Cai, W., 175
Calliari, E., 54
Calderia, k, 4—5
Cameron, J., 450
Cameron, N. J., 295, 317, 351
Campbell, B. M., 425
Cannariato, K., 476
Cant, D. J., 275
Cantonati, M., 300—305, 317
Canuti, P., 181—183
Cao, P., 296
Caron, J. M., 296
Carpenter, T. H., 73
Carril, A., 134
Carvajal, L. F., 134—135
Carvalho, L., 295, 317, 351
Case, M., 146
Catalan, J., 295, 317
Cautley, P. T., 111
Cecchini, S., 452—453
Cerdà, A., 36
Cerveny, P. F., 273
Chakraborty, P. P., 276
Chamoli, A., 296
Chandimala, J., 73
Chandra, A., 423—424
Chandrapala, L., 61, 82—83
Chandrasekar, N., 191—192
Chandrasekaran, R., 191—192
Chang, C.-L., 172—173
Chang, J., 296
Chang, X. G., 108—110

Chang, Y. -P., 171–173, 317
Chappellaz, J., 439, 440f
Charles, C. B., 295, 317
Charles, D. F., 295
Charvátová, I., 458
Chatterjee, S., 450
Chattopadhyaya, U., 113–114
Chaturvedi, A., 479
Chaturvedi, K. K., 430–431
Chaturvedi, S. K., 177–178, 191–192, 195, 208–209, 414–415
Chaudhari, S. K., 431–432
Chaudhri, H. S., 11
Chauhan, P. R., 113–114
Chaumeil, P., 296–297
Chen, C. -T., 172
Chen, C. T. A., 171
Chen, G., 4–5
Chen, L., 79
Chen, S. H., 172–173, 296–297, 317, 327–337
Chen, X., 296–297, 300, 305, 317, 327, 337
Chen, Y. -M., 172
Cheng, H., 296, 476
Cheng, J., 296
Chengappa, R., 411
Cheong, D., 305, 337
Cherry, N. J., 73–76
Chesner, C. A., 106
Chhetri, G. B., 125
Chhetri, L. K., 124
Chhinh, N, 28
Chhogyel, N., 125–126
Chhun, C., 23
Chiba, T., 327
Chimi, R., 125–126
Chindarkar, N., 34
Ching, B. T., 106
Chinnusamy, V., 430–431
Chinvanno, S., 28–29
Chitra, V., 430–431
Cho, A., 305, 327, 337
Cho, P. Y., 305, 317, 327
Choi, T. S., 327
Choudhury, R., 124
Chris, C., 51–52
Christl, M., 457
Chu, G., 305, 317, 337
Chuang, P. P., 172–173
Chun, J., 25, 26f
Clark, G. A., 105–106
Clarke, L. E., 287–288
Clarkson, C., 113–114

Cleland, S., 79
Clemens, S. C., 476
Clements, G. R., 110
Clews, P., 79
Clow, G. D., 468
Cobo, M. J., 425
Codilean, A. T., 466–467
Cohen, R., 34
Cohen, P. A., 438b
Cole, E. E., 317
Cole, R. D., 463
Collings, H. D., 106–108
Collins, D., 82–86
Collins, E. S., 208
Collins, M., 175
Compagnucci, R. H., 134
Conde, C., 134
Cook, E. R., 124, 296
Cooper, M., 260
Corbert, G. B., 110
Corlett, R., 21–22
Cornet, V., 296–297
Correa, G. J., 4–5
Correa-Metrio, A., 296
Cournoyer, L., 296–297
Courtillot, V., 479
Cox, G. M., 438b
Cradock-Henry, N., 24–25
Cramb, R.,, 37
Cranbrook, E. O., 105–106, 111–112
Crawford, R. M., 295
Crespo, O., 124
Creveling, J. R., 438b
Crosby, A.W., 36
Crossetti, L. O., 305
Crowthera, A., 113–114
Cubie, D., 54
Cuccurullo, C., 425
Cumming, B. F., 295–296, 317
Curnoe, D., 105–110, 113–114
Curray, J. R., 216, 218–219, 225–227, 275, 284
Curtis, G. H., 108–110
Cushman, J. A., 209, 237

D

Dahlbom, B., 105–106
Dairaku, K., 146, 147f, 161–166
Damon, P., 445f, 457–458
Damon, P. E., 457–458
Dapper, M. D., 112–113
Darby, D. A., 455

Dargusch, P., 423–424
D'Arrigo, R., 20–21
D'Arrigo, R. D., 296
Darshika, D. T., 85f–88f
Daryin, A., 337
Das, M., 259–260, 474–475
Dash, C., 473
Datta, A., 137
Davey, K. A., 317
Davis Jr., G. E., 480
Davis, R. B., 317
Davis, S. M., 2, 4–5
Davison, G. W. H., 111
Dawa, T., 125–126
Dayal, J. S., 430–431
de Andrade, A. M., 124–125
Dearing, J.A., 51
De Raaf, J. F. M., 275
De Vries, H., 453–454, 457–458
de Wolf, H., 327–337
Deb, P., 27
Debaveye, J., 112–113
deBeaulieu, J. L., 305, 337
Decet, F., 300–305, 317
Deepa, P. J. J. V., 191–192
Deino, A., 106
Delgado, J., 25
deMenocal, P. B., 177
Demisse, G., 132
Demissie, M., 177–178
Dendup, N., 124
Deng, F.M., 34
Denis, A., 27
Denys, L., 296–297
Deser, C., 11, 18–20
deSouza Junior, E., 124–125
Detemmerman, V., 84–85
Devkota, A., 137
Dhaila, B. S., 296
Dhal, S. P., 473
Dharani, K., 191–192
Dharma, A., 106
Dhingra, H., 461
Dias, J.M., 36
Dickinson, W. R., 285–287, 289
Diehl, J. F., 106
Diekmann, B., 337
Dikpati, M., 450
DiNezio, P., 18–20
Ding, Y., 124, 296
Ding, Z., 461–462
Ditchfield, P., 113–114

Dixit, S. S., 296
Doblas-Reyes, F. J., 88–90
Doezema, J., 34
Donald, A., 79
Donald, F., 295, 317
Dong, Z., 18–20
Donnadieu, Y., 438b
Dorji, R., 125
Dorji, S., 125
Dorji, T., 121–124
Dorji, T. Y., 125
Dorji, W., 125–126
Dorofeyuk, N., 337
Douglas, M. S. V., 337
Doyle, M., 134, 136
Drake, R., 106
Duan, Z., 430–431
Dubey, A. K., 113–114
Dubey, R., 414–415
Dulham, R., 227
Dumka, R. K., 464–465
Dunbar, R. B., 446, 447f
Dung, N. C., 476
Dungan, J. L., 88–90
Dunn, F. L., 105–106, 110–111
Dunne, J., 469
Duplessy, J. C., 247
Duthie, H. C., 295
Dutt, S., 476
Dutta, K., 470
Duží, B., 51

E

Earl, G. W., 106–108
Easterling, D. R., 84–85
Eastham, J., 18, 27–28
Ebbers, T., 27
Eddy, J. A., 441, 443–444, 452
Edirisinghe, M., 84
Edmonds, J., 1, 88
Edwards, A., 244
Edwards, R. L., 296, 476
Egeson, C., 451–452
Eggleston, S., 439, 440f
Ekundayo, T. C., 432
Elderfield, H., 260
El Hassani, A., 472
Elisa, F., 54t–55t
Ellerman, F., 443, 450
Ellwood, B. B., 472
Elmore, D., 468–469

Emmel, F. J., 216, 218−219, 227
Endo, H., 146, 153
Endo, K., 327
Eriyagama, N., 82−83
Erwin, D. H., 438b
Escobar, J., 296
Estes, L.D., 51
Evan, A. T., 4−5
Evans, M. N., 455−456, 476−477

F

Fagiolo, G., 51
Fairchild, I. J., 475f
Fairservice Jr., W. A., 409−410
Falconer, H., 111
Falk, G.C., 51
Farooqui, A., 296
Fazzuoli, M., 181−183
Federica, C., 54t−55t
Fekadu, H., 131
Fernando, I. M. K., 71
Fernando, K., 82−83
Fernando, T. K., 82−84
Ferriz-Mas, A., 458
Finkel, R., 468
Finkel, R. C., 468, 470
Finkelstein, S. A., 300−305
Fischer, H., 439, 440f, 457
Fleitmann, D., 296
Flemming, B. W., 192, 194t−195t
Flower, R. J., 295−296, 317, 351
Foland, K. A., 290−291
Folland, C. K., 85
Fontugne, M. R., 251
Foster, G. L., 473
Fousiya, A. A., 470−471, 472f
Fox, H. E., 247
Franc, A., 296−297
Frank, D., 454f
Frerichs, W. E., 210−211
Frics, M., 458b
Friedrich, M., 471b
Frierson, D. M., 4−5
Fritts, H. C., 476−477
Fritz, S. C., 305
Fröhlich, C., 33−34, 470
Fu, Q., 2, 4−5
Fu, X., 80−81
Fu, Y. -X., 108−110
Fujita, K., 124
Fujita, M., 146, 153
Fukuyama, Y., 146

Fuller, L., 475f
Fuller, R.A., 51
Fülöp, R. H., 466−467
Furu, P., 52

G

Gabrieli, J., 300−305, 317
Gabunia, L., 108−110
Gage, K. S., 452−453
Gaiser, E. E., 300
Gallego, M. C., 454
Gallet, Y., 444t−445t, 457−458, 459f
Galli, M., 452−453
Ganachaud, A., 175
Ganai, M., 6
Ganssen, G., 260
Gao, C., 296−297, 300−305, 317
García, J. A., 454
Gare, A.E., 36
Garg, K. C., 427−431
Gasse, F., 300−305, 317, 327
Gatti, E., 105−106, 112−113
Gaur, A. S., 413
Gaur, R., 291
Gaur, V. K., 227
Gemenne, F., 36
Geremu, G., 177, 195
Getty, A. G., 108−110
Ghimiray, M., 125
Ghosh, S. K., 273−274, 276, 282, 285, 289−291
Ghosh, S. S., 405−406
Gibbard, P., 105−106, 112
Gibson, C. E., 327, 337
Gibson, J., 296−297
Gijbels, R., 112−113
Gilbert, P.R., 52
Gleason, B., 84−86
Gleissberg, W., 452
Glodniok, S., 305, 327, 337, 351
Gnevyshev, M. N., 450
Gobbett, D., 110−111
Goh, H. M., 105−110, 113−114
Gomes, C., 67
Gong, Z. J., 305
Gooch, P., 36
Gopal, V., 296
Gosaye, B., 178, 195
Gosliner, T. M., 245
Gosselin, M. -P., 296−297
Goswami, B. N., 6−7, 76
Goswami, S., 430−431
Gouveia, A., 225

Govil, P., 178, 209, 296, 474–475
Gowan, E. J., 4
Graf, H. -F., 113–114
Grau, H. R., 131
Gregersen, L. E., 425
Greibe Andersen, J., 52
Grise, K. M., 2, 4–5
Grootes, P. M., 443–444
Grosch, C. E., 455
Grover, A., 52
Grover, H., 34
Grünbühel, C., 37
Grytnes, J. A., 295, 317
Gu, Z., 305, 317, 337
Gualda, G. A., 106, 112–113
Guggemoos, P., 124–125
Guilyardi, E., 175
Gujar, A. R., 191–192
Gulati, A., 423–424
Guoxiong, W., 299–300
Gupta, A. K., 208–209, 259–260, 461, 474–476
Gupta, H. K., 411–412
Gupta, I. D., 430–431
Gupta, M. C., 113–114
Gurung, D. R., 124
Gutiérrez, J. M., 19f–20f

H

Hadgu, G., 132
Hagemann, S., 113–114
Hagg, W., 124–125
Hahne, J., 300
Hai, C., 299–300
Hajdas, I., 455–456
Halberg, F., 452
Hale, G. E., 443, 450
Hale, R. E., 251
Hall, R., 327
Hall, R. I., 295, 305, 317, 327
Hallock, P., 238, 245
Halpert, M. S., 73
Hamada, A., 146–147
Hameed, S., 82–84
Hammond, J., 423–424
Hamza, M., 51
Han, J., 305, 317, 337
Hansen, C. J., 440–441
Hansen, F., 444t–445t
Hao, Q., 317, 337
Hapuarachchi, H. A. S. U., 73, 74f
Haq, B. U., 405–406

Haraguchi, T., 327
Harashina, K., 51
Hargan, K. E., 300–305
Harpending, H. C., 108–110
Harris, C., 113–114
Hart, A. M., 245
Hartz, F., 54
Harzhauser, M., 455
Hasan, M., 33–34
Hasan, M. K., 125
Hashimi, N. H., 178, 208–209, 247–251, 407–408, 411, 418
Hashimoto, S., 17–18
Hashino, M., 145
Haslam, M., 113–114
Hasler, P., 295
Hathaway, D. H., 440
Hatwar, H. R., 79
Haworth, E. Y., 305, 317
Hayashi, Y., 61, 65, 73, 83–84
Haylock, M., 84–86
Haynes, L. L., 439f
Heffernan, O., 4–5
Heikkilä, U., 457
Hejda, P., 458
Helle, G., 477
Hendon, H., 79
Hendon, H. H., 79–80
Henley, B., 439
Henriques, P. J., 208–209, 407
Henry, M., 216, 218–219, 227
Herath, H. R. C., 85f–88f
Herath, S., 22, 27, 29
Herbert, T. D., 465f
Herrera, F., 425
Herrera-Viedma, E., 425
Herron, M. M., 479
Herzschuh, U., 296
Hibbard, K., 1
Hibbard, K., 88
Hibino, K., 146–147, 161–166
Higashino, M., 146
Higgitt, D. L., 216
Hilgen, F. J., 461–462
Hill, J. E., 110
Hippler, D., 296
Hirano, J., 146, 147f
Hla, U., 209, 247
Hobday, A., 20
Hock, S. S., 110
Hodell, D. A., 461
Hoey, T. B., 216

Hoffecker, J. F., 105–106
Hoffman, P. F., 438b
Hoffman, S.M., 33–34
Hoffmann, E.M., 51
Hoffmann, S., 455–456
Hofmann, H. J., 468
Hofmann, W., 337
Hohenegger, J., 238, 245
Holifield, R., 53–54
Holle, R. L., 71
Hollifield, J., 34
Holmquist, J. R., 300–305
Hong, A., 299–300
Hong, C., 299–300
Hong, Y., 299–300
Hönisch, B., 439f
Hooijer, D. A., 110–111
Horton, B., 85
Horton, R., 20–22, 25, 27–28
Hossain, M.D., 51
Hossain, M.F., 54
Hossain, S.M., 41t–42t
Hoy, A., 124
Hoyos, N., 296
Hoyt, D. V., 441–442, 453b
Hsieh, M. L., 171–173, 317
Hu, L., 296
Hu, Y., 4
Huang, G., 251, 351
Huang, K., 296
Huang, M. -S., 172
Huang, S. Y., 172–173
Huang, W., 108–110
Huang, Y., 351
Huh, C. A., 296–297, 317, 327–337
Huq, S., 54
Hulot, G., 444t–445t, 457–458, 459f
Hulugalla, R., 61, 65, 73, 83–84
Humphrey, L. T., 105–106, 111–112
Hurd, G. A., 450
Hurley, M. A., 305
Hurtt, G. C., 88
Hustedt, F., 305, 317
Hutton, C.W., 51
Huyer, S., 425

I

Ibrahim, N., 106
Ibrahim, Y. K., 105–106, 111–112
Iizumi, T., 146, 147f
Immerzeel, W. W., 124
Inceoglu, F., 444t–445t

Ingham, F. T., 110–111
Ingole, B. S., 208–209
Ishidaira, H., 145, 147–153
Ishihara, K., 146
Ishii, M., 146–147, 153, 161–166
Ishikawa, H., 146, 161–166
Ishizaki, N. N., 145–146
Islam, M.R., 33–34
Islam, N., 82–84
Ito, R., 146, 161–166
Iversen, J., 458b
Iwata, S., 124, 126–127
Izett, G., 106, 112–113

J

Al-Jabi, S. W., 432
Jacob, T., 108–110
Jacobsen, B. H., 444t–445t
Jacoby, G. C., 296
Jade, S., 227
Jadhav, S. K., 451f
Jaglan, S., 476
Jahan, M.U., 41t–42t
Jain, A. K., 290–291
Jain, N., 430–431
Jamason, P., 85
Jamoneau, A., 295
Jani, R. A., 473
Jannathulla, R., 430–431
Jantan, A., 112–113
Jasin, B., 112–113
Jat, M. L., 125
Jáuregui, E., 134
Javaraiah, J., 452–453
Javed, M., 50–51
Jayaratne, K. C., 67
Jayawardena, I. M. S. P., 67, 69f–70f, 71, 72f–74f, 73, 77, 79, 79f–88f, 81–82
Jenkins, T., 108–110, 113–114
Jenkins, J.C., 51
Jerram, A. D., 181–183
Jessen, K., 458b
Jessica, K., 51–52
Jha, D., 113–114
Ji, J., 472
Ji, M., 337
Jia, X., 79
Jian, L. K., 449b
Jiang, H., 351
Jiang, W., 317, 337
Jiangyu, M., 299–300
Jianping, L., 299–300

Jin, F. F., 175
Jochum, M., 175
Joh, G., 327
Joh, G. J., 305, 327
Johannes, O., 124
Johansen, J., 300
Johanson, D., 105–106
Johansson, A., 54
Johansson, C., 317
Johnson, N. M., 273
Johnston, V. E., 469
Jones, C. R., 110–111
Jones, P. D., 6, 85
Jones, S., 113–114
Jones, S. C., 113–114
Jones, V., 317
Jones, V. J., 295, 317, 351
Joos, F., 459f
Jorde, L. B., 108–110
Jose, J., 473
Joshi, A., 52
Joshi, L. M., 296
Joshi, S., 124
Jourdain, D., 27
Jouzel, J., 479
Joy, A. H., 443, 450
Juan, F., 299–300
Juch, D., 181–183
Juggins, S., 295–296, 300, 317, 351
Julca, A., 51
Julia, C., 51–52
Julian, P. R., 76–77, 80
Julius, M. L., 295
Juneng, L., 77

K

Kaba, M., 468, 469f
Kademani, B., 424, 431
Kadir, M. Z. A. A., 67
Kaesler, R. L., 245
Kahlert, M., 296–297
Kainuma, M., 88
Kale, V. S., 407–408
Kallestrup, P., 52
Kalugin, I., 337
Kamiguchi, K., 146–147
Kanada, S., 145, 161
Kane, R. P., 65, 73
Kang, I. -S., 80–81
Kang, S., 124
Kang, Y.H., 33–34

Kao, S. J., 171–173, 317
Kaphaliya, B., 431–432
Kapur, D., 423–424
Karekezi, C., 52
Karisiddaiah, S. M., 245, 251
Karl, T. R., 85
Karma, K., 126–127
Karma, T., 124
Karoff, C., 444t–445t
Kashima, K., 305, 327, 337
Katel, O., 124
Kathiroli, S., 411
Kathyayani, S. A., 430–431
Kato, T., 145, 161
Katsuki, K., 305, 337
Katsuki, K., 327
Kattel, G. R., 296–297, 300–305, 317
Katwal, T. B., 125
Kawahata, H., 473
Kawaler, S. D., 440–441
Kawamura, K., 464f
Kawase, H., 146–147, 153–157, 161–166
Kebedew, M. G., 178
Keck, F., 296–297
Keisling, B., 466–467
Keller, F., 181–183
Keller, W., 300–305
Kelman, I., 51
Kelly, M. G., 295
Kem, Z., 106
Kempadoo, K., 34
Kennedy, J. J., 6
Kermarrec, L., 296–297
Kern, A. K., 455
Kerr, H. A., 208
Khadkikar, A. S., 410
Khan, H., 209, 474–475
Khan, K., 414–415
Khanal, N. R., 124
Khare, N., 178, 208–209, 254, 258–260, 405–406, 474
Kharya, A., 273, 291
Khatri-Chhetri, A., 125
Khokhlov, A., 459f
Khokiattiwong, S., 296
Khurana, R., 125
Kieckhefer, R., 216, 218–219, 227
Kieffer, B., 181–183
Kienel, U., 300
Kiew, Y. M., 105–110, 113–114
Kiguchi, M., 22
Kilham, P., 337
Kilham, S. S., 337

Kim, J. C., 305, 337
Kim, Y. S., 327
Kingston, J. C., 317
Kinzig, A., 17
Kipfstuhl, S., 439, 440f
Kirk, M., 288
Kiss, K. T., 327
Kitoh, A., 146–147
Klaatsch, H., 105–106
Klaus, A., 124–125
Klein, S. A., 4–5
Klein Tank, A. M. G., 84–86
Kleinen, T., 113–114
Kliot, N., 35–36
Knight, M. D., 106
Kniveton, D., 33–34
Knudsen, K. L., 351
Knudsen, M. F., 444t–445t
Kobayashi, J. T., 110–111
Kochman-Kedziora, N., 300
Kociolek, P., 295
Kock, A., 295
Koenig, L. S., 124
Köhler, P., 439
Kokocinski, M., 295
Kolesa, L., 300
Kondo, R., 327
Konerding, V., 51
Kong, L., 20
Kong, X., 296
Kopczynska, E., 296–297
Korisettar, R., 113–114
Kornkanitnan, N., 296
Korsman, T., 327
Korup, O., 124–126
Koshy, J., 113–114
Kotla, S. S., 273, 291
Kotlia, B. S., 296, 464–465
Kovaltsov, G. A., 444t–445t, 457–458, 459f, 466f, 471b
Kozak, E., 300
Kraef, C., 52
Krahn, K. J., 305, 327, 337, 351
Kram, T., 88
Kramers, J., 296
Krammer, K., 300, 305, 327–337
Krasske, G., 317
Kreiser, A., 317
Krey, V., 88
Kripalani, R. H., 11
Krishnamurti, T. N., 88–90
Kromer, B., 455–456, 455f, 471b, 478f
Krovnin, A. S., 453b
Krstic, S. S., 305, 317
Krüger, K., 113–114

Krynine, P. D., 285–287
Kubik, P. W., 457
Kubota, J., 327
Kukla, G. J., 296
Kulaimi, N. A., 110
Kulkarani, A., 11
Kumar, A., 6, 476
Kumar, B., 474–475
Kumar, D., 227
Kumar, K. R., 61, 63–66
Kumar, L., 125–126
Kumar, P., 276
Kumar, R., 273–274, 276, 282–283, 285, 289–291, 430–431
Kumar, R. P., 6
Kumar, R. R., 430–431
Kumar, S., 430–431
Kumaran, K. P. N., 410
Kumararaja, P., 430–431
Kumaravel, V., 273–274, 291
Kumari, R., 430–431
Kumon, F., 305, 317, 327
Kurihara, K., 146, 161
Kurtarkar, S., 208–209
Kurtarkar, S. R., 473
Kutser, T., 296

L

Labeyrie, L., 247
Labracherie, M., 247
Lahr, M., 108–110
Lahr, M. M., 113–114
Laing, T. E., 337
Lal, B. B., 405–406
Lal, D., 466–467
Lal, K., 405–406
Laluraj, C. M., 479
Lamarque, J. F., 88
Lambeck, K., 247
Landais, Petrovay, K., 439
Landscheidt, T., 446, 447f
Lane, C., 113–114
Lange-Bertalot, H., 300–305, 317, 327–337
Langer, M. R., 245
Langway, C. C., 470, 479
Lapierre, H., 181–183
Larsen, K. M., 227
Last, W. M., 178
Lauterbach, S., 305, 327, 337, 351
Lavenex, S., 54t–55t
Lawrence, J., 24–25
Lawrence, K. T, 465f
Lawver, L. A., 216, 218–219, 227

Lázár, A.N., 51
Lazarev, V., 468, 469f
Lea, D. W., 234, 260, 473
Leamon, R. J., 446–447
Leblanc, K., 296–297
Leboucher, T., 295
Lécuyer, C., 473
Lee, C. -Y., 172–173
Lee, E.S., 35–36
Lee, H., 305, 317, 327
Lee, J. -Y., 80–81, 305, 317, 327, 337
Lee, S., 17–18
Lee, S. D., 305, 317, 327
Lee, T. -Q., 88–90, 172–173, 296–297, 317, 327–337
Lehman, S. J., 468
Lehmann, B., 468
Leira, M., 317
Lelwala, R., 71
Lengaigne, M., 175
Leng, M. J., 351
Leshnik, L. S., 409–410
Levitus, S., 214f
Leydesdorff, L., 432
Li, B., 113–114
Li, C., 79
Li, D., 351
Li, H. -C., 171–173, 296–297, 317, 327–337
Li, J., 317, 337
Li, K., 296
Li, L., 299–300
Li, W., 296
Li, W. -H., 108–110
Li, X., 9
Li, Y., 296–297, 300, 305, 317, 327, 337
Li, Y. L., 305
Li, Z., 124
Liangcheng, T., 299–300
Liao, M., 296
Licht, A., 459f
Licker, R., 51
Liebrand, D., 461
Liew, P. -M., 172–173
Li, H. C., 317
Lim, J., 305, 317, 327
Lim, T. T., 106, 111–112
Lin, S. -F., 172–173
Lindholm, R. C., 285–287
Line, J. M., 295, 300, 317
Linshy, V., 208–209
Linshy, V. N., 208–209
Lipper, L., 423–424
Lipps, J. H., 245
Lisiecki, L. E., 440f
Littler, M. M., 244

Liu, D., 296
Liu, E., 337
Liu, J., 4, 305, 317, 337
Liu, S., 296
Liu, T., 317, 337, 461–462
Liu, W., 296
Liu, X., 296
Liu, Y., 300–305
Liu, Z., 296, 465f
Liyanage, J. P., 71
Lloyd, J. M., 351
Lockyer, N., 442
Lockyer, W. J., 451–452
Lodders, K., 440–441
Lohmann, G., 4
Lone, A., 296
Lone, M. A., 296, 476
Long, C., 430–431
Long, X., 472
Loo, Y., 20
Loomba, M., 52
Lopes, F., 444t–445t, 457–458, 459f
López-Herrera, A. G., 425
Lorenzo, T., 17
Lotay, Y., 123–124
Lotter, A. F., 337
Lotti-Bond, R., 455–456
Lou, J. -Y., 171–172
Lourens, L. J., 461–462
Loveson, V. J., 191–192, 408, 414–415, 418
Lovholt, F., 414–415
Lowe, R. L., 296–297
Lowell, W. E., 480
Lu, H., 296, 305, 317, 337
Lu, J., 4–5
Lu, X. X., 216
Lu, Z., 430–431
Lu, Z. T., 466–467
Lücke, A., 477
Ludlow, F., 471b
Luhmann, J. G., 449b
Lund, S. P., 455
Luoto, M., 295
Lwin, C., 17–18

M

M Krishna, P., 6
Ma, L., 296–297, 300–305, 317, 454
Ma, L. H., 454–455, 455f
Malhotra, B., 52
Ma, X. Z., 468
MacDonald, G. M., 300–305
Mackay, A. W., 305

Mackensen, A., 259–260
Mackey, A. W., 317, 337
Madden, R. A., 76–77, 80
Magaña, V., 132
Magana, V. O., 296
Mahagaonkar, A., 124
Mahakur, M., 6
Mahanta, C., 124
Maharjan, A., 124
Maharjan, S. B., 124–125
Mahesh, R., 191–192
Mahmood, N., 414–415
Majid, Z., 105–108
Mäkelä, A., 67, 69f–70f, 71, 72f–73f
Malherbe, W., 295
Malhotra, B., 52
Malmgren, B. A., 61, 65, 73, 83–84, 208–209, 474–475
Mandal, U. K., 431–432
Mandic, O., 455
Mangini, A., 296
Manickaraj, S. D., 191–192
Manjunatha, B. R., 473
Mann, D. G., 295
Mann, M., 457
Mann, M. E., 482
Manou, D., 33–35, 54
Manton, M. J., 82–84
Manzoor, N., 82–84
Marchetto, A., 337
Markovic, S. B., 305, 317
Marschall, H. R., 473
Marshall, H., 296–297
Martin, E. R., 79
Martin, J., 79
Martin, M., 33–34
Martin, P. A., 234, 260
Martin, P.L.Martin-Puertas, C., 34, 444t–445t
Martiniello, M., 36
Masarik, J., 468, 469f
Mashiotta, T. A., 260
Mason, B. G., 106
Mason, I., 317
Masson-Delmotte, V., 439, 479
Massonet, S., 468, 469f
Mastrorillo, M., 51
Masui, T., 88
Matschullat, J., 124
Matter, A., 296
Mattey, D., 475f
Matthes, K., 444t–445t
Matthews, A. J., 79
Maung, K., 17–18

Mayer, B., 54
Mayewski, P. A., 113–114, 457
Mazari, R. K., 273–274
Mazumder, A., 178, 208–209, 237, 245, 474–475
McAndrews, J. H., 208
McCarthy, F. M. G., 208
Mccracken, K. G., 457–458
Mcdermott, F., 475f
Mcdonnell, T., 34
McGowan, S., 300, 327
Mcintosh, S. W., 446–447, 450
McLeman, R., 53
McKee, E. D., 289–290
McKeever, K., 300
McKie, R., 105–106
McManus, J. F., 247
McNamara, K. E., 423–424
McPhaden, M. J., 72
McPherson, J. G., 287–288
McQuistan, C., 54
Meade, R. H., 215–216
Medioli, F. S., 208
Meehl, G. A., 88–90, 446–447
Meeker, L. D., 113–114, 457
Meese, D. A., 457
Meinke, H., 79
Mélice, J. L., 461
Melton, F. S., 88–90
Mende, W., 442
Menezes, E. T., 407–408
Merbold, L., 423–424
Merh, S. S., 251, 407–408
Merla, G., 181–183
Mertens, A., 296–297, 300–305, 317, 327–337
Merz, B., 25
Mesa, O. J., 134–135
Metcalfe, S., 296–297, 337
Meugniot, C., 181–183
Meyer, K., 240–245
Miall, A. D., 276–281, 288
Michaelis, A., 88–90
Michel, E., 247
Mihr, A., 33–35, 54
Mikami, T., 61, 65, 73, 83–84
Milani, F., 458
Miller, H., 457
Minetti, J. L., 134
Mingram, J., 305, 317, 327, 337, 351
Minhas, P. S., 431–432
Mischke, S., 305, 327, 337, 351
Mishra, D., 430–431
Mishra, D. P., 113–114

Mishra, K., 418
Mishra, R., 209, 474–475
Misra, A., 52
Mitchell, F. J. G, 317
Mittal, N., 52
Mizuta, R., 146–147, 153, 161–166
Mogessie, A., 177–178
Mohammad, A. A. A., 111–112
Mohan, K., 474–475
Mohan, R., 296
Mohd Razif, F., 106
Mokssit, A., 84–85
Molden, D. J., 124
Mool etal, 126–127
Mooley, D. A., 450
Moore, D. G., 216, 218–219, 227
Mori, N., 146–147, 161–166
Morice, C. P., 6
Morse, D. L., 468
Moss, R., 1
Moy, C. M., 446–447
Mudelsee, M., 296, 476
Muhammad, R., 111
Muhammad, R. F., 105–106, 111–113
Muhammad, R. F. B. H., 105–106, 112–114
Mukal, M., 227
Mukhopadhyay, A., 37
Mukhopadhyay, B., 7
Mukhopadhyay, R., 6
Munro, M. A. R., 317
Murakami, M., 17–18
Muralidhar, M., 430–431
Murata, A., 146–147, 153–157, 161–166
Murphy, M. J., 71
Mursula, K., 470
Muscheler, R., 444t–445t, 455–456
Muttil, N., 124
Muzikar, P., 469

N

Nag, A., 71
Nagai, H., 124
Nagatomo, T., 153–157
Nagavel, A., 430–431
Nagayasu, K., 305, 317, 327
Naidu, P. D., 208–209, 296, 474–475
Naik, S. S., 479
Nair, M., 6
Nair, R. R., 208, 247–251, 259–260, 407–408, 418, 474
Naito, N., 124, 126–127
Nakakita, E., 146, 161–166
Nakano, M., 145, 161

Nakano, Y., 245
Nanda, A. C., 291
Nanni, T., 452–453
Narama, C., 124, 126–127
Naser, M.M., 53
Naselli-Flores, L., 305
Natawidjaja, D., 227
Nath, B., 405–406
Nautiyal, S., 51
Nayak, G. N., 208–209
Nayak, S., 145–146, 161–166
Neal, V. T., 446
Neff, U., 296
Negendank, J. F. W., 317, 337
Nehrbass-Ahles, C., 439, 440f
Nemani, R. R., 88–90
Nemec, W., 287–288
Nepal, S., 124
Neudorf, C., 113–114
Newman, D., 468
Ni, J., 296
Nicholson, S. B., 443, 450
Nicholls, R.J., 51
Nigam, R., 177–178, 207–209, 227, 237–238, 247–251, 254, 258–260, 404–411, 414–415, 418, 473–475
Nigam R., 296
Nilsen, T. H., 289–290
Nilsson, G. B., 105–106
Ninkovich, D., 106, 112–113
Nishiizumi, K., 468
Nishimori, M., 146, 147f
Nishimura, S., 105–106, 112–113
Nitecki, D. V., 105–106
Nitecki, M. H., 105–106
Noda, A., 146
Noga, T., 295, 300
Nolte, E., 468, 469f
Nora, G., 125–126
Nordlander, L., 54
Norman, K., 113–114
Norris, J. R., 4–5
Nosaka, M., 146, 153–157, 161–166
Nuñez, M., 134, 136
Nur, S. I. A. T., 111–112
Nwe, T., 21–22

O

Obradovich, J. D., 106, 112–113
Obreht, I., 305, 317
O'Brien, S. R., 457
Oeschger, H., 468, 470
Oeurng, C., 28

OfDell, C. W., 4–5
O'Gorman, P. A., 146
Oh, J. H., 11
Ohba, M., 146, 161–166
Oh'izumi, M., 153–157
Ohl, A. I., 450
Ojha, J. R., 177–178, 191–192, 195
Okada, Y., 146, 153, 161–166
Okoh, A. I., 432
Okumura, Y., 18–20
Okuno, J. I., 464f
Oliver, J. E., 145
Oliver-Smith, A., 51
Olsen, J., 444t–445t
Olsson, L., 54
Onstott, T. C., 106
Oo, N. W., 216
Oppenheimer, C., 105–106, 108–110, 112–114
Oppenheimer, M., 51
Oppenheimer, S., 108–110, 109f
Orrenius, P., 34
Ortiz, J. D., 455
Ortolani, S., 458
Osakada, Y., 161–166
Otani, H., 305, 317, 327
Otto-Bliesner, B., 296
Overpeck, J. T., 208–209

P

Pabón, J. D., 132
Padisak, J., 305
Padmakumari, V. M., 296
Paepe, P. D., 112–113
Pagani, M., 438–439, 439f, 462f
Pai, D. S., 79
Paillard, D., 439
Pajaczek, A., 300
Pajunen, V., 295
Pal, J., 113–114
Pal, J. N., 113–114
Pal, S., 431–432
Pälike, H., 461
Palmer, T. N., 296
Panchang, R., 208–209, 210f, 227, 236–238, 247, 251, 254, 259–260, 474
Panebianco, S., 50–51
Panizzo, V. N., 305
Pant, G. B., 61, 63–66
Pant, R. K., 296
Papenfus, T., 260
Pappas, T., 124–125

Pardo-Igúzquiza, E., 461
Parkash, B., 273
Parker, D. E., 85
Parker, L., 125–126
Parrenin, F., 439
Parthasarathy, B., 450
Parthiban, G., 112–113
Passy, S. I., 295
Paterne, M., 251
Paterson, A. M., 295, 300–305
Pathak, M. C., 407–408
Pati, P., 473
Patnaik, R., 273, 291
Patrick, R., 317
Pattan, J. N., 112–113
Paul, J., 227
Paul, V. J., 240–245
Pavan, V., 88–90
Peacock, B. A. V., 105–106, 110–111
Pearce, N. J., 106, 112–113
Pearce, N. J. G., 112
Peatman, S. C., 79
Pecher, A., 181–183
Peloquin, J., 296–297
Peltier, W. R., 247
Peng, L. C., 105–106, 111–112
Peng, T. H., 260
Penjor, T., 124
Peszek, L., 300
Peter, B., 466–467
Peterson, T. C., 84–86
Petraglia, M., 113–114
Petrick, C., 444t–445t
Petrovay, K., 481
Pettijohn, F. J., 285–287, 289
Phadtare, N. R., 296
Phartiyal, B., 296
Philander, S. G. H., 72
Philibert, A., 317
Phirun, N., 23
Pickup, G., 79
Pienitz, R., 296–297, 337
Piguet, E., 35–36
Pilgrim, G. E., 273–274, 274t
Piller, W. E., 455
Piontkovski, S., 296–297
Piper, P. J., 111
Pipitpukdee, S., 27
Pirazzoli, P. A., 251, 407
Plessen, B., 305, 327, 337, 351
Plummer, N., 85
Poch, B., 28

Polvani, L. M., 4–5
Poncelet, A., 36
Poornima, M., 430–431
Porter, M., 53–54
Porter, S. C., 296
Portes, A., 35–36
Possnert, G., 444t–445t
Postma, G., 287–288
Potapova, M., 295
Potter, P. E., 285–287, 289
Poulickova, A., 295, 337
Poulton, A. J., 296–297
Poveda, G., 134–135
Power, S., 175
Prabhakar, M., 431–432
Prabhat, P., 470–471, 472f
Pradhananga, S., 124
Prairie, Y. T., 317
Prajapati, S., 464–465
Praveen, S., 430–431
Premalal, K. H. M. S., 83–84
Price, K., 113–114
Priesnitz, A., 317
Procházka, D., 51
Puig, D., 54
Pumijumnong, N., 137
Punyadeva, N. B. P., 73
Punyawardena, B. V. R., 73–76
Pyle, D., 113–114
Pyle, D. M., 106

Q

Qin, Y., 300, 327
Quamar, M. F., 296
Quay, P. D., 443–444, 470
Queguiner, B., 296–297
Quinn, W. H., 446
Quintela, R. M., 134

R

Racca, J. M. J., 317
Racca, R., 317
Rachel, L., 51–52
Raffi, I., 461
Rafi, Z., 414–415
Rahman, M., 51
Rahman, M.F., 54
Rahman, M.M., 51
Rahaman, W., 470–471, 472f
Rahimzadeh, F., 84–86
Rahman, R., 51

Rahman, S.M.M., 41t–42t
Rahut, D. B., 125
Rai, A. K., 430–431
Rai, G. K., 430–431
Raiker, V., 216, 245–247
Raitt, R. W., 216, 218–219, 227
Raiverman, V., 273–274
Rajagopalan, G., 208–209, 247, 407–408, 418
Rajamanickam, G. V., 191–192
Rajan, R. S, 296
Rajesh, R., 430–431
Rajgopalan, G., 247–251
Rajguru, S. N., 407–408
Rajkumar, U., 430–431
Rajshekhar, C., 410
Raju, K., 218–219, 227
Rajwar, D., 431–432
Ram, S., 124–125
Ramakrishna, S. S. V. S., 6
Ramalingeswara Rao, B., 218–219, 227
Ramanathan, A. L., 124
Ramaswamy, V., 17–18, 216, 218–219, 220f, 245–247
Rame, H. P., 113–114
Ramesh, R., 208–209, 296, 473, 476
Ramirez-Gonzalez, G., 425
Ramprasad, K. A., 218–219, 227
Rana, S. S., 208–209, 251, 474–475
Ranasinghe, R., 20
Ranjan, A., 122–123
Rao, K. H., 216
Rao, K. N. N., 464–465
Rao, M. R., 291
Rao, N. R., 237
Rao, P., 17–18
Rao, P. S., 216, 218–219, 220f, 245–247
Rao, S. R., 409–410, 413
Rao, S. V. R., 430–431
Rao, T. P. S., 218–219, 227
Rao, V. P., 251, 473
Rao, T. N., 6
Rashidi, N., 105–106, 112
Rasmusson, E. M., 73
Rasul, G., 124
Rashid, M.M., 33–34
Rathi, G., 273–274
Ratnayake, S., 110
Ravindra, R., 178
Raviprasad, G. V., 209, 247
Ray, D. K., 209, 247
Raymo, M. E., 440f, 464f
Raynaud, D., 439
Rayner, N. A., 6

Raza, W., 476
Raza, W., 296
Razuvayev, V., 85
Reading, H. G., 275
Reason, C. J., 77
Reddy, E. P. K., 430−431
Reid, G. C., 441, 452−453
Reid, M. A., 295
Reille, M., 305, 337
Reimer, C. W, 317
Reimer, P. J., 234
Renaud, F., 51
Reineck, H. E., 276−281, 288
Ren, F., 79
Renberg, I., 295, 317, 327
Renberg, J. F., 317
Reolid, M., 461
Riahi, K., 88
Richard, G., 113−114
Richards, M., 425
Ridley, H. N., 111−112
Ridley Thomas, W. N., 251
Riedel, F., 337
Rimet, F., 296−297
Rinzin, S., 124
Rioual, P., 300, 305, 317, 327, 337
Rippey, B., 327, 337
Ro, S. R., 405−406
Roachanakanan, R., 137
Roberts, M., 113−114
Roberts, R., 113−114
Roberts, R. G., 105−106, 112−113
Robinson, M., 35
Robinson, R. A. J., 216
Rodbell, D. T., 446−447
Rodionov, S. N., 453b
Rodolfo, K. S., 215, 219, 225, 245−247
Rodríguez-Tovar, F. J., 461
Roe, F. W, 110−111
Rogers, A. R., 108−110
Ropelewski, C. F., 73
Rose, W. I., 106
Rosebery, J., 295
Rosen, P., 327
Rosignoli, F., 34
Roth, R., 459f
Round, F. E., 295
Roy, A. K., 273
Rudaya, N., 337
Ruehlemann, C., 113−114
Ruhland, K., 296, 317
Ruhland, K. M., 295, 300−305, 337

Rupper, S., 124
Russell, C. T., 449b
Rutter, N., 461−462
Rybak, M., 295

S

Saalim S. M., 296
Sabade, S. S., 11
Sabbe, K., 337
Sachsenhauser, H., 468, 469f
Saes, M. J., 461
Sagar, A., 424, 431
Sagar, N., 296
Sagri, M., 181−183
Sahak, I. H., 111−112
Saidin, H. M., 105−106
Saidin, M., 105−110, 112−114
Saing, C., 18
Saito, F., 464f
Saitou, H., 146
Saji, N. H., 76
Sakai, A., 124, 126−127
Saklani, U., 122−123
Salehin, M., 51
Salimun, E., 77
Salinger, M. J., 85
Sammonds, P., 34
Sandeep, K., 473
Sangode, S. J., 273−274, 282, 289−291, 296
Sankalia, H. D., 409−410
Santos, F.D., 36−37
Sapkota, T. B., 125
Saraswat, R., 208−209, 259−260, 414−415, 473
Saraswat R., 296
Saraswathy, R., 430−431
Saraswati, P. K., 245
Sarkar, A., 208−209
Sarnthein, M., 260, 351
Sartono, S., 106, 110
Sasaki, H., 146, 161−166
Sato, T., 146, 153, 305, 327
Satpathy, R. K., 209
Satta, N., 51
Satyal, G., 227
Savage, H. E. F., 110−111
Saw, C. Y., 105−110, 113−114
Sawalha, A. F., 432
Scafetta, N., 458
Scarpati, O. E., 134
Schaefer, J. M., 124, 466−467
Schatten, K. H., 441−442, 453b

Schiebel, R., 296—297
Schieber, J., 283
Schleser, G. H., 477
Schlosberg, D., 54
Schmidt, L., 37
Schmidt, M. W., 453—454
Schmidt, R., 337
Schmitt, J., 439, 440f
Schneider, T., 146
Schove, D. J., 456—457
Schumacher, C., 79
Schüssler, M., 455, 455f, 470, 478f
Schutt, B., 181
Schwabe, H., 447—448, 448f
Schwalb, A., 305, 327, 337, 351
Schwalbe, G., 105—106
Schwander, J., 439, 468
Schwarz, A., 305, 327, 337, 351
Schwenninger, J.-L., 113—114
Scott, D. B, 208
Scrivenor, J. B., 105—106, 112
Seckbach, J., 295
Seekell, D. A., 296
Sehgal, R. K., 273, 291
Seltzer, G. O., 446—447
Selvaraj, K., 171—173, 317
Seng, Y. K., 77
Sengupta, P. K., 465
Seong, Y. B., 473
Serna, Y., 296
Seto, K., 305, 327
Shackleton, N. J., 106, 112—113
Shah, U. P., 409—410
Shahidan, S. I., 105—110, 113—114
Shamsudduha, M., 34
Shankar, R., 473
Shanmugam, M., 430—431
Shanmukha, D. H., 208—209
Sharma, K. P., 137
Sharma, M., 282, 285, 287—291
Sharma, P. C., 431—432
Sharma, R. P., 273
Sharma, S., 52, 178, 282, 285, 287—291
Sharp, S. P., 110
Shaw, T. A., 3—5
Sheikh, M. M., 82—84
Shekhar, M., 474—475
Shekhar, S., 276
Shen, C. C., 296, 476
Shen, J., 296—297, 300, 305, 317, 327, 337
Sherry, S. T., 108—110
Shetye, S. R., 225

Shi, X., 4, 296
Shine, T., 35
Shiau, L. J., 296—297, 317, 327—337
Shibutani, Y., 161—166
Shiferaw, A., 132
Shimmield, G. B., 474—475
Shiogama, H., 146—147, 161—166
Shipe, R., 296—297
Shipton, C., 113—114
Showers, W., 455—456
Shrestha, A. B., 124
Shrestha, B. B., 137
Shrestha, F., 124—125
Shrestha, R., 27
Shrestha, S., 27, 137
Shrestha, U. B., 137
Shukla, A., 276
Shukla, J., 296
Shukla, S. K., 296
Shukla, U. K., 275, 282, 284—285, 287—291
Shulmeister, J., 296
Siddaiah, N. S., 273—274, 291
Siddiqui, T., 33—34
Sieh, K., 227
Sierra, E. M., 134
Sieveking, A. G. de, 106—108
Sieveking, G. de G., 106—108
Siever, R., 285—287, 289
Sigert, C., 300
Sigg, A., 468
Sikka, D. R., 4, 7
Silverman, S. M., 452f
Simões, J. C., 124—125
Simoniello, R., 444t—445t
Simpson, I. R., 2, 4—5
Simpson, N. P., 124
Singh, A., 20, 105—106, 112—114
Singh, A. D., 209, 474—475
Singh, A. K., 296
Singh, B., 430—431
Singh, D. S., 275, 284
Singh, G. P., 430—431
Singh, H., 209
Singh, I. B., 275—282, 284—285, 287—291
Singh, P., 209, 430—431
Singh, P. K., 430—431
Singh, R., 274—275, 277—281, 277f, 431—432
Singh, R. K., 474—475
Singh, V. K., 7
Sinha, A., 476
Sinha, S., 282
Sinkeldam, J., 296—297, 300—305, 317, 327—337

Siriwardhana, M., 73
Siscoe, G. L., 443–444
Siwakoti, M., 137
Sloan, L., 438–439, 439f, 462f
Smakhtin, V., 82–83
Smith, C. L., 475f
Smol, J. P., 295–296, 300–305, 317, 337
Snyder, C. W., 457–458
Soininen, J., 295
Solanki, S., 470
Solanki, S. K., 444t–445t, 455, 455f, 466f, 468f, 471b, 478f
Soliman, A., 455
Solovieva, N., 337
Somayajulu, B. L. K., 470
Son, S. W., 4–5
Sonett, C., 445f, 457–458
Sonett, C. P., 457
Sonnadara, D. U. J., 71
Sonnet, C. P., 457–458
Soodyall, H., 108–110
Soren, S., 430–431
Sorkhabi, R. B., 290–291
Spötl, C., 475f
Sree, K., 296
Sreejith, O. P., 79
Srikantha, H., 137
Srinivasa, R., 216, 245–247
Srinivasan, M. S., 207–208
Srivastava, A., 450
Srivastava, A. K., 105–106, 112–114
Srivastava, P., 275, 284
Staffelbach, T., 468
Stan-Sion, C., 468, 469f
Stanek-Tarkowska, J., 300
Stanley, D. J., 247–250
Stauffer, B., 468, 470
Stauffer, P. H., 105–106, 112–113
Stebich, M., 305, 327, 337, 351
Steenhuis, T. S., 178
Stefan, H. G., 146
Stefanova, L., 88–90
Stefels, J., 296–297
Steig, E. J., 468
Steinhilber, F., 454f, 457–458, 470
Steinmetz, S., 39
Stellmacher, R., 442
Stepaniak, D. P., 296
Stephenson, D. B., 88–90
Stevenson, A. C., 295, 300
Stevenson, M. A., 300, 327
Stevenson, A. C., 317
Stix, J., 273

Stocker, T. F., 439, 440f
Stoermer, E. F., 295
Stojanov, R., 51
Stoller, C., 470
Stone, J. R., 305
Stoneking, M., 108–110
Storey, M., 105–106, 112–113
Stouffer, R. J., 88–90
Strassburg, M., 20
Stringer, C., 105–106
Stuiver, M., 234, 443–444, 470
Stump, E., 290–291
Al-Subbary, A. A., 296
Subramanian, K. S., 247
Sucher, C. M., 468
Sudha, V., 191–192
Sudhakar, M., 296
Suess, H. E., 453–454
Sugai, T., 327
Sugimoto, S., 146, 161–166
Sukumaran, S., 430–431
Sumathipala, W. L., 73, 77, 79, 79f–84f, 81–82
Sun, J., 431
Sun, W., 296
Sundaresh, Gaur, A. S., 414–415
Sunderasen, D., 191–192
Sunderesh, A, 413
Suokhrie, T., 209, 296
Suppiah, R., 65, 71, 73
Suprijo, A., 108–110
Suribabu, D., 464–465
Suseela, G., 296
Suter, M., 470
Suzuki-Parker, A., 145–146
Svircev, Z., 305, 317
Svobodova, H., 305, 337
Swain, D., 22, 27, 29
Swain, D. K., 137
Swapan, M.S.H., 53
Swe, A., 216
Swe, L., 27
Swe Thwin, N., 216, 245–247
Sweileh, W. M., 432
Swisher, C. C., 108–110
Synal, H. A., 468, 469f
Szabo, K. E., 327

T

Tacconi, P., 181–183
Tadesse, T., 132
Tadono, T., 124

Tagipour, A., 84–86
Taha, A. J., 105–106
Tahirkheli, R. A., 273
Taib, N. I., 105–106, 112–114
Takahashi, K., 464f
Takayabu, I., 145–146, 161–166
Takemi, T., 146, 161–166
Takeuchi, K., 145, 147–153
Talib, K., 105–106, 112
Talling, J. F., 178–180
Taloor, A. K., 464–465
Tamang, A. M., 125
Tamiru, L., 131
Tandon, S. K., 283, 285, 289
Tangang, F. T., 77
Tarasov, P., 337
Tariq, M. A., 124
Tarique, M., 470–471, 472f
Tatzreiter, F., 245
Tauxe, L., 273
Taylor, J. C., 295
Taylor, K., 113–114
Taylor, K. E., 88–90
Tegan, B., 51–52
Tekaia, F., 327
ter Braak, C. J. F., 295, 300
Thamban, M., 251, 473, 479
Thambyahpillay, G., 63–66
Than Thi, N., 125–126
Thanasupsin, S., 27
Thapa, P., 124
Thébault, E., 459f
Theriot, E. C., 295
Thiemann, S., 181
Thin, N., 27
Thirumalai, K., 18–20
Thoeun, H., 23
Thomas, E., 438–439, 439f, 462f
Thompson, R., 295, 317
Thomson, A., 88
Thorp, T., 54
Thrasher, B. L., 88–90
Thwin, S., 17–18, 216, 218–219, 220f, 245–247
Tian, L., 296
Tibby, J., 295
Tidwell, A., 146
Tiffany M. A., 295
Tilahun, S. A., 178
Timlin, M. S., 446
Timmereck, C., 113–114
Timmermann, A., 175
Tiwari, R. K., 464–465

Tjia, H. D., 106–108, 111
Toepoel, V., 39
Tomas, R. A., 296
Tomiyama, N., 124
Tortajada, C., 122–123
Toushingham, A. M., 247
Tranvik, L. J., 296
Trenberth, K. E., 72, 296
Trewin, B., 84–86
Trimble, V., 440–441
Tripathi, H. K., 427–431
Tripathy, A., 413
Tserendash, N., 337
Tsering, K., 124
Tshen, L. T., 105–106, 110–112
Tshering, D., 124
Tshering, L., 125
Tshering, P., 124
Tsunematsu, N., 146, 147f
Tubbs, L. E., 468
Tuenter, E., 461–462
Tun, T., 216
Turck-Chiéze, S., 444t–445t
Turner, F., 305, 327, 337, 351
Tweedie, M. W. F., 105–106
Twickler, M. S., 113–114, 457

U

Ukita, J., 124
Ullah, A.A., 51
Ulrich, R. K., 452–453
Umarajeswari, S., 191–192
Ummenhofer, C., 20–21
Unuma, T., 161–166
Usoskin, I. G., 444t–445t, 455, 455f, 457–458, 459f, 466f, 470, 471b, 478f

V

Vaganov, E. A., 476–477
Val, P., 466–467
Van Alstyne, K., 240–245
Van Campo, E., 300–305, 317
Van Dam, H., 296–297, 300–305, 317, 327–337
Van de Vijver, B., 317, 327
Van de Vijver, B. V., 300, 327–337
Van De Wal, R. S., 461–462
Van der Kaars, S., 113–114
Van Dijk, A. E., 461
Vanhala, L., 54
Van Geel, B., 444t–445t
van Leeuwe, M. A., 296–297

Van Loon, H., 446–447
van Manen, F. T., 110
Van Meijl, H., 24–25
Van Vuuren, D. P., 88
van Wijk, M. T., 423–424
Vance, D., 209
Vanormelingen, P., 295
Vaquero, J. M., 443, 454
Varela, M., 296–297
Vargas, W. M., 134
Varghese, J., 218–219, 227
Vasselon, V., 296–297
Vaz, G. G., 247–251
Vázquez, M., 443
Vecchi, G., 175
Vedlitz, A., 34
Veerayya, M., 245, 251
Veettil, B. K., 124–125
Veh, G., 124–126
Vehovar, V., 39
Vekua, A., 108–110
Velez, M. I., 296
Venkatesham, K., 296
Ventra, D., 287–288
Verma, A., 430–431
Verma, D., 209, 474–475
Verma, K., 209
Verma, N., 430–431
Verpoorter, C., 296
Verschuren, D., 296
Vidhya, B., 191–192
Vijayan, K. K., 430–431
Villa, I., 296
Vinayachandran, P. N., 76–77
Vinoth, A., 430–431
Vir Singh, S., 430–431
Visher, G. S., 276–281, 288
Vithanage, J. C., 82–84
Vo, X., 20
Vora, K. H., 245, 246f, 251
Vos, H., 477
Vos, P. C., 327–337
Vyverman, W., 337

W

Wacker, L., 471b
Waddington, E. D., 468
Waelbroeck, C., 247
Walker, G., 53–54
Walker-Crawford, N., 54
Wagh, A. B., 407–408
Wagle, B. G., 245, 251
Wagle, N., 124
Wagner, M., 337
Wagner, S., 444t–445t
Waliser, D. E., 80–81
Walker, D., 106–108
Walker, R. G., 275
Wallace, J. M., 11
Walz, A., 124–126
Wang, B., 80–81, 296
Wang, B. Z., 79
Wang, G., 296–297, 300–305, 317
Wang, J., 295
Wang, L., 305, 317, 337
Wang, L.-C., 171–173, 296–297, 317, 327–337
Wang, Q., 4, 305, 327, 337
Wang, S., 300
Wang, W., 88–90
Wang, Y., 296
Wang, Y. S., 172–173
Wangchuk, K., 124
Wangchuk, S., 124–125
Wanner, H., 454f
Waring, W. S., 432
Warner, K., 51
Warrier, A. K., 473
Watanabe, S., 146, 153
Watanabe, T., 317, 327–337
Watson, J.E., 51
Waugh, D. W., 2, 4–5
Weber, H., 52
Webster, P. J., 296
Weerasekera, A. B., 71
Weesakul, S., 137
Weijian, Z., 299–300, 473
Weinlein, W. A., 453–454
Weir, G. W., 289–290
Weis, D., 181–183
Weldeab, S., 259–260
Wen, X., 296
Westaway, K. E., 105–106, 111–112
Wester, P., 124
Westgate, J. A., 106, 112–113
Wheeler, M. C., 79–81, 79f–82f
Wheeler, R. E. M., 409–410
White, J. W. C., 468
White, K., 113–114
Whitlow, S., 113–114
Whitlow, S. I., 457
Whitmore, T. J., 317
Whitton, B. A., 295
Widdicombe, C., 296–297

Willenbring, J. K., 466–467
Willermet, C. M., 105–106
Williams, E., 244
Williams, E. R., 472
Williams, G. C., 245
Williams, M. A. J., 113–114
Williams, P. A., 124
Wilson, L., 145
Wilson, P. A., 461
Wilson, R. M., 440
Win, S. L., 216
Winter, J. G., 295
Wirasantosa, S., 106
Witkowski, A., 296
Witon, E., 296
Witt, R., 305, 327, 337, 351
Wolf, J., 51
Wolfli, W., 470
Wolpoff, M. H., 108–110
Wolter, K., 446
Wood, R. B., 178–180
Worku, W., 132
Wray, J. L., 244
Wray, L., 106–108
Wright, R., 237, 473
Wright, W. E., 296
Wu, J. T., 172–173, 296–297, 317, 327–337
Wu, M.-H., 172–173
Wylie, C. R., 240–245
Wyrtki, K., 225

X

Xia, J., 425
Xiao, J., 296
Xiao, X., 296–297, 300, 305, 317, 327, 337
Xu, B., 317, 337
Xu, Z. X., 145, 147–153
Xue, B., 296–297, 305, 317, 327
Xuefeng, L., 473

Y

Yabuki, H., 124
Yadava, M. G., 296, 473, 476
Yadunath, B., 125–126
Yahiya, Z., 73
Yallop, M., 296–297
Yamada, K., 305, 327
Yamagata, T., 76
Yamagishi, H., 178–180
Yamanokuchi, T., 124
Yanai, M., 296

Yang, D., 305
Yang, D. Y., 305, 337
Yang, H., 296
Yang, T. N., 172–173
Yang, X., 296–297, 300, 305, 327, 337
Yanjun, C., 299–300
Yao, Z., 296
Yasunari, T., 77, 296
Yasutomi, N., 146–147
Yatagai, A., 146–147
Yaya, O., 20
Yeap, E. B., 111
Yellin-Bergovoy, R., 450
Yim, W. S., 351
Yim, W. W. S., 251
Yimin, L., 299–300
Yin, J. H., 4–5
Yirgu, G., 181–183
Yojoyama, T., 106
Yokozawa, M., 146, 147f
Yonga, G., 52
Yordanova, E., 245
Yoshino, M. M., 71, 73
Youbin, S., 299–300
Yu, G., 296
Yu, Q., 430–431
Yu, X., 296–297, 300–305, 317
Yuan, B. Z., 431
Yue, S., 145
Yuen, B., 20
Yun, W. T., 88–90
Yun, S. M., 305, 317, 327

Z

Zachos, J., 438–439, 439f, 462f
Zambrano-Gonzalez, G., 425
Zanchettin, D., 113–114
Zawadzki, J., 124–125
Zech, W., 305, 317
Zelinka, M. D., 4–5
Zeng, Q. C., 79
Zhai, F., 17
Zhang, C., 77
Zhang, E., 296–297, 305, 317, 327, 337
Zhang, G., 124
Zhang, H., 296–297, 305, 317, 327
Zhang, J., 430–431
Zhang, L. N., 79
Zhang, S., 124
Zhang, X., 84–86, 296
Zhao, J., 105–106, 111–112
Zhao, K., 296

Zhao, L., 124
Zheng, F., 430–431
Zheng, W., 296
Zheng, Z., 296
Zhengguo, S., 299–300
Zhuang, J., 17
Ziegler, M., 461–462
Zielinski, G. A., 113–114
Zilberman, D., 423–424
Zimale, F. A., 178
Zomer, R., 21–22
Zong, Y., 351
Zorita, E., 444t–445t
Zubair, L., 73
Zubel, R., 295
Zyoud, S. H., 432

Index

Note: 'Page numbers followed by "*f*" indicate figures, "*t*" indicate tables and "*b*" indicate boxes'.

A

A. trispinosa, 231
Abiotic stress, 430–431
Accelerator mass spectrometry (AMS), 234, 234t
Acidic condition, 319t–326t
Adaptation, 13–14
Agricultural Universities (AU), 424
Agriculture and water resources, 24–28, 26f
    Cambodia, 28
    India, 8–9
    Myanmar, 27
    Thailand, 27–28
Alkaline conditions, 327, 359t–360t
Alluvial deposits, animal remains from, 110–112
Alluvial plain, 217
Alpine meadows, 121
Animal science (Ani. Sci.), 426
Annual rainfall distribution, 63f
Antarctic climate changes, tropical teleconnection impacts, 9–11, 10f
Antarctic temperature changes, 439
Anthropogenic activities, 178
Aquatic ecosystems, 295
Archeo-tourism, 403
Archeological Survey of India (ASI), 409, 413
Archeology, 403
Associated relict fauna, 240–244
Asterix, 242t–243t
Asterorotalia trispinosa, 254–256, 474
Atifact industry, 106–108
Atmospheric circulation, 3–4, 3f
Atmospheric layer, 437
Atmospheric linkages of study area, 236–238
Auto cyclic and allocyclic processes, 273
Average temperature, 136–141, 138f–139f
Ayeyarwady continental shelf, 218
Ayeyarwady delta, 217–218

B

Bangladesh
    climate-induced migration, 35, 53–54
        adaptation example, 54, 54t–55t
        climate change, 43–44
        description of, 35–36
        initiation, 41, 41t–42t
        period of, 42–43, 43t
        reasons for, 45–46, 46f
        respondents behavior, 40–41, 40f
        triggers of, 41t–42t
        types of, 44–45, 45f
    climate justice, 35, 53–54
    methodology
        gathering information, 39
        qualitative approach, 39–40
        quantitative sample size distribution, 39t
        sample size using quantitative methods, 37
    qualitative analysis
        adaptation options, 49t–50t
        case studies, 46–48
        focus group discussions, 46–48
        income generation relocation, 47–48
        in-depth interviews, 46–48
        informant interviews, 46–48
        interviews, 46–48
        participatory rural appraisal, 46–48
        preserve existence, 46–47
        workshops, 46–48
    quantitative analysis, 40–46
    society and climate experience transformation, 36
    study area, 36–37, 38f
Benthic and Planktonic foraminifera, 252–254
Benthic foraminiferal distributions, 211
Beryllium, 467
Beryllium-10, 467
Bhutan, 122, 124–125
    agriculture, 125–126
    farming, 125–126
    GDP, 125
    geographies of, 126–127
    glaciers, 124–125
    global warming, 124–125
    hydropower, 122–124
    Southern foothills, precipitation on, 123–124
    strategic resource, 123
Bio-geo-chemical components, 465
Bipolar spots, 443
Boreal summer intra-seasonal oscillation (BSISO), 80–82
Brackish environment, 351–352
Brückner-Egeson-Lockyer Cycle, 451–452

C

Calcium and potassium, 197
Calcium carbonate, 407
Carbon dioxide, 477
Carbon neutral development, 122
Carbon-14, 467–468
Carbon-negative, 121
Catastrophic weather events, 131
Central Andaman Rift, 218–219, 227
Chamo Lake, 181
Channel system, 276
Chiropodomys gliroides, 111–112
Chronic diseases, 131
Circum-neutral condition, 327–337, 338t–341t
Clast-supported conglomerate, 282
Climate, 183, 218
   event, 256
   modeling, 135–136
   signals, 146
   variability, 131
   simulation, 146–147
   versus tectonics control, 290–291
Climate change (CC), 137–141
   adaptive actions, practices for, 28–29
   agriculture and water resources, 24–28, 26f
      Cambodia, 28
      Myanmar, 27
      Thailand, 27–28
   Antarctic climate changes, tropical teleconnection impacts, 9–11, 10f
   climate-resilient agriculture and water sectors, 29–30
      climate information, 29
      National institutional framework, 30
      research and development, capacities for, 30
   COVID-19 pandemic, 11
   future climate implications, 6–7
      greenhouse gases, 6
      rational view of, 6–7
   general atmospheric circulation, 3–4, 3f
   global earth system, 1
   greenhouse gas, 12–14
   heat wave, 12
   human-caused climate change, 2
   Indian agriculture, 8–9
   Indian context, 7
   national security, 7–8
   ocean heat, 12–14
      adaptation, 13–14
      displacement, 12–13
      regional climate risk, 14
   precipitations and teleconnection, projection/prediction of, 11
   South Asia, 8
   Southeast Asia, 18–24, 19f–20f, 21t
      Cambodia, 23–24
      Myanmar, 20–22
      Thailand, 22, 23f–24f
   temperature, 12–14
   tropical belt, 4–5
   tropical region, climate on, 4, 5f
Climate Data Tools (CDT), 132
Climate risk, 14
Climate-induced extreme events, 28
Climate-induced migration, Bangladesh, 35, 53–54
   adaptation example, 54, 54t–55t
   climate change, 43–44
   description of, 35–36
   initiation, 41, 41t–42t
   period of, 42–43, 43t
   reasons for, 45–46, 46f
   respondents behavior, 40–41, 40f
   triggers of, 41t–42t
   types of, 44–45, 45f
Climate justice, 35, 53–54
Climate-resilient agriculture and water sectors, 29–30
   climate information, 29
   National institutional framework, 30
   research and development, capacities for, 30
Climate-Smart Agriculture (CSA), 423–425
Climatic aberration, 400
Climatic fluctuation, 177–178
Climatic regime, 401
Climatic scenario, 177–178
Clustering analysis, 197–202
Coastal climate, 171
Coastal geomorphology, 218–219
Coastal hydrography, 219–225
Coastal zone, 218
Coastline, 215
Coco's Island, 219
Cold surge, 77
Conglomerates, 290–291, 411–412
Consecutive dry days (CDD), 86
Consecutive wet days (CWD), 86
Contemporary climate, 171–172, 172f
Coral sclerites, 236
Coralline algae, 244
Coronal mass ejections (CME), 441
Cosmic ray fluxes, 469
Cosmic ray spallation, 467
Cosmogenic isotopes, 466–470
Council of Scientific and Industrial Research (CSIR), 211–212, 412–413
Coupled Model Intercomparison Project 6 (CMIP6), 22

COVID-19 pandemic, 11
Crop science (Crp Sci.), 426
Cryptocrystalline, 285
Cultural heritage, 403
Cultural reserve, 177–178
Cyclotella meneghiniana, 300

D

Data collection and analysis, 227–236
Deep-water oxygenation, 209
Dendrogram, 201f
Depositional model, 287–289
Desert topography, 377–378
Dholavira, 414–417
Diatoms, 295
Displacement, 12–13
Dissolved oxygen, 225
Distal ash, 105–106
Drainage, 215–216

E

Earth Simulation Model (ESM), 113–114
Earth's climate, 437
Earth's orbital variations, 459–463
Eccentricity, 461
Echo sounding profiles, 227
Ecological inference, 297
Ecological processes, 141
Economic losses, 23–24
Eddy Cycle, 455–456
El Nino Southern Oscillation (ENSO), 20–21, 72–76, 74f, 446–447
Elemental ratios, 260
Elemental studies, 233–234
Elephas maximus, 110–111
Elephus namadicus, 110–111
Endemic taxa, 361
Endorheic drainage systems, 178–180
Energy transmission, 441–442
Enhancing National Climate Services (ENACTS) project, 132
ENSO, 175
Environmental hazard, 131
Erosional bases, 275–276
Erosional processes, 283
Eutrophic condition, 311t–316t
Excavation, 404
Exclusive Economic Zone (EEZ), 212, 227
Extensive water conservation system, 414–415
Extreme weather events, 12–13

F

First intermonsoon (FIM), 61, 65–66, 71, 73, 76, 90
Flash floods, 24–25, 126
Flood basalt, 181–183
Fluctuation curves, 407–408
Food security, 27
Foraminifera, 208–209, 406–407
Foraminiferal signature, 473–475
Fossil fuels, 439
Freshwater and marine, 298
Freshwater systems, 317

G

Galactic cosmic rays (GCR), 468
General atmospheric circulation, 3–4, 3f
Geo Forschung Zentrum (GFZ), 233–234
Geo-physical data, 227
Geochemical analysis of lake sediments, 196–197
Geochronology, 234–236
Geological periods, 298
Geology, 183–188
Geomagnetic field change, 466f
Geomorphology, 181–183
Geotechnical assessment, 386–397
Glacial-melt water, 124–125
Glacial–interglacial change, 260
Glacial–interglacials records, 438–439
Glacier melt, 124
Gleissberg cycle, 452–453
Global average surface temperature (GAST), 440f
Global Climate Models (GCMs), 18–20
Global dispersion, 106
Global earth system, 1
Global warming, 124–125
Globigerina bulliodes data, 446
Grand minima and grand maxima, 443–445
Grasslands, 111
Greenhouse effect, 137
Greenhouse gases, 6, 12–14, 437
Gross Domestic Product (GDP), 17
Ground penetrating radar (GPR), 404, 405f, 415
Gulf of Martaban, 218–219

H

Hale cycle, 450
Hallstatt/Bray cycle, 456–458
Heat wave, 12
Hemisphere, 443
Highlands, 61
Himalayan Frontal Thrust (HFT), 274–275
Himalayan orogeny, 273–274

Hockey Stick Curve, 482
Holocene, 209, 407
Holocene period, 177–178
Homogenous soil layer, 415
Horizon marker, 106
Horizontal and low-angle planar gravelly sandstone, 275–281
Horizontal planar beds, 288
Human-caused climate change, 2
Humid climate, 287–288
Hurricane, 131
Hydel dams, 123
Hydrocarbon deposits, 178
Hydroclimate datasets unavailability, 175
Hydrodynamic conditions, 191–192
Hydrodynamic energy conditions, 201–202
Hydrodynamic energy gradient, 201–202
Hydrodynamic regimes, 192
Hydrogen into helium, 440
Hydrogen peroxide (H_2O_2), 191
Hydrological condition, 178
Hydrological system, 124
Hydropower, 122–124

I

Ice cores, 479
Indian agriculture, 424
Indian Council of Agricultural Research (ICAR), 424
Indian National Science Academy (INSA), 412–413
Indian Ocean Dipole Mode (IOD), 20–21
Indian Ocean originate, 180
Indian Summer Monsoon (ISM), 296
Indo-Myanmar joint oceanographic studies, 211–212
Inductively coupled plasma mass spectrometer (ICP-MS), 476
Industrialization, 177
Inter-Tropical Convergence Zone (ITCZ), 183
Interannual and intraseasonal rainfall variability, 71–82
Interannual variability, 76
Intergovernmental Panel on Climate Change (IPCC), 17, 88, 146
Intermediate zone, 63–65
International Geological Correlation Program, 407
International Monetary Fund, 437
International tourism, 418–419
Intra-seasonal SWM rainfall variability, 80–82
Intradecadal paleoclimatic records, 251–256
Iron and titanium, 197
Irrigation, 29
Ivory indicators, 105–106

J

Japan
　climate signals, 146
　impacts, 161–166, 164f–165f
　Intergovernmental Panel on Climate Change (IPCC), 146
　precipitation, 147–149, 148f, 149t
　present climate over, 146–153
　projected climate change over, 153–161
　　precipitation, 153–157, 156f, 157t
　　temperature, 157–161, 160f, 161t, 162f
　regions of, 147f
　temperature, 149–153, 152f, 153t

K

Kinetic and biological fractionation, 479
Kuroshio Current, 171

L

La Nina, 72–76
Laminated mudstone, 283
Larger benthic foraminifera, 238
Lateral accretion (LA), 289–290
Latitudinal belt, 65
Lensoidal geometry, 282
Leopoldamys sabanus, 111–112
Lithofacies analysis, 275–283
Lithofacies association, 283–284
Long period average (LPA), 378
Long-term solar activity, 465–470
Lothal, 409–411

M

Madden-Julian oscillation (MJO), 76–79, 80f–81f
Magnetic cycle, 450
Magnetic field, 442–443
Magnetic susceptibility (MS), 471–473
Magnetostratigraph, 273–274
Magnetostratigraphic studies, 290–291
magnitude, 157–160
Main Ethiopian Rift (MER), 178–180
Malaysian peninsula
　alluvial deposits
　　animal remains from, 110–112
　Earth Simulation Model (ESM), 113–114
　geochemical attempts, 112–113
　human occupation in, 106–108, 107f
　migration, 108–110
　origin, 108–110
　quaternary faunas, 110–112
　settlement, 108–110
　volcanic ash, occurrence and source of, 112–113

YTT ash, 113–114
Mammalian fossils, 291
Marine
    and nonmarine, 405–406
    archeology, 418–419
    environments, 410
    sediments, 405
Marine diatom taxa, 357t–358t
Marine environment, 352
Marine heatwaves (MHWs), 20
Marine micro-organisms, 208
Mass spectrometers, 466–467
Material and methods, 188–191
Matrix-to-clast-supported conglomerate, 282
Mean proloculus size (MPS), 231, 473
Mesohalobous taxa, 351–352
Microfossils, 404
    with special reference to foraminifera, 405–406
Mid-fan gravelly sandstones, 284
Milankovitch theory, 459–461, 463
Mineralogical immaturity, 289
Monochromatic photograph, 443
Monsoonal patterns, seasonal shifts and changes in, 27
Morphometric analysis, 231
Multipolar spots, 443
Multitaper method (MTM), 454
Multivariate ENSO Index (MEI), 446
Myanmar, 17–18
Myanmar Ayeyarawaddy delta shel, 212–227

N

National Academy of Agricultural Sciences (NAAS), 425
National Initiative for Climate Resilient Agriculture (NICRA), 424, 428f
National Meteorological and Hydrological Services (NMHSs), 29
National security, 7–8
Natural ecosystems, 132–134
Natural gas bearing potentials., 209
Natural Resource Division (NRM), 431
Neodymium isotope, 209
Network Project on Climate Change (NPCC), 424
Niton XRF analyzer, 191
Nitrogen atoms, 467
Northeast monsoon (NEM), 61, 65, 90
Nuclear weapons, 467–468
Nutrient level, 305–317

O

Obliquity, 461–462
Ocean heat, 12–14
    adaptation, 13–14
    displacement, 12–13
    pathways, 14
    regional climate risk, 14
Oligohalobous taxa, 351
Oligotrophic conditions, 317
Olympus stereozoom microscope, 231
Organic matter, 191, 196f
Orography, 61
Oxygen concentration, 225
Oxygen isotopic ratios, 476
Ozone Depleting Substances, 122

P

Pacific Oscillation, 4
Palaeocurrent pattern, 284–285
Paleo-hydrological condition, 300
Paleobathymetric, 407–408
Paleocene-Eocene thermal maxima (PETM), 439f
Paleoclimate, 172
Paleoclimatic proxies, 208
Paleoclimatic reconstruction, 469
Paleoclimatic records
    from employing cosmogenic proxie, 470
    Indian subcontinent
        employing magnetic susceptibility as proxy, 473
        using foraminifera as proxy, 474–475
        using speleothem as proxies, 476
    periodicity in, 258–262
Paleoecology, 106
Paleozoic planation surface, 183
Past sea-level changes, 407–408
Permian period, 183
Petrography, 285–287
Photosynthesis, 467, 477
Physico-chemical environmental parameters, 407
Physiographic zones, 121
Physiography of Ethiopia, 178–181
Phytoplankton community, 295
Pinjor Formation, 276t
Plagioclase, 285
Planar cross-bedded gravelly sandstone, 281–282
Planktonic percentage, 407–408
Pleistocene, 457t
Plotting rose diagrams, 284
Polar ice caps, 479
Polar regions, 468
Precession, 462–463
Precipitation, 147–149, 148f, 149t
    and teleconnection, projection/prediction of, 11
    variability, 132–136, 134f–135f
Precipitation, 11
Prediction of solar cycles, 479–481

Projected climate change over, 153–161
    precipitation, 153–157, 156f, 157t
    temperature, 157–161, 160f, 161t, 162f
Proximal-to-mid-fan boundary, 290
Proxy records
    dating uncertainties for, 175–176
    stand testimony, 481

Q

Q-mode cluster technique, 197–200
Quaternary Period, 106

R

R-mode cluster data, 201
Radargram, 404
Radioactive isotopes, 466
Radiocarbon, 477
Rainfall anomalies, 73
Rainy climate, 171
Rakhine coast, 218
Ramsetu, 417–418
Regional climate risk, 14
Regional sea-level curve, 247
Relict benthic foraminifera, 238–251
Rice Bowl of Asia, 18
Ripple cross-bedded sandstone, 282
River Ayeyarwady, 215
River basins, 106
River fluxes, 260
River mouth, 198f
River Salween, 215
River Sittang, 215–216
River-borne terraces, 400
Rotalidium annectens, 474

S

Sagar Kanya, 227
Salinity, 224–225, 351–352
    level condition, 352, 353t–354t
Sand and silt fraction, 201–202
Sand lenses, 282
Sandstone and siltstone, 291
Schwabe cycle, 447–449
Sea level fluctuations off Myanmar, 251
Sea paleoclimatic records, 438–439
Sea surface temperatures (SSTs), 17, 452–453
Sea-level fluctuations and marine archeology, 409–418
Seasonal average rainfall, 64f
Second intermonsoon (SIM), 61, 65, 71, 73, 79
Sediment Depositional Environment, 178

Sediment distribution on Ayeyarwady Shelf, 219
Sediment processing, 230–231
Sediment sample collection, 227
Sediment texture analysis, 190–195
Sediment's textural parameters, 178
Sedimentary basin, 178
Sedimentation, 289–290
Shared socioeconomic pathways (SSPs), 23f–24f
Short-term solar activity, 463–465
Silicon and aluminum, 197
Silt-sized grained pyroclastic material, 112
Singular spectrum analysis (SSA), 454
Siwalik molasse, 273
Slope stability assessment, 382–385
snow-capped mountains, 121
Sodium hexametaphosphate, 190–191
Soft coral sclerites and caryophyllids, 240–244
Soil health, 424
Solar activity, climatic manifestations of, 441–442
Solar cycles, 441–463
South America
    catastrophic weather events, 131
    Climate Data Tools (CDT), 132
    climate modeling, 135–136
    climate variability, 131
    data and methodology, 132
    Enhancing National Climate Services (ENACTS) project, 132
    GCM, 136
    map of, 133f
    precipitation variability, 132–136, 134f–135f
    temperature variability, 136–141
        average temperature, 136–141, 138f–139f
        climate change, 137–141
        variability prediction, 137–141
South Asia, 8
Southeast Asia, 17–24, 19f–20f, 21t
    Cambodia, 23–24
    Myanmar, 20–22
    Thailand, 22, 23f–24f
Southern foothills, precipitation on, 123–124
Southern Oscillation Index (SOI), 446
Southwest monsoon (SWM), 65, 71, 73, 80–82
Spatial and temporal variations of rainfall, 63–66
Spatial distribution of elements in lake, 196–197
Species extinction, 131
Spectroheliograph, 443
Speleothem record, 475–476
Speleothems, 475
Sri Lanka
    annual rainfall distribution, 63f
    boreal summer intra-seasonal oscillation (BSISO), 80–82

consecutive dry days (CDD), 86
consecutive wet days (CWD), 86
El Nino Southern Oscillation (ENSO), 72–76, 74f
first intermonsoon (FIM), 61, 65–66, 71, 73, 76, 90
future climate change projections, 88–95
    annual rainfall, 88–92
    heatwave occurrence, 92–95, 96f, 98f–99f
    seasonal rainfall, 88–92
interannual and intraseasonal rainfall variability, 71–82
interannual variability, Indian Ocean Dipole on, 76
Intergovernmental Panel on Climate Change (IPCC), 88
intra-seasonal SWM rainfall variability, 80–82
ITCZ, 65
lightning, spatial and temporal variations of, 67–71, 70f
Madden-Julian oscillation (MJO), 76–79, 80f–81f
northeast monsoon (NEM), 61, 65, 90
observed local trends, 82–87
    climatic extremes, 84–87, 85f
    extreme rainfall events, 89t
    spatial distribution maps, 87f
seasonal average rainfall, 62t, 64f
seasonality of, 61
second intermonsoon (SIM), 61, 65, 71, 73, 79
southwest monsoon (SWM), 65, 71, 73, 80–82
spatial and temporal variations of rainfall, 63–66
temperature variations, 66, 68f
topography map, 61, 62f
Stable isotopic ratios, 262
Stable oxygen isotopic ratios, 231–232
Statistical analysis of elements in sediments, 197–202
Strategic resource, 123
Strontium, 470
Strontium isotopic ratios, 470–471
Sub-surface distribution of foraminifera, 252–254
Sub-surface samples, 231
Subject matter divisions (SMDs), 426
Submerged Dwarka, 412–414
Subsurface sediment dating, 235–236
Sun
    as eternal driver of the Earth's climate, 437–439
    structure and nature, 442–443
    the external driver of Earth's climate, 440–441
Sunspot, 439
Surface sediment dating, 234
Sustainable Development Goals (SDGs), 30
Swampy bodies, 111

T

T-test statistical analysis, 231
Taiwan, 175–176
    contemporary climate in, 171–172, 172f
    critical issues remaining, 175–176
    ENSO, 175
    hydroclimate datasets unavailability, 175
    LGM, 175
    proxy records, dating uncertainties for, 175–176
    typhoons, 175
ENSO, 175
hydroclimate datasets unavailability, 175
hydroclimate of, 171
hydroclimate variability, 172–175
late holocene paleoclimatic records, 174f
LGM, 175
map of, 173f
proxy records, dating uncertainties for, 175–176
typhoons, 175
western tropical Pacific (WTP), 171
Tatrot Formation, 291
Tectonic and geological history, 225–226
Tectonic setting of study area, 225–227
Temperature, 12–14, 149–153, 152f, 153t
    variations, 66, 68f
Temperature extremes, 153
Temperature threshold, 161
Temperature variability, 136–141
    average temperature, 136–141, 138f–139f
    climate change, 137–141
    variability prediction, 137–141
Temporal patterns, 73–75
Tephra, geochronology of, 113
Terrain effects, 77
Thermal ionization mass spectrometer (TIMS), 476
Thermal neutrons, 467
Tides and current circulation pattern, 225
Total foraminiferal number (TFN), 252
Total solar irradiance (TSI), 444, 464
Transparent plastic sheet, 191
Tree rings, 476–479
Triangular stones, 411f
Triangulated Irregular Network (TIN), 182f
Tropical belt, 4–5
Tropical climate, 18
Tropical maritime, 61
Tropical rainfall, 72
Tropical region, climate on, 4, 5f
Tropical storm, 132–134
Troposphere and stratosphere, 467
Turbulent flow, 282
Typhoons, 175

U

Ultraviolet irradiance (EUV), 440
Unimodal pattern, 66
Unipolar spots, 443

V

Variability prediction, 137−141
Vertebrate fauna, 111−112
Vienna Pee Dee Belemnite (VPDB), 231−232
Vienna Standard Mean Ocean Water (VSMOV), 477
Volcanic ash, occurrence and source of, 112−113
Volcanic plug belt, 181−183

W

Warming climate, 12−14
Water stress, 125−126
Water turbulence, 337
Web of Science (WoS), 425, 426f

X

XRF analysis, 191

Y

YTT ash, 113−114

Z

Zoogeographic investigations, 237
Zooplankton, 473
Zooxanthellae, 238

Printed in the United States
by Baker & Taylor Publisher Services